THE MATERIALITY OF THE SKY

THE MATERIALITY OF THE SKY

The proceedings of the
Twenty-second Annual Conference of the
Société Européenne pour l'Astronomie dans la Culture (SEAC)
held in Malta
22–26 September 2014

Edited by Fabio Silva, Kim Malville,
Tore Lomsdalen and Frank Ventura

SOPHIA CENTRE PRESS

© Sophia Centre Press 2016

First published in 2016.

All rights reserved. No part of this publication may be reproduced
or utilised in any form or by any means, electronic or mechanical,
including photocopying, recording, or by any information storage
and retrieval system, without permission in writing from the Publishers.

Sophia Centre Press
University of Wales, Trinity St David
Ceredigion, Wales SA48 7ED, United Kingdom.
www.sophiacentrepress.com

Cover Photo: © Daniel Cilia
Book and Cover Design: Jennifer Zahrt

ISBN: 978-1-907767-09-8

British Library Cataloguing in Publication Data.
A catalogue card for this book is available from the British Library.
Printed in the UK by Lightning Source.

Contents

Contents   v

List of Figures   viii

List of Tables   xiii

Introduction: The Materiality of the Sky   xv
Frank Ventura

Contributors   xxv

Maltese Archaeoastronomy

Reading Messages from the Past: Interpreting Symbols   3
Possible Archaeoastronmical Significance in Malta
Frank Ventura

The 'Oracle Holes' of the Maltese Prehistoric Temples: 21
An Investigation of their Astronomical/Solar Alignments
Tore Lomsdalen

Inclusion and Exclusion of Sunlight and Moonlight From   35
Temples of the Ġgantija and Tarxien Phases
John Cox

Cosmology and Cosmovision

Cosmovisions Put Upon a Disk: Another View of the Nebra Disk   47
Michael A. Rappenglück

Astronomy and Landscape in Carthago Nova   65
Juan Antonio Belmonte, José Miguel Noguera Celdrán,
A. César González-García & Andrea Rodríguez-Antón

A Status Report: A Review of Research on the Origins   79
and Diffusion of the Belief in a Sky Bear
Roslyn M. Frank

In Search of Päivätär, the Finnish Solar Goddess   89
Marianna Ridderstad

The Nordic Calendar and the Great Midwinter Sacrifice at Old Uppsala   99
Göran Henriksson

Fire from the Heavens: The Idea of Cosmic Fire across Archaic Cultures   111
Michael A. Rappenglück

Astronomical Orientations

On the Orientation of the Historic Churches of Lanzarote: 125
When Human Necessity Dominates over Canonical Prescriptions
Alejandro Gangui, A. César González García, Mª Antonia Perera Betancort & Juan Antonio Belmonte

Orientation of Roman Camps and Forts in Britannia: 135
Reconsidering Alan Richardson's Work
Andrea Rodríguez-Antón, Antonio César González-García &
Juan Antonio Belmonte

Evidence for the Existence of Solar and Lunar Alignments in Western Scotland: 145
The Contrasting Nature of Backsights, Foresights and Alignments
Thomas Gough

Architecture, Illumination and Cosmology: 155
the Arles-Fontvieille Monuments, Archaeoastronomy and Megalithic Studies
Morgan Saletta

An Ethnoastronomy Study on the Astronomical Orientation and Astral Decoration 163
of the Stone Granaries (Hórreos) of Vilaboa (Galicia, Spain)
Fátima Braña Rey & Ana Ulla Miguel

Connections: 177
The Relationships between Neolithic and Bronze Age Megalithic Astronomy in Britain
Gail Higginbottom & Roger Clay

Winter Solstice at the Iberian Cave-Sanctuary of La Nariz 189
César Esteban & José Ángel Ocharan Ibarra

Raising Awareness of Light Pollution by Simulation of Nocturnal Light 197
of Astronomical Cultural Heritage Sites
Georg Zotti & Günther Wuchterl

New Findings at the 'Petre De La Mola' Megaliths 205
L. Lozito, F. Maurici, V. F. Polcaro, and A. Scuderi

Astronomy and Culture in Historical Times

Sirius (al-'Abūr) Proper Motion as Recorded in the Arabic Star Mythology 213
Flora Vafea

The Stones of Penas de Rodas: 223
Can the 'Spell of Archaeo-astronomy' Create a Contemporary Sacred Place?
Benito Vilas Estevez

The Sphere in Antiquity 233
Mª Pilar Burillo-Cuadrado

North and South America

Houses of the Sun and the Collapse of Chacoan Culture — 245
J. McKim Malville & Andrew Munro

Astronomy and the Ceque System of Cusco — 255
Steven R. Gullberg

The Temple of the Inscriptions in the Spiritual Landscape at Palenque — 267
Stanisław Iwaniszewski

Egypt, the Mediterranean and Asia

A Comparative Study of Megalithic Monuments in Sardinia and Beyond — 291
A. César González-García, Mauro P. Zedda & Juan A. Belmonte

Archaeoastronomy In Sicily: Megaliths And Rocky Sites — 301
Andrea Orlando

The Tall Gnomon Of Guo Shoujing:
An Astro-Archaeological Analysis — 317
Vance Tiede

The Sophia Centre — 327

Figures

Page 9, *Fig. 1.1*: *Ceiling painting in red ochre at the Ħal Saflieni hypogeum, Malta showing disks, spirals and coils, Heritage Malta.*

Page 10, *Fig. 1.2*: *Ceramic shard with 'solar wheel' decoration from Ħaġar Qim temple, Malta, Heritage Malta.*

Page 11, *Fig. 1.3*: *Large Stone with damaged low relief carvings of a bull and sow with thirteen appendages from the Tarxien Middle temple.*

Page 12, *Fig. 1.4*: *Limestone slab with star and moon symbols from Tal-Qadi temple, Malta.*

Page 13, *Fig. 1.5*: *Large floor slab with holes at the SE end of the concave facade of the Tarxien West temple, Malta.*

Page 16, *Fig. 1.6*: *Stone with presumed cup-marks lying against the outer wall of the Mnajdra Middle temple, Malta.*

Page 27, *Fig. 2.1*: *The apse with the adjoining external Oracle Room and the two smaller holes to the right.*

Page 28, *Fig. 2.2*: *Room M from the west with the altar in the back and the oracle hole on the front left hand side; lower right insert, the oracle hole seen from inside the room.*

Page 29, *Fig. 2.3*: *The summer solstitial sunrise seen from the area inside the temple where the slab is illuminated as a waning moon.*

Page 30, *Fig. 2.4*: *Photos taken by Lomsdalen on 23 December 2012 of the winter solstitial sunrise.*

Page 36, *Fig. 3.1*: *Location of Temple Period Sites, after Trump 2002.*

Page 37, *Fig. 3.2*: *Orientation of Multi-Chambered Temples according to phase.*

Page 38, *Fig. 3.3*: *Direction of far-southerly moonrise and of midwinter sunrise observed from Ta'Ħaġrat South Temple.*

Page 40, *Fig. 3.4*: *Sunrise 29 December 2012 observed from north side of the interior of (Evans') room 1 at Mnajdra South Temple and in cross-jamb view through the entrance.*

Page 41, *Fig. 3.5*: *Setting path of object at declination twenty-nine degrees south constructed over a star-field viewed across (Evans') room 9 and photographed 17 December 2009 from waist between rooms 14 & 17, Tarxien North Temple.*

Page 48, *Fig. 4.1*: *a) The Nebra Disk at an exhibition in the Neanderthal Museum in Mettmann, Germany.*

Page 52, *Fig. 4.2*: *a) A fragment of a stone plate, originally probably a disk, from Tal Qadi, Malta, ca. 3300-3000 BCE. It shows engravings of astral objects 'stars', crescent) and an abstract division into sectors.*

Page 56, *Fig. 4.3*: *a) Rosette shaped icon from Mesopotamian region and Jemed Nasr period 2,900–2,400 BCE: the seven dots originally are related to ^{d}Sibittum (^{d}Vll-bi), having a triple meaning: the Seven Sages, the seven sons of Enmešarra, and the Pleiades.*

Page 58, *Fig. 4.4*: *a) An iron sword from a site near Allach-Untermenzing, Munich, 5th century BCE (Iron Age).*

Page 60, *Fig. 4.5*: *a) Christ in Majesty* (Maiestas Domini), *Abraham Clemetsson Albo (1764–1841).*

Page 67, *Fig. 5.1*: *The peculiar topography of Cartagena in antiquity.*

Page 68, *Fig. 5.2*: *A snapshot (a) of the temple of Republican era on the top of Cerro del Molinete (ancient Arx Hasdrubalis, H1) before partial reconstruction*

Page 69, *Fig. 5.3*: *A hypothetical* **porticus duplex** *that was built to the southwest of the Augusteum could be the source of a series of antefixes with representations of Capricorns.*

Page 72, *Fig. 5.4*: *Summer solstice sunrise from the gate of the sanctuary of Atargatis (in H1), a sector where the sacred area of a female Punic deity was possibly located.*

Figures   ix

Page 74, *Fig. 5.5: The principal buildings and part of the recovered street grid of Roman Carthago Nova over imposed to a cardinally orientated modern city aerial view.*

Page 75, *Fig. 5.6: Sunset at summer solstice on the highest peak of Sierra Espuña as observed from the sacellum of Iuppiter Stator (Jupiter Tonante) to the southeast of the original urban area.*

Page 102, *Fig. 8.1: Sunset along the big mounds on 8 February 1993.*

Page 103, *Fig. 8.2: On the earliest date for the Great Midwinter Sacrifice, 28 January, the sun set straight over the highest point on Tunåsen and on the last date, 26 February, at the position of the burial cairn in the middle mound (Aun's?), as observed from the menhir at Tingshögen (Thing Mound).*

Page 105, *Fig. 8.3: The southern part, first excavated ca. 70 m, of the about 850 m long row of postholes discovered in May 2013 beneath the old main road leading out from Old Uppsala toward the north is seen in the upper left photo.*

Page 106, *Fig. 8.4: The depiction in the lower part of the figure shows the position of the upper limb of the rising Midwinter full moon, on 27 January 180 CE, for a distant zero horizon beyond the hill.*

Page 108, *Fig. 8.5: Sunrise at Old Uppsala on 6-7 February, the day in the middle of the winter when the sun rose along the more than 600 m long southern row of posts.*

Page 127, *Fig. 10.1: Two churches in Lanzarote with unique features: (a) the church of Nuestra Señora de las Mercedes (Our Lady of Mercy) in Mala; ands (b) the chapel of San Rafael, located alone and isolated on a plateau overlooking El Jable on the outskirts of the village of Teguise.*

Page 129, *Fig. 10.2: (a) Orientation diagram for the churches and chapels of Lanzarote; and (b) Diagram of winds for Arrecife Airport in Lanzarote, illustrative of the prevailing winds on the islands*

Page 130, *Fig. 10.3: Declination histogram for the chapels and churches of Lanzarote.*

Page 131, *Fig. 10.4: Map showing the geographical location of all the measured churches (indicated by ellipses), together with the orientation of the axis of the buildings towards the apse.*

Page 133, *Fig. 10.5: (a) Diagram showing the azimuth of the churches and chapels versus the most probable date of construction (or its first mention in the sources); and (b) Same as the above diagram, but with extended area close to the solar azimuth range.*

Page 139, *Fig. 11.1: Declination histograms towards west (top right) and east (top left).*

Page 141, *Fig. 11.2: Map of Britannia with the 93 camps and forts in the sample. Triangles indicate settlements to the south of Hadrian's Wall, and asterisks those to the north of Hadrian's Wall.*

Page 142, *Fig. 11.3: Declination histograms towards west (left) and east (right) of the data set of Roman camps and forts to the north (top) and south (bottom) of Hadrian's Wall.*

Page 147, *Fig. 12.1: Explanation of a Lunar Band for the 'standstill' north showing the relevant declinations*

Page 148, *Fig. 12.2: Lunar alignment, Port Ellen, Islay (Map ref. NR 3715 4559).*

Page 149, *Fig. 12.3: The combined lunar bands in the north with alignments.*

Page 151, *Fig. 12.4: Solar foresights.*

Page 152, *Fig. 12.5: Examples of lunar foresights found.*

Page 156, *Fig. 13.1: Sunlight striking the rear wall of the Hypogée de la Source, 23 September 2011.*

Page 158, *Fig. 13.2: Cut and plan of the Grotte de Cordes (de Fondouce, 1873).*

Page 167, *Fig. 14.1: Horreos in Santa Cristina and Santo Adrián of Cobres, 2014.*

x   Figures

Page 167, *Fig. 14.2: Santa Cristina and Santo Adrián of Cobres, from Googlemaps, 2014.*

Page 167, *Fig. 14.3: Hórreos of Santa Cristina and Santo Adrián (Vilaboa, Spain). Magnetically uncorrected Azimuths distribution of their largest longitudinal direction; from 0º N to 180º S, (in degrees, plus/minus 1º of uncertainty).*

Page 168, *Fig. 14.4: Number of hórreos of Santa Cristina and Santo Adrián (Vilaboa, Spain) per Azimuth (from 0º N to 180º S) with an average magnetic correction of 3ºfor the whole (245) sample. In degrees, plus/minus 1º of uncertainty.*

Page 170, *Figs. 14.5, 6, 7, and 8: Wind distributions.*

Page 171, *Fig. 14.9: Gamble with star, moon and circle.*

Page 171, *Fig. 14.10: Gamble with moon and hexa-petal rosette.*

Page 181, *Fig. 15.1: This is the 3D rendering of the landscape around the classic site of Uluvalt on Mull.*

Page 182, *Fig. 15.2: 3D landscapes of slabs and single menhirs, where the centre of the landscape is north. The red, vertical lines indicate the direction of the alignment of the site where it touches the horizon. In order from the top: two classic sites of (a, top) Torran, Argyll NM87880488; (b, middle) Rowanfield, Argyll NM82059585; and (c, bottom) Cillchriosd, Mull NM37735348.*

Page 185, *Fig. 15.3: 3D landscapes of Late Neolithic stone circles, classic sites. In order from the top: (a, top) Callanish, Isle of Lewis, Scotland; NB21303300; (b, middle) Castlerigg, Cumbria, England; NY29142363 and (c, bottom) Swinside, Cumbria, England; SD17168817.*

Page 186, *Fig. 15.4: 3D landscapes of stone circle reverse sites. In order from the top: (a) Stenness, Orkney, Scotland; HY30671252 and (b) Druids Circle, Gwenydd, Wales; SH72297466.*

Page 190, *Fig. 16.1: Cave-sanctuary of La Nariz (Moratalla, Murcia). The insert on the left indicates its location in the Iberian Peninsula.*

Page 191, *Fig. 16.2: Left, plan of the cave-sanctuary of La Nariz; UE 1 and UE 2 are the two main cavities. Right, photos of the interior of the cavities UE 1 and UE 2 obtained from their entrances; the location of the carved basins UC 2 and UC 3 are indicated with arrows.*

Page 192, *Fig. 16.3: Left: Horizon visible from the carved basins of UE 1 (up) and UE 2 (down). Some landmarks are marker with letters. A circle indicates the size of the solar disc. Centre and right: Sunset at a date very close to winter solstice as seen from the interior of UE 1 and UE 2, respectively.*

Page 193, *Fig. 16.4: The four images on the top show the illumination of basin UC 2 (in the interior of UE 1) as the winter solstice sunset progresses. The lower photo shows the carved basin UC 2 just a few minutes before sunset.*

Page 200, *Fig. 17.1: A fisheye photograph taken near the Großmugl public observing site on 5 November 2008 2:21CEDT (top) and simulation of the scene with the new illumination layer in Stellarium 0.13.1 (bottom). North is at top. ss*

Page 202, *Fig. 17.2: A simulated night scene of Valletta Harbour, Malta, created by combination of day and night photographs. Screenshot from Stellarium 0.13.0 with the new nocturnal illumination layer..*

Page 206, *Fig. 18.1: Laser scan section of the megaliths: the observing point (a), the viewfinder oriented to the meridian (b) and the gallery pointing to the winter solstice sunset (c) are marked. Blue arrows show pictures of these artificial features.*

Page 207, *Fig. 18.2: Light effects at the «Petre de la Mola» at winter solstice: A) noon B) sunset. Both pictures have been taken from the same observing point: the break on the platform on the North side of the megalith shown in Fig. 18.1.*

Page 207, *Fig. 18.3: The meridian and the winter sunset directions indicated by the petroglyph.*

Page 208, *Fig. 18.4: The petroglyph indicating the cardinal direction and the basin sited in the third modified flat rock.*

Page 208, *Fig. 18.5: The artificial basin and carving on the third flat rock.*

Page 215, *Fig. 19.1: Graph showing the variance of the galactic latitude of Sirius (α CMa) within 54 millennia.*

Figures xi

Page 217, *Fig. 19.2: Graph showing the variance of declination of the stars Canopus, Sirius, Procyon, Betelgeuse and Rigel, as calculated with the software Voyager 4.5. The values come from Table 19.2.*

Page 218, *Fig. 19.3: Positions of the stars of the myth for years 49,850 BCE and 35,360 BCE.*

Page 219, *Fig. 19.4. Positions of the stars of the myth for years 24,168 BCE and 12,076 BCE.*

Page 220, *Fig. 19.5: Positions of the stars of the myth for year 815.*

Page 224, *Fig. 20.1: The stones of Penas de Rodas.*

Page 225, *Fig. 20.2: Natural mark at the top of the stone, marked in red colour.*

Page 226, *Fig. 20.3: 'Supposed' semicircular theatre.*

Page 226, *Fig. 20.4: Marks of the wedges in other stones.*

Page 235, *Fig. 21.1: Bronze sphere found in Mainz, Germany. Photo by Pilar Burillo-Cuadrado.*

Page 236, *Fig. 21.2:* Farnese Atlas. *Detail of the intersection of the Celestial Equator with the Ecliptic. Photo by Pilar Burillo-Cuadrado.*

Page 237, *Fig. 21.3: Left, detail of the sphere held by Helios found in Pompeii, Italy; and on the right the sphere of the* Farnese Atlas *showing the intersection of the equator and the ecliptic.*

Page 239, *Fig. 21.4: Left, detail of the Celestial Sphere held by the* Muse Urania *in the Naples Museum, Italy; and on the right* Urania *on the Sarcophagus of the Ostia Museum, Italy.*

Page 240, *Fig. 21.5: The* Orbis Terrarum. *Basilica S. Maria Sopra Minerva in Rome. Photo by Pilar Burillo-Cuadrado.*

Page 246, *Fig. 22.1: Great Houses at Chaco Canyon. The Late Bonito Great Houses are identified in boxes, in which DSSR and JSSR indicate horizon markers for December solstice sunrise and June Solstice sunrise, respectively.*

Page 247, *Fig. 22.2: Storage floor area added to Chaco Canyon Great Houses.*

Page 248, *Fig. 22.3: Sunset observed from Casa Chiquita. Photo by G. B. Cornucopia.*

Page 249, *Fig. 22.4: Kin Kletso June Solstice Sunrise (photo by G. B. Cornucopia).*

Page 249, *Fig. 22.5: December solstice sunrise near Roberts Small Pueblo.*

Page 250, *Fig. 22.6: Horizon Survey Results at Headquarters Site A.*

Page 257, *Fig. 23.1: Rowe's Cusco ceque numbering system.*

Page 260, *Fig. 23.2: Ceques of the four suyus of Cusco. The sun temple, Coricancha, is at the centre.*

Page 261, *Fig. 23.3: The opening in the northeast cave of An 3:4 (Lacco) is aligned for the June solstice sunrise.*

Page 262, *Fig: 23.4: The altar in the inner chamber of the southeast cave of An. 3:4 (Lacco) is illuminated through a light-tube at the time of the zenith sun.*

Page 264, *Fig. 23.5: Tangential lines at a small un-numbered huaca between Ch. 1:2 (Kenko Grande) and An. 3:4 (Lacco) guide the eye to the horizon for sunrises and sunsets at solstice and equinox.*

Page 268, *Fig. 24.1: The Temple of the Inscriptions, Structures XIII ('Red Queen') and XII.*

Page 274, *Fig. 24.2: The Sarcophagus Lid*

Page 278, *Fig. 24.3: The Plan of the Temple of the Inscriptions*

Page 279, *Fig. 24.4: Schematic plan of the temple atop the Temple of the Inscriptions.*

Page 282, *Fig. 24.5: The Temple of the Cross as seen from the top of the Temple of the Inscriptions.*

xii   Figures

Page 283, *Fig. 24.6:. Western horizon as seen from the top of the Temple of the Inscriptions.*

Page 294, *Fig. 25.1: Left, dendrogram for the orientation of megalithic monuments of Sardinia. Right, map showing the results of the k-means analysis. Adapted from González-García et al. (2014).*

Page 295, *Fig. 25.2: Extended dendrogram for the orientation of megalithic groups in the Western Mediterranean.*

Page 298, *Fig. 25.3: Declination histograms for several groups of monuments in Sardinia (dark grey shade) compare with a homogeneous population of the same number of elements (light grey). TdG stands for Tombe di Giganti and N for Nuraghe.*

Page 302, *Fig. 26.1: The plan of the Sese number 31 (left); and the orientations diagram of passages in the Sesi of Pantelleria (right).*

Page 304, *Fig. 26.2: Map of the Sicily with the various markers that indicates the archaeological sites involved in the archaeoastronomical studies.*

Page 308, *Fig. 26.3: On the top: one of the rocky tubs present on the Water's Rock; on the bottom: the azimuths of Rocca Novara measured from the altar (green line) and the Tower (red line).*

Page 309, *Fig. 26.4: The so-called Temple of Diana on the Cefalù Rock (from Houel, 1785).*

Page 312, *Fig. 26.5: Planimetric sketch of the north wall (from Mauceri, 1896).*

Page 318, *Fig. 27.1: From left to right, Guo Shoujing, Sun & Shadow at Meridian Transit, Grooved Template Scale (facing south).*

Page 319, *Fig. 27.2: Shadow Definer; Observing Table; Ying Fu Principle; Reflected starlight on water; Reconstructed Horizontal Scale.*

Page 320, *Fig. 27.3: Left and right, Solstice Altitudes; center, Shadow Cast on Template Scale at Meridian Transit.*

Page 321, *Fig. 27.4: Left – Tall Gnomon, Gaocheng (NB: Shadow on scale's mid-point at Midwinter, 26 Dec 2009 ≈ 11:00; Right – Site of Yuan Observatory, Dadu (Beijing), 11 Dec 2003 ≈ 11:00. North = top.*

Page 325, *Fig. 27.5: Transits of Venus: top left, 3 December +1360/1200 Local (-23° dec), Beijing; top right, 8 June +2004 (+23°dec) Helioscope, Yokohama Science Center (Itsuo Inouye, AP); bottom, 29 October +1244/1200 Local (-27° dec), Gaocheng (Starry Night Pro Plus-6).*

Tables

Page 42, *Table 3.1: Orientation of the Main Axis in Multi-Apse Temples with a note on Observational Features.*

Page 70, *Table 5.1: Data on the orientation of ancient structures at Carthago Nova (37°36' N; 0°58' W).*

Page 85, *Table 6.1: Cross-cultural comparisons of the tenets of bear ceremonialism.*

Page 100, *Table 8.1: Important events in early Swedish History in accordance with the eight-year cycle.*

Page 109, *Table 8.2: Age of the posts at Old Uppsala.*

Page 128, *Table 10.1: Table showing orientations for the chapels and churches of Lanzarote.*

Page 140, *Table 11.1: Declination values and corresponding dates of the peaks obtained in eastern and western directions for the whole sample.*

Page 141, *Table 11.2: Declination values and corresponding dates of the peaks obtained at the south and the north of Hadrian's Wall towards eastern and western direction, respectively*

Page 150, *Table 12.1: Possible calendrical sites on Mull and Islay*

Page 157, *Table 13.1: Azimuthal orientations of the Arles-Fontvieille monuments.*

Page 173, *Table 14.1, part 1 and 2: Data collected in Cobres parishes for an initial sample of 9 hórreos with 'astral decoration'.*

Page 215, *Table 19.1: Variance of the galactic latitude of Sirius (α CMa) within 54 millennia.*

Page 217, *Table 19.2: The variance of declination of the stars Canopus, Sirius, Procyon, Betelgeuse and Rigel.*

Page 247, *Table 22.1: Bonito Phase Great Houses.*

Page 248, *Table 22.2: Late Bonito Great Houses.*

Page 262, *Table 23.1: Astronomical and cosmological characteristics of huacas on ceques by suyu.*

Page 279, *Table 24.1: Basic orientations of the temple located atop the Temple of the Inscriptions.*

Page 280, *Table 24.2: Tomb orientations at Palenque.*

Page 296, *Table 25.1: Data considered for the cluster analysis for Sardinian and western Mediterranean groups.*

Page 323, *Table 27.1: Yuan znang/chi, Metric Linear Unit.*

Page 323, *Table 27.2: Yuan Shih Dec & Tall Gnomon Locations.*

Page 324, *Table 27.3: Tall Gnomon Minimum Declinations at Gaocheng & Beijing.*

Top: The SEAC happy family on excursion in front of the Ħaġar Qim Temple. Photo: Daniel Cilia.

Left: Donation of conference surplus to the Archaeological Society of Malta (ASM). Photo: James Moffett. L to R, Dr Reuben Grima SEAC 2014; Mr Tore Lomsdalen SEAC 2014 Secretary; Professor Frank J. Ventura SEAC 2014 Chair; Ms Patricia Camilleri President ASM; Ms Ann Gingell Littlejohn Hon. Sec. ASM; Professor Nicholas Vella Vice President ASM.

INTRODUCTION: THE MATERIALITY OF THE SKY

Frank Ventura

Introduction

From the earliest times, human beings have been driven by basic needs to procure food and water, shelter and defence, and communication with other members of the group. This led to the extensive exploration and exploitation of the organic and physical resources in the landscape for food and clothing, tools and construction, and caves for shelter, among others. Human beings living in coastal areas, rivers and lakes could also exploit the resources of the seascape or waterscape and use it to explore other territories. The skyscape was different as it was beyond the reach of man and could not be manipulated. Yet, the innate imagination of human beings could not be unmoved by the sun and the moon, which dominate the day and night and apparently move in a well-ordered fashion, and the stars, which provide a splendid canopy on clear dark nights. Although celestial objects could not be handled and exploited in a tangible manner, man's creativity sought to understand them, to find some use for them in relation to his needs and activities and to generate ideas about their nature and their meaning for humanity.

In other words, skyscapes were 'metaphysically appropriated through projection whereby intangible material culture is mapped onto the heavens'.[1] One must add, however, that concepts, patterns, myths and other creations of this intangible culture could be transformed into material culture, including iconography, calendars, orientations and other human creations. These manifestations constitute the materiality of the sky and bear witness to man's interest in the sky. In other words, our hypotheses about the relationship between man and the skyscape can be corroborated by research on the materiality of the sky, which was the theme of the Twenty-second Annual Conference of the *Société Européenne pour l'Astronomie dans la Culture* (SEAC – European Society for Astronomy in Culture), held between the 22nd and 26th September 2014 in Malta at the Valletta Campus of the University of Malta. During the five days of the conference, participants discussed various aspects of man's fascination with the sky and its relation to his culture.

While it is expected that the understanding and use of celestial objects may have been similar among various peoples worldwide, ideas about their nature and meaning must have been very different depending upon the different cultures that emerged in different geographical contexts.[2] These hypotheses will just be fanciful ideas unless specific research is conducted to discover supporting evidence in the material culture of prehistoric and proto-historic peoples and in the artefacts, written and oral records of later literate cultures. The papers in this volume illustrate the variety in research activity generated in this field of study across a broad spectrum of ages, cultures and geographical regions. They also provide a good sample of the categories of material evidence, methodological issues and approaches to archaeoastronomy outlined by Ruggles, whose analysis provides a useful device for grouping the papers into categories as suggested hereunder.[3]

---

[1] T. Darvill, 'Afterward: Dances Beneath a Diamond Sky', in *Skyscapes. The Role and Importance of the Sky in Archaeology*, ed. F. Silva and N. Campion (Oxford: Oxbow Books, 2015), pp. 141.
[2] J. McKim Malville, 'Meaning and Intent in Ancient Skyscapes – An Andean Perspective', in *Skyscapes. The Role and Importance of the Sky in Archaeology*, ed. F. Silva and N. Campion (Oxford: Oxbow Books, 2015), p. x.
[3] C. L. N. Ruggles, 'Nature and Analysis of Material Evidence Relevant to Archaeoastronomy', in *Handbook of Archaeoastronomy and Ethnoastronomy*, ed. C. L. N. Ruggles (New York: Springer, 2015), pp. 353–372.

Orientations

The measurement and interpretation of orientations remains a major focus of attention in archaeoastronomy and the following papers in the present publication reflects this interest. In his study of the orientations of the five megalithic hypogées at Arles-Fontvieille (~3500–3000 BCE), **Morgan Saletta** suggests that the intention was not the targeting of the sun, the moon or the stars but 'seasonal illumination' or the light and shadow effects at certain significant days of the year which were connected with the cosmological principles held by the Late Neolithic culture that built the monuments. Comparing these effects to similar hierophanies at Newgrange and Maeshowe, he goes on to suggest that the cosmological principles spread by contact and diffusion from France to Ireland and Scotland. This hypothesis has already attracted a number of studies but its corroboration can only be considered as work in progress.

Within the same timeframe, **John Cox** distinguished between the orientations of the Maltese temples of the Ġgantija phase (3600–3000 BCE), which are orientated towards the southeast, and the later temples of the Tarxien phase (3000–2500 BCE), which have orientations towards the southwest generally. Furthermore he suggests a transitional phase for temples with orientations that fall in between these two. He suggests that these differences in the preferred orientation were driven by changes in tradition and ritual practice, which required either the inclusion or the exclusion of illumination from the winter solstice sunrise or sunset and from the moon at the major standstill south. These interesting suggestions require further study but they illustrate the proper approach we need to adopt to move beyond the collection and analysis of orientations towards an understanding of what motivated the builders to choose particular orientations and not others.

Turning to the Bronze Age monuments in Scotland, **Thomas Gough** quite rightly stresses the need of precise measurements and a detailed understanding of motions of the sun and the moon if we wish to resolve the dispute arising from Alexander Thom's claims of precise lunar and solar alignments in Scotland and their rebuttal by Clive Ruggles. Gough provides evidence garnered since 2007 from a very careful reassessment of the Early Bronze Age standing stones in several regions in western Scotland which show precise lunar alignments in Argyll and lunar and solar alignments at Mull and Islay. Also in Scotland, **Gail Higginbottom and Roger Clay** studied many Bronze Age freestanding stone monuments erected between 1400 and 900 BCE and noted correspondences between their astronomically significant orientations and landscape features for monuments. They then investigated when and from where these patterns were first introduced. Surprisingly, they noticed the same patterns in monuments that were erected two millennia earlier in the Neolithic, which led them to argue that the findings demonstrate cosmological continuity over millennia in spite of the drastic material and social changes which must have occurred between the Neolithic and the Bronze Ages.

Leonardo Lozito et al. introduce readers to evidence relating to an archaeological site of interest for archaeoastronomy on a mountain in the Lucanian Dolomites of Southern Italy, which was in use during to the Neolithic and Bronze Ages. The site consists of an imposing group of rocks clearly modified by man to produce alignments with winter solstice sunset and the meridian, both of which are indicated by a petroglyph on the rock surface at the observation point. Other observation points are suggested due to the presence of carved basins on the rocks at these points. The authors hypothesise that this was a sacred site utilised for calendric and ceremonial purposes. Focusing on Sicily, the largest island in the Mediterranean Sea, **Andrea Orlando** notes its significant number of prehistoric and historic monuments which have been almost neglected by archaeoastronomers for many years. However, the author recalls work carried out by Nissen, Koldewey and Puchstein, and also Penrose, in the late 19th and early 20th centuries on the orientations of the Greek temples in Sicily and the hiatus until 1992 when Hoskin et al. measured the orientations of the *sesi* on the nearby island of Pantelleria. He then gives an account of the resurgence of interest in this field of study, especially since 2012 when investigations of various promising sites, including megalithic temples, dolmens, cyclopean walls and other structures, have been taking place in collaboration with archaeologists.

In contrast to Sicily the nearby island of Sardinia has been the focus of several studies of the orientations of its hundreds of megalithic monuments ranging from nuraghe, *tombe di giganti*, *domus de janas*, and dolmens. In a reappraisal of the data at hand and new data obtained in the last years, **César González-García et al.** carried out a cluster analysis of the orientations which confirmed the differences between the northern and southern halves of the island, especially concerning the *tombe di giganti*. They then went on to compare the orientations of the groups of monuments in Sardinia with other groups of contemporaneous megalithic monuments on the shores of the central Mediterranean. The result was that the north-south differences noted in Sardinia are related to the orientation customs on the north and south shores of the Western Mediterranean. This conclusion led them to suggest that Sardinia was at the crossroads of diffusion for orientation customs – a hypothesis which requires a review of the diffusionist paradigm which archaeologists had rejected years ago. Turning further west to the region of Murcia in southwest Spain, **César Esteban and José Ángel Ocharan Ibarra** present an investigation of the orientations of a Late Bronze Age cave sanctuary, combining astronomy, archaeology, religious beliefs and rituals with references from ancient authors. The cave has dual cavities of approximately the same length and orientation, both of which have a water spring and a carved basin at the inner end. One cavity is oriented towards the winter solstice sunset over a distant mountain range while the other is possibly aligned with the major lunar standstill or maybe even with the setting of Venus. The authors explain the motivation for these orientations with interesting information about the ritual and the deity worshipped in this sanctuary and make a comparison with similar activities in contemporary sanctuaries and others of a later age.

Moving on to a much later age and a larger scale, **Belmonte et al.** investigated the orientations of the city Carthago Nova established by the Carthaginians, who opted for a series of astronomical alignments related to their main deities. When the city passed into Roman rule, Julius Caesar refounded the city and his successor Augustus established another orthogonal grid with clear astronomical orientations. The rich fusion of orientations, astronomy, religion, landscape archaeology, history and the tradition of planning the orientations of cities in this investigation provide an impetus for further studies of ancient cities where the original street grid and the surrounding landscape can still be observed or deduced.

Keeping to the same era, **Andrea Rodríguez-Anton et al.** analyse the orientation data of 93 Roman camps and forts in Britannia collected by Alan Richardson by means of a protractor, which they complement with estimates of horizon altitudes using a Digital Terrestrial Model. They then use a statistical model to compare the distribution of the resulting declinations with a set of homogeneous distributions. From this analysis they conclude that the observed orientations are non-random, that there is a concentration around orientations towards sunrise or sunset positions on particular dates. These dates could be related to feast days of the Roman deities worshipped in Britannia or dates related to the Roman warfare season. Interestingly, this pattern is observed on either side of Hadrian's wall and is similar to the trend observed in Hispania. Once more, this preliminary investigation could serve as a stimulus for more precise fieldwork in Britannia and in other regions of the Roman Empire.

Crossing over to the other side of the Atlantic, **Andrew Munro and Kim Malville** noted meaningful differences between the Great Houses of the Bonito Phase of Chaco Canyon New Mexico built before and after the drought of 1090—1100 CE. Besides differences in building technique, a significant characteristic of the Great Houses built after the drought was that they were located at places from where solstice sunrise or sunset could be observed at notable horizon features. The authors discuss alternative explanations for these differences and relate them to the eventual changes in the power structure of Chacoan society.

Delving deep into the Inca world, **Steven Gullberg** carried out a re-assessment of the 328 *huacas* (Inca shrines) in the Cusco basin of Peru which are set along 41 *ceques* or lines organized in four groups: three with nine *ceques* each and a fourth group with 14 *ceques*. Gullberg found that the *ceques* did not function as straight line sites directed towards astronomical events on the horizon as had been proposed in the 1970s. However, in many cases the orientations of *ceques* correlated with sunrises and sunsets over the component *huacas*, light and shadow effects, light tubes and also with cosmological aspects including water springs and channels, and carved stairs, among others. Gullberg's discussion of the complex interactions between sacred sites, astronomical events and cosmological aspects highlights the difficulties faced by other researchers who attempt to determine the motivation behind the orientations of monuments in Neolithic, Bronze Age and other contexts where the anthropological information is simply non-existent.

The investigation of orientations of more recent structures has attracted the attention of **Alejandro Gangui et al.**, who measured the orientations of 30 churches built before 1810 and a few later buildings, which constitute almost all of the Christian sanctuaries on the island of Lanzarote in the Canary Archipelago. The analysis produced an unexpected two-fold pattern: the majority of the churches were oriented towards the east (or west), as generally observed in the Christian world, while a substantial proportion had an anomalous orientation towards the north-northeast. The fact that the latter orientation correlates well with the direction of the prevailing winds provides a plausible reason for the trend but the real reason remains an open question. **Fátima Braña Rey** and **Ana Ulla Miguel** introduce us to an ethnoastronomy study of 245 stone granaries (*hórreos*) of Vilaboa, NW Spain, dating from the end of 17th century and later, some of which have carved motifs representing the sun, moon and stars. A preliminary study of the orientations does not show any particular pattern while a more detailed ethnographic investigation of nine *hórreos* revealed that people in the neighbourhood did not know the meaning of the celestial motifs. Indeed from interviews it transpired that the orientations may depend on wind direction and other meteorological phenomena rather than on astronomy, and that the motifs could be symbols relating to social status.

Calendars and time reckoning

Many cultures around the world felt the need to organize ritual events on days which correspond to astronomical phenomena. **Göran Henriksson** studied the method used in Old Uppsala, Sweden, to regulate the date of the Great Midwinter Sacrifice which took place every eight years depending on the full moon which occurred between 28 January and 26 February. Using historical records and carrying out extensive calculations of all full moons from 200 CE to 1200 CE, he identified 852 as a date when the sacrifice took place. He then studied the farmers' rule for determining the phases of the moon in advance and another method involving the determination of the sacrificial dates by solar observations. The latter method gained support from archaeological evidence when two very long rows of postholes were accidentally discovered in 2013 whose astronomical orientations corresponded with very significant alignments with the sun and the full moon. With this evidence at hand, the author could establish a rule based on astronomical observations to determine the date of the sacrifice in advance.

In an easterly context and a later age, **Vance Tiede** investigated the unusual dimensions of the Tall Gnomon of Guo Shoujing in Henan Province ins China which was restored in recent years. The gnomon is 40 *chi* (Chinese feet; approximately 9.75m) high with a horizontal grooved template scale 128 *chi* (~31.19m) long. A simple calculation and the historical record shows that the template scale is very significantly longer than needed for observing the meridian passage of the sun, even when uncertainties about the length of the *chi* are taken into account. The author tackles this anomaly by exploring the possibility that the gnomon was used to observe the meridian passage of the moon at its major standstill south, the nocturnal culmination of bright planets or stars, and the meridian passage of Venus in daylight. Interestingly, Tiede also seeks possible links of these observations with traditional Chinese cosmology.

Introduction: The Materiality of the Sky   xix

Constellations, asterisms and their associated mythologies and iconography

The constellations patterns perceived by the Egyptians, the Mesopotamians, the Greeks, the Chinese and other ancient cultures clearly reflect their geographical and cultural contexts. This observation implies that an understanding of the motivation behind the naming and the meaning of the constellations in different cultures requires a cross-disciplinary approach. **Roslyn Frank** presents an interesting example of this approach in her overview of the origins and diffusion of 'sky bears' in the northern hemisphere. She shows how hunter gatherers and other cultures projected the figure of a bear, or a bear-human hunter associated with 'bear ceremonialism'. The author discusses how this constellation has been enculturated across the northern hemisphere from the point of view of two contrasting paradigms: the diffusionist paradigm and the environmental niche paradigm. Frank argues for cognitive continuity to explain the observed and reported similarity in the cosmology connecting bears and humans across cultures in Europe, Eurasia and North America.

In a similar vein, but taking a wider, global perspective **Michael Rappenglück** reviews the cosmic fire motif across cultures and epochs. Archaic cultures considered several sources as the origin of cosmic fire, including the sun, the moon, the planet Venus, stars and other celestial bodies. The aurora, lightning, meteors, comets and other sky phenomena were considered as sources of transient fire. In several cultures fire was considered male, antagonistic but also complementary to water which was female. Between them fire and water accounted for creation, transformation, destruction and regeneration. The myth of the theft of fire from the heavens by a hero, beliefs about fire drilling, the ritual of kindling of fire at the beginning of a time cycle, and purification and cremation rituals also cut across many cultures. This study again illustrates the interaction of celestial and meteorological phenomena, religious beliefs and rituals which contribute to the cosmology of various cultures.

Starting from the Ancient Greek geometrical approach to an understanding of celestial motions, **Francisco Burillo and Pilar Burillo** review the symbolism acquired by the sphere by quoting several examples of representations of it in mosaic, marble, bronze, paintings and on coins. A sphere with circles on it was meant to represent the cosmos, the ecliptic, the celestial equator and latitudes. However, other images of the sphere were meant to represent perfection and, by analogy, divinity. Thus the authors show that the symbol of the sphere is found in images of Zeus, Urania and in Christian art.

Marianna Ridderstad explores the role of the solar goddess of the Baltic Finns, Päivätär, which was probably one of a long tradition of Iron Age female solar deities in the Baltic region. The author collates a mine of relevant information from rites mentioned in folklore, runic poetry incantations, and medieval staff runic calendars combined with archaeological, archaeoastronomical and ethnographic evidence. She also suggests that, with the arrival of Christianity in Finland, the major feast of the Virgin Mary was denoted by solar symbols which were attributes of the solar goddess. Understandably in this northern region, the interest in the sun as a source of fertility and regeneration and associated traditions were maintained irrespective of the change in religious beliefs. **Flora Vafea** refers to an Arabic myth connecting the constellation Orion representing a bride, and the bright stars Canopus, Sirius and Procyon which represented relatives: a brother and two sisters, respectively. The myth relates that Canopus and Sirius fled to the south of the galactic equator and became invisible while Procyon remained in the northern hemisphere. The author explores the possibility that this was not just a fanciful myth but that it reflects reality. The investigation produces a surprising result since the different proper motions of the stars can be used to show that the myth reflects a reality that occurred in a long distant past.

Frank Ventura discusses the general challenges that must be addressed when interpreting artefacts as symbols of celestial objects and events, especially in the context of prehistoric and preliterate cultures. He then suggests criteria, such as cultural continuity, that can be used to evaluate alternative interpretations to avoid relativism. The author then illustrates the use of these criteria in a discussion of the validity of interpretations of selected artefacts from the Late Neolithic temples of Malta which may symbolize the sun, the moon and the stars. On a similar tack, **Michael Rappenglück** addresses the challenges of determining the meaning and purpose of the iconography on ancient artefacts, with particular reference to the Nebra Disk. The author questions the methodologies that have been used to interpret the symbols on the disk and the rash fixing of hypotheses as the accepted truth. He then discusses in detail the various emblems on the disk and their possible meanings. Quoting many examples of similar emblems in various artefacts from different cultures, the author suggests that the iconography of the disk is emblematic, referring to myth, magic and ritual, although a simple astronomical meaning cannot be ruled out. For example, Rappenglück notes that the emblems on the Nebra Disk seem to reflect the tradition of decorating the shields of ancient Greek warriors with significant symbols. In particular, Homer's description of the shield of Achilles refers to representations of the earth, sea and sky, with the sun, the moon and the constellations at the centre, as well as the presence of several other representations of other elements.

Another warning about misrepresentations of the astronomical significance of sites emerges from the paper by **Benito Vilas Estevez,** who investigated the claim by uninformed investigators that a grouping of massive stones in Galicia, Spain, was a Roman observatory. The author shows that the impressive stones are simply natural boulders which are grouped without human intervention and the claimed orientations are simply random. From structured interviews the author finds that the false claims are increasingly being believed especially by foreigners. This is a clear example of why many archaeologists are sceptical about the claims of archaeoastronomy.

Beliefs, rituals and the sky

The two papers that fall under this heading illustrate the different methodologies that can apply for an investigation of beliefs and ritual which complement and corroborate the archaeoastronomical evidence. In a Late Neolithic context, the author of the first paper relies mainly on the archaeological evidence and a comparative survey of later cultures in other countries. In the second paper, the author reaches his conclusions from textual, iconographic and archaeological evidence. The first paper refers to six of the known temples of the Maltese Islands and an underground burial site with perforations in the walls which archaeologists interpret as oracle holes. **Tore Lomsdalen** reviews this explanation in the light of literature from other cultures and suggests that the oracle interpretation makes sense in the context of the religious beliefs, traditions and the cosmology of the culture concerned. He notes that the holes are positioned on the right-hand side of the monuments and may have involved the passage of objects or some form of intangible communication from one side to the other. An alternative purpose for at least two of the oracle holes may have been to let in sunlight at the solstices for reasons that require further investigation. In a study of a later culture, Stanisław **Iwaniszewski** investigates the Temple of the Inscriptions at Palenque, Mexico, which represents the underworld. It is also the memorial and mausoleum of Mayan ruler Pakal (615-683 CE) whose soul was supposed to enter the realm of the sun through a special conduit in the burial chamber. The author studies references to meteorological events, celestial bodies and Pakal's ancestors in the long hieroglyphic inscription and the iconography as well as the astronomical alignments of the monument and its visual connections to later eighth century structures in the vicinity.

Finally, a warning and the provision of a tool for mitigating the effects of the rapidly increasing level of illumination in inhabited areas and the insensitive lighting around monuments, particularly those of archaeoastronomical interest, which may further alienate the general public about the importance of the skyscape in past and modern cultures. **Georg Zotti and Günther Wuchterl** highlight this difficulty and recommend two approaches. The first is to continue raising public awareness about the negative effects of excessive nocturnal illumination and the second involves the provision of software tools to site managers to help them reach reasonable decisions when designing lighting for architectural features and other monuments. The authors describe and illustrate the use of a new feature in Stellarium planetarium software which permits site managers to foresee site-dependent lighting effects of various sources of illumination.

In conclusion

This brief summary of the papers in this volume serves to stimulate interest in reading the original papers and illustrates the range of interests displayed during the SEAC 2014 conference. Besides the interesting presentations, the programme included excursions to the temples at Ħaġar Qim and Mnajdra, the enigmatic cart-ruts at Dingli, and the old city of Mdina for all participants. Some participants opted for a visit to observe the (near) equinox sunrise aligned with the central axis of the Mnajdra South temple. Despite an overcast sky, almost miraculously, the sun made a glorious appearance at the right moment, to the delight of participants who had opted for an early morning rise. Another optional excursion included a visit to the Ħal Saflieni hypogeum, an underground burial site of the Temple Period carved out of living rock in a form that reflects the architecture of the temples. The General Assembly at the end of the conference was marked by the award of the prestigious Carlos Jaschek award to Kim Malville for his very significant contributions to archaeoastronomy. This was a notable event since it is only the fourth time that this award has been made since its institution in 2006. To finish off an intense week, there was a relaxed post-conference tour to Gozo, visiting the twin temples of Ġgantija as well as the nearby hypogea known as the Xagħra Circle. After an open-air lunch consisting entirely of local organic produce, a visit to the small but impressive fortified Cittadella city closed off the week.

The success of this conference depended directly on the commendable contributions of the presenters and indirectly on several other persons. In this respect, a word of thanks must go to the staff at the Valletta Campus of the University of Malta for their administrative support, Heritage Malta for their support and offer of free entrance to museums to participants, the staff of the Malta Tourism Authority and the staff of the Ministry of Finance the Economy and Investment for their logistics and financial support, Mr George Barbaro Sant of the Alberta Group for assistance throughout the project, the scientific committee for their efficient evaluation of the proposed abstracts, and to members of the local organizing committee, especially Tore Lomsdalen for his unstinting efforts to attract the conference to Malta and to be involved in all practical details and actions to ensure the smooth running of the event. We would also like to thank the Sophia Centre Press for their wonderful work bringing this volume to fruition, particularly the extensive editing work by Kathleen White, and the gorgeous layout skills of Jenn Zahrt.

Bibliography

Darvill, T. 'Afterword: Dances Beneath a Diamond Sky'. In *Skyscapes. The Role and Importance of the Sky in Archaeology*, edited by F. Silva and N. Campion, pp. 140–148. Oxford: Oxbow Books, 2015.
McKim Malville, J. 'Meaning and Intent in Ancient Skyscapes – An Andean Perspective'. In *Skyscapes. The Role and Importance of the Sky in Archaeology*, edited by F. Silva and N. Campion, pp. viii–xvi. Oxford: Oxbow Books, 2015.
Ruggles, C. L. N. 'Nature and Analysis of Material Evidence Relevant to Archaeoastronomy'. In *Handbook of Archaeoastronomy and Ethnoastronomy*, edited by C. L. N. Ruggles, pp. 353–372. New York: Springer, 2015.

Top: Kim Malville receives the Jashek Memorial Award from the SEAC president Michael Rappenglück. Photo: Tore Lomsdalen.

Bottom: A well-deserved lunch break during the post-conference tour to Gozo consisting entirely of local biological organic products. Photo: Tore Lomsdalen.

Facing page: Equinox sunrise tours to the Mnajdra Temple on 24 September 2014. Photos: Tore Lomsdalen.

Next page: Equinox sunrise tours to the Mnajdra Temple on 25th September 2014. Top photo by Frank Ventura; bottom photo by Alejandro Gangui.

Contributors

Juan Antonio Belmonte is a staff astronomer at the Instituto de Astrofísica de Canarias (IAC) in La Laguna (Tenerife, Spain). As PI of the Orientatio ad Sidera Project, he has been engaged in extensive research on Cultural Astronomy around the world, notably in the Mediterranean region.

Dr Fátima Braña Rey has a Sociology Degree from Universidad Complutense in Madrid, Spain, and a PhD in Applied Social Anthropology from Santiago de Compostela University, where she did a case study whose objective was the Museum of the Galician People. Dr Braña Rey's speciality is social patrimony and she has participated in different projects, including the Ronsel Project about immaterial heritage of Galician Region, and those at Galician ethnographic museums such as the Museum of Galician People, the Ethnology Museum of Ribadavia, the Museum of Wine in Ourense and others.

She was an invited professor at Vigo University in areas of Sociology and Anthropology, and has participated in different research programs of the autonomous government, Santiago de Compostela University and Vigo University. She has also taken part in the Ethnography Commission of the Galician Cultural Department of the Galician Autonomous Government. In addition, her writing is published in various journals and she presents her work at conferences on the topics of anthropology and patrimony, the significance of museums and the methodology of researching the contents of museums. Currently Dr Braña Rey is an Associate Professor in the Social Anthropology department of Universidade de Vigo.

Pilar Burillo-Cuadrado, University of Zaragoza, Ciencias de la Antigüedad, Campus de Teruel. Spain. e-mail: pburillo@unizar.es

Roger Clay is Emeritus Professor at the School of Physical Sciences of the University of Adelaide, Australia.

John Cox was born in 1947 and educated at the University of York and the University of Westminster. He is an artworker with a particular interest in archaeoastronomy and astronomical cartography. In 1976 he curated the exhibitions 'Peruvian Ground Drawings', 'Megalithic Sites', and 'Wind and Water' shown at the Institute of Contemporary Arts, London. In collaboration with Richard Monkhouse he has assisted the development of the astronomical imaging programme 'Starry8' designed to produce astronomical maps from digital catalogues. Images and maps produced by this programme have been exhibited at the London Planetarium, The Royal Greenwich Observatory (1988), and the Harvard Smithsonian Center for Astrophysics (2002). With Richard Monkhouse he has had published an atlas showing spectral types (*Colour Star Atlas*, George Philip, 1991) and an atlas showing stars and galaxies in simulated 3D (*3-D Atlas of Stars and Galaxies*, Springer-Verlag, 2000). In 2006-2007 he coordinated observations of moonrise from prehistoric temples in Malta and Gozo. At present he is working on star atlases that include a new figuring of Ptolemy's catalogue.

César Esteban has a PhD in Astrophysics and is a lecturer at the University of La Laguna and researcher at the Instituto de Astrofísica de Canarias. Although his main research topics are the chemical composition of ionized nebulae and chemical evolution of the Universe, he devotes part of his time to archaeoastronomy. He has carried out fieldwork in the Iberian Peninsula, Canary Islands, North Africa, the islands of Polynesia and Micronesia and the Valley of Mexico. In recent years, he has developed research projects on pre- and protohistoric sites in south and eastern Spain in collaboration with archaeologists.

Benito Vilas Estevez has a degree in History from the University of Santiago de Compostela and an MA in Archaeology by the University of Santiago de Compostela. He is currently pursuing the MA in Cultural Astronomy and Astrology at the University of Wales Trinity Saint David.

Roslyn M. Frank, Professor Emeritus at the University of Iowa, has done extensive research in the areas of archaeo- and ethnoastronomy as well as ethnomathematics. Two of her recent articles are included in *The Handbook of Archaeoastronomy and Ethnoastronomy*, one on the origins of 'Western' constellations and the other on the skylore of the indigenous peoples of northern Eurasia. In her research she has attempted to expand the notion of cultural astronomy by applying the tools of archaeological ethnology to the study of landscape and skyscape. For the past two decades she has been engaged in a project called 'Hunting the European Sky Bears'. That project has resulted in the publication of a series of articles that bring together ethnographic and linguistic materials which, in turn, have been applied to documenting and reconstructing what is a earlier pan-European cosmology grounded in the belief that humans descended from bears. Her other areas of research are cultural linguistics, cognitive linguistics, ethnography and anthropological linguistics with a special emphasis on the Basque language and culture. More information is available at: http://uiowa.academia.edu/RoslynMFrank.

Alejandro Gangui holds a PhD in Astrophysics from the International School for Advanced Studies, Trieste, Italy. Currently he is an Independent Researcher in CONICET, the National Scientific and Technical Research Council of Argentina, working at IAFE, the Institute for Astronomy and Space Physics. He is also Professor of Physics at the University of Buenos Aires, Argentina.

A. César González-García is a Ramón y Cajal Fellow at the Instituto de Ciencias del Patrimonio (Incipit) of the Spanish CSIC at Santiago de Compostela. He currently is the Secretary of the European Society for Astronomy in Culture (SEAC), discipline in which he is investigating.

Dr Thomas Gough holds an honours degree in chemistry (St. Andrews University, 1967) and a PhD in chemistry (Edinburgh University, 1971). During this time he built a number of telescopes for observation of variable stars, culminating in a 16" Newtonian reflector.

From 1971–74 he was employed at the Royal Observatory Edinburgh (R.O.E.) on a 'Site Testing Project', which involved searching for a new site in the Northern Hemisphere for an astronomical observatory. Gough proposed the Island of La Palma in the Canary Islands in 1972 and following more extended testing the site was chosen for the Roque de los Muchachos Observatory, a multinational observatory for use mainly by UK, Spain and Scandinavian countries.

From 1975–1980 and 1984–1997, Gough taught science at school level. From 1980–1984 he circumnavigated the globe in a 40 ft. yacht via the Panama Canal, Australasia and South Africa. This was before the electronics era and so ocean navigation was entirely by sextant. In 2006 he began a systematic investigation in limited regions in Scotland to reassess the possible existence of prehistoric alignments for the sun and moon using menhirs as backsights and natural horizon features as foresights, and field work began in May 2007.

Steven Gullberg was born in Jamestown, New York (USA) in 1956, and has a BS from the University of the State of New York and a MA of Liberal Studies (Ancient Astronomy) from the University of Oklahoma. In 2010 he graduated with a PhD in Astronomy from James Cook University and is a member of the International Astronomical Union (IAU). He is currently an Assistant Professor of Interdisciplinary Studies at the College of Liberal Studies, University of Oklahoma where he teaches astronomy and is the lead faculty member for the college's Liberal Studies degree program. His research interests lie primarily with Incan and Babylonian astronomy and he has authored a number of associated research papers.

Göran Henriksson was born in 1943. In 1983 he received a PhD in Astronomy at the Astronomical Observatory, Uppsala University. In 1985 he developed a computer program for precise calculations of positions for the sun, moon and other celestial objects, useful during the last six millennia. It was used in investigations of the Neolithic grooves on Gotland, the passage graves in Västergötland, the Bronze Age rock-carvings in Sweden and the sacrificial Iron Age calendar in Old Uppsala. In 1992 cooperation started with Mary Blomberg, Department of Ancient History, Uppsala University. All the palaces and peak sanctuaries on Crete were measured and Minoan astronomy was discovered. Absolute chronologies have been established from solar eclipses in the Near East, Egypt and China.

In 2011, he combined his knowledge in Archaeoastronomy and Modern Cosmology and made an accurate calibration of the lunar secular acceleration in longitude from 33 total solar eclipses back to 3653 BCE. This made it possible to verify Einstein's predicted precession of the geodesic in the earth-moon system. It became even possible to identify an extra lunar secular acceleration predicted in a new theory of gravity with a massive graviton. In 2016 he determined the mass of the Graviton.

Gail Higginbottom is a Landscape Archaeologist whose research has a broad remit and includes Cultural Landscapes, GIS and philosophy. Specialisms include European Prehistory and Cultural Astronomy, where she investigates the reasons behind why monuments were erected and what places were chosen to build them and why. She has also worked and researched in cultural heritage management in the UK and Australia. She is currently a visiting research fellow at both the Australian National University and The University of Adelaide, as well as a Fellow of the Society of Antiquaries of Scotland.

Dr Stanislaw Iwaniszewski is Professor-Researcher of Archaeology at the Escuela Nacional de Antropología e Historia – Instituto Nacional de Antropología e Historia, Mexico City, and a former SEAC (1999-2005) and ISAAC (2007-2014) President. In 2015, he made habilitation with the monothematic cycle of publications on the theoretical and methodological aspects of research in archaeoastronomy. From 2016 he has served as Managing Editor of the *Journal of Cultural Astronomy*.

Tore Lomsdalen has a Masters Degree in Cultural Astronomy and Astrology from the University of Wales Trinity Saint David, UK. He is currently affiliated with the Department of Classics and Archaeology at the University of Malta, where he undertaking a PhD on the research topic Cosmology in Prehistoric Malta. He is a member of the European Society for Cultural Astronomy (SEAC) and the Italian Society for Archaeoastronomy (SIA); in addition, he was the secretary of the Local Organisation Committee of the SEAC 2014 Conference at the University of Malta Valletta Campus and is a member of the editing committee of the SEAC 2014 proceedings publication.

Leonardo Lozito is national vice president and regional president of Apulia and Basilicata of the Archaeological Groups of Italy. He works at the Ministry of Heritage and Culture, in the Superintendence of Apulia and in the Superintendence of Basilicata. He was involved in numerous archaeological excavations, especially in Basilicata, resulting in the publication of results of the investigated sites in many scientific articles. Leonardo also deals with the protection of mobile works (paintings, statues, various furnishings) and their containers (such as churches, ethnographic collections, etc.) and conducts outreach to a wide audience via exhibitions, conferences and educational activities.

J. McKim (Kim) Malville. During the International Geophysical Year Kim spent a year in the Antarctic where he studied the aurora australis. He obtained his BS in physics from Caltech and his PhD in astrophysics from the University of Colorado. Kim is presently Professor Emeritus in the Department of Astrophysical and Planetary Sciences at the University of Colorado and Tutor at the University of Wales, Trinity Saint David. His primary research areas are solar physics and archaeoastronomy. In 1997 he was a member of the team that revealed the world's oldest known megalithic astronomy at Nabta Playa near Abu Simbel in southern Egypt. In 2003 he was involved in the rediscovery of the Inca ceremonial centre of Llactapata, previously lost in a cloud forest above Machu Picchu. In 2014 he received the Carlos Jaschek award of SEAC. Books he has authored or edited include *A Feather for Daedalus, Prehistoric Astronomy of the Southwest, Canyon Spirits: Beauty and Power in the Ancestral Puebloan World* (with John Ninnemann and Steve Lekson), *Ancient Cities, Sacred Skies: Cosmic Geometries and City Planning in Ancient India, Pilgrimage: Sacred Landscapes and Self-Organized Complexity, Chimney Rock: the Ultimate Outlier*, and *Machu Picchu's Sacred Sisters: Choquequirao and Llactapata* (with Gary Ziegler).

Ferdinando Maurici. After graduating in literature from the University of Palermo, he received a Master's degree in medieval archaeology at the University of Barcelona, where he studied with Manuel Riu, a PhD in medieval history and, again at the University Barcelona, where he was awarded a doctorate in Christian Archaeology. He was a fellow of the Alexander von Humboldt Foundation and an adjunct professor at the Universities of Bamberg and Frankfurt an der Oder. Author of many scientific papers and books, he is presently executive manager of the Office for Cultural Heritage of the Sicilian Region and contract professor of Christian Archaeology at the University of Bologna and of Medieval Archaeology at the LUMSA.

Dr Ana Ulla Miguel holds a Bachelor in Physics and a PhD in Astrophysics. Between 1992 and 1997, she held various research and teaching posts in national and foreign institutions, including the University of Tromsø (Norway), the Laboratory for Space Astrophysics and Fundamental Physics (LAEFF, Madrid), ESRIN-ESA (Italy), the Niels Bohr Institute in Copenhagen and the IAC (Canary Islands). In October 1997 she joined the Applied Physics Department of the University of Vigo (Spain), where she became, in January 2001, an Associate Professor.

Dr Miguel is a member of the International Astronomical Union (IAU), the Spanish Society of Astronomy (SEA) and of the Royal Spanish Physical Society (RSEF). Her research interest includes late stages of stellar evolution, exoplanets, the (ESA) Gaia satellite and Cultural Astronomy, among others. So far, she has co-supervised 5 PhD works and has authorised or co-authorised over 100 publications, including both peer reviewed papers and workshops contributions. Between March and June of 2016, she was a visiting researcher at the Astronomy Unit of Queen Mary University of London.

Andy Munro holds an MSc in Astronomy from Swinburne University, and a BS (Hon) in Political Science from Charter Oak College. He obtained his PhD from James Cook University in 2012 working with Kim Malville. His thesis was 'The Astronomical Context of the Archaeology and Architecture of the Chacoan Culture'. He continues extensive fieldwork in the Chaco Culture National Historical Park and outlying Chacoan Great Houses. He is interested in carefully documenting evidence for the uses of visual astronomy by Ancestral Puebloans with the goal of enhancing our understanding of their culture. His findings provide support for the idea that groups of people with varied cultural traditions collaborated at Chaco Canyon, and that Chaco may have operated as regional ritual and pilgrimage centre with consistently powerful overtones of astronomy. He supervises astronomy graduate student projects for Swinburne University of Technology (Australia), and manages a consulting firm in metropolitan Washington, DC.

José Miguel Noguera Celdrán is Professor of Archaeology at the Universidad de Murcia in Spain, where Roman History and Archaeology is his main research interest. In the last two decades, he has been engaged in the excavations of Punic and Roman Cartagena, including those sites discussed in the article.

José Ángel Ocharan Ibarra was born in Bilbao (Bizkaia, Spain) in 1968. He is a professional archaeologist with an undergraduate degree in history (University of Murcia, Spain), where he specialized in Prehistory, Archaeology and Ancient History. His postgraduate diploma is in Geography and History Education (Universidad of Murcia), with a speciality in Archaeological Drawing and Submarine Archaeology. He also holds a Masters Degree in Professional Archaeology and Heritage Management, and is a PhD student in Prehistory and Archaeology (University of Alicante).

Ocharan Ibarra is the President of the Asociación Cultural de Estudios Protohistóricos y Arqueología and a research member of both the Instituto Alavés de Arqueología and the Association of Archaeological Illustrators and Surveyors (University of Reading, Whiteknights, UK). His research focuses on Prehistoric Cultural Manifestations, Rock art, Religion and Symbolism in the Prehistoric Iberian Peninsula.

He has attended, organized and lectured in nearly fifty national and international conferences. He is actively involved in research Archaeology and has taken part in multiple Archaeological interventions, both national and international, thirty of which he has conducted himself. An avid traveller, he currently lives in Murcia (Spain), where he conducts the Archaeological interventions in the Sanctuary Cave of La Nariz (Moratalla, Murcia).

Andrea Orlando holds an astrophysics degree from the Institute of Interplanetary Space Physics (INAF/Rome) and a PhD in nuclear and particle astrophysics from the Superior School of the University of Catania and the Astrophysical Observatory of Catania (2012), with a thesis titled 'Multispacecraft observations of Coronal Mass Ejections'. Orlando has presented over 100 public lectures emphasizing the importance of education and culture as integral elements of a scientific researcher. Since 2012, he has been a member of the Italian Society of Archaeoastronomy (SIA), and has pioneered archaeoastronomy research in Sicily. In 2015 he organized the fifteenth conference of the SIA in collaboration the Department of Humanities/University of Catania. He also promoted the first comprehensive study of archaeoastronomy into the Valley of the Temples in Agrigento, in collaboration with the Politecnico of Milan (Italy) and the University of Waikato (New Zealand). Orlando is the founder and president of the Institute of Sicilian Archaeoastronomy, a cultural association that deals with the study of archaeological sites in Sicily with archaeoastronomical value. He is the creator and organizer of cultural events about ancient astronomy, including 'In Search of Astronomy and Ancient Music' and 'Stones & Stars - Archaeoastronomy Festival'. He is also author and/or co-author of numerous articles and scientific publications, nationally and internationally.

Mª Antonia Perera Betancort is chief archaeologist in the Heritage Unit from Cabildo de Lanzarote.

Francesco Polcaro. Born in 1945, graduated in Mechanical Engineering and Mathematics (General Relativity) and with a PhD in Aerospace Engineering, Francesco is an Associate Researcher at the Institute of Space Astrophysics and Planetology (INAF, Rome, Italy), a member of the Astronomy and Cultural Heritage (ACHe) Centre at the University of Ferrara (Italy) and a Research Associate at the Dublin Institute of Advanced Studies in Ireland. He is author of more than 300 articles and 3 books on Stellar Spectroscopy, High Energy Astrophysics, Aerospace Technologies, Historical Astronomy and Archaeoastronomy. Currently, his main fields of study are the evolution of very high mass stars, the astronomical contents of Bronze Age monuments and the collection and interpretation of information of astrophysical interest from ancient documents.

Michael A. Rappenglück was born 1 December 1957 in Karlsruhe, Germany. From 1977–1983 he studied philosophy, logic, theory of science, Christian philosophy and theological propaedeutic at Ludwig-Maximilian University, Munich, and was awarded a MA in 1984. From 1984–1997 he studied the history of science (astronomy) and in 1998 was awarded his PhD. From 2001-2009 he served as vice president of the German Society for Scientific Research of Symbols, and from 2003–2004 he served as vice president of SEAC, from 2005–2011 as secretary, and since 2011 as president.

Since 2006 Michael has been vice president of the Chiemgau Impact Research Society (CIRT), and was cofounder of the Society for Archaeoastronomy, Germany (GfA), serving as secretary from 2008–2011 and since 2011 as president. His memberships include: 'History of Surveying' Federation of German Surveyors (VDV); Working group of history of astronomy, Astronomical Society, Germany; ISAAC; Astronomical Society, Germany; Working Group on Astronomy and World Heritage (IAU, UNESCO); and the German Society for the Scientific Research of Symbols.

Current affiliations include: executive director and headmaster of the Adult Education Center, Gilching; head of public observatories in Gilching and Fürstenfeldbruck; educational consultant for the city of Gilching; peer reviewer for professional journals; scientific expert in national and international broadcastings and telecasts; scientific consultant for planetarium shows; and member of editorial boards, including the *Journal of Skyscape Archaeology* and *Scientific Culture*.

Dr Marianna Ridderstad obtained her PhD in astronomy (archaeoastronomy) in 2015. She also holds a MSc degree in theoretical physics (2002) and a Lic.Phil. in astronomy (astrophysics) 2011 from the University of Helsinki. In 2008, together with archaeologist Dr Jari Okkonen from the University of Oulu, she launched Finland's first archaeoastronomical project, which examined the orientations of the large Neolithic stone enclosures known as 'Giants' Churches'. Her other interests in archaeoastronomy have included investigating the astronomical symbolism and mythology of Finnish and Minoan cultures, as well as measuring the orientations of prehistoric stone cairns and Christian churches and graves in Finland. She also has taught astrobiology at the University of Helsinki since 2006 and written many popular scientific articles on archaeoastronomy in Finnish.

Andrea Rodríguez-Antón is PhD Student at the Department of Astrophysics of La Laguna University (Spain) and a member of the IAC. She is doing her PhD on the customs of orientation of cities and fortresses in the Roman Empire with special emphasis in Hispania.

Morgan Saletta holds Master's degrees in Cultural Anthropology from the Ecole des Hautes Etudes en Sciences Sociales and in Heritage Studies from the Muséum National d'Histoire Naturelle, Paris. He completed his PhD in the History and Philosophy of Science at the University of Melbourne.

Alberto Scuderi is national vice president and regional president for Sicily of the Archaeological Groups of Italy and Associate Member of the Italian Institute of Prehistory and Proto-History (Florence). He organized and directed many archaeological excavations in Sicily and in various parts of the world, such as in Romania (Adamclisi), Oman, Tunisia (Hergla).

He has to his credit dozens of excavations in collaboration with some of the most illustrious archaeologists, such as Professor Sebastiano Tusa, and with foreign universities such as that of Goteborg (Sweden), Oslo (Norway) and Illinois (USA). He is the author of many archaeological publications and has received several awards for his continued scientific contributions to archaeology of Western Sicily.

Fabio Silva is a Marie-Skłodowska-Curie Fellow at the Institut Català de Paleoecologia Humana i Evolució Social (Spain) and a tutor at the Sophia Centre for the Study of Cosmology in Culture, University of Wales Trinity Saint David (United Kingdom), where he is responsible for a postgraduate module titled 'Skyscapes, Cosmology and Archaeology'. His research interests centre on how humans perceive their environment (skyscape and landscape) and use that knowledge to time and adjust their social, productive and religious behaviours. His skyscape research has mostly focused on Neolithic Portugal, though he has also done fieldwork in the United Kingdom and Malta. His books include *Skyscapes: The Role and Importance of the Sky in Archaeology* (edited with Nick Campion, Oxbow Books, 2015). He has a PhD in Astrophysics (2010) and an MA in Cultural Astronomy and Astrology (2012).

Vance Tiede of Astro-Archaeology Surveys in Guilford, CT (USA) is an astro-archaeologist specializing in GIS and remote sensing applications regarding astronomical orientation in ancient monumental architecture worldwide. He has conducted field surveys at sites in China, Greece, Guatemala, Iceland, Italy, Ireland, Mexico, the United Kingdom and the United States (see http://yale.academia.edu/VanceTiede). Current research interests include geo-locating the (est. 16+) RAF/USAAF unexploded bombs (IX 1943) inside the Pompeii Archaeological Zone (*Scavi di Pompei*) with cluster analysis of its 160+ bomb craters. He is a graduate of The Johns Hopkins University (BA, History) and Yale University (MA, Archaeological Studies). Tiede is a member of the Archaeological Society of Connecticut, Historical Astronomical Division of the American Astronomical Society, and *Société Européenne pour l'Astronomie dans la Culture* (SEAC).

Flora Vafea was born in Piraeus, Greece, in 1958. She holds a degree in Mathematics from the National and Kapodistrian University of Athens (1981), a MA from Cairo University (1998) with a thesis titled 'Topics in Ancient Egyptian Mathematics', and a PhD in the History of Sciences & Technology from University Paris 7 (2006), with a thesis titled 'Les traités d'al-Ṣūfī sur l'astrolabe'. Vafea has been a teacher in Secondary School for 35 years in Greece, Cairo and Khartoum; she was a Visiting fellow at the *Bayerische Akademie der Wissenschaften* in Munich (2014), and a researcher in the Library *Dār al-kutub* in Cairo (2014-15). She speaks Modern and Ancient Greek, Arabic, English, French, German, Spanish, Ancient Egyptian and Latin. Her publications related to the History of Science include *The astronomical instruments in Saint Catherine's Iconography; From the Celestial Globe to the Astrolabe; Abd al-Raḥman al-Ṣūfī: Study of the Lunar and Solar Eclipses with the Astrolabe*; and *The Mathematics of Pyramid Construction in Ancient Egypt* among others. She has presented numerous papers at international conferences in UK, Germany, Greece, Egypt, Hungary, Peru and Malta. She has also worked in collaboration with the *PAL* project (since 2013), the research team on the Antikythera Mechanism (2007–2013), and the Planetarium and *Alexploratorium* of Alexandria Library (2009–2013), and is involved in the construction of astronomical instruments following Arabic manuscripts.

Frank Ventura was the first head of the Department of Mathematics, Science and Technical Education in the Faculty of Education at the University of Malta. He is currently the chairman of the MATSEC Examinations Board of the University of Malta which is responsible for national examinations at the end of compulsory education and pre-university levels. His interest in archaeoastronomy goes back to the late 1970s and since then Ventura has published several papers and articles on the archaeoastronomy of the Neolithic temples of the Maltese Islands and the Punic tombs in Malta which have appeared in local and international publications. He was also deeply involved in the organization of the INSAP 2 and SEAC 2014 conferences, both of which were held in Malta. He has also contributed to publications on science education, environmental education, educational assessment and areas relating to astronomy and archaeology.

Kathleen White is a writer, editor and researcher with a background in academia, journalism, PR, corporate communications, business management and education. She holds a BSc in classical piano performance and art history and a Postgraduate Diploma in Cultural Astronomy and Astrology. Her areas of interest include cultural cosmology and astrology, intellectual history, classical music, sound and music in ritual magic, and energetic herbalism and bodywork practices. She is the owner of InkHorn Editing, based in Bournemouth, Dorset in the UK.

Günther Wuchterl is an astronomer, Director of the Kuffner-Sternwarte for the Kuffner-Obs.-Society (Verein Kuffner-Sternwarte), and Co-Investigator of the CoRoT and PLATO 2.0 missions that search for planets outside the solar system.

Dr Jennifer Zahrt researches the history, philosophy and epistemology of astrology, with a special focus on the German cultural realm. Her doctoral thesis, titled *The Astrological Imaginary in Early Twentieth Century Germany* (University of California, Berkeley, 2012), concerns the ways in which astrology shaped philosophy, literature and the arts during the Weimar Republic. Her work appears in *Sky and Symbol* (Sophia Centre Press, 2013), *Verdant Gnosis: Cultivating the Green Path*, Vol. 2 (Rubedo Press, 2016) and *Table Talk* (Counterpoint, 2014); her translation work includes *Zoroaster's Telescope* (Ouroboros Press, 2013) and selected alchemical writings of Paracelsus, in *Occlith* 1 (Viatorium Press, 2016). She is the Creative Director of the Sophia Centre Press, and she lives in Seattle, WA.

Mauro P. Zedda was born in Isili, Sardinia in 1962. He is a land surveyor, having majored in Cultural Heritage at the University of Sassari in Italy. Zedda has published a dozen books and a long series of articles as a writer and as co-author with Italian and European scholars on the orientation of the prehistoric and early historic monuments of Sardinia. He has recently measured the orientation of all the Romanesque churches of Sardinia and a statistically significant sample of those of Corsica and Tuscany, and is about to proceed to publication. He is currently completing his PhD thesis with professor Anna Depalmas on 'Analysis and distribution of different types of nuraghe'.

Georg Zotti is a computer scientist and astronomer, a researcher at LBI ArchPro and member of the Stellarium development team. His research in virtual archaeology includes improving Stellarium for research and demonstrations in historical and cultural astronomy.

MALTESE ARCHAEOASTRONOMY

READING MESSAGES FROM THE PAST: INTERPRETING SYMBOLS OF POSSIBLE ARCHAEOASTRONOMICAL SIGNIFICANCE IN MALTA

Frank Ventura

ABSTRACT: The study of the megalithic temples of Malta from an archaeoastronomical perspective has focussed on temple orientations and their significance. Following this effort, archaeologists have recognised that the sky could have been an important element in the cosmology of the culture of the temple builders. If this model is correct then we should expect to find some reference to the sun, moon and stars and related events in the extensive repertoire of artefacts retrieved from the temple sites, including ceramic artefacts, relief carvings, statues and statuettes, and paintings on the ceiling of the hypogeum site at Ḥal Saflieni. The search for such evidence entails the interpretation of geometric, figurative and abstract designs as symbols of celestial objects and events. Certainly, this exercise is fraught with many difficulties. This does not mean, however, that we should not face the challenges and take the risk of not getting it right since a critical assessment of the symbolism should lead to a better understanding of the culture. The aim of this paper is to discuss the general challenges that must be addressed when interpreting artefacts and motifs as symbols of celestial objects and events, especially in the context of prehistoric cultures; the criteria that can be used to evaluate alternative interpretations so as to avoid relativism; the validity of the interpretations of some artefacts as symbols of astronomical significance for communities of the Temple Period in Malta.

Introduction

The megalithic temples of Malta are recognized by archaeologists as extraordinary architectural monuments of the Late Neolithic which demonstrate the builders' skill in planning, constructing and decorating complex buildings with limited resources. This achievement is even more remarkable when one considers that the structures date from as early as 3600 BCE and continued in use until around 2500 BCE when the temple culture came to a mysterious end. Fortunately, several structures have survived complete destruction, despite the ravages of time, weathering, human action and neglect. These precious remains, along with the artefacts retrieved from them, provide us with tangible evidence from which we can obtain an insight into the culture that created the monumental buildings. Naturally, the task of eliciting an intangible and complex characteristic such as culture from the material evidence requires careful interpretation. Given the subjective nature of interpretation, it would not be surprising if investigators from different theoretical backgrounds – archaeologists, anthropologists and scientists - arrive at different conclusions. Indeed, the following are some examples of divergent conclusions relating to the temples of Malta reached by different researchers:

- the structures were built by giants/by the Phoenicians/by prehistoric people;
- the monuments are temples/multi-purpose structures;
- the buildings had a roof /no roof;
- many of the statutes and statuettes represent the goddess of fertility/do not represent the goddess, indeed their gender is purposely not indicated;
- entrance to the temple was restricted to an elite/not restricted;
- the temple builders lived in isolation and had scarce contact with the outside world/were in contact with neighbouring islands.

Interestingly, only the first of the above-mentioned conflicting views has been resolved definitively when carbon dating of organic remains amply confirmed the prehistoric origin of the temples.[1]

Equally divergent views have been expressed about the temple builders' possible interest in the sky and how this is displayed in the structures they built and furnished. Starting from as early as 1840, the military officer commissioned with the clearing of the complex temple of Ħaġar Qim, commented that various circumstances and observations, which he did not specify, led him to believe that the temple builders had an express interest in the motions of the sun, moon and planets.[2] Such speculation continued sporadically over the years with Italian archaeologist Luigi Ugolini and photographer Gerald Formosa supporting the astronomical hypothesis while the eminent archaeologist J. D. Evans, architect Renato Laferla and archaeologist David Trump rejecting it.[3] Subsequently, systematic attempts to investigate the temples from an astronomical point of view began in the late 1970s and continue to this day. Almost invariably, the main focus of these studies has been on temple orientations and their interpretation. This work has led to a general agreement on their orientations but different interpretations of the builders' intentions for orientating their temples.[4] Again, this is understandable because orientations can be measured fairly accurately and objectively while interpretations are intrinsically subjective, dependent on the conceptual schema of the investigator and difficult to corroborate.

What has this effort achieved? It is safe to surmise that on a popular level there is now a general belief that the builders orientated the Mnajdra South temple intentionally to mark the equinox and the solstices. On a professional level, at least some archaeologists have taken note of such analyses and included the celestial domain as an important element in their models of the cosmology of the temple culture.[5] As far as archaeoastronomers are concerned, it is still debatable whether the orientations demonstrate the builders' interest in sunrise at the solstices and the equinox, the moon, or in the Pleiades and the Crux-Centaurus star clusters, or in both the sun and the stars.[6]

The purpose of the present paper is to widen the research focus by suggesting that if the temple orientations display the builders' interest in the sky then there should be some evidence of this interest in the iconography that has survived from that age. Fortunately, both the early clearing of some of the sites and the later scientific excavations have unearthed an extensive repertoire of statues, statuettes, decorated

---

[1] Colin Renfrew, *Before Civilization: The Radiocarbon Revolution and Prehistoric Europe* (London: Pimlico edition, 1999); C. Malone, S. Stoddart, and G. Clark 'Dating Maltese Prehistory', *Mortuary customs in prehistoric Malta: Excavations at the Brochtorff Circle at Xagħra (1987-94)*, ed. C. Malone, S. Stoddart, A. Bonanno, and D. Trump (Cambridge: McDonald Institute Monographs, 2009), pp. 341–46.
[2] J. G. Vance, 'Description of an ancient temple near Qrendi, Malta', *Archaeologia* XXXIX (1842): pp. 227–240, plates XXIII – XXVIII.
[3] Luigi Ugolini, *Malta: Origini della Civiltà Mediterranea* (Roma: La Libreria dello Stato, 1934), pp. 137–38; Gerald Formosa, *The megalithic monuments of Malta* (Vancouver, Canada: Skorba Publishers, 1975); J. D. Evans, *Malta* (London: Thames and Hudson, 1959), p. 125; and David Trump, *Malta: Prehistory and Temples* (Malta: Midsea Books, 2002), p. 199.
[4] Frank Ventura, 'Temple Orientations', in *Malta before History: The World's Oldest Free-standing Stone Architecture*, ed. Daniel Cilia (Malta: Miranda Publishers, 2004), pp. 307–325; Renato Laferla, 'La struttura, l'ubicazione e l'evoluzione dei templi cuprolitici nelle isole maltesi', in *L'Architettura à Malta* (Roma: Atti del XV Congresso di Storia dell'Architettura, Malta, 1970), pp. 195–212.
[5] S. Stoddart, A. Bonanno, T. Gouder, C. Malone, and D. Trump, 'Cult in an Island Society: Prehistoric Malta in the Tarxien Period', *Cambridge Archaeological Journal* 3, no. 1 (1993): pp. 3–19; S. Stoddart, 'The Maltese Death Cult in Context', in *Cult in Context: Reconsidering Ritual in Archaeology*, ed. David A. Barrowclough and Caroline Malone (Oxford: Oxbow Books, 2007), pp. 54–60; Caroline Malone, 'Ritual, Space and Structure – The Context of Cult in Malta and Gozo', in Barrowclough and Malone, *Cult in Context: Reconsidering Ritual in Archaeology*, pp. 23–34.
[6] J. Cox, 'Moonrise over Malta', *Astronomy and Geophysics* 49, no. 1 (2009): pp. 1.7–1.8; J. Cox and T. Lomsdalen, 'Prehistoric cosmology: observations of moonrise and sunrise from ancient temples in Malta and Gozo', *Journal of Cosmology* 9 (July 2010): pp. 2217–2231.

pottery, low relief sculpture, decorated stone, and rare paintings which can be examined.[7] Since practically all of the retrieved artefacts come either from the temple sites or from the two major underground burial sites of the Temple Period, it is reasonable to assume that they were of special significance for the activities that occurred at the sites, whether religious, social or political. In other words, the artefacts must have had a particular meaning for the people who created them. Interestingly, the scholarly study of prehistoric art has endured for several decades and continues, as demonstrated by a recent paper by Robb, who conducted a quantitative analysis of 2011 prehistoric art traditions from various countries across Europe, including Malta, and ranging from 40,000 BCE to 0 CE.[8] However, rather than exploring meaning, Robb was interested in the significance of how the amounts of art fluctuated over the period, and whether there were long-term patterns in the kinds of art that were made and in the social uses of art. In a detailed study of the passage tombs of the Irish and British Neolithic, Robin sought for meaning in the location of the different carved motifs and the artefacts, such as pendants, pins and beads, as well as the architecture of the tumuli and the thresholds in the tombs.[9] His conclusion was that the whole ensemble represented a system of shared beliefs and rituals of the passage from life to death among the Neolithic societies of the region. In the Maltese context, Grima's insightful analysis of architecture and iconography of five temples is particularly relevant.[10] This included a careful consideration of the different floor levels of the courts and adjacent apses of five temples, the low relief iconography of the screening slabs that delineate the courts and other carved images within the apses. The observations led Grima to argue that this evidence together with the location of the temples between embarkation points on the coast and the interior of the island suggests a cosmology of land and sea and the possibility of ritual activity connected to a metaphoric journey occurring within the temples. Furthermore, Malone considered about 250 figurative and other objects from the Maltese temples and burial sites and categorised them by attributes such as form, nature, meaning and portability as well as by the context of discovery within temples or burial sites.[11] From the patterns that emerged, Malone concluded that the collective imagery suggests a multi-layered cosmos including the subterranean, marine, land and upper layers inhabited by different creatures and possibly spirits. However, these studies by archaeologists hardly ever mention the sky and celestial objects. The challenge now is to try to identify correctly the art and artefacts that could signify the Maltese temple builders' interest in the sky and to decipher the original meaning of this material culture.

Challenges

At this point, it may be pertinent to ask: why have we neglected this facet of archaeoastronomy in Malta? A possible reply comes from a comment by Hodder to archaeologists when he stated that '... perhaps because of the enormous difficulty of making sense of the fragmentary data from long-gone societies, perhaps because of the difficulty of saying anything with any degree of certainty about the past, most ar-

---

[7] Anthony Pace, 'The Artistic Legacy of Small Island Communities: The Case of the Maltese Islands', in *Maltese Prehistoric Art 5000 – 2500 BC*, ed. A. Pace (Malta: Fondazzjoni Patrimonju Malti/The National Museum of Archaeology 1996), pp. 1–12.

[8] John Robb, 'Prehistoric art in Europe: a deep-time social history', *American Antiquity*, 80, no. 4 (2015): pp. 635–654.

[9] Guillaume Robin, 'Spatial structures and symbolic systems in Irish and British passage tombs: the organization of architectural elements, parietal carved signs and funerary deposits', *Cambridge Archaeological Journal* 20, no. 3 (2010): pp. 373–418.

[10] Reuben Grima, 'An iconography of insularity: a cosmological interpretation of some Images and Spaces in the Late Neolithic Temples of Malta', *Papers from the Institute of Archaeology* 12 (2001): pp. 48–65.

[11] Caroline Malone, 'Metaphor and Maltese Art: Explorations in the Temple Period', *Journal of Mediterranean Archaeology* 21, no. 1 (2008): pp. 81–109.

chaeologists prefer to become absorbed in data and method'.[12] Obviously, this comment is also relevant to archaeoastronomers who study prehistoric monuments and who prefer to adhere to measurements and calculations and avoid the risk of expressing views on the meaning of other available evidence dreading that they can be refuted at a later date. However, as Hodder remarks even 'what we measure and how we measure it are theoretical' and therefore subjective.[13] An example of how this comment applies to the discipline of archaeoastronomy can be deduced from the measurements of the orientations of the Maltese temples.

The temples are enclosed structures with a defined entrance, a central corridor with, typically, two pairs of semi-oval chambers or apses on each side and either a semi-oval central apse or a roughly rectangular niche at the rear end. The temples also have a concave facade which delineates an open-air oval court in front where the community could congregate for celebrations and ritual. Given this general description of the temple plans, which orientations should be measured? Different assumptions lead to different replies. The plans suggest that the main focus of interest is the innermost central apse or niche – where one would expect the 'high altar' to be situated – so that one should measure the orientation of the corridor, which coincides with this axis of symmetry. Accordingly, the abovementioned studies have measured the orientations of the central axes from the interior looking out towards the sky above the visible horizon, assuming the builders' interest in celestial objects. However, several studies have measured other orientations which involve the cross-jamb illumination at sunrise or sunset of what are considered significant areas of the temples. Cross-jamb illumination refers to the lighting by the sun's rays entering at an angle rather than perpendicularly through the doorway assuming that also particular structures on the sides of the central corridor had some special significance. Special attention has been given to illumination at the solstices and the equinoxes but also on cross-quarter days. On the other hand, archaeologists have suggested that the orientations in the opposite direction were also important. In fact, the congregation in the outer court would have been looking into the temples whose axes happen to be in the general direction of Sicily and nearby islands from where the builders' ancestors arrived and from where they obtained exotic materials such as obsidian, flint and red ochre.[14] Furthermore, the Ħaġar Qim temple has six doorways leading to the exterior which requires a decision on whether to measure the orientation of all entrances or just the one at the centre of the concave facade.

The next question concerns the method of measurement of the orientations. Methods may involve the use of a wide range of resources including published plans and a protractor, or the use of a prismatic compass and a clinometer, a theodolite, a GPS device, remote sensing and other means. Besides being prone to personal errors, all of these methods have a theoretical basis which is accepted as unproblematic or simply not questioned. Interestingly, the theoretical assumptions underlying decisions on which orientations to measure and how to measure them do not deter researchers from conducting their work and reaching conclusions using established methods of procedure.

The task of deciphering the meaning of artefacts, however, seems more daunting and risky leading to a reluctance to follow this interpretive approach. This is understandable since the theoretical assumptions involved seem to have a less solid basis than those concerning the measurement of orientations. Despite these difficulties, the search for meaning could also lead to very rewarding results. Indeed, through an understanding of the symbolism of the material artefacts not only can we provide further support for the conclusions reached from the study of the temple orientations but, more importantly, we can potentially understand the culture of the builders who produced them. This approach could contribute towards a

---

[12] Ian Hodder, *Theory and Practice in Archaeology* (London and New York: Routledge, 2005) e-book, p. 2.
[13] Hodder, *Theory and Practice*, p. 4.
[14] Stoddart et al., 'Cult in an Island Society', pp. 16–17.

fusion of archaeoastronomy and archaeology similar to the 'kind of synthesis between science studies and archaeology which is implicit in the interpretive and linguistic approach to landscape archaeology and more architectural approaches to the performance of space' suggested by Turnbull.[15] However, accepting that proceeding along this path is fraught with difficulties, it is important to identify the challenges that will be met, how best to face them and how to arrive at credible and persuasive conclusions.

In our interpretations of the meaning of ancient artefacts we cannot avoid using our western scientific conceptual framework which has been developing tortuously over hundreds of years. This framework embraces a wide range of simple and higher order concepts and powerful over-arching theories. Almost certainly, our modern conceptual framework is very different from the conceptual framework of the people of the temple culture. Yet, despite these differences we must somehow understand the other in their own terms and empathise with their culture if we wish to elicit and understand the intended meaning of the artefacts which we study. Hodder refers to this major challenge as the double hermeneutic.[16] A second challenge concerns cultural continuity and change. The beliefs, ritual and customs of the temple people may have changed over the one thousand year duration of the temple period. In fact, there is evidence that the temple structure and building techniques evolved and the incidence of decorative elements inside the temples increased and these changes may have been driven by cultural changes. On the other hand, there is also evidence of continuity, for example, in representations of obese figures, in a variety of sizes, which could represent ancestors or the deity. Another challenge is presented by the possibility that researchers from different disciplines and schools of thought using different conceptual schemes can ascribe different meanings to the same artefact. Additionally, arriving at a correct interpretation is complicated by the uniqueness of most of the relevant artefacts. Indeed, in most cases only unique specimens exist and this precludes the confirmation of the interpretation by reference to a pattern obtained from larger samples, such as the contexts in which several examples of an artefact are found.

Rising to the challenge

With the possibility of arriving at alternative and conflicting interpretations of the symbolism of the same artefacts, it is necessary to establish criteria that help us to determine their strengths and weaknesses and to decide which to accept and which to reject.

Criteria for accepting or rejecting interpretations of symbols.
1. Fitness within the whole context of what is known about the culture is possibly the best test of a good interpretation. A strong interpretation should fit very well in a narrative which describes the temple period and its achievements. Such an interpretation would match with the underlying knowledge, technological skills and artistic skills that were required for producing the monuments and the various artefacts. In other words, a valid interpretation is corroborated by internal evidence from the culture which produced the artefact. Hodder refers to this fit as 'wholeness',[17] while Renfrew and Bahn explain that in the interpretation of symbols 'it is the assemblage, the ensemble, that matters, not the individual object in isolation'.[18]
2. Interpretations that involve concrete processes such as observation and recording of observations are considered stronger than interpretations that imply the use of higher order concepts and abstract thinking. To take a fictitious example, in a prehistoric context, the interpretation of carvings of crescents and various

---

[15] David Turnbull, 'Performance and Narrative, Bodies and Movement in the Construction of Places and Objects, Spaces and Knowledges: The Case of the Maltese Megaliths', *Theory, Culture & Society* 19 (2002): pp. 125–134.
[16] Hodder, *Theory and Practice*, p. 139.
[17] Hodder, *Theory and Practice*, p. 22.
[18] Colin Renfrew and Paul Bahn, *Archaeology: Theories, Methods and Practice* (Fourth Edition) (London: Thames and Hudson, 2004), p. 394.

forms of circles as representations of phases of the Moon is more acceptable than an interpretation which claims that the carvings demonstrate knowledge of how sunlight is reflected from the Moon at various points in its orbit around the Earth.
3. In the case of unique artefacts where internal evidence is lacking, interpretations that can be supported by external evidence can be considered stronger than interpretations which stand alone and which are specific to the culture under study. Thus, we can accept well-established interpretations of analogous artefacts from other cultures as supporting evidence. Such analogies should be even more acceptable if the physical environment, stage of cultural development and other relevant factors both local and foreign are comparable since these factors generate comparable experiences.
4. Interpretations that lead to testable hypotheses or predictions are preferred to sterile interpretations that cannot be tested.

Armed with these thoughts, it is now instructive to discuss some prehistoric artefacts of the temple period which could have an astronomical significance. They will be divided into two groups: artefacts with presumed symbols of the sun and the moon, and artefacts with presumed symbols of the stars. Almost all of the interpretations mentioned hereunder have appeared in the literature but they have not been discussed properly and in most cases not cited in later studies. These examples illustrate the challenges of applying the suggested criteria for accepting or rejecting interpretations and possibly for inspiring new and more persuasive interpretations.

Sun and Moon Symbolism
Solar disks and the tree of life
The underground burial site at Ħal Saflieni, which is carved into the living rock with similar architectural features to the temples, contains rare examples of painting in red ochre on the ceiling and walls of two of the chambers. Both paintings show a composition of randomly arranged parts of coils or spirals which contrasts sharply with the spirals carved with remarkable symmetry in the temples above ground. One of them shows a number of definite disks of varying sizes interspersed among the coils and spirals (Fig. 1.1). This curious painting has led to various interpretations which agree that it had some symbolic – possibly religious – meaning but disagree about its actual meaning. Not long after the excavation of the site, Zammit reported that some had considered the arrangement of coils and spirals as a representation of the tree of life.[19] The disks were explained variously as the fruit of the tree, astral representations or even the cup-and-ring markings characteristic of Neolithic art. Ridley expanded on the interpretation of the tree of life by suggesting that the spirals represented the force and energy of nature and the disks stood for the sun, the life-giving force itself.[20]

These explanations address two points: the meaning of the coils and part spirals, and the meaning of the disks. Interestingly, Zammit supported the interpretation of this painting by analogy when he mentions that A. H. Sayee – professor of Assyriology at the University of Oxford – was among those who agreed that it represented the tree of life, a symbol largely used by the early Sumerians in their symbolic representations. However, he did not choose between different interpretations of the disks. Ridley's interpretation does not account for the numerous disks and it assumes an understanding of the dependence of plant growth on the sun, which reflects a modern scientific conceptual framework. Furthermore, this interpretation does not fit the context of the burial site unless another assumption is made that the symbol of the sun as the life-giving force is somehow related to belief in the afterlife.

---

[19] T. Zammit, *The Neolithic hypogeum at Ħal-Saflieni at Casal Paula-Malta: A short description of the monument with plan and illustrations* (Second edition) (Malta: Empire Press, 1935).
[20] Michael Ridley, *The megalithic art of the Maltese Islands* (Dorset: Dolphin Press, 1976), pp. 23–24.

Reading Messages from the Past: Interpreting Symbols of Possible Archaeoastronomical Significance in Malta

Fig. 1.1: Ceiling painting in red ochre at the Ḥal Saflieni hypogeum, Malta showing disks, spirals and coils, Heritage Malta.

Devereux has suggested an alternative interpretation of the painting by considering that the chamber in which it is found has special acoustic characteristics which amplify a deep voice by resonance.[21] He noted that the painting on the ceiling started from above a niche in the wall which seemed to be the cause of resonance. There were traces of three disks in the niche, which indicated a connection with the painting, and the size of the spirals and the disks become smaller the further they were from this niche. For Devereux, the painting could be a form of acoustic notation signalling sound and its amplitude – possibly the voices of gods or the ancestors. This interpretation fits the context better but it implies the abstract representation of sound, which is intangible, and raises the question of the why the users of the hypogeum felt the need to complement the sense of hearing by visual means.

Solar wheel decoration
The decoration on a unique black pottery shard recovered from the Ħaġar Qim temple many years ago and which was exhibited and published for the first time in recent years has been interpreted as an astronomical symbol. This curious decoration consists of a circle divided into eight approximately equal sectors with a rough dot close the centre of each sector. There are also short projections on the circumference which give the impression of 'rays' emanating from the circle (Fig. 1.2). Given the 'spokes' which divide the circle into sectors and the presumed 'rays', this decoration has been described as a 'solar wheel'.[22]

---

[21] Paul Devereux, 'A ceiling painting in the Ḥal-Saflieni hypogeum as acoustically-related imagery: a preliminary note', *Time and Mind: The Journal of Archaeology, Consciousness and Culture* 2, no. 2 (2009): pp. 225–232.
[22] Pace, 'Artistic Legacy', p. 9.

Fig. 1.2: Ceramic shard with 'solar wheel' decoration from Ħaġar Qim temple, Malta, Heritage Malta.

This decoration is unique in the repertoire of artefacts from the temple period and the validity of its sun symbolism can be supported by considering whether it fits in the archaeological context at Ħaġar Qim and by comparison with analogous symbols in other cultures. The solar connection at Ħaġar Qim was first noted by Formosa who showed how sunlight illuminated specific areas in the temple at the summer solstice sunrise and sunset.[23] Additionally, a study of solar illumination in the temple throughout the year has led Vassallo to conclude that Ħaġar Qim was built to mark time. Though these observations are certainly interesting, supporting evidence is needed to confirm that they were intentional.[24] There are also several instances of analogous symbols which have been interpreted as representations of the sun in foreign sources but they do not have dots within the sectors. Thus although the solar wheel interpretation of the symbol seems reasonable, it does not provide a satisfactory explanation for these curious dots. Interestingly, Prendergast[25] noted that among repertoire of motifs in Irish passage tombs, a complex motif that generally included a decorated semicircle with a radial pattern of lines terminated or punctuated by dots or circles occurs only at tombs that are either aligned towards or face the sun during its apparent annual or diurnal motion.

Sun and moon symbolism at Tarxien
Two of the four temples at the Tarxien site are richly decorated with spirals and animal representations and one of them encloses about one half of a larger than life anthropomorphic statue. Particularly interesting are the large but poorly preserved low relief carvings of a bull and a companion animal with thirteen appendages under its belly on a huge megalith and another bull carving on an adjacent megalith

---
[23] Formosa, *The megalithic monuments*, pp. 19–20.
[24] Mario Vassallo, 'Ħaġar Qim: a leading marker of Neolithic time', *The Sunday Times*, 13 February 2011, pp. 52–53.
[25] Frank Prendergast, 'Linked landscapes: Spatial, Archaeoastronomical and Social Network Analysis of the Irish Passage Tomb Tradition' (Unpublished PhD thesis, University College, Dublin, 2011), section 5.9.

Fig. 1.3: Large Stone with damaged low relief carvings of a bull and sow with thirteen appendages from the Tarxien Middle temple, highlighted by means of a rough outline to bring out the main details; photo, Frank Ventura.

in one of the side chambers in the central temple (Fig. 1.3). The companion animal on the first megalith has been interpreted as the representation of a sow and the appendages as piglets. Clearly, this interpretation stems from the idea that these representations symbolise fertility. However, Cutajar has suggested an astronomical interpretation. For him, the bull is a symbol of the sun and the companion animal is a cow, symbolising the moon, while the thirteen appendages represent the thirteen lunar cycles in a solar year.[26] This interpretation assumes that the temple builders measured the passage of time in terms of the solar year and lunar cycles and depicted the comparable length of the solar year and thirteen lunar cycles by means of the relief carving. The assumption of knowledge of the approximate length of the solar year is reasonable if we accept the temple builders' interest in observing and recording the interval of days from a winter or summer solstice to the next winter or summer solstice. But what could have been the motivation of showing the number of lunar cycles that fit into a solar year? In this case, the interpretation of these relief carvings as representations of fertility fit the context better than the astronomical interpretation because other decorative elements in the temple such as representations of phalli support the fertility interpretation.

Symbols related to the Stars

The Tal-Qadi stone

The Tal-Qadi stone is undoubtedly the artefact which presents the best evidence of the temple builders' interest in the sky. This unique fragment of globigerina limestone (29cm long by 24cm wide and 5cm thick) has inscribed radial lines dividing the surface into five sectors. Four sectors contain what are generally accepted as representations of six- and seven-pointed stars, symmetrically arranged in the wider part of the sectors and short lines at the narrower end. The other sector is empty except for a D-shaped figure

---

[26] Dominic Cutajar, 'Two relief-carvings of Chalcolithic Malta', in *Archaeology and Cult in the Ancient Mediterranean*, ed. A. Bonnano (Amsterdam: B.R. Gruner, 1986), pp. 163–67.

Fig. 1.4: Limestone slab with star and moon symbols from Tal-Qadi temple, Malta; photo, Frank Ventura.

which has been interpreted as the crescent moon (Fig. 1.4). Considering that the slab was discovered within the temple, Pace has suggested that it could have served as a votive offering. This interpretation accepts that the inscribed figures represent the stars but it implies that the composition is simply decorative and not an attempt to represent star positions in the sky. Other interpretations go further and try to identify an astronomical intention behind the production of this artefact. Some of these attempts may have been inspired by the fact that the Tal-Qadi temple is situated on the edge of a wide alluvial plain which ends in the nearby Salina Bay. In prehistoric times, this bay extended further inland and the temple would have been only a short distance from the seashore.[27] Presumably, the community in this area would have included seafarers who needed knowledge of the sky for safe navigation in the open sea. Support for the navigation hypothesis comes from the observation that if the stone is turned by 180 degrees, the upturned D-shaped figure can be seen as representing some form of sea-craft.[28]

The division of the slab into sectors around a radiant point has led to an alternative suggestion that it is an attempt to divide the night sky into regions of identifiable groups of stars thus producing a star chart. If this interpretation is correct, it is now practically impossible to identify the star groups especially when we consider that the artefact may be only a small part of a larger, rectangular or maybe circular slab. Yet there has been at least one suggestion that the symbols in three sectors represent the stars of Taurus, the Pleiades and Perseus, while the single star in the other sector represents Sirius or a bright star of the constellation Orion.[29] Some support for the star chart interpretation can be deduced by analogy from

---

[27] Nick Marriner et al., 'Geoarchaeology of the Burmarrad ria and early Holocene human impacts in western Malta', *Paleogeography Paleoclimatology Paleoecology* 339-341 (2012): pp. 52–65.

[28] Anthony Pace, 'The Buġibba, Tal-Qadi Cluster', in *Malta before History: The World's Oldest Fee-standing Stone Architecture*, ed. Daniel Cilia (Malta: Miranda Publishers, 2004), pp. 162–63.

[29] Peter Kurzmann, 'Die neolithische Sternkarte von Tal-Qadi auf Malta' ('The Neolithic star map at Tal-Qadi in Malta'), *Archaeology Online*, 2014, available at http://www.archaeologie-online.de/magazin/fundpunkt/sonstiges/2014/die-neolithische-sternkarte-von-tal-qadi-auf-malta. [accessed 26 July 2016].

Reading Messages from the Past: Interpreting Symbols of Possible Archaeoastronomical Significance in Malta

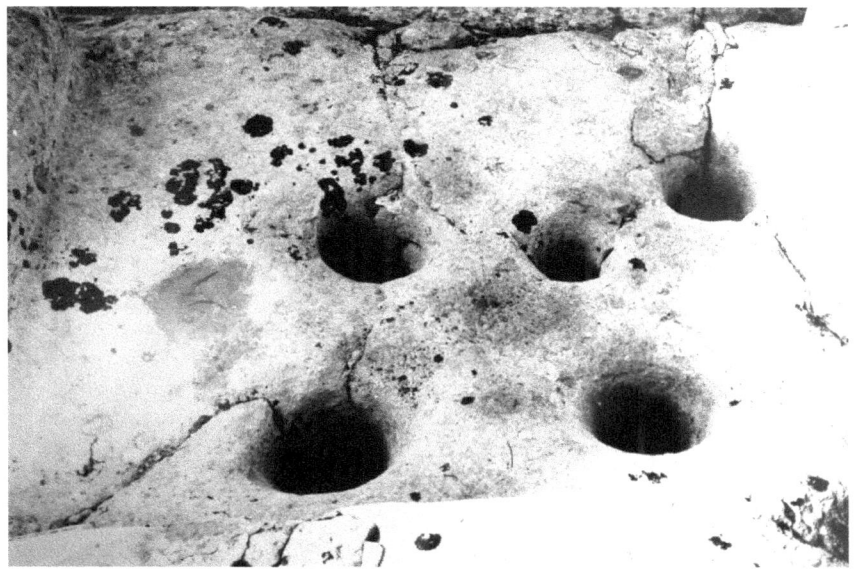

Fig. 1.5: Large floor slab with holes at the SE end of the concave facade of the Tarxien West temple, Malta; photo, Frank Ventura.

the small circular clay tablet K8538 from the library of Ashurbanipal, which is exhibited in the British Museum, London. The tablet, which is known as a planisphere, depicts the night sky divided into eight equal sectors in which the stars are represented by large and small holes in the clay with accompanying text in cuneiform.[30] Although, also in this case, there are different interpretations of the symbols and the text, the most quoted interpretation is that it shows the night sky of 3-4 January 650 BCE over Nineveh and the cuneiform text describes the constellations and includes astrological or magical text. Of course, the Tal-Qadi stone is about two thousand years older but, assuming that there was the required motivation, the skill needed for representing the night sky by means of star symbols grouped in sectors was not beyond the ability of the temple builders as it only requires the organisation of observations and the marking of the stone surface. Another suggestion has been that the decoration represents the record of a particular event, which again is difficult to identify. The feasibility of this interpretation arises if the D-shaped figure on the slab is a rough representation of the partially eclipsed sun rather than the moon. A deep partial eclipse of the sun visible from Malta during the period 3000 BCE to around 2500 BCE could have made such a great impression on the people of Tal-Qadi that they recorded it on stone. In this interpretation, the star symbols just represent the sky and not any particular star group.

Crux Pattern of Holes at Tarxien

The concave facade of the southwest temple at Tarxien is unique because of the large rectangular stone slabs at each end which could have acted as the floors of cubical recesses for use in some ritual. There are only scant remains of the slab at the west end but the slab at the east end is practically intact with five holes dug right through (Fig. 1.5). The excavator of the temple reported that the intriguing pattern in which these holes are set had led some to suggest that they represented the distinctive Southern Cross

---

[30] British Museum *Cuneiform Texts from Babylonian Tablets, etc. In the British Museum*. Part XXXIII (50 plates). (London: Department of Egyptian and Assyrian Antiquities, British Museum, 1912), Plate X.

(Crux) star group which was visible from Malta during the whole Temple Period.[31] The pattern is certainly suggestive of the asterism but this interpretation was disregarded soon after it was put forward. However, it has gained some credence after Foderà Serio et al. showed that the best fit for the orientations of the main temples was that they were aligned with the rising and procession of the Crux-Centaurus configuration of stars.[32] Interestingly, the investigation of the orientations of the monuments of other prehistoric cultures in the Central and Western Mediterranean region may serve to support this interpretation.

Thus, Hoskin has argued quite convincingly that the taulas on the island of Menorca in the Western Mediterranean and possibly also the pre-talayotic sanctuaries of the nearby larger island of Mallorca were also aligned with this group of stars.[33] Similarly, a major conclusion reached by Zedda and Belmonte was that a large percentage of the orientations of 452 circular nuraghes throughout Sardegna may have been oriented to the group of stars, consisting of members of the Southern Cross followed by Rigil Kent and Hadar, the brightest stars of Centaurus.[34] In view of this research and other possible evidence from central Italy, Magli suggested that the Crux and Centaurus stars could have been connected with an astral religion which spread across the Mediterranean over a long period of time.[35]

In contrast to this astronomical interpretation, archaeologists have interpreted similar holes as libation holes for liquid offerings. However, these holes are usually found in the threshold slabs at the entrances of several temples. Additionally, the excavator of the Tarxien temples recorded the presence of over a hundred small stone balls found nearby and surmised that they might have been connected to the five holes in the slab. He then suggested that the balls could have been used in some sort of ritual game for the purpose of divination.[36] Later, in the landmark survey of the prehistoric antiquities of the Maltese Islands, J. D Evans referred to this structure as the so-called 'divination block' although he and added that more probably it was a shrine. These interpretations do not exclude the astronomical hypothesis but neither do they support it.[37]

Standing Stone with cup-marks

Another intriguing artefact is the 1.46m high orthostat with curious marks which is currently lying against the outer wall of the Mnajdra Middle temple. Most probably, this stone is not in its original location in the temple and, perhaps for this reason, it has not attracted the attention of archaeologists as it has never been mentioned in the archaeological literature. This is surprising since it has four different types of artificial marks, two of which need some explanation. The most notable of these marks is a series of about 25 pits or cupules with an average diameter of 3.6cm which can be described as cup-marks. Most of them are on the upper right hand side of the stone. The other marks include a roughly inscribed cross at the top, an area of decoration consisting of small shallow circular pits at the middle, and a less prominent series

---

[31] T. Zammit, *The Neolithic Temples at Ħal-Tarxien, Malta* (Malta: Empire Press, 1929), p. 13.

[32] G. Foderà Serio, M. Hoskin, and F. Ventura, 'The orientations of the Temples of Malta', *Journal for the History of Astronomy* 23 (1992): pp. 107–19.

[33] M. Hoskin, *Tombs, Temples and their Orientations: A new perspective on Mediterranean prehistory* (Bognor Regis: Ocarina Books, 2001), pp. 37–52.

[34] Mauro Zedda and Juan Antonio Belmonte, 'On the orientations of the Sardinian nuraghes: some clues to their interpretation', *Journal for the History of Astronomy* 35 (2004): pp. 85–107.

[35] Giulio Magli, *Mysteries and Discoveries of Archaeoastronomy: From Giza to Easter Island* (New York: Copernicus Books, 2009), p. 62.

[36] T. Zammit, 'Third Report on the Ħal-Tarxien excavations, Malta (read 24th June 1920), Oxford Society of Antiquaries', *Archaeologia* LXX (1920): pp. 179–200.

[37] J. D. Evans, *The Prehistoric Antiquities of the Maltese Islands* (London: Athlone Press, 1971), p. 118.

of marks snaking down towards the bottom of the stone (Fig. 1.6). The cross and the sinuous marks at the bottom may be of recent origin but the small pitted area in the middle is probably prehistoric since it is similar to the pitted decoration at the top end of the east pillar flanking the entrance to the inner apse of the Mnajdra East temple. The interpretation of the presumed cup marks clearly presents the most interesting challenge from the archaeoastronomical point of view. The fact that so far this is the only stone with cup-marks found in the Maltese temples exacerbates the difficulty of interpreting them. So, before attempting an interpretation, it is important to establish that all the marks are man-made and not due to natural erosion. The next step is to establish that the marks can be considered as cup-marks or cupules by comparing them to similar marks on stones in Scotland, Ireland, England, Germany and other countries, where cup-marks on stones are an established phenomenon, even though the archaeological contexts in which they are found may be very different from the one at Mnajdra. The final and most difficult step is to determine whether the marks and their pattern are connected to the sun, moon or stars, keeping in mind that in most cases the astronomical interpretations of cup-marks or cupules on stones in foreign countries have been controversial. Among the few cases that provide strong supporting evidence for an astronomical interpretation is the careful study by Douglas Scott of the standing stones with cup-marks and cup-and-ring marks found at Kilmartin Glen, Scotland. Scott found that there are good correlations between stone alignments and the rising and setting of the midwinter sun and the south major standstill moon.[38] At an earlier date, Ruggles and Burl measured the azimuths of 11 stones with cup-marks from the centres of recumbent stone circles at various places in Aberdeenshire in Scotland. They found that the values of seven of the declinations corresponded well with rising or setting of the major standstill of the moon and another three declinations were within one degree of the minor standstill of the moon.[39] This study, which was supported by the evidence from the archaeological context, led the investigators to conclude that if the cup-marked stones (at least in the recumbent stone circles) have an astronomical significance, then we should undoubtedly undertake investigations of cup-marks elsewhere. In a completely different archaeological context, Edwards and Belmonte (2004) also arrived at a well-supported interpretation of cup-marks in their re-assessment of megalithic astronomy of Easter Island. In this case, besides studying the orientations of the ceremonial platforms (*ahu*) topped with statues (*moai*), they were guided by ethnographic evidence to investigate places where celestial bodies were observed by a community of astronomer-priests. One of these places was an outcrop in a remote peninsula on the island known as the 'rock for seeing the stars'. They not only found the rock, but close-by they found a boulder with ten cupules which corresponded to a presumed 'star map' that had been reported earlier. On investigation, the authors agreed that the cupules could be a representation of *Mataraki* (the Pleiades), which was an important asterism of the culture of Easter Island as corroborated by the other archaeological and ethnographic information.[40]

With reference to the Mnajdra stone, perhaps the most intriguing pattern of its cup-marks is that found towards the bottom left hand side of the stone (circled in Fig. 1.6) which consists of a cupule of more than average size accompanied by seven smaller cupules arranged in a roughly circular fashion around it. This pattern has a curious resemblance to a similar pattern on the Nebra disk. This does not imply any association between the two artefacts. Neither does it mean that an interpretation which fits for the

---

[38] Douglas Scott, 'Recent astronomical observations at Kilmartin Glen, Argyll, Scotland', *Antiquity* 84, no. 324 (June 2010), available at http://antiquity.ac.uk/projgall/scott/324 [accessed 25 July 2014].
[39] C. L. N. Ruggles and H. A. W. Burl, 'A new study of the Aberdeenshire recumbent stone circles, 2: Interpretation', *Archaeoastronomy* 16, no. 8 (1985)): pp. S25–S60.
[40] E. R. Edwards and J. A. Belmonte, 'Megalithic astronomy of Easter Island: a reassessment', *Journal for the History of Astronomy* 35 (2004): pp. 421–433.

Fig. 1.6: Stone with presumed cup-marks lying against the outer wall of the Mnajdra Middle temple, Malta; photo, Frank Ventura.

pattern on the Nebra disk would also fit the Mnajdra cup-marked stone. However, it provides further incentive for a closer investigation of this stone and its possible connection with the skyscape during the Temple period.

Conclusion

In conclusion, let us return to the concerns set out at the beginning with two questions: is it worth following this approach of trying to interpret the meaning of artefacts which may have an astronomical significance and taking the risk of being wrong? How can we minimize the uncertainty which is intrinsically involved in this process?

A possible answer to the first question can be found in an observation made by Thomas Kuhn, one of the most influential philosophers of science of the 20th century, about a change in worldview accompanying revolutions in science:

... when paradigms change, the world itself changes with them. Led by the new paradigm, scientists adopt new instruments and look in new places. Even more important, during revolutions scientists see new and different things when looking with familiar instruments in places where they have looked before. It is rather as if the professional community had been suddenly transported to another planet where familiar objects are seen in a different light and are joined by unfamiliar ones as well.[41]

These remarks can be adapted to our knowledge of the culture of the builders of the Maltese temples. By embracing the archaeoastronomy paradigm, a number of artefacts that were either neglected or overlooked have been brought to the fore. The interpretative attempts presented here may generate an interest in incorporating the celestial dimension in understanding the cosmology of the temple builders. Thus, for example, quite likely the 'solar wheel' sherd would have remained hidden in storage had it not been interpreted in the light of the new paradigm. Likewise, the rows of drilled holes in the Mnajdra East temple which were first described in 1913 were neglected for 80 years until they were interpreted as a tally probably related to the heliacal rising of a sequence of stars starting with the rising of the Pleiades.[42] An example of an artefact that has been overlooked is the stone with 'cup-marks' found leaning against the outer wall of the Mnajdra Middle temple whose interpretation still needs to be determined. The other examples mentioned in this paper show how the archaeoastronomy paradigm can apply to different types of artefacts.

The problem posed by the second question on how to minimize the intrinsic uncertainty of the interpretive approach is more difficult to resolve. Besides searching for more supporting evidence for our interpretations, a possible way forward is to adopt a critical approach in formulating interpretations, engage in scholarly debates on the strengths and weaknesses of alternative interpretations and broader interdisciplinary perspectives.

Bibliography

British Museum. *Cuneiform Texts from Babylonian Tablets, etc. In the British Museum*. Part XXXIII (50 plates). London: Department of Egyptian and Assyrian Antiquities, British Museum, 1912.
Cox, John. 'Moonrise over Malta'. *Astronomy and Geophysics* 49, no. 1 (2009): pp. 1.7–1.8.
Cox, John, and Tore Lomsdalen. 'Prehistoric cosmology: observations of moonrise and sunrise from ancient temples in Malta and Gozo'. *Journal of Cosmology* 9 (July 2010): pp. 2217–2231.
Cutajar, Dominic. 'Two relief-carvings of Chalcolithic Malta'. In *Archaeology and Cult in the Ancient Mediterranean*, edited by Anthony Bonnano, pp. 163–7. Amsterdam: B.R. Gruner, 1986.
Devereux, Paul. 'A ceiling painting in the Ħal-Saflieni hypogeum as acoustically-related imagery: a preliminary note'. *Time and Mind: The Journal of Archaeology, Consciousness and Culture* 2, no. 2 (2009): pp. 225–232.
Edwards, E.R., and J. A. Belmonte. 'Megalithic astronomy of Easter Island: a reassessment'. *Journal for the History of Astronomy* 35 (2004): pp. 421–433.
Evans, J. D. *Malta*. London: Thames and Hudson, 1959.
Evans, J. D. *The Prehistoric Antiquities of the Maltese Islands*. London: Athlone Press, 1971.
Foderà Serio, Giorgia, Michael Hoskin, and Frank Ventura. 'The orientations of the Temples of Malta'. *Journal for the History of Astronomy* 23 (1992): pp. 107–19.
Formosa, Gerald. *The megalithic monuments of Malta*. Vancouver, Canada: Skorba Publishers, 1975.
Grima, Reuben. 'An iconography of insularity: a cosmological interpretation of some Images and Spaces in the Late Neolithic Temples of Malta'. *Papers from the Institute of Archaeology* 12 (2001): pp. 48–65.
Hodder, Ian. *Theory and Practice in Archaeology* [e-book]. London and New York: Routledge, 2005.

---

[41] T. S. Kuhn, *The Structure of Scientific Revolutions* (Third Edition) (Chicago and London: The University of Chicago Press, 1996), p. 111.
[42] F. Ventura, G. Foderà Serio and M. Hoskin, *Journal for the History of Astronomy* 24 (1993): pp. 171–183; Hoskin, *Tombs, Temples and their Orientations*, pp. 31–36.

Hoskin, Michael. *Tombs, Temples and their Orientations: A new perspective on Mediterranean prehistory*. Bognor Regis: Ocarina Books, 2001.

Kuhn, T. S. *The Structure of Scientific Revolutions* (Third Edition). Chicago and London: The University of Chicago Press, 1996.

Kurzmann, Peter. 'Die neolithische Sternkarte von Tal-Qadi auf Malta' ('The Neolithic star map at Tal-Qadi in Malta'), *Archaeology Online*, 2014. http://www.archaeologie-online.de/magazin/fundpunkt/sonstiges/2014/die-neolithische-sternkarte-von-tal-qadi-auf-malta.

Laferla, Renato. 'La struttura, l'ubicazione e l'evoluzione dei templi cuprolitici nelle isole maltesi'. In *L'Architettura à Malta. Atti del XV Congresso di Storia dell'Architettura, Malta*, pp. 195–212. Roma, 1970.

Magli, Giulio. *Mysteries and Discoveries of Archaeoastronomy: From Giza to Easter Island*. New York: Copernicus Books, 2009.

Malone, Caroline. 'Metaphor and Maltese Art: Explorations in the Temple Period'. *Journal of Mediterranean Archaeology* 21, no. 1 (2008): pp. 81–109.

Malone, Caroline. 'Ritual, Space and Structure – The Context of Cult in Malta and Gozo'. In *Cult in Context: Reconsidering Ritual in Archaeology*, edited by David A Barrowclough and Caroline Malone, pp. 23–34. Oxford: Oxbow Books, 2007.

Malone, Caroline, S. Stoddart, and G. Clark. 'Dating Maltese Prehistory'. In *Mortuary customs in prehistoric Malta: Excavations at the Brochtorff Circle at Xagħra (1987-94)*, edited by C. Malone, S. Stoddart, A. Bonanno, and D. Trump, pp. 341–46.. Cambridge: McDonald Institute Monographs, 2009.

Marriner, Nick, Timothy Gambin, Morteza Djamali, Christophe Morhange, and Mevrick Spiteri. 'Geoarchaeology of the Burmarrad ria and early Holocene human impacts in western Malta'. *Paleogeography Paleoclimatology Paleoecology* 339-341 (2012): pp. 52–65.

Pace, Anthony. 'The Artistic Legacy of Small Island Communities: The Case of the Maltese Islands'. In *Maltese Prehistoric Art 5000 – 2500 BC*, edited by A. Pace, pp. 1-12. Malta: Fondazzjoni Patrimonju Malti/The National Museum of Archaeology, 1996.

Pace, Anthony. 'The Buġibba, Tal-Qadi Cluster'. In *Malta before History: The World's Oldest Free-standing Stone Architecture*, edited by Daniel Cilia, pp. 158–65. Malta: Miranda Publishers, 2004.

Prendergast, Frank. 'Linked landscapes: Spatial, Archaeoastronomical and Social Network Analysis of the Irish Passage Tomb Tradition'. Unpublished PhD thesis, University College, Dublin, 2011, section 5.9.

Renfrew, Colin. *Before Civilization: Thee Radiocarbon Revolution and Prehistoric Europe*, (London: Pimlico edition, 1999)

Renfrew, Colin, and Paul Bahn. *Archaeology: Theories, Methods and Practice* (Fourth Edition). London: Thames and Hudson, 2004.

Ridley, Michael. *The megalithic art of the Maltese Islands*. Dorset: Dolphin Press, 1976.

Robin, Guillaume. 'Spatial structures and symbolic systems in Irish and British passage tombs: the organization of architectural elements, parietal carved signs and funerary deposits', *Cambridge Archaeological Journal* 20, no. 3 (2010): pp. 373–418.

Robb, John. 'Prehistoric art in Europe: a deep-time social history'. *American Antiquity*, 80, no. 4 (2015): pp. 635–654.

Ruggles, C. L. N., and H.A.W. Burl. 'A new study of the Aberdeenshire recumbent stone circles, 2: Interpretation'. *Archaeoastronomy* 16, no. 8 (1985)): pp. S25–S60.

Scott, Douglas. 'Recent astronomical observations at Kilmartin Glen, Argyll, Scotland'. *Antiquity* 84, no. 324 (June 2010). http://antiquity.ac.uk/projgall/scott/324.

Stoddart, S., A. Bonanno, T. Gouder, C. Malone, and D. Trump. 'Cult in an Island Society: Prehistoric Malta in the Tarxien Period'. *Cambridge Archaeological Journal* 3, no. 1 (1993): pp. 3–19.

Stoddart, Simon. 'The Maltese Death Cult in Context'. In *Cult in Context: Reconsidering Ritual in Archaeology*. Edited by David A. Barrowclough and Caroline Malone, pp. 54–60. Oxford: Oxbow Books, 2007.

Trump, David. *Malta: Prehistory and Temples*. Malta: Midsea Books, 2002.

Turnbull, David. 'Performance and Narrative, Bodies and Movement in the Construction of Places and Objects, Spaces and Knowledges: The Case of the Maltese Megaliths'. *Theory, Culture & Society* 19 (2002): pp. 125–134.

Ugolini, Luigi. *Malta: Origini della Civiltà Mediterranea*. Roma: La Libreria dello Stato, 1934.

Vance, J. G. 'Description of an ancient temple near Qrendi, Malta'. *Archaeologia* XXXIX (1842): pp. 227–240, plates XXIII – XXVIII.

Vassallo, Mario. 'Ħaġar Qim: a leading marker of Neolithic time'. *The Sunday Times*, 13 February 2011, Malta, pp. 52–53.

Ventura, Frank. 'Temple Orientations'. In *Malta before History: The World's Oldest Free-standing Stone Architecture*, edited by Daniel Cilia, pp. 307–325. Malta: Miranda Publishers, 2004.

Ventura, F., G. Foderà Serio, and M. Hoskin. 'Possible Tally stones at Mnajdra, Malta'. *Journal for the History of Astronomy* 24 (1993): pp. 171–183.

Zammit, T. *The Neolithic Temples at Ħal-Tarxien, Malta*. Malta: Empire Press, 1929.

Zammit, T. *The Neolithic hypogeum at Ħal-Saflieni at Casal Paula-Malta: A short description of the monument with plan and illustrations* (Second edition). Malta: Empire Press, 1935.

Zammit, T. 'Third Report on the Ħal-Tarxien excavations, Malta (read 24th June 1920) Oxford Society of Antiquaries'. *Archaeologia* LXX (1920): pp. 179–200.

Zedda, Mauro, and Juan Antonio Belmonte. 'On the orientations of the Sardinian nuraghes: some clues to their interpretation'. *Journal for the History of Astronomy* 35 (2004): pp. 85–107.

THE 'ORACLE HOLES' OF THE MALTESE PREHISTORIC TEMPLES: AN INVESTIGATION OF THEIR ASTRONOMICAL/SOLAR ALIGNMENTS

Tore Lomsdalen

ABSTRACT: Several scholars have suggested that the small openings that may be observed in various Maltese megalithic temples (3600 BCE–2500 BCE) may be tied to oracular practices. The evidence for such an interpretation is critically reviewed. The surviving material evidence is also examined on site. An alternative hypothesis is put forward. It is argued that solar alignments and cardinal orientations may also be a significant factor in accounting for these features.
Keywords: Malta, Prehistory, Temple culture, Archaeoastronomy, Oracle culture.

1. Introduction

The aim of this paper is to investigate critically whether the so-called 'oracle holes' in the Maltese temples were an integral feature within a sacred or religious cultural context. The paper will also investigate to what extent the perforations in the temple walls have any significant astronomical orientations: alignments towards celestial bodies or cardinal points related to seasons and the solar year.

The word 'oracle' implies divine inspiration and contact through a form of direct communication with a god through a professional seer and interpreter of divine will, often in the context of religious rituals, sacred cult performances and mystery religions.[1] It is important to distinguish between prophetic mechanisms which are interpreted as visible signs of the gods, such as flights of birds, sudden sounds, falling stars or entrail divination of sacrificed victims, and divine presence or appearance with a close physical proximity of gods to men.[2] This paper will concentrate on the second issue, namely if there might have been a ritual or ceremonial leader or master who performed religious acts connecting heaven and humans.

An essential question for the evaluation of whether oracular performances took place inside the Maltese temples is to ask if these monuments existed in the context of a sacred space where cult, rituals and religious ceremonies were conducted; in essence, were they actually temples. Secondly (and equally) essential is the question of a solar-influenced oracular culture; to what extent did the sky and celestial bodies influence the sacredness of the monuments themselves and rituals and ceremonies conducted there, in term of their performance and in terms of the time when they could occur.

The methodology for this investigation includes fieldwork, site survey, horizon astronomy, observation and research in relevant literature related to the Maltese Temple Period and oracular culture in general.

2. Early oracular, omen and prophecy cultures

The first recorded forms of divination and priestly seers have been verified in Mesopotamia; Babylonia and Assyria practised many forms of oracles. By the second millennium BCE, they predominantly depended on mechanical devices and visible signs, for example, the flight of birds, entrails of sacrificed animals, abnormal human and animal births and forms of lecanomancy.[3] According to Henriksson and

---

[1] Bernard C. Dietrich, 'Oracles and Divine Inspiration', *Kernos* 3 (1990): p. 159.
[2] Dietrich, 'Oracles', p. 162.
[3] Dietrich, 'Oracles', p. 165.

Blomberg, if the Minoan's (about 2000 BCE) knowledge of the motions of the heavenly bodies were acquired at the peak sanctuaries and if this knowledge had important consequences for religion (cult calendar) and the economy (agriculture, navigation), then it is also likely to have affected other areas of society in significant ways; they further suggest that if the Mycenaeans used a lunisolar calendar, their source is likely to have been the Minoans.[4]

Rochberg agrees with Dietrich that the origin of written or scholarly divination goes back to the Old Babylonian period (second millennium BCE).[5] Huber, however, has advanced the hypothesis that some lunar-eclipse omens may refer to historical events from specifically the Akkadian and Ur III periods, implying that celestial divination existed already in the third millennium BCE.[6] Koch-Westenholz claims that eclipses were very important celestial omens in Babylonia and Assyria and an intense watch was kept when their occurrences were expected.[7] That the Babylonians were keen observers of celestial events is not only evidenced by their many omens related to planetary movements, but also the earliest known evidences for the zodiacal division of the year and the institution of lunar calendars.[8] Neugebauer asks whether the 'Metonic cycle' was Meton's own invention or of Babylonian origin; however, he seems to imply that it may derive from the Mesopotamian lunisolar calendar.[9] According to Neugebauer, Babylonian mathematical astronomy had much influence on Hellenistic science and Greek astronomers like Hipparchus, who used Babylonian methods and experiences.[10] Campion asserts that it is clear from the records that accumulated empirical celestial observations were being used to match political events to astronomical patterns.[11]

The Temple of Pythian Apollo at Delphi, probably the most well known oracular centre and a contemporary icon of oracular practices, was constructed in 550 BCE. According to Liritzis and Castro, to schedule Delphic festivals and find the proper day for a prophecy involved careful, exact calculations carried out by learned priests and philosophers.[12] The Delphic calendar was a lunar-solar-stellar one and the temple itself was oriented towards the seasonal celestial motion of Lyra and Cygnus, two of Apollo's favourite constellations.[13] Furthermore, Castro et al. suggest this has also been proved for the Apollo Temples at Didyma and Hierapolis.[14]

Although the Maltese Temple Period was contemporaneous with the Babylonian and early Hellenistic

---

[4] Göran Henriksson and Mary Blomberg, 'Evidence for Minoan Astronomical Observations from the Peak Sanctuaries on Petsophas and Traostalos', *Opuscula Atheniensia* XXI, no. 6 (1996): p. 110.
[5] Francesca Rochberg, *The Heavenly Writing: Divination, Horoscopy, and Astronomy in Mesopotamian Culture* (Cambridge: Cambridge University Press, 2004), p. 65.
[6] Peter J. Huber, 'Dating by Lunar Eclipse Omens with Speculations on the Birth of Omen Astrology', in *From Ancient Omens to Statistical Mechanics: Essays on the Exact Science Presented to Asger Aaboe*, ed. J. Berggren and B. R. Goldstein (Copenhagen: Copenhagen University Library, 1987), pp. 3–13.
[7] Ulla Koch-Westenholz, *Mesopotamian Astrology*, Vol. 1 (Copenhagen: Museum Tusculanum Press, 1995), p. 105.
[8] Otto Neugebauer, *Astronomy and History: Selected Essays* (New York: Springer-Verlag, 1983), pp. 383–85.
[9] Neugebauer, *Astronomy*, pp. 382–83.
[10] Neugebauer, *Astronomy*, p. 383.
[11] Nicholas Campion, *The Dawn of Astrology: A Cultural History of Western Astrology*, Vol. 1, *The Ancient and Classical Worlds* (London: Continuum, 2008), p. 48.
[12] Ioannis Liritzis and Belén Martin Castro, 'Delphi and Cosmovision: Apollo's Absence at the Land of the Hyperboreans and the Time for Consulting the Oracle', *Journal of Astronomical History and Heritage* 16, no. 2 (2013): p. 184.
[13] Liritzis and Castro, 'Delphi', p. 184.
[14] Belén Martin Castro, Ioannis Liritzis. and Anne Nyquist, 'Oracular Functioning and Architecture of Five Ancient Apollo Temples through Archaeoastronomy: Novel Approach and Interpretation', *Nexus Network Journal, Architecture & Mathematics* DOI 10.1007/s00004-015-0276-2(2015).

period, there is no written evidence that any omen-based, oracular or prophecy culture existed, based on the simple fact that that this prehistoric culture has left us no written record; writing was first introduced by the Phoenicians around 700 BCE.[15] All indications of religious rituals, belief systems and cosmology in prehistoric Malta are evidenced through the archaeological record and the physical layout of the temple sites and consequently subject to modern bias, perception and interpretation. However, that ritual performances may have occurred is supported by archaeological finds of quantities of human and animal bones, offering bowls, libation holes in thresholds, stone tools, ornaments, figurative artefacts, statues, carved stones and art. There are also series of altar-like arrangements and evidence of apertures that could control access and visibility to certain areas within the temple compound. In front of the entrances to both Hagar Qim and Mnajdra South temples are two holes bored through the rocky ground, which may possibly have been used for tying up sacrificial animals.[16] Malone further proposes that the temples show indications of being places of highly structured ceremonial ritual practice which may reflect an equally complex social structure.[17] Renfrew suggests that Temple-Period Malta possessed a chiefdom society where the chief had a socio-economic role and received, as dues or gifts, a significant part of the produce from each group and area.[18]

3. Oracular culture and the Maltese temples

That the Maltese megalithic structures were actually temples is indisputably based on circumstantial evidence.[19] However, most scholars seem to agree with Trump's statement: 'There can now be very little argument but that they really were temples'.[20] The same reasoning applies regarding whether they were sacred spaces for rituals and ceremonies. In assuming they were so, Malone offers the following justification: 'Malta presents some of the most sophisticated designed architectural ritual spaces furnished with symbolic iconography and material culture in early western Europe'.[21] Furthermore, this paper's author has more extensively analysed the sacredness of Maltese temples in another publication.[22] A temple by definition implies sacredness, rituals, religious feasts and festivals performed by ceremonial officiators in front of a congregation; furthermore, Barrowclough suggests that 'Practices performed in these buildings would necessarily bear some relation to those of contemporary religions'.[23]

In 1882, Caruana remarks that in the right apse of the Mnajdra South Temple that several of the megaliths were pierced through suggesting they were 'holes like the Keltic monoliths'.[24] Zammit in his

---

[15] Anthony Bonanno, *Malta: Phoenician, Punic, and Roman* (StaVenera, Malta: Midsea Books Ltd, 2005), p. 6.

[16] Tore Lomsdalen, *Sky and Purpose in Prehistoric Malta: Sun, Moon and the Stars of the Temples of Mnajdra*, Vol. 2, Sophia Centre Master Monographs (Ceredigion, Wales, UK: Sophia Centre Press, 2014), pp. 48–49; Themistocles Zammit, *The Copper Age Temples of Hagar Qim and Mnajdra: With Plans and Illustrations* (Valetta: Facsimile Edition, 1927), pp. 26–27.

[17] Caroline Malone, 'Ritual, Space and Structure - the Context of Cult in Malta and Gozo', in *Cult in Context: Reconsidering Ritual in Archaeology*, ed. David A. Barrowclough and Caroline Malone (Oxford: Oxbow Books, 2007), p. 23.

[18] Colin Renfrew, *Before Civilization: The Radiocarbon Revolution and Prehistoric Europe* (London: Pimlico, 1973), pp. 170–71.

[19] Giulio Magli, *Mysteries and Discoveries of Archaeoastronomy from Giza to Easter Island* (New York: Copernicus Books, 2009), p. 49.

[20] David H. Trump, *Malta: An Archaeological Guide* (London: Faber and Faber Ltd., 1972), p. 24.

[21] Malone, 'Ritual', p. 27.

[22] Tore Lomsdalen, 'Cult, Ritual, Sacred Space and the Sky in the Prehistoric Temples of Malta' (paper presented at The Marriage of Heaven and Earth: Images and Representations of the Sky in Sacred Space, Bath Spa, UK, 2014), in press.

[23] David A. Barrowclough, 'Putting Cult in Context: Ritual, Religion and Cult in Temple Period Malta', in *Cult in Context: Reconsidering Ritual in Archaeology*, ed. David A. Barrowclough and Caroline Malone (Oxford: Oxbow, 2007), p. 47.

[24] A. A. Caruana, *Report on the Phoenician and Roman Antiquities In the Group of the Islands of Malta* (Malta: Government

1927 publication was the first to identify the two rectangular windows at the South Temple of Mnajdra as part of oracular chambers.[25] He also considered the rectangular hole at Hypogeum Hal-Saflieni to be part of an oracular chamber.[26] Trump seems convinced that there were oracle holes at Mnajdra South and Tarxien Temples, and mentions an Oracle Room at Hypogeum Hal Saflieni and possibly an oracle-like hole at Hagar Qim.[27] Skeates, on the other hand, seems more reluctant to accept them as 'oracle holes' and uses inverted commas whenever mentioning the term.[28] Skeates further suggests that the 'oracle holes' at Mnajdra and Tarxien may be later features, and it is unclear what passed through these holes; nevertheless he proposes they may have been used for religious communication, ritual practices and/or sensory experiences.[29] Zammit argues that oracular rooms form a prominent feature of the Maltese megalithic sanctuaries, indicating that all these places of worship were built with great forethought and hosted rites in the islanders' religious life.[30] Zammit further claims that oracular chambers appear to be areas in which an inspired priest was supposed to respond to the inquiries of worshippers through communication with the spiritual world.[31]

3.1 Oracular temple sites

Malone refers to five temple sites with so-called oracle holes; this paper's author suggests there may be six in total. As noted by Malone, they are all placed to the right-hand side of the temple entrance. This is supported by site surveys conducted by the present author. Hertz argues that some authors believe the differentiation of right and left is effectively explained by the rules of religious orientation and sun worship.[32] Based on Hertz's dualistic scheme – the perception that left is weak and right is strong – Malone has developed a model for laterality in prehistoric Malta.[33] The concept that the prehistoric Maltese had such a notion of left and right could partially explain why oracle holes were placed on the right-hand side of a temple entrance.

3.1.1 Xaghra Circle

In the detailed, extensive report of the 1987–1994 excavation of the hypogeum at Brochtorff Circle (renamed Xaghra Circle), Malone and Stoddart argue that this underground temple had many elements which suggest a complex of beliefs and rituals, reinforced by rules relating to positions of right, left and ahead, including 'oracle holes' built into the wall of the right-hand side of the entrance.[34] The so-called

---

Printing Office, 1882), p. 15.

[25] Zammit, *Copper Age Temples*, p. 28.

[26] Themistocles Zammit, *The Neolithic Hypogeum at Hal-Saflieni at Casal Paula-Malta: A Short Description of the Monument with Plan and Illustrations*, 3rd edition (Valletta: Empire Press, 1935), p. 43, Fig. 11.

[27] David H. Trump, *Malta: Prehistory and Temples*, Photography Daniel Cilia (Malta: Midsea Books, 2002).

[28] Robin Skeates, *An Archaeology of the Senses: Prehistoric Malta* (Oxford: Oxford University Press, 2010).

[29] Skeates, *An Archaeology of the Senses*, p.170; Skeates, *An Archaeology of the Senses*, p. 209.

[30] Karl Ing. Mayrhofer, *The Prehistoric Temples of Malta and Gozo: A Description by Prof. Sir Themistocles Zammit* (Ing. Karl Mayrhofer, 1995), p. 35.

[31] Zammit, *The Neolithic Hypogeum*, p. 27.

[32] Robert Hertz, 'The Pre-Eminence of the Right Hand: A Study in Religious Polarity', in *Right and Left: Essays on dual symbolic classification*, trans. Rodney Needham and Claudia Needham, ed. Rodney Needham (Chicago: University of Chicago Press, 1973), pp. 335–57, pp. 3–31.

[33] Malone, 'Ritual', pp. 31–33.

[34] Caroline Malone and Simon Stoddart, 'Conclusions', in *Mortuary Customs in Prehistoric Malta: Excavations at the Brochtorff Circle at Xaghra (1987-94)*, ed. Simon Stoddart, Caroline Malone, Anthony Bonanno, and David Trump, with Tancred Gouder and Anthony Pace (Cambridge: McDonald Institute Monographs, 2009), p. 372.

'chapel', which could conceivably have formed an oracle room, is placed immediately to the right at the bottom of the easterly oriented entry steps.[35] Based on the assumption that the eastern entrance of the funerary complex was aligned to the equinoctial sunrise, it is nevertheless difficult to make any inferences on whether the 'chapel' could have been a part of any solar ritual or ceremonies. However, the author of this paper suggests that this assumption should not be completely excluded and could be an area for further research.

3.1.2 Ggantija

Based on the author's on-site observations the Ggantija temples have a number of holes in the temple wall of various sizes and shapes. However, to suggest that some of these are oracle holes would currently be purely speculative; more examination is needed to verify whether they fit the criteria and arguments on what constitutes an 'oracle hole'. Some holes inside the temple give more the impression of being used as rope holes for closing off certain areas.

Nevertheless, Malone and Stoddart argue that there is an oracle hole on the right-hand side of the entrance to Ggantija Temple, without giving further details.[36] Personal correspondence from Malone and Grima refers to an altar-like arrangement of the inner right-hand apse (no longer *in situ*) of the south temple.[37] The only indications that such a possible oracle hole ever existed can been seen in Brochtorff's drawings and pictorial documentation from the nineteenth century.[38] Further speculation about the solar or celestial alignments of this possible oracle hole must wait until further facts are at hand. From Brochtorff's drawings it is also difficult to establish whether the hole was a feature built into or penetrating the temple wall.

Vassallo suggests that Ggantija South Temple's entrance itself has a cross-jamb offset illumination at the winter solstice.[39] This event has also been observed and photographically documented by the present author on 27 December 2010; at the moment of sunrise it was clouded, but cleared up after some time. Another astronomical event related to this temple is made by Cox, who claims it is aligned with the far-southerly Moonrise.[40]

3.1.3 Hal Saflieni Hypogeum

Zammit was the first to name the hypogeum at Hal Saflieni a 'sanctuary' and to relate one of the apses to an oracle room.[41] This room is situated on the middle level and on the right-hand side of the entrance. In the chamber is an oval-shaped niche, decorated with red ochre discs, which may have been used as an echo amplifier to transmit sounds throughout the temple.[42] That the so-called oracle hole possesses special acoustic effects was tested by Grima, Cortis and the present author on 28 December 2010.

Skeates maintains that it is unlikely that the 'Oracle Chamber' at the Hypogeum Hal Saflieni was intentionally created by its builders as its extraordinary acoustics are enhanced today by the fact that the

---

[35] Malone, 'Ritual', p. 33.
[36] Malone and Stoddart, 'Conclusion', p. 372.
[37] Tore Lomsdalen, 16 June 2014.
[38] Reuben Grima, *The Archaeological Drawings of Charles Frederick De Brocktorff*, ed., design and photography Daniel Cilia (Malta: Midsea Books and Heritage Malta, 2004), p. 62.
[39] Mario Vassallo, 'Sun Worship and the Magnificent Megalithic Temples of the Maltese Islands', *The Sunday Times of Malta*, 23 January 2000, pp. 40–1, 30 January 2000, pp. 44–5, and 6 February 2000, pp. 36–7.
[40] John Cox, 'Observations of Far-Southerly Moonrise from Hagar Qim, Ta' Hagrat and Ggantija Temples from May 2005 to June 2007', *Cosmology Across Cultures, ASP Conference Series*, Vol. 409 (2009).
[41] Themistocles Zammit, *Malta: The Islands and Their History*, 2nd edition (Valletta: The Malta Herald, 1929), pp. 60–65.
[42] Anthony Pace, *The Hal Saflieni Hypogeum: Paola* (Malta: Heritage Malta, 2004), p. 40.

catacomb has been almost totally cleared of deposits; these would originally have reduced its resonance.[43]

The oracle hole itself is oriented to about 160°/340° degrees, twenty degrees from the north/south cardinal direction; it thereby does not qualify for any equinoctial or solstitial solar risings or settings. On the other hand, the oracle chamber may have cosmological symbolism, represented by its many drawings in red ochre, a metaphor for the blood of life and death and a potent symbol associated with ancestral creation and the past.[44] The uniform design of interlinking spirals may represent trees of life, symbolizing life.[45] Red ochre may also be a metaphor for the sun, another life force symbol.[46] When it comes to the hypogeum's solar alignments, Pace suggests the monument's orientation would have allowed the best light penetration around the summer solstice, particularly during the afternoons and sunset hours and that this effect may have been intentional.[47] Vassallo, on the other hand, observed that the main axis of the hypogeum is oriented towards azimuth 121° from north; this coincides with the position of the winter solstice sunrise during the Temple Period.[48]

3.1.4 Tarxien

Gozo's two temples, Xaghra Circle and Ggantija, one underground and the other aboveground, are in each other's vicinity; so, too, are the Hal Saflieni and the Tarxien Temples on Malta. Barrowclough refers to and agrees with Zammit that the hole in one of the apses of the East Temple was part of an adjoining, external Oracle Room where small objects such as greenstone pendants could be dropped through a channel by a priest-like figure to worshippers.[49] Twelve greenstone pendants were excavated from within and around the Oracle Room area.[50] By the time of the author's visit to the site in 2014/15, two U-formed holes had been noticed, one through the apse wall and the other in front of the oracle hole on the outside, which may have been used for passing small objects as suggested by Zammit and Barrowclough. Further to the right of this hole (seen from inside), Zammit observes that two round holes pierced through the temple wall, 'a few inches apart' and about 10 cm in diameter, but concludes their purpose is not evident.[51] Evans agrees with Zammit's evaluations of the perforations in the temple wall and maintains that the apse with the oracle hole to the left is evidence that some kind of 'oracle' is very plausible (Fig. 2.1).

Although it is impossible to be certain of the exact nature of the room's use, Evans nevertheless states, 'We may be sure at least that some kind of "priestcraft" was the underlying reason for such elaborate arrangements'.[52] Trump, on the other hand, describes the apse as having 'Two oracle holes open to the right'.[53] Trump's statement implies that these, rather than the ones suggested by Zammit and Evans, were the oracle holes (see Fig. 2.1).

---

[43] Skeates, *An Archaeology of the Senses*, pp. 209–10.
[44] Christopher Tilley, *The Materiality of Stone: Explorations in Landscape Phenomenology: 1* (Oxford: Berg, 2004), pp. 140–43.
[45] Marija Gimbutas, *The Living Goddesses* (Berkeley, CA: University of California Press, 2001), p. 61.
[46] Campion, *The Dawn of Astrology*, p. 5.
[47] Anthony Pace, *The Hal Saflieni Hypogeum: 4000 BC - 2000 AD*, ed. Anthony Pace (Malta: National Museum of Archaeology, Museums Department, Malta, 2000), pp. 12–15.
[48] Mario Vassallo, 'Sun Worship in the Hypogeum: The Guiding Light for the Dead', *Times of Malta*, 23 and 30 January 2000, 6 February 2000.
[49] Barrowclough, 'Ritual', pp. 51-52; T. Zammit, 'Third Report on the Hal-Tarxien Excavation, Malta', *Archaeologia* 70, no. 1918-20 (1920): p. 182.
[50] Barrowclough, 'Putting Cult in Context', pp. 50–51.
[51] Zammit, 'Third Report on the Hal-Tarxien Excavation', p. 182.
[52] John D. Evans, *The Prehistoric Antiquities of the Maltese Islands: A Survey* (London: The Athlone Press University of London, 1971), p. 132.
[53] Trump, *Malta: Prehistory and Temples*, p. 123.

Fig. 2.1: The apse with the adjoining external Oracle Room and the two smaller holes to the right. Photo: Lomsdalen.

Based on a site survey by the present author, the oracle hole suggested by Zammit does have an eastern orientation when viewed from inside out but its exactness is subject to where the measurement is taken within the apse. Measured from the centre of the ingress of the apse of the East Temple, it indicates an azimuth of about 90°. The oracle hole is in the right-hand apse from the temple entrance.

3.1.5 Tas-Silg

An exceptional case of the continued use of architectural structures is a late Neolithic megalithic building, parts of which were used for ritual purposes. Evidently still well-preserved, it was transformed into the cell of a Phoenician-Punic temple dedicated to Astarte, later to the Roman goddess, Hera-Juno, and finally into a baptistery during the Byzantine period.[54] The Neolithic part of Tas-Silg has a threshold with three libation holes oriented towards the east.[55]

In temple no. IV is the rectangular formed room M (Fig. 2.2). On the east side of the room there is a nearly four meters long altar-like arrangement with an oracle hole opposing it, giving the room an impression of being part of a 'shrine'.[56] Based on personal observations and measurements in 2014 and 2015, the north/south aligned vertical orthostat (now partly broken) includes the oracle hole with an east/west orientation. The slab with the hole has, from outside the room, an unworked natural surface, while on the inside, is carefully dressed with rope holes for closure, supporting the theory that this may have been a special and important room.

---

[54] Alberto Cazzella and Giulia Recchia, 'Tas-Silg: The Late Neolithic Megalithic Sanctuary and Its Re-Use During the Bronze Age and Early Iron Age', *Scienze dell'Antichita* 18 (2012).

[55] Cazzella Alberto and Giulia Recchia, 'Revisiting Anomalies: New Excavations at Tas-Silg and a Comparison with the Other Megalithic Temples I Malta', *Accordia Research Papers* 10 (2004–2006): p. 64.

[56] Cazzella and Recchia, 'Tas-Silg: The Late Neolithic Megalithic Sanctuary', p. 21.

Fig. 2.2: Room M from the west with the altar in the back and the oracle hole on the front left hand side; lower right insert, the oracle hole seen from inside the room. Photo: Lomsdalen.

Tas-Silġ is a complicated site for which to establish the temple's main entrances. According to plans of Cazzella and Recchia,[57] indications are that the temple has both an eastern and western entrance. Based on the eastern entrance to the Neolithic temple, the oracle hole and the 'shrine' would be on the right-hand side. It is questionable if the equinoctial sunrise or sunset would illuminate the hole itself, and its cardinal orientation may have a more symbolic or cosmological meaning. This site deserves further studies regarding these features.

3.1.6 Ħagar Qim
Formosa is the first known person to photographically document the summer solstitial sunrise through the so-called oracle hole at Ħagar Qim in 1974.[58] Trump calls this hole 'possibly an oracle like Mnajdra's'.[59] Skeates connects this 'oracle hole' – which penetrates the temple wall – with an angle to an adjoining external shrine.[60] Mayr suggests this arrangement served as a kind of altar.[61] The penetration of the light of the summer solstice sunrise has been documented by Vassallo in various publications.[62]

From fieldwork in 2011 the present author has also observed how the light from the summer solstice sunrise penetrates the oracle hole and illuminates a formation of a crescent waxing moon onto an altar-like arrangement inside the temple apse due to the concave formation of the oracle hole and the slab placed outside the hole (see Fig. 2.3).

---

[57] Cazzella and Recchia, 'Tas-Silġ: The Late Neolithic Megalithic Sanctuary', p. 19
[58] Gerald J. Formosa, *The Megalithic Monuments of Malta* (Vancouver: Skorba, 1975), pp. 20–21.
[59] Trump, *Malta: Prehistory and Temples*, p. 144.
[60] Skeates, *An Archaeology of the Senses*, p. 168.
[61] Albert Mayr, *Die Vorgeschichtlichen Denkmäler Von Malta* (München: Verlag der k. Akademie, 1901), pp. 667–68.
[62] Vassallo, 'Sun Worship', pp. 44–45; Vassallo, 'Sun Worship in the Hypogeum'; Mario Vassallo, 'Ħagar Qim's Layout Shows Yearly Movements of the Sun', *The Sunday Times*, 6 February 2011, pp. 52–53.

Fig. 2.3: The summer solstitial sunrise seen from the area inside the temple where the slab is illuminated as a waning moon (see insert, lower left). Photo: Lomsdalen.

Elaboration on the exactness of the illumination at the time of sunrise would require a study of its own. However, based on preliminary observation and calculations of the first illumination onto the slab inside the temple apse, the sun had a declination of about +21.35°; according to Agius and Ventura the sun rose at declination +24.0° during the Maltese Temple Period.[63] Based on the assumption that the oracle hole and the slab in front of it had the same layout during the Temple Period as today, it is highly questionable that the rising sun at the summer solstice would illuminate a waning moon inside the temple; the sun at Hagar Qim would rise nearly two degrees or about three to four solar discs further north. Based on flashlight tests by the author it does not seem to be the oval perforation of the oracle hole itself that forms the waning moon, but the form of the slab outside that creates this special effect. Evans, on the other hand, does not give any clear indication whether the slab outside the oracle hole is an original or a later feature.[64]

3.1.7 Mnajdra

The Mnajdra site consists of three individual temples: the small trefoil, the middle, and the lower, also named the South Temple. In the latter are two oracle holes in the right-hand apse after the main entrance.[65] The smaller and more finely elaborated hole has only an entrance from the back side of the altar; the other has a portal entrance with rope holes for closure from inside the temple apse itself. The room had two altars; one collapsed in 1990. The altar arrangements have been named a shrine by Zammit, who further suggests that both rooms are oracular chambers.[66] Skeates also refers to these two apertures as

---

[63] George Agius and Frank Ventura, *Investigation into the Possible Astronomical Alignments of the Copper Age Temples in Malta* (Malta: University Press, 1980), p. 13.
[64] Evans, *Antiquities*, pp. 81–82.
[65] Trump, *Malta: Prehistory and Temples*, p. 150.
[66] Zammit, *The Copper Age Temples*, pp. 27–28; Zammit, *Malta: The Islands and Their History*, p. 54.

'oracle holes'; however, he suggests they are later features of the Tarxien phase.⁶⁷ This chronology agrees with Lomsdalen's theory of the building sequence which also dates them to the Tarxien period.⁶⁸

The South Temple has a clearly defined eastern orientation with a central axis azimuth of 92.7°, horizon altitude of 3.72° and a declination of 0.0°.⁶⁹ Ventura et al. argue that the temple may have been intentionally aligned to the heliacal rising of the Pleiades, which rose at declination 0° during the Temple Period, rather than the equinoctial sunrise.⁷⁰ As the temple stands today, it is also aligned to both summer and winter solstice sunrises, a common feature that Lomsdalen argues was used throughout all construction phases of the Mnajdra Temple.⁷¹

Lomsdalen further argues that the smaller oracle hole is aligned to the winter solstice sunrise and the larger one to both summer and winter solstice sunrises.⁷² Ventura was the first to observe this latter event through a small hole in the temple façade, and that the sun actually rose over the winter solstice posthole (a possible marker for the winter solstitial sunrise which was discovered in 1981 by Ventura in collaboration with Agius).⁷³ Both these occurrences have been confirmed photographically by Lomsdalen (Fig. 2.4). There are no clear indications whether this hole is human made, a result of natural erosion, if the façade slab was placed intentionally to produce this sunrise effect, or if it happened by pure chance.

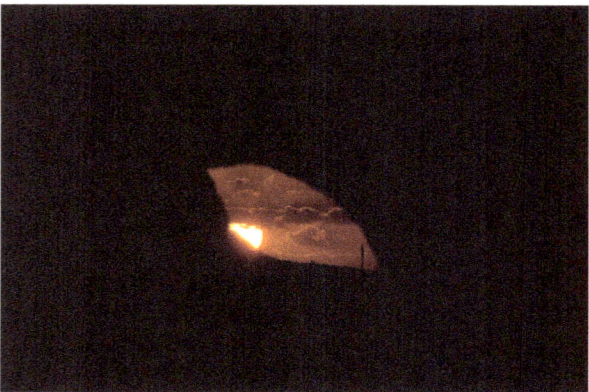 

Fig. 2.4: Photos taken by Lomsdalen on 23 December 2012 of the winter solstitial sunrise. The left one is taken through the hole in the façade wall, showing the sun illuminating the frame of the oracle hole, while on the right, the rising of the sun as observed from the oracle hole close to the winter solstice horizon posthole.

In personal communication between Grima, Ventura and the author the issue of whether the slab was placed there by chance or intention was addressed in the following manner:⁷⁴ a straight line from the illumination point of the oracle hole, through the hole in the façade to the winter solstice sunrise – and allow-

---

⁶⁷ Skeates, *An Archaeology of the Senses*, p. 170.
⁶⁸ Lomsdalen, *Sky and Purpose in Prehistoric Malta*, pp. 145–60.
⁶⁹ Giorgia Fodera Serio, Michael Hoskin, and Frank Ventura, 'The Orientations of the Temples of Malta', *Journal for the History of Astronomy* 23 (1992): p. 115.
⁷⁰ Frank Ventura, Giorgia Fodera Serio, and Michael Hoskin, 'Possible Tally Stones at Mnajdra, Malta', *Journal for the History of Astronomy* 24 (1993): p. 176.
⁷¹ Lomsdalen, *Sky and Purpose in Prehistoric Malta*, pp. 145–60.
⁷² Lomsdalen, *Sky and Purpose in Prehistoric Malta*, pp. 128–29.
⁷³ Agius and Ventura, *Investigation*, fn 14, p. 32.; Ventura et al., 'Possible Tally Stones', pp. 172–73.
⁷⁴ Tore Lomsdalen et al., Archaeoastronomy, June 2014.

ing an error of two degrees (one degree on each side of this line) with the hole in the façade – indicates a six in one thousand chance that the alignment is due to chance: 2/360 (degrees of a circle) = 1/180 = 0.006. Adding the fourth alignment point to this line, the horizon posthole, brings the apparent probability of chance to an absolute minimum that has to be corroborated further.

4. Conclusion

Based on the first research objective of this paper, whether the so-called 'oracle holes' in the Maltese temples were an integral part of a sacred or religious cultural context, the indications are that they may have been so. This is due, not only to the physical layout and placement of the oracle holes and chambers within the temple compound, but due to indications that there also seem to have been rituals or performances enacted. As mentioned earlier, Zammit, Skeates and Barrowclough refer to objects being passed through these holes from a priestly figure to worshippers; however, based on Renfrew's chiefdom society, objects being transferred the other way cannot be excluded. Furthermore, the depicted iconography and cosmological representations, the many retrieved artefacts symbolizing elements of religious ceremonies or rituals in both the hypogea and the temples aboveground for the living, is in accord with Malone's suggestion that the Maltese temples were places of highly structured, ceremonial ritual practice. Hypotheses regarding sounds passed through these holes, due to lack of the written word, are obviously pure speculation, but possible. Nevertheless, the use of sound as part of rituals and feasts or for prophecies, giving secular or spiritual advice or a comparable forerunner to religious confessions, are all possible but not provable. With reference to the Tarxien Temples, Evans also concludes that the 'oracular' idea fits with the psychological implications of the whole construction and draws parallels to the so-called oracle holes at the Mnajdra South Temple.[75]

Regarding the second research question – if the oracular culture was solar influenced and part of a coherent sun-worshipping worldview, cosmology or religion – can be inferred from the concept of left/right dualism (if not factual, at least symbolic or metaphoric) suggested by Hertz and Malone. All extant oracle holes in the above-ground temples included in this field study indicate orientations or alignments towards specific cardinal directions and the sunrise at either the equinoxes or solstices. Whether this was intentional or happened by chance is, again, hypothetical. However, based on various scholarly research, both in astronomy and archaeoastronomy reflected in the archaeological material, it cannot be completely excluded that the builders of the oracle holes may have been influenced by celestial phenomena, especially the sunrise on specific dates during a solar year. We may only speculate whether this may have been a calendric indicator for religious or secular feasts and ceremonies.

Whether oracular practices did exist in prehistoric Malta cannot yet be scientifically verified. Nevertheless, based on indications from archaeological excavations and astronomical observations, it cannot be excluded that a solar influenced – factual or symbolic – form of oracular performance may have been a part of a religious or ritual temple culture. Further data collection is needed to reach more substantial conclusions.

Acknowledgments
First of all I would like to thank Heritage Malta for allowing me access to the prehistoric temples sites. Without their continuous help and support this research would not be possible. Further, I am grateful to Professor Frank Ventura and Dr Reuben Grima, both from the University of Malta, for constructive discussions and valuable inputs.

---

[75] Evans, *Antiquities*, pp. 132–3.

Bibliography

Agius, George, and Frank Ventura. *Investigation into the Possible Astronomical Alignments of the Copper Age Temples in Malta*. Malta: University Press, 1980.

Barrowclough, David A. 'Putting Cult in Context: Ritual, Religion and Cult in Temple Period Malta'. In *Cult in Context: Reconsidering Ritual in Archaeology*, edited by David A. Barrowclough and Caroline Malone. Oxford: Oxbow, 2007.

Bonanno, Anthony. *Malta: Phoenician, Punic, and Roman*. StaVenera, Malta: Midsea Books Ltd, 2005.

Campion, Nicholas. *The Dawn of Astrology: A Cultural History of Western Astrology*, Vol. 1, *The Ancient and Classical Worlds*. London: Continuum, 2008.

Caruana, A. A. *Report on the Phoenician and Roman Antiquities In the Group of the Islands of Malta*. Malta: Government Printing Office, 1882.

Castro, Belén Martin, Ioannis Liritzis, and Anne Nyquist. Oracular Functioning and Architecture of Five Ancient Apollo Temples through Archaeoastronomy: Novel Approach and Interpretation'. *Nexus Network Journal, Architecture & Mathematics* DOI 10.1007/s00004-015-0276-2 (2015).

Cazzella Alberto, and Giulia Recchia. 'Revisiting Anomalies: New Excavations at Tas-Silg and a Comparison with the Other Megalithic Temples I Malta'. *Accordia Research Papers* 10 (2004-2006): pp. 61–70.

Cazzella, Alberto, and Giulia Recchia. 'Tas-Silg: The Late Neolithic Megalithic Sanctuary and Its Re-Use During the Bronze Age and Early Iron Age'. *Scienze dell'Antichita* 18 (2012): pp. 12–38.

Cox, John. 'Observations of Far-Southerly Moonrise from Hagar Qim, Ta' Hagrat and Ggantija Temples from May 2005 to June 2007'. In *Cosmology Across Cultures, ASP Conference Series*, Vol. 409 (2009): pp. 344-48.

Dietrich, Bernard C. 'Oracles and Divine Inspiration'. *Kernos* 3 (1990): pp. 157–74.

Evans, John D. *The Prehistoric Antiquities of the Maltese Islands: A Survey*. London: The Athlone Press University of London, 1971.

Fodera Serio, Giorgia, Michael Hoskin, and Frank Ventura. 'The Orientations of the Temples of Malta'. *Journal for the History of Astronomy* 23 (1992): pp. 107–19.

Formosa, Gerald J. *The Megalithic Monuments of Malta*. Vancouver: Skorba, 1975.

Gimbutas, Marija. *The Living Goddesses*. Berkeley, CA: University of California Press, 2001.

Grima, Reuben. *The Archaeological Drawings of Charles Frederick De Brocktorff*, edited, designed and with sphotography by Daniel Cilia. Malta: Midsea Books and Heritage Malta, 2004.

Henriksson, Göran, and Mary Blomberg. 'Evidence for Minoan Astronomical Observations from the Peak Sanctuaries on Petsophas and Traostolos'. *Opuscula Atheniensia* XXI, no. 6 (1996): pp. 99–114.

Hertz, Robert. 'The Pre-Eminence of the Right Hand: A Study in Religious Polarity'. In *Right and left: Essays on dual symbolic classification*, translated by Rodney Needham and Claudia Needham, edited by Rodney Needham, pp. 3–31. Chicago: University of Chicago Press, 1973, pp. 335–57.

Huber, Peter J. 'Dating by Lunar Eclipse Omens with Speculations on the Birth of Omen Astrology'. In *From Ancient Omens to Statistical Mechanics: Essays on the Exact Science Presented to Asger Aaboe*, edited by J. Berggren and B. R. Goldstein, pp. 3–13. Copenhagen: Copenhagen University Library, 1987.

Koch-Westenholz, Ulla. *Mesopotamian Astrology*, Vol. 1. Copenhagen: Museum Tusculanum Press, 1995.

Liritzis, Ioannis, and Belén Castro. 'Delphi and Cosmovision: Apollo's Absence at the Land of the Hyperboreans and the Time for Consulting the Oracle'. *Journal of Astronomical History and Heritage* 16, no. 2 (2013): pp. 184–206.

Lomsdalen, Tore. 'Cult, Ritual, Sacred Space and the Sky in the Prehistoric Temples of Malta'. Paper presented at The Marriage of Heaven and Earth: Images and Representations of the Sky in Sacred Space, Bath Spa, UK, 2014. In press.

Lomsdalen, Tore. *Sky and Purpose in Prehistoric Malta: Sun, Moon and the Stars of the Temples of Mnajdra*. Vol. 2, Sophia Centre Master Monographs. Ceredigion, Wales, UK: Sophia Centre Press, 2014.

Magli, Giulio. *Mysteries and Discoveries of Archaeoastronomy from Giza to Easter Island*. New York: Copernicus Books, 2009.

Malone, Caroline. 'Ritual, Space and Structure - the Context of Cult in Malta and Gozo'. In *Cult in Context: Reconsidering Ritual in Archaeology*, edited by David A. Barrowclough and Caroline Malone. Oxford: Oxbow Books, 2007.

Malone, Caroline, and Simon Stoddart. "Conclusions." In *Mortuary Customs in Prehistoric Malta: Excavations at the Brochtorff Circle at Xaghra (1987-94)*, edited by Simon Stoddart, Caroline Malone, Anthony Bonanno, and David

Trump, with Tancred Gouder and Anthony Pace, pp. 361–84. Cambridge: McDonald Institute Monographs, 2009.

Mayr, Albert. *Die Vorgeschichtlichen Denkmäler Von Malta*. München: Verlag der k. Akademie, 1901.

Mayrhofer, Karl Ing. *The Prehistoric Temples of Malta and Gozo: A Description by Prof. Sir Themistocles Zammit*: Ing. Karl Mayrhofer, 1995.

Neugebauer, Otto. *Astronomy and History: Selected Essays*. New York: Springer-Verlag, 1983.

Pace, Anthony, ed. *The Hal Saflieni Hypogeum: 4000 BC - 2000 AD*. Malta: National Museum of Archaeology, Museums Department, Malta, 2000.

Pace, Anthony. *The Hal Saflieni Hypogeum: Paola*. Malta: Heritage Malta, 2004.

Renfrew, Colin. *Before Civilization: The Radiocarbon Revolution and Prehistoric Europe*. London: Pimlico, 1973.

Rochberg, Francesca. *The Heavenly Writing: Divination, Horoscopy, and Astronomy in Mesopotamian Culture*. Cambridge: Cambridge University Press, 2004.

Skeates, Robin. *An Archaeology of the Senses: Prehistoric Malta*. Oxford: Oxford University Press, 2010.

Tilley, Christopher. *The Materiality of Stone: Explorations in Landscape Phenomenology: 1*. Oxford: Berg, 2004.

Trump, David H. *Malta: An Archaeological Guide*. London: Faber and Faber Ltd., 1972.

Trump, David H. *Malta: Prehistory and Temples*. Photography Daniel Cilia. Malta: Midsea Books, 2002.

Vassallo, Mario. 'Hagar Qim's Layout Shows Yearly Movements of the Sun'. *The Sunday Times*, 6 February 2011, pp. 52–53.

Vassallo, Mario. 'Sun Worship and the Magnificent Megalithic Temples of the Maltese Islands'. *The Sunday Times of Malta*, 23 January 2000, pp. 40–1, 30 January 2000, pp. 44–5, 6 February 2000, pp. 36–7.

Vassallo, Mario. 'Sun Worship in the Hypogeum: The Guiding Light for the Dead'. *Times of Malta*, 23 and 30 January 2000, 6 February 2000.

Ventura, Frank, Giorgia Fodera Serio, and Michael Hoskin. 'Possible Tally Stones at Mnajdra, Malta'. *Journal for the History of Astronomy* 24 (1993): pp. 171–83.

Zammit, T. 'Third Report on the Hal-Tarxien Excavation, Malta'. *Archaeologia* 70, no. 1918-20 (1920): pp. 181–200.

Zammit, Themistocles. *Malta: The Islands and Their History*, 2nd edition. Valletta: The Malta Herald, 1929.

Zammit, Themistocles. *The Copper Age Temples of Hagar Qim and Mnajdra: With Plans and Illustrations*. Valetta: Facsimile Edition, 1927.

Zammit, Themistocles. *The Neolithic Hypogeum at Hal-Saflieni at Casal Paula-Malta: A Short Description of the Monument with Plan and Illustrations*, 3rd edition. Valletta: Empire Press, 1935.

INCLUSION AND EXCLUSION OF SUNLIGHT AND MOONLIGHT FROM TEMPLES OF THE ĠGANTIJA AND TARXIEN PHASES

John Cox

ABSTRACT: Temple sites in Malta and Gozo were constructed in two main and overlapping phases, Ġgantija and Tarxien. Some details of their construction are unresolved. A number are orientated to admit the light of sunrise and moonrise through the entrance. Two temples of the Ġgantija phase are aligned with the far southerly moonrise at major standstill. A number of Ġgantija phase temples allow winter sunrise and a far southerly moonrise to the left side of the interior; this is termed an 'offset illumination' to the left. While one temple of the Tarxien phase admits the same offset illumination to the left, other temples of the Tarxien phase might exhibit other effects. Mnajdra South Temple admits sunrise and moonrise in ways contrary to that seen in other temples.

Prehistoric temples in Malta and Gozo, free standing stone buildings with a central corridor and opposed apses, date from a 'Temple Period' 3600 to 2500 BCE. The period is divided into two main phases or periods. A 'Ġgantija Phase' of about 3600 to 3100 is named after Ġgantija South Temple. Temple enclosures in the Ġgantija phase have a simple ground plan in a clover-leaf shape of three or five apses. A 'Tarxien phase' of about 3200 to 2500 BCE is named after a construction at Tarxien. At Tarxien the apse at the head is scaled down and the ground plan made complicated, with one temple butted into another. Tarxien is smaller than Ġgantija, and the stonework at Tarxien in particular is more decorative, with geometric bas-relief carvings and mirroring patterns. Some buildings of the Ġgantija phase were restructured and otherwise modified into the Tarxien phase, and there is a continuity of use across the whole temple period. The Ġgantija overlaps with the Tarxien around the middle of the Temple Period, and for the purpose of the present essay, temples built in the middle of the Temple Period, or first built in one phase and then modified in the next are assigned to a 'middle phase'.

Some temples are found in a complex of two or more individual enclosures. Some complexes are at the centre of a geographical cluster with outlying and smaller temples that have been lost. An example is given by Kordin III, the only surviving temple from at least three and possibly five megalithic enclosures identified in the immediate area in 1892.[1]

Two features of temple structure open to question are the shape of the entrance to a temple enclosure and how much of the interior was roofed over. This essay assumes the conventional idea that the entrances were formed by two uprights and a lintel set into a façade. However, this essay entertains the idea that some parts of some temples, in particular some courtyard areas and perhaps some part of some apses were left unroofed.

The idea of a partly-roofed structure is counter to the idea of an entirely roofed-over structure favoured (among others) by Trump.[2] It revives the model favoured by Zammit,[3] and in the specific area of Evans' room 1 at Tarxien it found recent support from Reuben Grima.[4] A partly unroofed model answers practical difficulties, how was the interior illuminated for cleaning, and how did some temples (such as Ġgantija South Temple) manage fire and smoke from the hearths? But in the temples that have an orientation that admits the sunrise and the moonrise it requires a conceptual distinction between a formal illumination at

---

[1] N. C. Vella, *Prehistoric Temples at Kordin III* (Santa Verena: Heritage Books, 2004), p. 8.

[2] D. H. Trump, *Malta, Prehistory and Temples* (Malta: Heritage Books, 2002), p. 196.

[3] T. Zammit, *The Copper Age Temples of Hagar Qim and Mnajdra* (Malta: Altaprint, 2000), p. 31.

[4] In a suggestion made during a presentation given at the SEAC conference in Valletta on the 23rd September 2015.

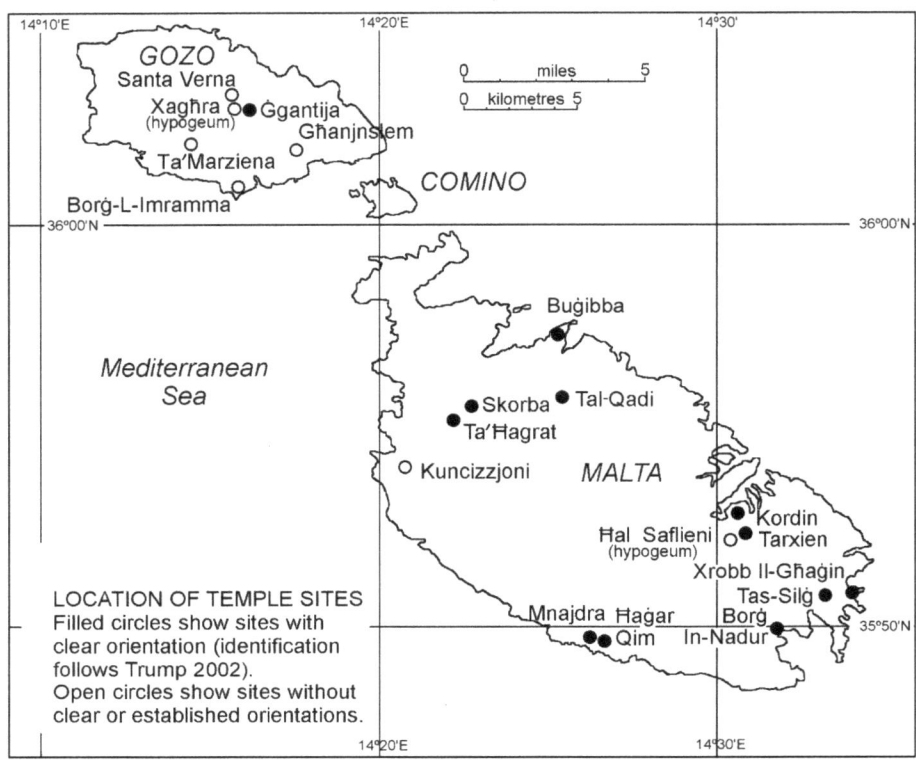

Fig. 3.1: Location of Temple Period Sites, after Trump 2002.

moonrise and sunrise observed through the doorway, and a casual illumination by sun and moon shining into an unroofed area by day and by night.

The layout of a temple can be compared with the layout of two hypogea used for burial and ritual practice. At the Brochtorff Circle (also called the 'Xagħra Circle'), natural caves were used for burials before the Temple period, and were subsequently enlarged in the Ġgantija and Tarxien periods. At Xagħra stonework and partitions were constructed underground that reproduced the form of the overground temples. At Ħal Saflieni (ca. 3200-3000 BCE) the architectural forms of a built temple were cut out of the living rock.

Because the hypogea served to house the bones of the dead they were literally part of an afterlife and underworld. The layout and furniture of the hypogea (screens, thresholds, shelving, water basins, figurines and exotic materials) informs an interpretation of the layout of the temples. In a 'lateral ritual layout' of an 'idealized temple', 'plant and animal images' are located left facing into the temple, 'animal remains [or] feasting debris' located right.[5]

One way the temples might have been used is in an employment by specialists in ritual.[6] Such

---

[5] Caroline Malone and Simon Stoddart, 'Conclusions', in *Mortuary Customs in prehistoric Malta, Excavations at the Brochtorff Circle at Xagħra (1987-1994)*, ed. Caroline Malone, Simon Stoddart, Anthony Bonanno and David Trump (Cambridge: McDonald Institute Monographs, 2009), p. 372.

[6] Caroline Malone, 'Ritual, Space and Structure - The Context of Cult in Malta and Gozo', in *Cult in Context, Reconsidering*

specialists might have endeavoured to commune with the natural world, with the spirit world and with the ancestors, and might have attempted to mediate between those worlds and the life of the larger community. In that specialized use the temples could have been used in two ways, in a more or less private way to assist that communion, and in a more or less public display of mediation.

Regarding each temple enclosure as an individual case (and excluding the hypogea) there are at least twenty-one temple enclosures well surveyed or sufficiently well surveyed to show an axis though a multi-apse structure whose orientation can be measured or estimated. The largest collection of surveys is found in Evans.[7] Within the sample of twenty-one seen in Figure 3.2, temples facing more or less south are identified with both main phases.

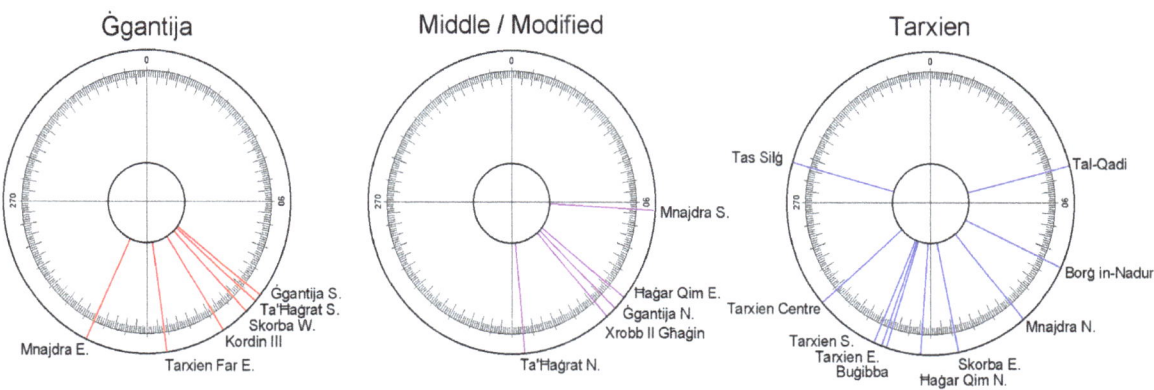

Fig. 3.2: Orientation of Multi-Chambered Temples according to phase. A list showing orientation and phase is found in Table 3.1, below. By 'Middle' is meant a period at about the middle of the Temple Period, ca. 3100 BCE. By 'Modified' is meant structures with Ġgantija origins that were comprehensively re-modelled into and in the Tarxien.

Where a temple faced more or less due south most of the interior would remain dark. Sunlight and moonlight would light up parts of the inside of the doorway at a winter sunrise and sunset and at a southerly moonrise and moonset, and there would be some light into the entrance through meridian passage, sunlight more so in the winter, moonlight more so in the summer, but direct light would not penetrate into the temple interior (unless some part of the interior remained unroofed). The dark character of these south facing structures would produce the character of a cave or underground chamber (and, it might be conjectured, in that sense mirrored the hypogea).

Only three temples have an orientation exactly towards an astronomically significant moonrise or sunrise. A near-coincidence between the orientations of Ġgantija South, Ta'Ħaġrat South and Ħaġar Qim East with a far southerly moonrise at major standstill was first identified by Agius and Ventura,[8]

Ritual in Archaeology, ed. David A. Barrowclough and Caroline Malone (Oxford: Oxbow Books, 2007), p. 24.

[7] J. D. Evans, *Prehistoric Antiquities of the Maltese Islands; A Survey* (London: Athlone Press, 1971).

[8] George Agius and Frank Ventura, 'Investigation into the Possible Astronomical Alignments of the Copper Age Temples in Malta', *Archaeoastronomy, The Bulletin of the Center for Archaeoastronomy College Park, MD* IV, no. 1 (University of Malta: 1980; University of Maryland, MD: 1981), pp. 10-21.

but the multi-phasal nature of the apparition was not identified. Following an observation at Ġgantija by Cox on 25 May 2005, an informal team made repeated observations in 2006 and 2007 from each of the three temples. Observations were intended to test the orientation of the temples and to explore how easily the observation of a far southerly moonrise ('FSMR') was made. Counting an observation from any of the three temples as a single observation, thirty-four observations were attempted altogether, thirty at night, and four (of a near full moon) before sunset. In fifteen observations the moonrise was visible on the horizon.[9] Of the four observations made before sunset, none succeeded in observing the Moon until it had risen.

What this series of observations suggests is that the observation of a far-southerly moonrise can be made to the horizon with about a fifty percent chance of success once (and sometimes twice, one before maximum, one after) every tropical month in the period January to June for two years at the maximum of the 18.61 year lunar cycle. The sequence of moonrises seen in darkness begins with a last crescent seen rising before dawn in late winter and it ends with a full moon seen rising before dusk in early summer.

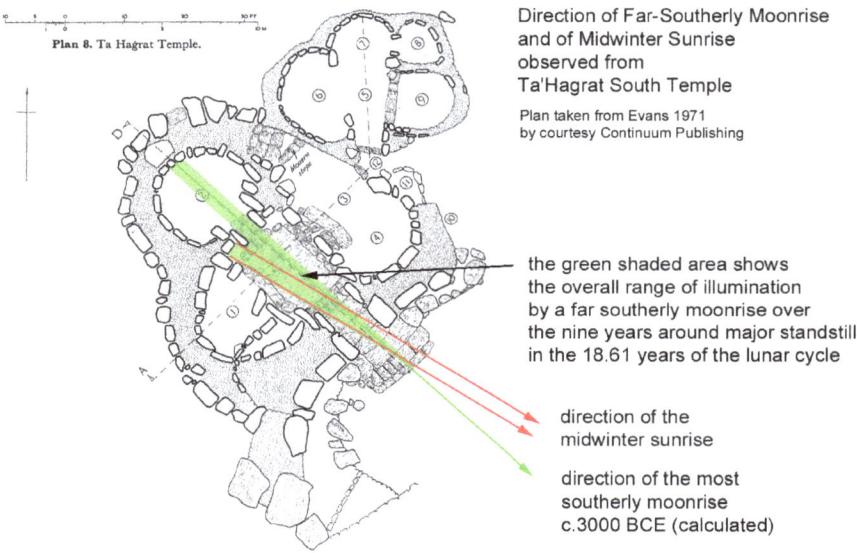

Fig. 3.3: Direction of far-southerly moonrise and of midwinter sunrise observed from Ta Ħaġrat South Temple. The direction of one or more southerly or far southerly moonrises will in coincide with the direction of a midwinter sunrise every tropical month for nine years of the 18.61 year cycle, but only those in the first half of the year are likely to be visible at rising.

While the orientations of Ġgantija South and Ta Ħaġrat South were found to be close to the direction of a far southerly moonrise in period, the orientation of Ħaġar Qim East was found to be about two

---

[9] J. Cox, 'Observations of Far-Southerly Moonrise from Hagar Qim, Ta'Hagrat, and Ggantija Temples from May 2005 to June 2007', *Cosmology Across Cultures*, Astronomical Society of the Pacific Conference Series, Vol. 409, ed. Jose Alberto Rubino, Juan Antonio Belmonte, Francisco Prada, and Antxon Alberdi (San Francisco: Astronomical Society of the Pacific, 2009), pp. 344-349.

degrees south. Although the three temples are orientated in a direction eight and ten degrees south of the position of a midwinter sunrise, the entrances are sufficiently wide to let in the light of the sunrise cross-jamb through the entrance. Observed at Ta´Ħaġrat South Temple the midwinter sunrise ('MWSR') shines through the entrance and lights up a niche to the left side of the waist of the temple (Fig. 3.3).[10] As the sun rises the illumination moves to centre, shrinks in size, and retreats through the entrance. At Ta´Ħaġrat South Temple the sunrise at midwinter has a special niche to accommodate it, a formal inclusion. For the rest of the year the sun rises north of the entrance and sunlight is not allowed to come in through the doorway, shut out apart from a glancing illumination of the south end of the entrance.

At Ġgantija South Temple (observed by Cox on 15 December 2008) the midwinter sunrise reaches the left hand side of the courtyard and corridor, but there is no obvious niche or obvious stone slab to receive it. The later date temple that was built next to it, Ġgantija North Temple (late Ġgantija or early Tarxien, in Figure 3.2 shown as 'middle') has an orientation several degrees south of its neighbour, too far south to admit a far southerly moonrise along the main axis, but sufficiently east for a midwinter sunrise to shine through the entrance and light up a stone slab placed to the left of the waist.

For the purposes of this essay this cross-jamb illumination to the left side of a temple at sunrise is called an 'offset illumination to the left'. From a study of plans it looks as if Skorba West (Ġgantija phase), and Xrobb Il Għaġin (middle of temple period) might show an offset illumination to the left by a midwinter sunrise, but the writer knows of no practical observation. The effect at Xrobb il-Għaġin is likely to remain theoretical because of that site's perilous condition. An offset illumination to the left at midwinter sunrise has been observed at Mnajdra North Temple by Albrecht,[11] and observed and photographed by Lomsdalen.[12] Mnajdra North Temple is an interesting case because it may be the only example of a pure Tarxien structure to contain this effect, and based on Evan's plan it will be complicated by a parallel illumination through the secondary entrance to the west of the main entrance.

The curtain wall at Ħaġar Qim Main Building East Temple indicates Ġgantija phase, but the interior seems to have been extensively re-arranged in the Tarxien. A peculiar feature is a second entrance at the northern end of the main axis, giving a view to the north-west. The view along the main axis south-east to the horizon is obscured by a modern wall, but approximates to a position about two degrees south of an object rising at a declination of 29°S,[13] the Moon's maximum topocentric declination south in 3000 BCE. When risen, a far southerly moon shines straight down the main axis and out of the north end. In the corollary view north-west the moonset at declination 29°S. would be to a point on the horizon about two degrees left of centre (constructed from an observation of moonset at declination 27.45°N to the horizon by John Cox and Michael Spiteri on 26 March 2007), so the moonset is always excluded from the main axis.

The offset illumination by midwinter sunrise at Ħaġar Qim East Temple is made to a niche to the left of the waist of the temple. A pedestal and a bas-relief carving of a volute were found in front of the niche when the structure was first excavated. Another feature is an artificial horizon. An outlying structure, the East Building, stands in the way of a natural horizon in the direction of sunrise, and the sun has to clear this obstacle before it can light up the niche. The illumination at sunrise does not make a perfect fit; it is cut off at the top and constrained to the right of the niche.[14]

---

[10] K. Albrecht, *Malta's Temples,* trans. Amanda Loughran (2001; Nadur, Gozo: de Bono Press, 2002), p. 77.

[11] Albrecht, *Malta's Temples,* p. s63.

[12] Lomsdalen, *Sky and Purpose in Prehistoric Malta,* Fig. 5.3.

[13] Cox, 'Observations'.

[14] Mario Vassallo showed photographs of the illumination effect at the SEAC conference in Valletta on the 23rd September 2015.

The third building with a close fit to a significant astronomical direction is the Tarxien phase Mnajdra South Temple. This temple faces the sunrise at equinox, a direction that coincided with the rising point of the Pleiades in period.[15] Mnajdra South Temple is unique among the temples of Malta and Gozo because it allows sunrise through the doorway through the whole of the year. Towards midsummer and towards midwinter the cross-jamb view becomes increasingly restricted and short-lived.

The cross-jamb view out of the temple in the direction of the midwinter sunrise is just over one solar diameter wide (Fig. 3.4). Because the midwinter sunrise was further south in period, and because the stones at the entrance will have been abraded by weather and by the handling of visitors in the present era, the view at midwinter will have been restricted to a view close to extinction.

Mnajdra South temple works in a way that is contrary to the other temples that let in sunrise and moonrise. It admits the sunrise over the whole year or over a period close to the whole year. It takes the winter sunrise to the right instead of the left. Other temples favour the midwinter sunrise but at Mnajdra South the midwinter sunrise is viewed close to extinction. Other temples let in the moonrise at the maximum of the cycle and exclude it at the minimum: Mnajdra South lets in the moonrise for the whole range north and south for the nine years around the minimum of the 18.61 year cycle, but excludes the extreme rising points for the nine years around the maximum.

Two temples constructed in the Tarxien phase have orientations north of east-west. Both are difficult to interpret: Tal Quadi is ruined and irregular, Tas Silġ was overbuilt and its remains are ambiguous: it might have faced the other way, it might have faced two ways. The orientation of Tal Quadi is towards the sunrise around the 20th of April and the 20th August, and Tas Silġ would have faced the sunset on much the same dates. In complement both structures would have faced a full moon about 20th February and 20th October. Whether this arrangement had astronomical or seasonal significance is open to question.

Fig. 3.4: Sunrise 29 December 2012 observed from north side of the interior of (Evans') room 1 at Mnajdra South Temple and in cross-jamb view through the entrance.

---

[15] F. Ventura, G. F. Serio, and M. Hoskin, 'Possible Tally Stones at Mnajdra, Malta', *Journal for the History of Astronomy* 24 (1993): pp.171-183, p. 178.

Fig. 3.5: Setting path of object at declination twenty-nine degrees south constructed over a star-field viewed across (Evans') room 9 and photographed 17 December 2009 from waist between rooms 14 & 17, Tarxien North Temple.

Further discovery of astronomical significance in Tarxien structures needs special pleading. Ventura recounts that in 1974 Gerald Formosa noticed a midsummer sunset that shone through the entrance to a chamber (Evans' room 10) at the south of the Ħaġar Qim complex and into a porthole niche (Evans' niche zeta).[16] The effect relies on the intervening chambers remaining unroofed, but a persuasive feature is that intervening walls are at the same level as the natural horizon.

From Tarxien Centre Temple there is a view towards a far southerly moonset. The South Temple creates an artificial horizon. The sightline depends on having no roof over the intervening (Evans') room 9. It might be that this observation has not been knowingly made in the modern period, but it can be demonstrated by taking a photograph of a star-field from the waist of Tarxien North Temple and superimposing the path of an object at declination 29° South (Fig. 3.5). The structure would also allow the offset illumination of a niche to the right of the waist by a midwinter sunset, and by the southerly moonset for nine years of the cycle.

---

[16] Frank Ventura, *'Temple Orientations'*, in *Malta before History*, ed. Daniel Cilia (Slema, Malta: Miranda Publishers, 2004), pp. 306-326, pp. 308-9.

Table: Orientation of the Main Axis in Multi-Apse Temples							
Temple	phase	Evans	A.& V.	S.H.& V.	Cox	Albrecht	Lomsdalen
		1971	1981	1992	2001	2002	2014
Ġgantija S.	Ġg	129	128*	125.5	129	128	128
	aligned FSMR observed, offset illumination MWSR observed but no target zone						
Kordin Temple W (III)	Ġg	148	149.6	149.6		144	
	conjectured dark						
Mnajdra E.	Ġg	[212]	207.2	204.0	206	207	200
	conjectured dark						
Skorba W.	Ġg	135		134.9	~138.0	134	137
	candidate for offset illumination MWSR, no observation						
Ta'Ħaġrat S.	Ġg	133	130.6	131.0		130	131
	aligned FSMR observed, offset illum. MWSunRise observed						
Tarxien Far E.	Ġg	171	(170)	168.5			182
	conjectured dark						
Ġgantija N.	MID	134	133*	134.5	135	132	137
	offset illumination MWSR observed						
Ħaġar Qim E.	MID	132	128.7	129.0	131	128	128
	FSMoonR obsrvd misaligned 2 degrees, offset illum. MWSR over artificial horizon observed						
Mnajdra S.	MID	[103]	92.7	92.7	94	92	92
	aligned EQSunRise observed, offset MSSR observed, near extinction MWSunR observed						
Ta'Ħaġrat N.	MID	178	172.7			173	
	conjectured dark						
Xrobb il Għaġin	MID	137	140*			140	
	candidate for offset illum. MWSunR, but no observation, site hazardous and possibly lost						
Ħaġar Qim N.	Tx	183	186.0	186.0		180	184
	conjectured dark						
Mnajdra N.	Tx	[148]	138.1	138.5	139	138	140
	offset illumination MWSR observed						
Skorba E.	Tx	170	168.5	168.5		168	172
	conjectured dark						
Tal-Qadi	Tx	072	76*			76	
	aligned to object rising with declination about 12°N						
Borġ in-Nadur	Tx	110				120	
	large difference in measure of orientation, no observation						
Buġibba	Tx	201	201*	193.5		201	
	conjectured dark						

Temple	phase	Evans	A.&V.	S.H.&V.	Cox	Albrecht	Lomsdalen
Tarxien S. (I)	Tx	204	200.2	201.0		205	204.5
	possibity of unroofed courtyard, no current theory						
Tarxien Centre. (II)	Tx	230	230.1			229	227
	possible unroofed area to SW, if so then align MWSunS, offset FSMoonS constructed						
Tarxien E. (III)	Tx	200	198.7	198.0		202	202
	conjectured dark						
Tas Silġ	Tx				286		
	aligned to object setting with declination about 11° N						
* taken from Agius & Ventura 1980							
Orientation of compass rose in Evans' plans of Mnajdra is in error, bearings shown in square brackets.							
Orientation of Tas Silġ from bearings taken Cox 2007.							

Table 3.1: Orientation of the Main Axis in Multi-Apse Temples with a note on Observational Features.

In summary it can be suggested that dark temples, which is to say temples that excluded direct internal illumination by sun and by moonlight, were constructed in all phases. In the Ġgantija phase two temples (Ġgantija S., Ta'Hagrat S.) were orientated on the most southerly moonrise at major standstill with the additional characteristic of admitting a midwinter sunrise in offset illumination to the left. In later phases a small number of temples were constructed to exclude axial illumination by the Moon but to admit the midwinter sunrise in an offset illumination to the left. At the end of the Ġgantija phase and into the Tarxien a number of variant structures were constructed, including light temples, temples facing sunrise and sunset without obvious significance, and convoluted arrangements with astronomical significance at Ħaġar Qim and at Tarxien but without obvious proof. Mnajdra South Temple stands apart as a middle period to Tarxien phase temple whose accommodations of sunrise and moonrise are contrary to all other cases.

Bibliography

Albrecht, Klaus. *Malta's Temples*. Translated by Amanda Loughran. 2001; Nadur, Gozo: de Bono Press, 2002.

George Agius and Frank Ventura. 'Investigation into the Possible Astronomical Alignments of the Copper Age temples in Malta'. *Archaeoastronomy, The Bulletin of the Center for Archaeoastronomy College Park, MD* IV, no., pp. 10-21. University of Malta: 1980; University of Maryland, MD: 1981.

Cox, John. 'The Orientations of Prehistoric Temples in Malta and Gozo'. *Archaeoastronomy, The Journal of Astronomy in Culture* XV1 (2004): pp. 24-37.

Cox, John. 'Observations of Far- Southerly Moonrise from Hagar Qim, Ta'Hagrat, and Ggantija Temples from May 2005 to June 2007'. In *Cosmology Across Cultures*, Astronomical Society of the Pacific Conference Series , Vol. 409, edited by Jose Alberto Rubino, Juan Antonio Belmonte, Francisco Prada, and Antxon Alberdi, pp 344- 349. San Francisco: Astronomical Society of the Pacific, 2009.

Lomsdalen, Tore. *Sky and Purpose in Prehistoric Malta*. Ceredigion: Sophia Centre Press, 2014.

Malone, Caroline, and Simon Stoddart. 'Conclusions'. In *Mortuary Customs in prehistoric Malta, Excavations at the Brochtorff Circle at Xaghra (1987-94)*, edited by Caroline Malone, Simon Stoddart, Anthony Bonanno and David Trump, pp. 361-384. Cambridge: McDonald Institute Monographs, 2009.

Malone, Caroline. 'Ritual, Space and Structure - The Context of Cult in Malta and Gozo'. In *Cult in Context, Reconsidering Ritual in Archaeology*, edited by David A. Barrowclough and Caroline Malone, pp. 23-34. Oxford: Oxbow Books, 2007.

Evans, J. D. *The Prehistoric Antiquities of the Maltese Islands; A Survey*. London: Athlone Press, 1971

Grima, Reuben. 'Landscape and Ritual in Late Neolithic Malta', In *Cult in Context, Reconsidering Ritual in Archaeology*, edited by David A. Barrowclough and Caroline Malone, pp. 35-49. Oxford: Oxbow Books, 2007.

Serio, Giorgia Foderà, Michael Hoskin, and Frank Ventura. 'The Orientations of the Temples of Malta'. *Journal for the History of Astronomy* 23, part 2 (1992): pp. 107-119.

Trump, David H. *Malta, Prehistory and Temples*. Malta: Midsea Books, 2002.

Ventura, Frank. 'Temple Orientations'. In *Malta before History*, edited by Daniel Cilia, pp. 171-183. Slema, Malta: Miranda Publishers, 2004.

Ventura, Frank, Giorgia Fodera Serio, and Michael Hoskin. 'Possible Tally Stones at Mnajdra, Malta'. *Journal for the History of Astronomy* 24 (1993): pp. 171-183.

Vella, Nicholas C. *The Prehistoric Temples at Kordin III*. Malta: Heritage Books, 2004.

Zammit, Themistocles. *The Copper Age Temples of Hagar Qim and Mnajdra*. Malta: Altaprint, 2000.

COSMOLOGY AND COSMOVISION

COSMOVISIONS PUT UPON A DISK: ANOTHER VIEW OF THE NEBRA DISK

Michael A. Rappenglück

ABSTRACT: This study provides insights into the primarily symbolic intentionality and uses of the Nebra Disk. It will be argued that, in spite of the disk's fascinating artwork, the Nebra Disk clearly reflects ancient pictorial traditions and motifs, and hence it does not support the farfetched conclusions drawn previously by most researchers. Finally, this chapter brings into focus the problem of developing suitable methodologies in the field of cultural astronomy as well as the difficulties that arise when hypotheses and interpretations put forward by official institutions become the generally accepted way of thinking about a given artefact.

Since the discovery of the Nebra Disk in 1999 scientists and non-professionals have hypothesized about its meaning and purpose. They have often claimed that a definitive resolution to the problem has been achieved. In reality, to date (2015) dozens of interpretations have been put forward and it is impossible to cite them all here. The majority of the researchers who have treated this topic assume that the Nebra Disk was employed to count or measure something and that its use had calendric, astronomic, or mathematic implications.[1] In other words, they argue that it was a scientific instrument. That approach is a specifically modern one ascribing to archaic cultures the same interest in mathematical precision and computational rigor that characterizes the work of scientists today.

Thinking about methodology
When attempting to judge the value of the remarkable number of hypotheses that have been put forward, one has to begin by doing a careful iconographical and semiotic analysis, based on archaeological and ethnological data. Such an analysis requires, first, bringing into view phenomenological aspects of the disk as well as perceptual psychology. Once this is done, the next step is to determine the validity of the hypothesis explored concerning the enormous knowledge base, allegedly hidden in the disk's design.[2] Astronomic phenomena can serve as subsidiary subject matter, the secondary focus of a main object (*quasi-astronomical*) or they can serve as the primary focus and, as such, be tied to different subsidiary matters (*exact-astronomical*). The question is whether the disk should be described as a quasi-astronomical object, e.g. a symbolic or emblematic one, or as an exact-astronomical one, e.g. as an instrument that functioned as a calendric or other kind of measuring device.

The proposed development of the iconography
The Nebra Disk (Fig. 4.1a) was found in a stony chamber located within a younger, ring-shaped rampart on top of Mittelberg (252m; ɸ: 51°17′02″N, λ:11°31′12″E).

---

[1] Harald Mellar and François Bertemes, eds., *Der Griff nach den Sternen. Wie Europas Eliten zu Macht und Reichtum kamen* (Halle [Saale]: Landesamt für Denkmalpflege und Archäologie Sachsen-Anhalt, Landesmuseum für Vorgeschichte, 2010).

[2] M. A. Rappenglück, 'Palaeolithic Stargazers and Today´s Astro Maniacs – Methodological Concepts of Cultural Astronomy focused on Case Studies of Earlier Prehistory', in *Ancient Cosmologies and Modern Prophets*, ed. Ivan Šprajc and Peter Pehani, Slovene Anthropological Society, Anthropological Notebooks XIX, supplement, 2013, pp. 85–103.

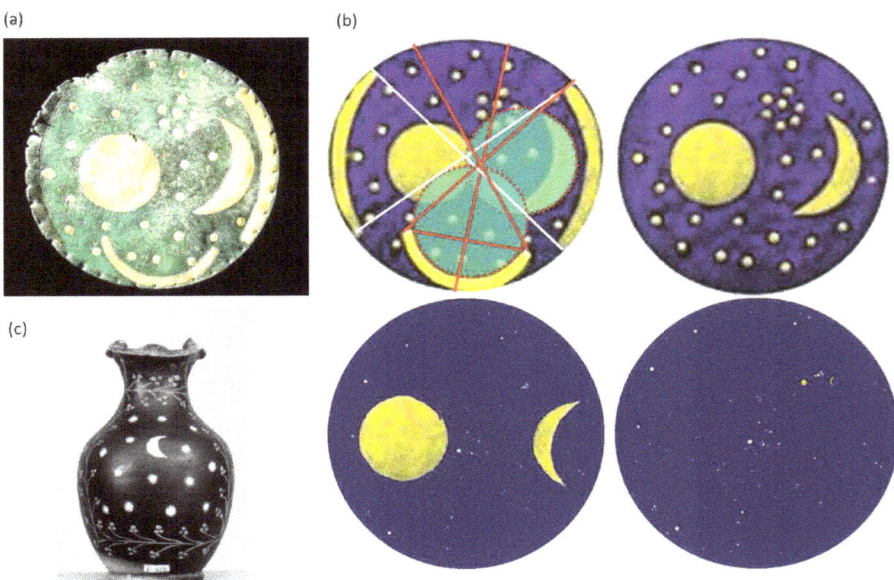

Fig. 4.1: a) The Nebra Disk at an exhibition in the Neanderthal Museum in Mettmann, Germany, 2007 © Michael A. Rappenglück. b) The disk's design is 'abstract' rather than 'concrete', © Michael A. Rappenglück. c) A distribution of stars around a crescent, similar to the Nebra Disk's design, can be recognized on this Gnathian oinochoe *(wine jug), 4th century BCE, © Trustees of the British Museum.*

It is still unclear whether this is a hoard or a grave.[3] The disk (Ø ca. 32cm; thickness, centre 4.5mm, edge 1.7mm; 2kg), is made of a copper-tin alloy (2.5% tin). The copper comes from the Mitterberg (Mühlbach, Hochkönig, Eastern Alps, in Austria) while the gold stems from mines in Cornwall (England).[4] Originally, the object had a tinge of dark brown or black. The corrosion layer of malachite causes the greenish stain seen on it today. Different analyses have shown five phases of development of the disk's iconography and that three artisans worked on it.[5] Thus, the disk's iconography, too, was repeatedly changed during the time that it was in use. There are significant adaptions: superimpositions and dislocation of elements. This suggests that the disk's design was adapted for another purpose or that the three artisans who worked on it, one after the other, were not aware of the real sense of the design or that it no longer had significance for them. The exact number of the small disks on the objects portrayed seems to have been particularly unimportant for the manufacturer.

---

[3] H. Mellar, 'Nebra: Vom Logos zum Mythos – Biographie eines Himmelsbildes', in Mellar and Bertemes, eds., *Der Griff nach den Sternen*, pp. 23–73; E. Pernicka, 'Archäometallurgische Untersuchungen am und zum Hortfund von Nebra', in Mellar and Bertemes, eds., *Der Griff nach den Sternen*, pp. 719–734.

[4] Anja Ehser, Gregor Borg, and Ernst Pernicka, 'Provenance of the gold of the Early Bronze Age Nebra Sky Disk, central Germany: geochemical characterization of natural gold from Cornwall', *European Journal of Mineralogy* 23, no. 6 (2011): pp. 895–910.

[5] Daniel Berger, Roland Schwab and Christina-Heinrich Wunderlich, Technologische Untersuchungen zu bronzezeitlichen Metallziertechniken nördlich der Alpen vor dem Hintergrund des Hortfundes von Nebra', in Mellar and Bertemes, eds., *Der Griff nach den Sternen*, pp. 751–777.

Dating

It is stated that all objects (swords, two flanged axes, a chisel and pieces of spiral bracelets) belong to the same hoard and that the tiny remains of soil adhering to the objects coincide with the soil of the Mittelberg, near Nebra.[6] A microscopic analysis however made evident problems with the different patinas found on the objects, indicating that they might not belong to the same assembly.[7] That challenges efforts that have been made to date the disk using the associated finds: hitherto the probable time line for the burial of the object had been indirectly determined by C^{14} dating of ca. 0.6mg carbon from a tiny piece of birch bark on the hilt of one of the accompanying swords, which gave 1639–1401 cal. BCE (2σ)[8]. For the swords and flanged axes stylistic criteria was utilized. The stylistic similarity of these swords to swords from Hungary confirmed that time-period. However, it also is possible that the swords do not belong to exactly the same time period as the disk itself. Moreover, the disk does not contain any carbon suited for C^{14} dating.

The copper of the disk comes from the ancient mine of Bischofshofen (Hochkönig Mountain, Austria). The production there started at least 18th century BCE, but an older age (19th century BCE or earlier) cannot be excluded.[9] The production continued up to 800 BCE.[10] With respect to the calibrated C^{14} age of the associated finds, the period of usage is limited to about 200–400 years.

The lateral arc and the tiny golden disk, located near it, consist of gold but a gold which has much more tin than the other gold applications. This implies that these inlays were inserted at another time. A manufacturing of the gildings in the course of several phases is very probable, making the differences in their composition a time-factored element.[11] Moreover, this also suggests that each of the three craftsmen who worked on the disk at different times was using a different charge of gold. While the first of the three was a well-trained artisan, the other two were not so qualified and did not respect the work of their predecessor.[12] As a consequence and contrary to the established interpretation regarding the complete design, one can hardly accept the idea of a well-planned depiction, a purposely intended cosmovision, or that the design was intended to be an illustration of a single datable sky event or celestial process.[13] Changes made to the design very probably reflect alteration in the underlying concepts.

The disk is no measurement device

Some have suggested that the Nebra Disk could have been used as a kind of measurement device. There is, however, essentially no evidence for that theory. Its weight of about 2kg undercuts the idea that it was a hand-held device that could be comfortably grasped and then manipulated to obtain exact measurements of data. In short, the weight of the disk goes against its practicality as a hand-held instrument. Thus, to

---

[6] Mellar, 'Nebra: Vom Logos zum Mythos – Biographie eines Himmelsbildes', pp. 35–42.
[7] Without a microscopically analysis problems with all objects' patina had been stated, see Peter Schauer, 'Kritische Anmerkungen zum Bronzeensemble mit „Himmelsscheibe"', angeblich vom Mittelberg bei Nebra, Sachsen-Anhalt. *Archäologisches Korrespondenzblatt* 35, no.3 (2005): pp. 323–328, 559; Indeed microscopically analysis by Rupert Gebhard (Director of the Bavarian State Collection, Munich; pers. communication) makes it evident that the patina of the objects differ and that the assembly it is not a hoard.
[8] Meller, 'Nebra: Vom Logos zum Mythos – Biographie eines Himmelsbildes', p. 56, fn. 43.
[9] Mellar, 'Nebra: Vom Logos zum Mythos – Biographie eines Himmelsbildes', pp. 62–63.
[10] Normann Herz, *Geological Methods for Archaeology* (Oxford: Oxford University Press, 1997), p. 242.
[11] H. Mellar, 'Nebra: Vom Logos zum Mythos – Biographie eines Himmelsbildes', pp. 59–70.
[12] Berger, Schwab, and Wunderlich, 'Technologische Untersuchungen zu bronzezeitlichen Metallziertechniken nördlich der Alpen vor dem Hintergrund des Hortfunds von Nebra', pp. 753–757.
[13] Mellar, 'Nebra: Vom Logos zum Mythos – Biographie eines Himmelsbildes', pp. 59–70.

justify that thesis, it must be conceptualized as having been horizontally fixed on a stable mounting. One might think that the 39 holes along the edge of the disk could have served that purpose. However, they are only 3mm (!) in diameter. A few bigger fasteners or curved brackets would have been sufficient. However, there is no evidence for that. The lateral arcs must be understood in an abstract and symbolical sense. Moreover, there is no sign of a central mounting and there are no rear and front sights. The only possibility for measuring something would have been to put pins into the 39 holes at the edge. But one of the problems with that alternative is that they were drilled in such a way that they affected or at least penetrated the two golden arcs on the outer edges of the disk.[14] Therefore, the assumption that the object was intended to be used as a measurement device can be excluded. Rather there is strong probability that the holes allowed the disk to be fixed horizontally or later also vertically on some type of mount for ritual purposes.[15] Similarly used objects have been found from south Eastern Europe to Scandinavia, 1600 BCE and later. There is additional evidence for the suggestion that the holes were utilized in some fashion as a way to mount the disk (Fig. 4.2d).

The iconography revisited: the disk's design seems to be primarily symbolic
When viewed closely, the decorative elements on the disk can be interpreted as representing a kind of geometrical construction: apart from its circular shape, the use of lines, circles and relative measures is evident.[16] However, that does not justify the myriad of interpretations that see in these elements elaborate mathematical and astronomical patterns. Despite the fact that the Nebra Disk might have some possible astronomical content, e.g. the lateral arcs, the geometrical construction is due to the manufacturing process, a process that also took into account metrological rules and aesthetic principles of the time-period in question.

In this sense, the structural organization of the elements making up the disk's design should be considered 'abstract', rather than 'concrete'. The 'abstraction' is evident if we assume the crescent moon or the sun, and the Pleiades (6+1 pattern) are among the elements depicted and then compare the depicted field of view as well as the apparent sizes of the three objects with reality (Fig. 4.1b). Moreover, if the Pleiades are in fact depicted on the disk, their portrayal does not correspond to the way they appear in the sky to a viewer. Rather they are portrayed in an abstract fashion.

Finally, statistical analysis proves that the 25 small golden disks, which often are identified with certain stars and asterisms, not having the 6+1 pattern, follow a random distribution. They are equidistant to the bigger objects, and were not arranged intentionally.[17] Such a distribution of stars around a crescent can be recognized on a Gnathian *oinochoe* (wine jug) from the 4th century BCE (Fig. 4.1c).[18] Moreover, a calendar count, arbitrarily using the small golden disks and the 6+1 pattern for hypothesizing different time units, without any evidence of such in the design, fails to take into consideration the need for

---

[14] Mellar, 'Nebra: Vom Logos zum Mythos – Biographie eines Himmelsbildes', pp. 69.
[15] Paul Gleirscher, Zum Bildprogramm der Himmelsscheibe von Nebra: Schiff oder Sichel, *Germania* 85, no. 1 (2007): p. 24–26.
[16] Heiko Breuer, 'Untersuchung der Maßverhältnisse der Himmelsscheibe von Nebra', in *Der Griff nach den Sternen. Wie Europas Eliten zu Macht und Reichtum kamen*, ed. Harald Mellar and François Bertemes (Halle (Saale): Landesamt für Denkmalpflege und Archäologie Sachsen-Anhalt, Landesmuseum für Vorgeschichte, 2010), pp. 91–96.
[17] Wolfhard Schlosser, 'Die Himmelsscheibe von Nebra – Astronomische Untersuchungen', in Mellar and Bertemes, eds., *Der Griff nach den Sternen*, pp. 913–933; Emilia Pásztor and Carl Roslund Carl, 'An interpretation of the Nebra disc', *Antiquity* 81 (2007), pp. 267–278; H. Breuer, 'Untersuchung der Maßverhältnisse der Himmelsscheibe von Nebra', p. 95.
[18] British Museum 1873,0820.312AN191821.

iconographical consistency and therefore is pure speculation.[19] In addition, the idea that originally the designer would have used the Nebra Disk for synchronizing the solar and lunar year with a so-called Pleiades Intercalation Rule is clearly refuted.[20] In sum, all of these discrepancies serve to confirm that the disk has an 'abstract' and not a 'concrete' design.

Indeed, the disk's design could be viewed as been conceptualized, deliberately, as a purely symbolic artefact.

Cross-cultural comparisons: other examples of emblematic and iconographic designs
Further proof of the symbolic nature of the design of the Nebra Disk can be brought into focus using cross-cultural comparisons. By comparing examples of cosmic shields, drums, mirrors and similar objects, equipped with symbols/signs of the sky, the earth and the underworld from different cultures and epochs, one can glean further information about concepts and purposes embedded in the pictorial traditions of earlier periods. For the most part, the makers of such objects designed them for several interrelated purposes: for getting and maintaining supernal and mundane power, for the purpose of avoiding harm and other types of negative influences, i.e., apotropaic magic; and for the psychological empowerment of the owner or manipulator of the object. In some cases, the object depicted was a complete *imago mundi*. In rare cases, calendric and orientation knowledge was involved.

There exists a fragment (29 x 24 x 5cm) of a stone plate, probably originally a disk, from Tal Qadi, Malta (Fig. 4.2a), ca. 3300-3000 BCE, which shows engravings of astral objects 'stars', a crescent) and an abstract division into sectors. That object's design is comparable to that of the Nebra Disk, but is much older.

The disk (ca. 25cm) of the Trundholm 'Sun Chariot' (Denmark, Nordic Bronze Age, ca. 1800-1600 BCE), is interpreted as the sun or full moon with the decoration on both sides representing a calendric count. It is, however, not an iconographical composition that integrates sun, moon, stars, and other objects on one single circular object. The designs of shaman drums, on the other hand, often combine astral bodies (moon, sun, Venus, single stars, star clusters, Milky Way, rainbow, and many other objects).[21] Notwithstanding that fact, most of

---

[19] Rahlf Hansen, 'Sonne oder Mond? Verewigtes Wissen aus der Ferne', in Mellar and Bertemes, *Der Griff nach den Sternen*, pp. 953–962; Mellar, 'Nebra: Vom Logos zum Mythos – Biographie eines Himmelsbildes', pp. 59–60.
[20] Manfred Feller and Johannes Koch, 'Geheimnis der Himmelsscheibe doch nicht gelöst?', available at http://home.arcor.de/manfred_feller/Himmelsscheibe [accessed 15 January 2015].
[21] Garrick Mallery, *Picture-Writing of the American Indians* (Washington, DC: Government Printing Office, 1894), pp. 515, 516, fig. 721, 722; Lisha Li, 'The Symbolization Process of the Shamanic Drums Used by the Manchus and Other Peoples', *North Asia. Yearbook for Traditional Music* 24 (1992): pp. 52–80.

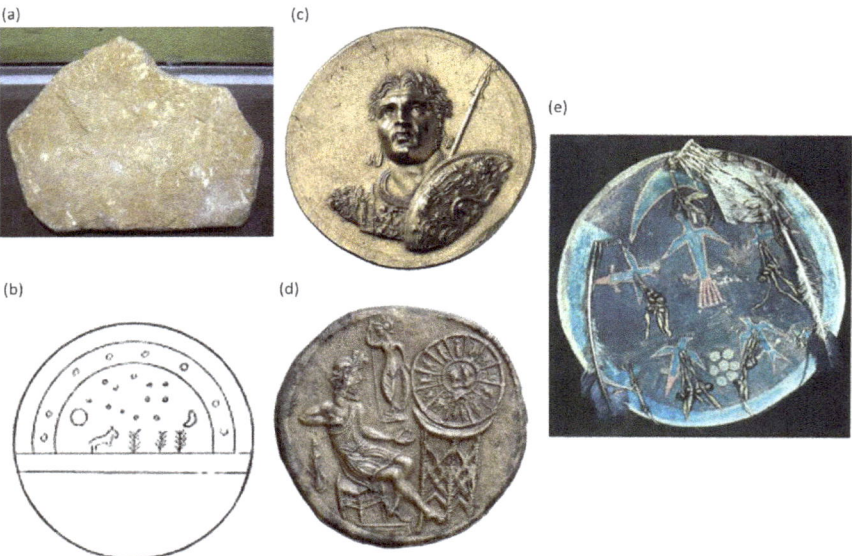

Fig. 4.2: a) A fragment (29 x 24 x 5cm) of a stone plate, originally probably a disk, from Tal Qadi, Malta, ca. 3300-3000 BCE. It shows engravings of astral objects 'stars', crescent) and an abstract division into sectors. National Museum of Archaeology, Malta © Michael A. Rappenglück. b) A Mongolian shaman's drum showing the three cosmic strata. In the upper part: the sky, a rainbow, the region of the stars. On the left and right sides: sun and moon (adopted from Mallery 1894: p. 515, fig. 721). c) A coin from Abukir, Egypt, 1st half of 3rd c. CE. The front side shows a bust of Alexander the Great, crowned by a diadem, together with a spear and a shield. The latter depicts Gaea, the heads of Helios and Selene, surrounded by five zodiacal constellations. © Münzkabinett, Staatliche Museen zu Berlin, r. Saczewski. d) A coin (Divus Traianus), 355-395/423 CE. On its backside, it shows Achilles' cosmic disk with depictions of the zodiac and busts of Selene/Luna and Helios/Sol on a folding table in front of Vulcan. © Münzkabinett, Staatliche Museen zu Berlin, r. Saczewski. e) The shield (Ø 49.5 cm) of chief Little Rock, Southern Cheyenne, ca. 1865. It shows a 6+1 cluster (very probably the Pleiades) associated with a crescent, circle with four triangular peaks, from which birds fly, along with a big bird in the centre, which goes up from the Moon in direction of the Pleiades. © Detroit Institute of Art.

them are datable to the 19th century or a few centuries earlier, their iconography follows much older traditions. A modern artwork of the Nunavut artist Kenojuak Ashevak, Cape Dorset, 1993, depicting the life-world (earth, sea, sky) on a disk, shows a remarkable resemblance to the main iconographic objects found on the Nebra disk.[22] The type of the figures, their relative sizes and positions are very comparable: The waxing moon is a little bigger than the sun. However, the illuminated part is not facing the sun, as would be astronomically correct. The distribution of the stars is random, although one can recognize some ring-shaped clusters positioned at the edge of the disk. Specific asterisms are not identifiable. In contrast to the Nebra Disk, no stars are shown in the unlighted area of the crescent and the Pleiades are not depicted.

Many ancient cultures handed down depictions of the waxing moon in an astronomically incorrect fashion, turned away from the sun.[23] Only a few illustrations accurately display the relation between

---

[22] John MacDonald, *The Arctic Sky: Inuit Astronomy, Star Lore, and Legend*, (Ontario: Royal Ontario Museum, 1998), front cover.

[23] Michael Oppitz, 'Drawings on Shamanic Drums', Anthropology and Aesthetic 22 (1992): p. 75; Roland Gröber,

the sun and the crescent moon, e.g., the circular depiction of the sky with sun, moon and stars in an ancient Hebrew manuscript from 1277–1324 CE (The British Library, London (MS 11639, fol. 517c) or an illustration of the Planisio Bible (ca. 1362) in the Vatican Apostolic Library (Vat. Lat. 3550, fol. 5v), displaying the sun, the crescent moon, the stars (randomly distributed), and some zodiacal signs. The astronomically incorrect depiction of a major crescent moon embracing a minor sun is a popular motive in art, handicrafts and elsewhere (Fig. 4.4b).

The representation of stars in the unlit area of the moon is a common symbolic tradition in different cultures. It is contrary to the widespread opinion that keen sky observers manufactured the Nebra Disk. A real sky watcher would not have made such a mistake. This error clearly indicates the at least partly emblematic meaning of the disk. In 2003, that tradition still existed on a shopping bag from Italy (Fig. 4.3g).

Depictions of a single luminary (sun, Venus, another planet, a star with several rays) put inside or slightly outside the unlighted area of a crescent icon commonly occur, e.g., on coins from the Roman Empire (27 BCE–476 CE).[24] Often the crescent is depicted horizontally and as a little bit bigger than the full disk that is slightly embedded in it. In a field of column #18 in Göbekli Tepe (Anatolia, Turkey), ca. 10,000 BC (PPNP/A), a ring shaped object inscribed in a crescent like figure has been carved out.[25] To date, this is the oldest known example for such an icon.

The Archaeological State Collection in Munich has an iron sword from a site near Allach-Untermenzing, Munich, 5th century BCE (Iron Age). Its iconography is partially similar to the Nebra Disk design (Fig. 4a):[26] On the front side are depicted a small golden disk, a golden crescent figure, separated by a golden vertical line, and an irregular set of silvery inlays. The backside displays the marquetry of an arc with a button on each end (probably a torc) and a disk with a triskelion.

The iconographic depiction of the Pleiades
The majority of researchers identify the rosette-shaped figure of the seven small plates with the open cluster of the Pleiades and, in principle, that hypothesis is possible. Cultural traditions from all over the world and across time provide evidence for equating the pattern of a pentagonal or hexagonal set of points around a dot in the centre with the Pleiades: comparatively recent depictions of that pattern are well known. For example, there is the Canadian Plains Indian Tipi (Canada), which has painted on it the star cluster as a 6+1 pattern (Fig. 4.3c).[27] The Milingimbi (Australia) identified the Pleiades with the wives (6+1 pattern) of three fishers (Orion belt stars) sitting in their celestial canoe.[28] Earlier, a Carolingian manuscript from the 9th century CE, based on the *Phainomena* of Aratus (ca. 315 BCE or 310

'Zur Stellung von „Sonne" und „Mond" auf der Himmelsscheibe von Nebra', *Megalithos* 3 (2004): pp. 88–91; Ursula Thiemer-Sachse, 'Zur „Stellung" von Sonne und Mond in Abbildungen aus dem indigenen Amerika. Ideen und Bemerkungen', *Megalithos* 3 (2004): pp. 151–154.

[24] William Ridgeway, 'The Origin of the Turkish Crescent', *The Journal of the Royal Anthropological Institute of Great Britain and Ireland* 38 (1908): pp. 241–258.

[25] Klaus Schmidt, *Sie bauten die ersten Tempel. Das rätselhafte Heiligtum der Steinzeitjäger*, (München: C.H. Beck, 2006), p. 172, fig. 80.

[26] Wolfgang David, 'Die Zeichen auf der Scheibe von Nebra und das altbronzezeitliche Symbolgut des Mitteldonau-Karpatenraumes', in Mellar and Bertemes, eds., *Der Griff nach den Sternen*, pp. 481–482, fig. 44.

[27] Ed C. Krupp, Seven Sisters', *Griffith Observer* 55, no. 1 (1991), p. 9.

[28] Roslynn D. Haynes, 'Astronomy and the Dreaming: The Astronomy of the Aboriginal Australians', in *Astronomy across Cultures. The History of Non-Western Astronomy*, ed. Helene Selin (Dordrecht: Kluwer Academic Publishers, 2000), p. 80, fig. 11.

BCE– 40 BCE), 3rd century BCE, illustrates the Pleiades as seven women (6+1 pattern).[29] Coins of Roman times show a combination of the rosette (6+1) pattern with the crescent moon. A silver Denarius, issued by Hadrian (76–138 CE), Rome, 125–128 CE, displays seven stars in a matrix-pattern (2 x 3; 1), partially embedded in a horizontal lying crescent (Fig. 4.3e).[30] Another silver Denarius minted by Hadrian from the same time period shows on one side a horizontal positioned crescent with seven stars (6+1 pattern), partially in its unlit area (Fig. 4.3f).[31] In particular, there exist multiple readings of the design on different coins, if seven stars are associated with the crescent: the seven planets (five planets, including sun and moon; but concerning the moon that raises the question why then an additional crescent was added), the Pleiades, or the Big Dipper asterism (the Latin *septemtriones*).[32] Although the combination of the rosette star pattern with the crescent corresponds very well to the iconography of the Nebra Disk, in the case of the Roman coins the interpretation of the stars is different: The seven 'star' design refers to the turn of eras, the renewing of the world and the new world ruler (the desired upcoming of the *saeculi felicitas*). The crescent and seven stars upon the Denarius of Septimus Severus indicate the same desired epochal change. Thus, it was an emblematical design emphasising the emperor's cosmic power. The same could have been the case of the meaning iconographically embedded in the design of the Nebra Disk.

The oldest and at the same time best verifiable examples for signifying the Pleiades by a rosette icon comes from Mesopotamian art, e.g., Jemed Nasr 2900-2400 BCE (Fig. 4.3a).[33] There, however, the seven dots originally referred to ᵈSibittum (ᵈVll-bi) and had a triple meaning: the Seven Sages, the seven sons of Enmešarra, and the Pleiades. The iconic image always had some kind of astral meaning, but it did not designate exclusively a distinct astronomical object, e.g., the Pleiades. In fact, the mantic meaning was important: seven stones, fitting well in the palm of a hand, had been used in throwing oracles. They signified the seven cities: Ur, Nippur, Eridu, Kullab, Keš, Lagaš, and Šuruppak. The goddess Ishtar (Venus) ruled the numbers. Since 1800 BCE, the seven-point cluster primarily referred to the Pleiades.

All over the world, for the most part people have modelled their iconographic depictions of the Pleiades on the natural perception of the cluster, arranging the six or seven stars into the pattern of a two-array matrix, a V-shaped figure, or a small wagon. There are many well-known examples starting from the Upper Palaeolithic to modern times. The Pleiades can be identified as a component of a wall painting in the cave of La Tête du lion (France), Gravettian (30,5–22ka BP), and as a set of six dots that appear floating above the aurochs (#18) in the cave of Lascaux (France), Magdalenian (18–10kaBP). These are the hitherto oldest examples for that notation (Fig. 4.3i).[34] Incidentally, both of these examples make clear that the depiction of the Pleiades on the Nebra Disk, if they are really meant, is not the oldest one and is not a 'concrete' one. A sand painting of the Kumeyaay people, southern California (USA), which is made during the boy's initiation ritual, images the Pleiades shaped like the Little Dipper or Great Dipper (Fig. 4.3d).[35] However, often one can find multiple depictions of the 6+1 pattern, frequently

---

[29] Krupp, 'Seven Sisters', p. 1.
[30] British Museum 1950, 1006. 419.
[31] British Museum 1931, 1014. 29.
[32] Harold Mattingly and Edward A. Sydenham, *The Roman Imperial Coinage. Vol. II: Vespasian to Hadrian* (London: Spink & Son, 1926), pp. 324, 328, 431; Andreas Alföldi, 'Der Neue Weltherrscher der Vierten Ekloge Vergils', *Hermes* 65, no. 4 (1930): p. 373, LXV, 24, 25.
[33] E. Douglas Van Buren, 'The Seven Dots in Mesopotamian Art and their Meaning', *Archiv für Orientforschung* 13 (1939–1941): pp. 277–289, and fig. 2.
[34] Michael A. Rappenglück, 'Possible astronomical depictions in Franco-Cantabrian Palaeolithic rock art', in *Handbook of Archaeoastronomy and Ethnoastronomy*, ed. C. L. N. Ruggles (New York: Springer Science+Business Media, 2014), pp. 1205–1212.
[35] Krupp, 'Seven Sisters', p. 15.

alternating with 4+1 or 7+1 sets, on various objects, where no explicit astronomical interpretation is possible. Therefore, they served a pure decorative purpose or it has to be assumed that such depictions had a different meaning. Mesopotamian cylinder seals,[36] a seal from Amman (early 8th century BCE; moreover, the pattern is associated with crescents),[37] the Phaistos Disk, Crete (ca. 1700 BCE),[38] or the depiction of the Euphorbus plate from about 610 BCE are further evidence that such images are purely decorative.[39]

There are, nonetheless, other instances showing multiple rosettes with a clear astronomical reference. Many such patterns appear, dispersed between the constellations, on the sky globe, Mainz, Roman-Germanic Central Museum (Fig. 4.3h), 150–220 CE, Asia Minor.[40] While the Pleiades are a matrix-like pattern in Taurus, one of the multiple rosettes like sets (6+1) could be recognized below Lepus. In the Budolfi church in Aalborg (Denmark) several lime paintings on the walls and the ceiling from about 1500–1520 CE show sets of 6+1 points, distributed randomly among the stars, which indicate the celestial realm (Fig. 4.3b).[41] From this example, one can infer that the pattern does not simply signify the Pleiades, but the idea of star clustering, derived from observing visible Open Clusters, and it was used as a general indication of the celestial realm.

With respect to the Nebra Disk, it is necessary to look at the 6+1 rosette as an iconographic element in the composition of disk-shaped objects. The multiple depictions of that pattern on the Phaistos Disk were already mentioned, but there is no distinct interpretation of the icon possible. The shield of chief Little Rock, Southern Cheyenne, ca. 1865, shows a 6+1 cluster associated with a crescent, circle with four triangular peaks, from which birds fly up. It has a big bird in the centre that is going up from the moon in the direction of the Pleiades (Fig. 4.2e).[42] The birds, flying up from the peaks, are connecting the earth to the sky. The shield (Ø 49.5cm), painted black, blue-green and red, is quite comparable to the Nebra Disk. Although there is no genuine ethnographic record interpreting the 6+1 patterns, they most probably signify the Pleiades.[43]

---

[36] Van Buren, 'The Seven Dots in Mesopotamian Art and their Meaning', pp. 277–289; Felix Blocher 'Gestirns- und Himmelsdarstellungen im alten Vorderasien von den Anfängen bis zur Mitte des 2. Jt. v. Chr.', in Mellar and Bertemes, eds., *Der Griff nach den Sternen*, pp. 973–987.
[37] Siegfried H. Horn, 'A Seal from Amman', *Bulletin of the American Schools of Oriental Research* 205 (1972): pp. 43–45, and fig. 1.
[38] Jean Faucounau, 'L'énigme du disque de Phaistos : où en est-on aujourd'hui ?', *L'Antiquité Classique* 67 (1998): pp. 262–263, fig. 1, 2, p. 265, fig. 3, #12.
[39] Tyler Jo Smith and Dimitris Plantzos, eds., '*A Companion to Greek Art*', Vol. I (Chichester: John Wiley & Sons, 2012), pp. 92–93.
[40] Ernst Künzl, Maiken Fecht, and Susanne Greiff. *Ein römischer Himmelsglobus der mittleren Kaiserzeit: Studien zur römischen Astralikonographie*. Mainz: Römisch-Germanisches Zentralmuseum, 2003.
[41] Jørgen Orlien and Viggo Petersen, *Domkirken i Aalborg. Budolfi kirke – fra begyndelsen til i dag* (Aalborg: Budolfi Sogns Menighedsråd, 2012).
[42] Imre Nagy, 'A Typology of Cheyenne Shield Designs', *Plains Anthropologist* 3, no. 147 (1994): pp. 31; Michael Kan Curator and William Wierzbowski, 'Notes on an Important Southern Cheyenne Shield', *Bulletin of the Detroit Institute of Arts* 57, no. 3 (1979): pp. 124–133.
[43] It could be also the Big Dipper (personal communication E. C. Krupp).

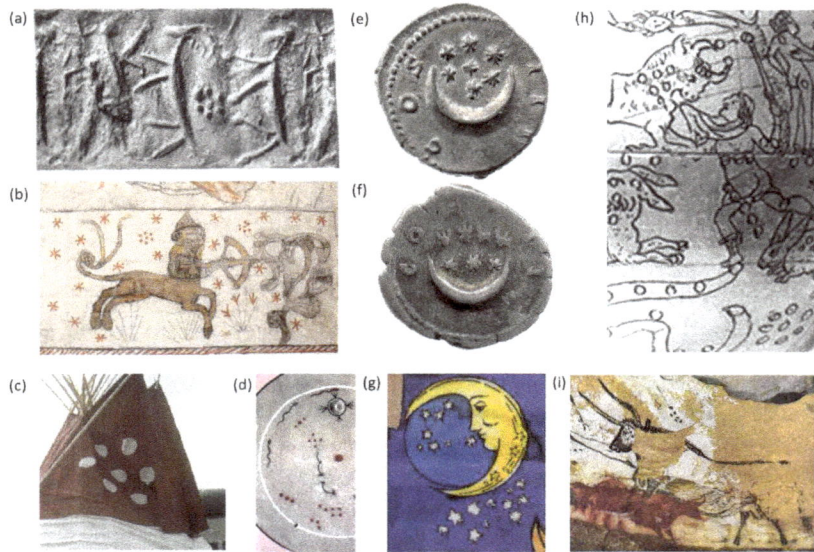

Fig. 4.3: a) Rosette shaped icon from Mesopotamian region and Jemed Nasr period 2,900–2,400 BCE: the seven dots originally are related to ᵈSibittum (ᵈVll-bi), having a triple meaning: the Seven Sages, the seven sons of Enmešarra, and the Pleiades. © E. Douglas Van Buren, 1939–1941: b) In the Budolfi church in Aalborg (Denmark), several lime paintings on the walls and the ceiling from about 1500–1520 CE show sets of 6+1 points distributed randomly among the stars, which indicate the celestial realm. © Västgöten, CC BY-SA 3.0: Budolfi Church in Aalborg Denmark. c) Depictions of the Pleiades as 6+1 pattern at a Tipi tent of the Canadian Plains Indian. © Ed. C. Krupp, 1991: p. 9, 15. d) A sand painting of the Kumeyaay people, Southern California (USA), made during a boy's initiation ritual, shows the Pleiades shaped like the Little Dipper or Great Dipper. © Ed. C. Krupp, 1991: p. 9, 15. e) A silver Denarius, issued by Hadrian (76–138 CE), Rome, 125–128 CE, displaying seven stars in a matrix-pattern (2 x 3; 1), partially embedded in a horizontal lying crescent. © Trustees of the British Museum. f) A silver Denarius minted by Hadrian (76–138 CE), Rome, 125–128 CE. On one side it shows a horizontal positioned crescent with seven stars (6+1 pattern), partially in a shaded area. © Trustees of the British Museum. g) Stars in the unlighted area of the crescent moon. In 2003, that tradition still existed on a shopping bag from Italy. © Roland Gröber. h) Multiple rosettes (7+1 and others): many such patterns appear on the sky globe, dispersed among the constellations, 150–220 CE, Asia Minor; Mainz, Roman-Germanic Central Museum. © Michael A. Rappenglück. i) A set of six dots, very probably the Pleiades, appears floating above the aurochs (#18) in the cave of Lascaux (France), Magdalenian (18–10ka BP). © Michael A. Rappenglück.

The lateral arcs as bows of dawn and dusk

According to some researchers, the lateral curved inserts (Fig. 4.1b, top left), originally placed at the left and the right borders represent arcs of horizon (82°), between winter and summer solstice sunrise and sunset at the Mittelberg, Ziegelroda, near Nebra (φ = 51°).[44] While the curved insert on the right edge is still there, the other one which was on the left edge has fallen off. The right insert would need to be related to the sunset area, because it is close to the inclined crescent. Only that interpretation, which works for every location at 51°N (including Stonehenge [51°10′44″N, 1°49′35″W]), close to the ancient Cornwall mines, remains invariant concerning all the contradictory hypotheses. That feature, however, was not originally intended by the first artisan and belongs to the design introduced by the second artisan.

A cross-cultural example might be the case of Little Rock's shield where the four triangular peaks

---

[44] Schlosser, 'Die Himmelsscheibe von Nebra – Astronomische Untersuchungen', pp. 926–928.

extending to the disk's centre can be correlated with the 'world corners'.[45] They indicate the points of sunrise and sunset at winter and summer solstice. Similar shields from the Southern Cheyenne show the four corners of the world on an arc, together with astral bodies (crescent, sun, Venus, and thunderbirds). A painted animal skin belonging to the Skidi-Pawnee (USA) displays a complete map of the sky (66 x 46 cm): Milky Way, asterisms, pole star, and the twilight arch at dawn and at dusk bows of dawn and dusk, painted in red (Fig. 4.4c). Hence, the lateral arcs on the Nebra Disk could have signified the twilight arc at dawn and at dusk as well as those of the horizon. While one interpretation is related to measurements and therefore much more 'astronomical', the other one has an emblematic meaning.

Iconography of the third arc
During the Nebra Disk's third phase of manufacture, the artisan added another arc, made of gold coming from a third source. This object is much more curved when compared to both of the lateral arcs. The third arc shows a double longitudinal partition following the curve. On each edge (concave/convex) a series of 100 lateral incisions is visible. In total, there are 200 cuts. The shape of the curved arc along with the location chosen for it was such that none of the small golden 'star' circles were covered. The third arc opens onto, that is, faces the 6+1 pattern. The three holes drilled into the arc were perforations made in the object prior to its being attached to the disk. This suggests that this artisan was unaware of the meaning that might have been attached to the third arc.

That the icon symbolizes the Milky Way can be excluded for typically this celestial feature is depicted as a wavy, curved or circular shape, without any multiple longitudinal partitioning.[46] A golden seal ring from Mycenae (Fig. 4.4d), 15th century BCE, e.g., displays the sun and moon, albeit in an astronomically incorrect fashion, positioned above a concave and wavy longitudinal tripartite Milky Way.[47]

According to another interpretation, the notches on the edges of the third arc on the Nebra Disk could signify rain. The icon with its lateral divisions then would indicate a rainbow with its coloured zones. A comparable pictograph can be recognized at a wall picture of the Chamber of the Laguna Pueblo, Western Pueblo (USA), early 20th century CE, displays a tripartite rainbow beneath a sun, close to a crescent moon, and three stars (Belt of Orion), and other mythical figures (Fig. 4.4f).[48] As a daylight object, a nighttime symbolism is not very probable. However, if understood as an emblematic icon, it would fit nicely to the assembly of celestial symbols.[49]

The third arc is identified most frequently as a (solar) barque.[50] Nevertheless, ancient representations of ships (boats) from Scandinavian, Egypt, or Minoan culture (Fig. 4.4e) from about the Nebra Disk's period, show sterns and superstructures.[51] Sometimes researchers refer to the incisions, which most

---

[45] Curator and Wierzbowski, 'Notes on an Important Southern Cheyenne Shield', pp. 124–133; I. Nagy, 'A Typology of Cheyenne Shield Designs', pp. 31.

[46] Krupp, Ed C., 'Spilled Milk'. *Griffith Observer* 57, no. 12 (1993), pp. 2–18; Schlosser, 'Die Himmelsscheibe von Nebra – Astronomische Untersuchungen', p. 928.

[47] Krupp, 'Spilled Milk', p. 5.

[48] Elsie Clews Parsons, 'Isleta, New Mexico', *Bureau of American Ethnology* 47 (1929–1930): pp. 193–166, 210 plate 17, p. 267.

[49] Schlosser, 'Die Himmelsscheibe von Nebra – Astronomische Untersuchungen', p. 928.

[50] Regine Maraszek, 'Ein Schiff auf dem Himmelsozean – Zur Deutung des gefiederten Goldbogens auf der Himmelsscheibe von Nebra', in Mellar and Bertemes, eds., *Der Griff nach den Sternen*, pp. 488–500.

[51] Arvid Göttlicher, *Kultschiffe und Schiffskulte im Altertum* (Berlin: Gebr. Mann, 1992), p. 155, 158, Abb. 82, pp. 161-163, Abb. 84; Thomas Guttandin, Diamantis Panagiotopoulos, Hermann Pflug, and Gerhard Plath, *Ausstellungskatalog Inseln der Winde - Die maritime Kultur der bronzezeitlichen Ägäis* (Heidelberg: Institut für Klassische Archäologie der Ruprecht-Karls-Universität, 2011).

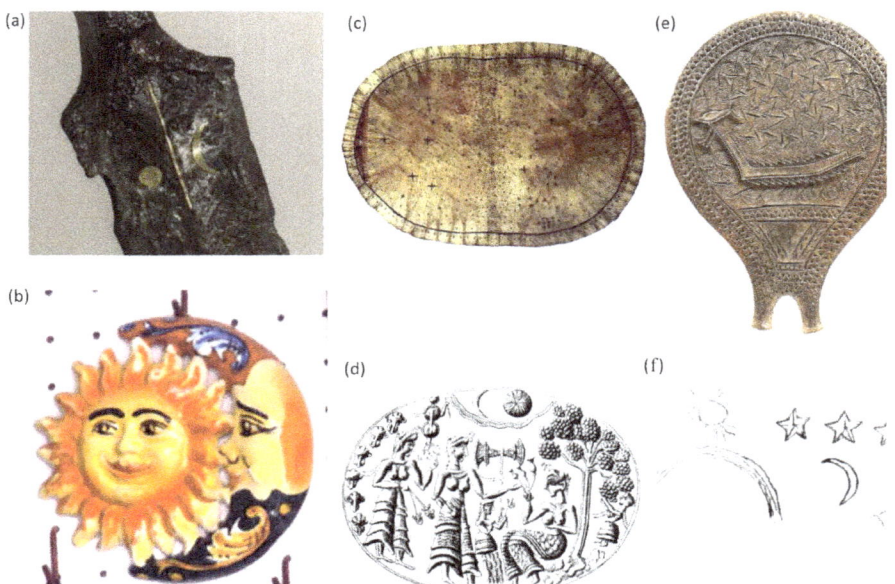

Fig. 4.4: (a) An iron sword from a site near Allach-Untermenzing, Munich, 5th century BCE (Iron Age). On the front side a small golden disc, a golden crescent figure, separated by a golden vertical line, and an irregular set of silvery inlays are depicted. The backside displays the marquetry of an arc with a button on each end (probably a torc) and a disc with a triskelion. Archaeological State Collection, Munich. © Michael A. Rappenglück. (b) The Moon embracing the Sun: an artwork on display at a shop in Taormina, Sicily 2014. © Michael A. Rappenglück. (c) A painted animal skin manufactured by the Skidi-Pawnee (USA) displaying a complete map of the sky (66 x 46 cm): Milky Way, asterisms, pole star, and the bows of dawn and dusk, painted in red. © Ed C. Krupp, 1993: p. 5. (d) Golden seal ring from Mycenae, 15th century BCE, displaying the sun and moon, depicted in an astronomically incorrect fashion, and above a concave and wavy longitudinal tripartite Milky Way (adopted from Evans, 1901). e) A skillet (size: 0.28cm), made of clay, showing an incised depiction of a ship, from a cemetery at Chalandriani, Syros, Early Cycladic II (2800–2300 BCE). Archaeological Museum, Athens, P 4974. © Michael A. Rappenglück. f) A wall picture of the Chamber of the Laguna Pueblo, Western Pueblo (USA), early 20th century CE, displays a tripartite rainbow beneath a sun, close to a crescent moon, and three stars (Belt of Orion), and other mythical figures (adopted from Elsie Clews Parsons, 1929–1930: 210, plate 17, 267).

probably were produced in the process of inlaying, as paddles. That would imply that the barque in question had about 2 x 100 oarsmen, unrealistic for such a simple boat.

The only interpretation left would be to conceive the arc as a (crescent-like) sickle. The so-called button sickles, e.g., from Preußlitz or Merseburg (Saalekreis, Sachsen-Anhalt), feature a very similar design. Three longitudinal ribs on the objects were needed to stabilize the tool. Ritual sickle deposits are typical for Central Germany, 1550–900 BCE.[52] In ancient traditions, copper sickles are related to the Moon, the separation of sky from Earth, and creation.[53] They were believed to have been ritually buried, deep in the Earth after the creational act, according to the Hittite *Song of Ulikumi* (1620-1220 BC) as well as later, in Hesiod's *Theogony* (Hes. Th. 154-210; LCL 57, p. xxxiv-xxxv; 700 BCE). Moreover, the poet also talks about the Pleiades as an important time signal of the farmer's almanac in *Works and*

---

[52] Gleirscher, 'Zum Bildprogramm der Himmelsscheibe von Nebra: Schiff oder Sichel?'), pp. 23–93.
[53] Volker Haas, *Geschichte der Hethitischen Religion (Handbook of Oriental Studies: Section 1; The Near and Middle East 15)*. Leiden, New York, Köln: Brill, 1994, p. 79 fn. 2, pp. 94–96, 123–124, 150–151, 175.

Days (Hes. WD 383–404, 571–581, 609–640; LCL 57, pp. 290–291, 304–305). That fits very well with the proposed emblematic design of the disk.

Cosmic shields of Ancient Greece

When attempting to interpret the Nebra Disk, the shield devices of ancient Greek culture, starting with Mycenaean Civilization (1600–1100 BCE) up to the Hellenistic period (323–146 BCE), are especially interesting. Some of them (and armours, too) had been decorated with (silver and golden) stars, with the moon, and the sun for apotropaic purposes and as expressions of celestial power.[54] Most commonly, authors describe the shield of Achilles in some detail. According to Homer's *Iliad*, it showed the sun, moon, stars and the three cosmic spheres.[55] The shield of the gigantic Tydeus displayed the full moon in the centre, surrounded by the stars indicating the starry sky.[56] A coin (Fig. 4.2c), Abukir, Egypt, on its front side shows the bust of Alexander the Great, crowned by a diadem, together with a spear and a shield. The latter depicts Gaea, the heads of Helios and Selene, surrounded by five zodiacal constellations. The coin dates from the first half of 3rd century CE.[57]

In addition, a coin (Divus Traianus), 355–395/423 CE on its backside shows Achilles' cosmic disk, with depictions of the zodiac and the busts of Selene/Luna and Helios/Sol on a folding table in front of Vulcan (Fig. 4.2d).[58] A figure of this type demonstrates clearly the custom of displaying such emblematic cosmic shields fixed on a special curved bracket. This probably was what was done with the Nebra Disk, too.

A Swedish counterpart of the Nebra Disk: a design from the 19th century

Since 1930 the museum of art (Hallands Konstmuseum) in Halmstad, Sweden, has displayed a remarkable painted wall hanging from Småland, dating from the late 18th century.[59] It comes from a local folkloristic tradition of painting biblical motives (*Bonadsmålning*) in southern Sweden (Småland, Halland, Västergötland, and Blekinge).

Though the tradition already existed in the Late Middle Ages lime paintings of churches, its climax was between 1750 and 1850. Abraham Clemetsson Albo (1764–1841) painted a peculiar Christ in Majesty (*Maiestas Domini*) using egg tempera on the wall hanging (Fig. 4.5a): A disk shows Christ above a globe. This figure is surrounded by a 16-rayed sun (eight red and eight white spouts), the waxing moon within a field of randomly distributed stars, and a group of six plus one stars above the crescent (Fig. 4.5b): Seven winged faces and 18 dots (5, 5, 5, and 3) in a red painted ring border the main scene. The pictorial elements depicted in this wall hanging are undoubtedly the ones that reveal the most similarity to those found in the Nebra Disk's design. This also strongly supports the possibility that the design of the Nebra Disk

---

[54] Rudolf Eisler, *Weltenmantel und Himmelszelt. Religionsgeschichtliche Untersuchungen zur Urgeschichte des antiken Weltbildes*, 2 Vol. (München: C. H. Beck, 1910), pp. 78–84, 302–318; George Henry Chase, 'The Shield Devices of the Greeks', *Harvard Studies in Classical Philology* 13 (1902): pp. 61–127; Nicolas Yalouris, 'Astral Representations in the Archaic and Classical Periods and Their Connection to Literary Sources', *American Journal of Archaeology* 84, no. 3 (1980): pp. 313–318. P. R. Hardie, 'Imago Mundi: Cosmological and Ideological Aspects of the Shield of Achilles', *The Journal of Hellenic Studies* 105 (1985): pp. 11–31.

[55] Hom. Il. 18.478-18.607, especially Hom. Il. 18. 483-18.489, in Homer, *Homeri Opera in five volumes* (Oxford: Oxford University Press, 1920).

[56] Aesch. *Seven*, 387-390, in Aeschylus, *Seven against Thebes*, trans. Herbert Weir Smyth (Cambridge, MA: Harvard University Press, 1926); Eisler, *Weltenmantel und Himmelszelt*, p. 314.

[57] Numismatic Collection, Berlin State Museums, #18200016.

[58] Numismatic Collection, Berlin State Museums, #18200659.

[59] HM 14682; Anna Holmström, 'Himlafenomen', in *Inzoomning-Bonadsmåleri* (Halmstad: Länsmuseet Halmstad, 2009), pp. 80–83.

Fig. 4.5: a) Abraham Clemetsson Albo (1764–1841) created a peculiar Christ in Majesty (Maiestas Domini) painted with egg tempera on a wall hanging: a disc shows Christ above a globe. He is surrounded by the 16-rayed sun (eight red and eight white spouts), the waxing moon within a field of randomly distributed stars, and a group of six plus one stars above the crescent: seven winged faces and 18 dots (5, 5, 5, and 3) in a red painted ring border the main scene. © Museum of Art's Halmstad (Hallands Konstmuseum), Sweden.) Detail of Fig. 4.5a: The Pleiades as 6 + 1 pattern. Museum of Art Halmstad (Hallands Konstmuseum), Sweden.

originated from a folkloristic conception, without an exact astronomical intention. That finding prompts several questions, which up to now remain unanswered: what really explains the extraordinary similarity of both unique astronomical disk-shaped depictions? Was there once upon a time a much older matrix that the Swedish painter was drawing on? Why is the very best object of iconographic comparison to the Nebra Disk so surprisingly recent?

Conclusions

The careful iconographic analysis of the Nebra Disk makes it evident that it was a symbolic object, designed to have emblematic significance, rather than as a scientific instrument. However, one cannot entirely rule out the possibility that there were a few modest, basic astronomical readings intended. The object refers primarily to cosmic myths, astral magic, ritual, and the astral power of armaments. It is quite remarkable that the designs of two objects, which belong to the 19th century (Shield of Little Rock and Clementsson Albo's wall hanging) so closely match the Nebra Disk's composition.

Acknowledgements
I am very grateful to Peter E. Blomberg for informing me of the painted wall hanging in the art museum in Halmstad and for contacting the officials there. I thank Roslyn Frank for improving the flow of the English text.

Bibliography

Alföldi, Andreas. 'Der Neue Weltherrscher der Vierten Ekloge Vergils'. *Hermes* 65, no. 4 (1930): pp. 369–384.

Aeschylus. *Seven against Thebes*. Translated by Herbert Weir Smyth. Cambridge, MA: Harvard University Press, 1926.

Berger, Daniel, Roland Schwab, and Christian-Heinrich Wunderlich. 'Technologische Untersuchungen zu bronzezeitlichen Metallziertechniken nördlich der Alpen vor dem Hintergrund des Hortfunds von Nebra'. In *Der Griff nach den Sternen. Wie Europas Eliten zu Macht und Reichtum kamen*. Edited by Harald Mellar and François Bertemes, pp. 751–777. Halle (Saale): Landesamt für Denkmalpflege und Archäologie Sachsen-Anhalt, Landesmuseum für Vorgeschichte, 2010.

Blocher, Felix. 'Gestirns- und Himmelsdarstellungen im alten Vorderasien von den Anfängen bis zur Mitte des 2. Jt. v. Chr'. In *Der Griff nach den Sternen. Wie Europas Eliten zu Macht und Reichtum kamen*. Edited by Harald Mellar and François Bertemes, pp. 973–987. Halle (Saale): Landesamt für Denkmalpflege und Archäologie Sachsen-Anhalt, Landesmuseum für Vorgeschichte, 2010.

Breuer, Heiko. 'Untersuchung der Maßverhältnisse der Himmelsscheibe von Nebra'. In *Der Griff nach den Sternen. Wie Europas Eliten zu Macht und Reichtum kamen*. Edited by Harald Mellar and François Bertemes, pp. 91–96. Halle (Saale): Landesamt für Denkmalpflege und Archäologie Sachsen-Anhalt, Landesmuseum für Vorgeschichte, 2010.

Chase, George Henry. 'The Shield Devices of the Greeks'. *Harvard Studies in Classical Philology* 13 (1902): pp. 61-127.

Curator, Michael Kan, and William Wierzbowski. 'Notes on an Important Southern Cheyenne Shield'. *Bulletin of the Detroit Institute of Arts* 57, no. 3 (1979): pp. 124–133.

David, Wolfgang. 'Die Zeichen auf der Scheibe von Nebra und das altbronzezeitliche Symbolgut des Mitteldonau-Karpatenraumes'. In *Der Griff nach den Sternen. Wie Europas Eliten zu Macht und Reichtum kamen*. Edited by Harald Mellar and François Bertemes, pp. 439–486. Halle (Saale): Landesamt für Denkmalpflege und Archäologie Sachsen-Anhalt, Landesmuseum für Vorgeschichte, 2010.

Ehser, Anja, Gregor Borg, and Ernst Pernicka. 'Provenance of the gold of the Early Bronze Age Nebra Sky Disk, central Germany: geochemical characterization of natural gold from Cornwall'. *European Journal of Mineralogy* 23, no. 6 (2011): pp. 895–910.

Eisler, Rudolf. *Weltenmantel und Himmelszelt. Religionsgeschichtliche Untersuchungen zur Urgeschichte des antiken Weltbildes*. 2 Vol. München: C. H. Beck, 1910.

Evans, Arthur J. 'Mycenaean Tree and Pillar Cult and its Mediterranean Relation'. *The Journal of Hellenic Studies* 21 (1901): pp. 99–204.

Faucounau, Jean. 'L'énigme du disque de Phaistos : où en est-on aujourd'hui ?'. *L'Antiquité Classique* 67 (1998): pp. 259–271.

Feller, Manfred, and Johannes Koch. 'Geheimnis der Himmelsscheibe doch nicht gelöst? Warum die angebliche Entschlüsselung der Himmelsscheibe durch R. Hansen und H. Meller falsch ist'. http://home.arcor.de/manfred_feller/Himmelsscheibe.

Gleirscher, Paul. 'Zum Bildprogramm der Himmelsscheibe von Nebra: Schiff oder Sichel?'. *Germania* 85, no. 1 (2007): pp. 23–93.

Göttlicher, Arvid. *Kultschiffe und Schiffskulte im Altertum*. Berlin: Gebr. Mann, 1992.

Gröber, Roland. 'Zur Stellung von „Sonne" und „Mond" auf der Himmelsscheibe von Nebra', *Megalithos* 3 (2004): pp. 88–91.

Guttandin, Thomas, Diamantis Panagiotopoulos, Hermann Pflug, and Gerhard Plath. *Ausstellungskatalog Inseln der Winde - Die maritime Kultur der bronzezeitlichen Ägäis*. Heidelberg: Institut für Klassische Archäologie der Ruprecht-Karls-Universität, 2011.

Haas, Volker. *Geschichte der Hethitischen Religion (Handbook of Oriental Studies: Section 1; The Near and Middle East 15)*. Leiden, New York, Köln: Brill, 1994.

Hansen, Rahlf. 'Sonne oder Mond? Verewigtes Wissen aus der Ferne'. In *Der Griff nach den Sternen. Wie Europas Eliten zu Macht und Reichtum kamen*. Edited by Harald Mellar and François Bertemes, pp. 953–962. Halle (Saale): Landesamt für Denkmalpflege und Archäologie Sachsen-Anhalt, Landesmuseum für Vorgeschichte, 2010.

Hardie, P. R. 'Imago Mundi: Cosmological and Ideological Aspects of the Shield of Achilles'. *The Journal of Hellenic Studies* 105 (1985): pp. 11–31.

Herz, Normann. *Geological Methods for Archaeology*. Oxford: Oxford University Press, 1997.

Hesiod. *Theogony. Works and Days. Testimonia.* Edited and translated by Glenn W. Most. Loeb Classical Library 57. Cambridge, MA: Harvard University Press, 2007.

Holmström, Anna. 'Himlafenomen'. In *Inzoomning-Bonadsmåleri*. Halmstad: Länsmuseet Halmstad, 2009.

Homer. *Homeri Opera in five volumes.* Oxford: Oxford University Press, 1920.

Roslynn D. Haynes. 'Astronomy and the Dreaming: The Astronomy of the Aboriginal Australians'. In *Astronomy across Cultures. The History of Non-Western Astronomy.* Edited by Helene Selin, pp. 53–90. Dordrecht: Kluwer Academic Publishers, 2000.

Horn, Siegfried H. 'A Seal from Amman'. *Bulletin of the American Schools of Oriental Research* 205 (1972): pp. 43–45.

Krupp, Ed C. 'Seven Sisters'. *Griffith Observer* 55, no. 1 (1991): pp. 2–16.

Krupp, Ed C. 'Spilled Milk'. *Griffith Observer* 57, no. 12 (1993): pp. 2–18.

Künzl, Ernst, Maiken Fecht, and Susanne Greiff. *Ein römischer Himmelsglobus der mittleren Kaiserzeit: Studien zur römischen Astralikonographie.* Mainz: Römisch-Germanisches Zentralmuseum, 2003.

Li, Lisha. 'The Symbolization Process of the Shamanic Drums Used by the Manchus and Other Peoples'. *North Asia. Yearbook for Traditional Music* 24 (1992): pp. 52–80.

MacDonald, John. *The Arctic Sky: Inuit Astronomy, Star Lore, and Legend.* Ontario: Royal Ontario Museum, 1998.

Mallery, Garrick. *Picture Writing of the American Indians.* Washington, DEC: Government Printing Office, 1894.

Maraszek, Regine. 'Ein Schiff auf dem Himmelsozean – Zur Deutung des gefiederten Goldbogens auf der Himmelsscheibe von Nebra'. In *Der Griff nach den Sternen. Wie Europas Eliten zu Macht und Reichtum kamen.* Edited by Harald Mellar and François Bertemes, pp. 488–500. Halle (Saale): Landesamt für Denkmalpflege und Archäologie Sachsen-Anhalt, Landesmuseum für Vorgeschichte, 2010.

Mattingly, Harold, and Edward A. Sydenham. *The Roman Imperial Coinage. Vol. II: Vespasian to Hadrian.* London: Spink & Son, 1926.

Mellar, Harald. 'Nebra: Vom Logos zum Mythos – Biographie eines Himmelsbildes'. In *Der Griff nach den Sternen. Wie Europas Eliten zu Macht und Reichtum kamen.* Edited by Harald Mellar and François Bertemes, pp. 23–73. Halle (Saale): Landesamt für Denkmalpflege und Archäologie Sachsen-Anhalt, Landesmuseum für Vorgeschichte, 2010.

Mellar, Harald, and François Bertemes, eds. *Der Griff nach den Sternen. Wie Europas Eliten zu Macht und Reichtum kamen.* Halle (Saale): Landesamt für Denkmalpflege und Archäologie Sachsen-Anhalt, Landesmuseum für Vorgeschichte, 2010.

Nagy, Imre. 'A Typology of Cheyenne Shield Designs'. *Plains Anthropologist* 39, no. 147 (1994): pp. 5–36.

Oppitz, Michael. 'Drawings on Shamanic Drums'. *Anthropology and Aesthetic* 22 (1992): pp. 62–81.

Orlien, Jørgen, and Viggo Petersen. *Domkirken i Aalborg. Budolfi kirke – fra begyndelsen til i dag.* Aalborg: Budolfi Sogns Menighedsråd, 2012.

Parsons, Elsie Clews. 'Isleta, New Mexico', *Bureau of American Ethnology* 47 (1929–1930): pp. 193–166, 210 plate 17, p. 267.

Pásztor, Emilia, and Roslund Carl. 'An interpretation of the Nebra disk'. *Antiquity* 81 (2007): pp. 267–278.

Pernicka, E. 'Archäometallurgische Untersuchungen am und zum Hortfund von Nebra'. In *Der Griff nach den Sternen. Wie Europas Eliten zu Macht und Reichtum kamen.* Edited by Harald Mellar and François Bertemes, pp. 719–734. Halle (Saale): Landesamt für Denkmalpflege und Archäologie Sachsen-Anhalt, Landesmuseum für Vorgeschichte, 2010.

Rappenglück, Michael A. 'Palaeolithic Stargazers and Today´s Astro Maniacs - Methodological Concepts of Cultural Astronomy focused on Case Studies of Earlier Prehistory'. In *Ancient Cosmologies and Modern Prophets.* Edited by Ivan Šprajc and Peter Pehani, pp. 85–103. Slovene Anthropological Society, Anthropological Notebooks XIX, supplement, 2013.

Rappenglück, Michael A. 'Possible astronomical depictions in Franco-Cantabrian Palaeolithic rock art'. In *Handbook of Archaeoastronomy and Ethnoastronomy.* Edited by Clive L. N. Ruggles, pp. 1205–1212. New York: Springer Science+Business Media, 2014.

Ridgeway, William. 'The Origin of the Turkish Crescent'. *The Journal of the Royal Anthropological Institute of Great Britain and Ireland* 38 (1908): pp. 241–258.

Schauer, Peter. 'Kritische Anmerkungen zum Bronzeensemble mit „Himmelsscheibe", angeblich vom Mittelberg bei Nebra, Sachsen-Anhalt'. *Archäologisches Korrespondenzblatt* 35, no. 3 (2005): pp. 323–328, 559.

Schlosser, Wolfhard. 'Die Himmelsscheibe von Nebra – Astronomische Untersuchungen'. In *Der Griff nach den Sternen. Wie Europas Eliten zu Macht und Reichtum kamen*. Edited by Harald Mellar and François Bertemes, pp. 913–933. Halle (Saale): Landesamt für Denkmalpflege und Archäologie Sachsen-Anhalt, Landesmuseum für Vorgeschichte, 2010.

Schmidt, Klaus. *Sie bauten die ersten Tempel. Das rätselhafte Heiligtum der Steinzeitjäger*. München: C.H. Beck, 2006.

Smith, Tyler Jo, and Dimitris Plantzos, eds. '*A Companion to Greek Art*', Vol. I. Chichester: John Wiley & Sons, 2012.

Thiemer-Sachse, Ursula. 'Zur „Stellung" von Sonne und Mond in Abbildungen aus dem indigenen Amerika. Ideen und Bemerkungen'. *Megalithos* 3 (2004): pp. 151–154.

Van Buren, E. Douglas. 'The Seven Dots in Mesopotamian Art and their Meaning'. *Archiv für Orientforschung* 13 (1939–1941): pp. 277–289.

Yalouris, Nicolas. 'Astral Representations in the Archaic and Classical Periods and Their Connection to Literary Sources'. *American Journal of Archaeology* 84, no. 3 (1980): pp. 313–318.

ASTRONOMY AND LANDSCAPE IN CARTHAGO NOVA

Juan Antonio Belmonte, José Miguel Noguera Celdrán, A. César González-García and Andrea Rodríguez-Antón

ABSTRACT: This paper presents a novel archaeoastronomical approach to the city of Carthago Nova, present day Cartagena, in the southeast of Spain (ancient *conventus Carthaginiensis, Hispania citerior*). The city was founded over a previous minor Iberian settlement by Hasdrubal, the brother-in-law of Hannibal Barca, ca. 229 BCE, as a twin to his motherland Carthage and capital of the Carthaginian domains in the Iberian Peninsula. The site was a very peculiar one with a local topography similar to that of the African metropolis. The town was founded on a small, well-protected peninsula, only open to land on its east side, which included a series of five hills. According to Polybius, these were devoted to the principal Punic divinities and to the legendary earlier founder of the city, Aletes. Preliminary landscape archaeology approaches on site had shown the relevance of the local orography for the location of the most sacred and relevant buildings in the different areas of the city,[1] where visibility played a most relevant role. When Hasdrubal chose and organized the site, he knew very well what he was doing. The highest spot dominating the site and the port was devoted to Eshmun (*Aesculapius* according to Polybius), as in Carthage itself. Another hill located on axis to the north of this one was devoted to Cronos, the Punic Baal Hammon. This last perhaps served as the node for a series of possible astronomical alignments. The most significant was the solsticial relationship to the sacred area of the *Arx Hasdrubalis* (Cerro del Molinete) to the west-south-west, where a sanctuary presumably dedicated to a female divinity (a later inscription mentions Atargatis) was erected. In 209 BCE, Scipio conquered the city and, exactly on the same spot, the Romans erected, in the Republican era, a sanctuary devoted to an unknown deity and with a monumental access, a splendid view and an orientation which fully justified the new Roman dominion over the city and its port, one of the best in the Mediterranean. Later on the city was re-founded under the title of *Colonia Urbs Iulia Nova Carthago* and especially under Caesar and Augustus a new orthogonal grid, with strong astronomical connections, was applied to the city. The urban plan even included *Mons Aletes*, the hill devoted to the legendary city founder to whom Augustus was perhaps assimilated. Hence, Carthago Nova can be considered an astronomical and topographic materialization of sacred space.

1. Introduction: an historical snapshot

Qart Hadašt (modern Cartagena, Spain) was founded in 229/228 BCE by the Carthaginian general Hasdrubal as the capital of the political and military protectorate set up by the Barca family of Carthage in southern Iberia in the third-century BCE. Protected by a deep bay, the Carthaginian – and later the Roman – city was located on a small peninsula – around 40 hectares in size – and was surrounded by the sea to the south-southeast (the Sea of Mandarache) and a large lagoon (Almarjal) to the north. It was linked to the mainland by an isthmus, over which the *via Heraclea* (later the *via Augusta*) was built. The peninsula was traversed by a number of small valleys running from the NE to the SW and from the SE to the NW and was surrounded by a belt of five hills of varying height and importance. The Iberians had previously settled some of these hills in the fourth and third centuries BCE. The city was in the centre of an ideal natural harbour, oriented towards northern Africa (and therefore Carthage), and in a nearly impregnable defensive position. It was strategically placed in a productive fishing area and rich in esparto-grass, and it was close to a mountain range abounding in lead and silver veins. These mines were intensively exploited during the Phoenician (Punic) and Roman Republic periods, and also during part of the first century CE.

---

[1] J. M. Noguera Celdrán, 'Carthago Nova: Urbs privilegiada del Mediterráneo occidental'. In *Hispaniae urbes. Investigaciones arqueológicas en ciudades históricas*, ed. J. Beltrán Fortes and O. Rodríguez Gutiérrez (Sevilla: Servicio de Publicaciones de la Universidad de Sevilla, 2012), pp. 121-190.

The written and archaeological sources combined provide us with an increasingly exact picture of the Phoenician city: it was fortified with strong walls built according to Hellenistic models, and it was topped with a sumptuous palace on the acropolis (*Arx Hasdrubalis*, today Cerro del Molinete, H1 in Fig. 5.1) and an agora. The terrain was prepared for construction, with the erection of retaining walls and terraces, to which the street outlined adapted. The archaeological record has indicated that domestic and industrial quarters coexisted and also that a number of sanctuaries to Punic deities were present (for example, on the summit of Cerro del Molinete), but it is still silent about the sacred areas which, according to Polybius, stood on the other hilltops. The divinities that this author associates with those sanctuaries must be regarded as an *interpretatio graeca* of the original Phoenician pantheon (see Fig. 5.1). Mount Asklepios (*Mons Aesculapii*, modern Cerro de la Concepción, H5), could have hosted a shrine to Eshmun; Kronos/Saturn could refer to Baal Hammon, a deity worshipped on Monte Sacro (H2), while Hephaistos (modern Despeñaperros hill, H4) could refer to Egyptian divinities worshipped by the Phoenicians such as Ptah or Bes. According again to Polybius, Cerro de San José hill (H3) could have been consecrated to an otherwise unknown deity, Aletes, presumably a mortal deified for his discovery of the silver mines and connected with the city *ager* and responsible for its wealth.[2]

After the Roman conquest of the city in 209/208 BCE, the victors realised the enormous potential of the city's strategic position and exploited the southeast of the Iberian Peninsula. The city founded by the Barcas became one of the main Mediterranean cities in the Peninsula: it was turned into the main operational base for the Roman navy and armies in Hispania. The new opportunities for trade soon attracted a large number of Italian merchants, although the majority of the population was made up of strongly-Hellenised artisans, workmen, traders and mariners, among which strong Semitic, North African and indigenous groups could also be detected. Throughout the second century BCE the city maintained a Phoenician character. The only major transformation in the city was the reconstruction of the walls, which followed Phoenician models. The intense construction and monumental activity of the city, which is attested in the archaeological record, took place in the late second and, especially, in the first centuries BCE. This activity was promoted by wealthy Italian mining and trading moguls. The city was equipped with a major harbour and commercial buildings decorated with frontal porticoes. Towards the end of the second century BCE, probably on the private initiative of Italian traders from Delos or their adjuncts, the Phoenician sanctuary in Cerro del Molinete (H1) was transformed. The so-called 'Sector 2' (presumably the sacred area) was furnished with a mortar pavement with inscriptions, including a large *tessellae*-made dedication to Atargatis (*Dea Syria*). This inscription included an ablative (*Salute*), which associated therapeutic faculties with this divinity. This is coherent with the results of the latest excavations, which have uncovered a *triclinium* (often called a *sacellum* or small sanctuary) that may have been used for health-related rituals such as the *incubatio*. The inscriptions were oriented towards the east, and their axis was aligned with a doorway and two steps that provided access to the room. The exact date of this entrance is unknown, but its opening could have coincided with the general renovation of the building. The doorway was loosely orientated towards the sanctuary of Baal Hammon on Monte Sacro (H2), which is located nearby.

The construction in the area adjacent to the sanctuary of Atargatis must also be dated to the final decades of the second or the early first centuries BCE. The new sanctuary involved the construction of several terraces and sturdy retaining walls, all of which faced the largest of the valleys running across the city. The sanctuary was now presided over by a temple built in the Italian tradition. Access

---

[2] J. M. Noguera Celdrán, 'Qart Hadast, capital bárquida de Iberia', in *Fragor Hannibalis. Aníbal en Hispania*, ed. M. Bendala Galán, M.ª Pérez Ruiz, and I. Escobar (Madrid: Comunidad de Madrid, 2013), pp. 134-173.

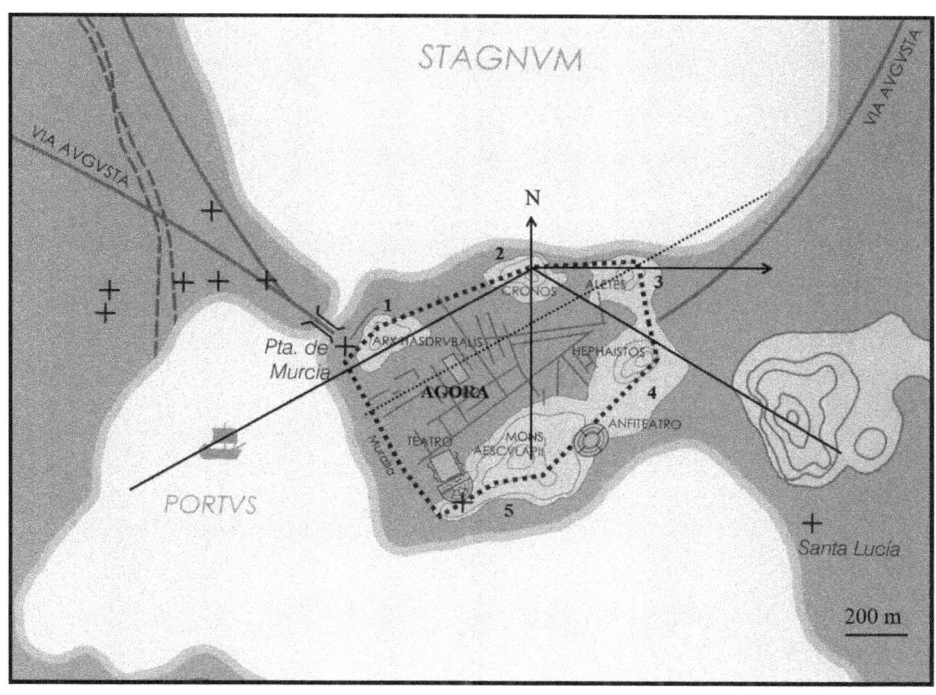

Fig. 5.1: The peculiar topography of Cartagena in antiquity. The diagram shows the location of the five main hills within the urban area: Arx Hasdrubalis (Hill 1, or H1 hereafter), Mons Cronos (Hill 2, H2), Mons Aletes (Hill 3, H3), Mons Hephaistos (Hill 4, H4) and Mons Aesculapii (Hill 5, H5). Either by chance or by deliberate selection, Mons Cronos, where the sanctuary of the main Punic deity Baal Hammon was presumably located, may have acted as a node of selective astronomical alignments (meridian, equinoctial or solstitial) with the other hills. The sector of the Agora could also be solstitially related to Mons Aletes. See the text for further discussion. Diagram by the authors, adapted from Ramallo Asensio (2011).

to the summit of the hill was achieved via a monumental staircase. The use of volcanic stone gave the complex a strong sense of constructive coherence, demonstrating that the whole project had been conceived as a discrete unit. Due to the lack of epigraphic or textual sources, we ignore the question of who was responsible for this vast operation (it can, however, plausibly be attributed to Italian-Delian *negotiatores*). The complex followed an eclectic style, and Italian,

Greco-Hellenistic and Phoenician formal and technical languages can be perceived. Due to the presence of hydraulic infrastructures, it is possible to speculate that it was dedicated to health-related and therapeutic 'Eastern' deities. The construction of this sanctuary on the top of a prominent orographic feature of the acropolis clearly indicates the status attained by the city. The temple, the staircase and the terraces completed a monumental façade presiding over the central valley. This façade was also clearly visible to the ships entering the harbour (see Fig. 5.2). The evidence dating to the modern period demonstrates the existence of a canal that provided access to the harbour. This canal avoided the so-called 'Laja' (a huge reef in the middle on the port mouth) which was located next to the quay where the Faro de Navidad now stands. A lighthouse may have been operational in this location from the Phoenician-Roman period. A photograph dated to the late nineteenth century, which was taken from a vessel that was sailing up this canal, provides us with a view of the city, the harbour and Cerro del Molinete (H1) to the left. With this evidence, it is easy to imagine the perspective presented to approaching ships from the mid-first century BCE: this would have included the terraced sanctuary, which for decades illustrated the Roman character of Hispania.

Fig. 5.2: A snapshot (a) of the temple of Republican era on the top of Cerro del Molinete (ancient Arx Hasdrubalis, H1) before partial reconstruction. It was located in a dominant position (b) facing the city and its port and could have served as a perfect topographic reference. However, the orientation of the building can offer another important clue since its perpendicular axis was orientated to sunrise on the day when the foundation of the city of Rome was commemorated. See the text for further details. Diagram by the authors.

An examination of the *IIvir quinquenales* sets the date of the colonial *deductio* of Carthago Nova to the year 54 BCE, or even the previous year, which was the year that Pompey's Spanish command began. Thus, the *deductio* would be related to the civil war between Pompeians and Caesarians. The colonial status triggered a new urban project for the city, which aimed to create an urban landscape expressive of *urbanitas* and civilisation. The initial stage of this project focused on the reconstruction of the old Republican walls, which took the best part of the second half of the century, and on the water supply. The Caesarian period (40-25 BCE) was witness to the construction of a new street layout, which also included the construction of sewers and pavements that were built with polygonal limestone blocks. The new project was resumed on a grand scale during the Augustan period, with the erection of a succession of open spaces and public and semi-public buildings in the regular city blocks that were located in the western sector of the peninsula, next to the harbour. This included the structures which are currently integrated in *Insula I* (area of Cerro del Molinete, H1), the *forum* and associated buildings, a hypothetical *porticus duplex* that was built to the southeast of the *forum* (this is the source of a series of antefixes representing Victories with Capricorns – see Figure 5.3 – and it followed the Augustan iconographic programme), the theatre with its peristyle, and the *Augusteum*.

The position of these structures clearly demonstrates that they were planned in unison in order to conform to the new urban orthogonal layout. This is further demonstrated by the homogeneity of the construction techniques and materials. Hence, at the beginning of the first century CE, Carthago Nova reached its apex in organic structuration and planning. It is the aim of this work showing how this planning reflected an interest in both local and the celestial landscapes.

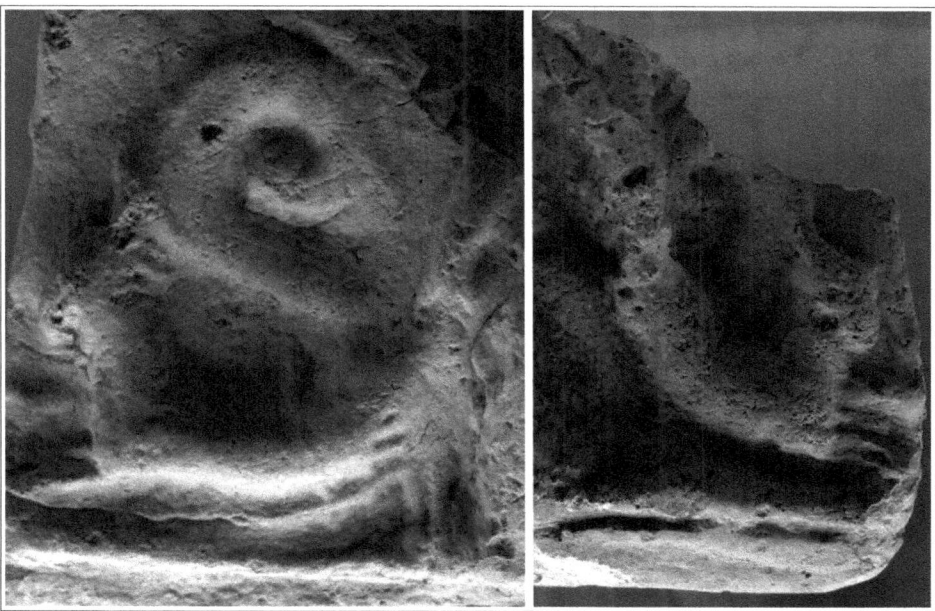

Fig. 5.3: A hypothetical porticus duplex that was built to the southwest of the Augusteum could be the source of a series of antefixes with representations of Capricorns. This zodiacal sign was adopted by Emperor Augustus and thus represented in coinage and other commemorative and symbolic objects. Images by the authors.

2. Discussion: Astronomy and Landscape

Our team is now conducting a research study to analyse the orientation of Roman settlements in Hispania while a first diachronic study was conducted at ancient Emerita Augusta (today Mérida).[3] Ancient Carthago Nova was a logical step further. In October 2013, the authors were invited by the "Cartagena, Puerto de Culturas" Consortium to perform a detailed archaeoastronomical analysis of the ancient remains discovered across the city in the last couple of decades. We were assisted by the local archaeologist Encarnación Zamora and by José Bonnet, president of the amateur astronomer association of the city (AAC), who had been the promoter of the initiative. The second author has been excavating in Cartagena for more than a decade and his archaeological advice was going to be fundamental for a best approach to the problem. The vast majority of the archaeological sites in the city and the most relevant orographic spots, including the five prominent hills within the urban area – even though archaeological remains are not self-evident on some of them – were visited, and orientation and topographic measurements obtained in most of them with high precision compasses and clinometers (see Table 5.1). The terrain in Cartagena is mostly limestone and hence no magnetic alterations are expected. However, some of the sites, such as the impressive *Augusteum* or the *Curia* were located in underground facilities, covered by modern con-

---

[3] See e.g. A. C. González-García, A. Rodríguez-Antón, and J. A. Belmonte, 'The orientation of Roman towns in Hispania: preliminary results', *Mediterranean Archaeology and Archaeometry* 14 (2014): pp. 107-19. For an early study on the topic, see G. Magli, 'On the orientation of Roman towns in Italy', *Oxford Journal of Archaeology* 27 (2008): pp. 63–71. For a diachronic case study, see A. C. González-García and L. Costa-Ferrer, 'The diachronic study of orientations: Mérida, a case study', in *Archaeoastronomy and Ethnoastronomy: Building Bridges Between Cultures*, ed. C. L. N. Ruggles, IAU Symposium 278 (2011): pp. 374-381.

Monument	a (°)	h (°)	δ(°)	Comments
Sanctuary of Atargatis (Cerro del molinete, H1)	82	1	6¾	Eastern gate
	262	3	−4¼	Opposite (western direction)
	171	0	−52	Southern 'gate': Canopus ?
	351	0	51	Hypothetical Northern Access
	61½	1½	23	Western wall altar limits: summer solstice and
	91¼	1	− 0½	Equinox sunrises
Republican temple (Cerro del Molinete, H1)	167	0½	−50¾	Temple axis; to the 'Cathedral'
	76¾	1	10¼	Perpendicular axis. Sunrise 4/18 GPC, 4/22 Julian Calendar in Caesar's time. 'Foundation of Rome': 4/21 in a certain year of the Roman Republic calendar.
	61½	2	23¾	North and south borders of the recent construction on Monte Sacro (*Mons Cronos*, H2): Summer solstice sunrise
	66	2	20½	
	71½	1	15	Cross at *Mons Aletes* (H3)
	136	5	−31	North and south borders of *Mons Aesculapii* (H5)
	140	5	−33½	
	121	1½	−23¼	Peña del Aguila
	125	4	−24½	El Calvario chapel: Sunrise at WS
	268	5½	1¾	La Atalaya
Castellum Aquae	175½	0	−52¾	At Cerro del Molinete (H1). Canopus ?
Temple at Forum	61½	2½*	23¾	Perpendicular axis (Axis of the Curia) Summer solstice sunrise
Temple of Isis and Serapis	151	B(2)	−42½	
'Barrio de las termas'	243½	B[0]	21	Cardus
	63½	B[0]	20½	Decumanus (23¾ for h=5°)
Roman Theater	60	[0]	23	Solstitial at flat horizon: varatio ?
	240	[0]	−23¾	
	150	23	−23¼	Cavea axis: winter solstice sunrise
Theatre's source shrine	338½	0	47	
Torre Ciega	108	2	−13½	

Table 5.1: *Data on the orientation of ancient structures at Carthago Nova (37°36' N; 0°58' W). Columns indicate the identification of the site, the azimuth (a) for different elements within a site, the angular height of the horizon (h) in that direction (B stand for a "blocked" view) and the corresponding declination (δ). Errors of ½° should be considered on both azimuth and angular height, reaching ¾° for the corresponding declination. Finally, some relevant comments are added. For the theater, the angular height was considered for a flat horizon. * The angular height of Mons Aletes (H3) from the Forum area was estimated from Cerro San José itself.*

crete buildings, without exterior view and data acquisition was thus impossible even with a GPS. For the analysis of these sites, we will recall on detailed excavation plans that were kindly made available to the authors by the Consortium.

Hasdrubal chose and organized the site for the new capital of the Punic empire in Hispania knowing very well what he was doing. *Qart Hadašt* (the New City, a nickname already borne by the north African metropolis) was erected as a minor scale replica of its mother town. The highest spot dominating the site and the port, Cerro de la Concepcion, was possibly devoted to Eshmun (*Aesculapius* according to Polybius 10, 10, 9)[4], as in Carthage itself. Another hill, Monte Sacro (H2, 'Sacred Mountain', its present name and perhaps a recall of a very ancient tradition) is located on axis to the north of this one was devoted to Cronos, the Punic Baal Hammon, and a sanctuary was presumably erected on the top. This spot could have served as the node for a series of possible astronomical alignments as shown in Figure 5.1. The most significant was the solstitial relationship to the sacred area of the *Arx Hasdrubalis* (Cerro del Molinete) to the west-south-west, where a sanctuary presumably dedicated to a female divinity, perhaps Tanit the supreme goddess of Carthage, was erected. As previously mentioned, exactly at the same place, in the said Punic sanctuary, a mosaic with an inscription dedicated to Atargatis was installed in the early Roman period (end of the second century BCE).

During the summer solstice of June 2014, our colleagues of the AAC were able to observe how the sun, as seen from that particular spot, climbed the northern slope of Monte Sacro (H2), finally abandoning the horizon on top of a modern building found at exactly the same place where the Cronos temple should have been located in Punic (and Roman) times. This is beautifully illustrated in Figure 5.4. This phenomenon is quite similar to some other places with similar features encountered in different locations throughout the Mediterranean, such as Hattusha, the capital of the Hittite Empire,[5] and could indeed have been important in the Iberian Peninsula during the Phoenician epoch.[6] Finally, Monte Sacro (H2) could have had an equinoctial relationship to *Cerro de San José (H3), consecrated to* Aletes, the mortal hero deified for his discovery of the silver mines responsible for the wealth of the city. Mons Aletes (H3) could also have acted as a reference point for the establishment of the *agora* of *Qart Hadašt* in the valley between the hills at a place where the summer solstice sunrise would have happened on this hill, in a certain parallel to the pair Arx Hasdrubalis-Mons Cronos (H1-H2). As we will see later, the Roman Forum followed a similar pattern.

In 209 BCE, Scipio conquered the city and, during the decades and centuries to come, the Romans reorganized the Punic city, erecting a series of important buildings which remains can be contemplated (and indeed measured) today. Table 5.1 shows the data obtained during our field campaign in Cartagena. Some buildings such as the foundations of the ruined temple of Isis and Serapis at the foot of Molinete (H1) and 'Torre Ciega' – a funeral monument in the outskirts of the city – do not show any evident archaeoastronomical connection, despite astronomical patterns having been discovered for these types of monuments and buildings elsewhere.[7] However, this was indeed the case for many others. On the acropolis of the city (the *Arx Hasdrubalis*, H1), the citizens of Carthago Nova erected at least two sanctuaries that, as shown in the introduction, have recently been excavated and restored. The first was the sanctuary of Atargatis and the second the so-called Republican sanctuary built in the Hellenistic and Italic tradition.

On the one hand, the sanctuary of Atargatis was perhaps the refurbishing of an older building, perhaps of Punic origin, for which two main accesses have been archaeologically reported so far. The main entrance, facing the inscription mentioning Atargatis, opened to an area where a masonry altar was

---

[4] See M. Koch, 'Aletes, Mercurius und das prönikisch-punische Pantheon in Neukarthago', *Madrider Mitteilungen* 23 (1982): pp. 101-13.

[5] A. C. González Garcia and J. A. Belmonte, 'Thinking Hattusha: astronomy and landscape in the Hittite lands', *Journal for the History of Astronomy* 42 (2011): pp. 461-494; see also J. A. Belmonte and A. C. González-García, 'The pillars of the Earth and the sky: capital cities, astronomy and landscape', *Journal of Skyscape Archaeology* 1 (2015): in press.

[6] C. Esteban and J. L. Escacena Carrasco, 'Archaeology of the sky. Astronomical orientations in Protohistoric buildings of the south of the Iberian Peninsula', *Trabajos de Prehistoria* 70 (2013): pp. 114-139.

[7] A. C. González-García, L. Costa Ferrer, M. Zedda, and J. A. Belmonte, 'The orientation of the Punic tombs of Ibiza and Sardinia', in *Light and Shadow in Cultural Astronomy*, ed. M. Zedda and J. A. Belmonte (Dolianova: Agora Nuraghica, 2007), pp. 47-56.

Fig. 5.4: Summer solstice sunrise from the gate of the sanctuary of Atargatis (in H1), a sector where the sacred area of a female Punic deity was possibly located since the foundation of Qart Hadašt, over a modern water deposit, was presumably built upon the foundations of the temple of Baal Hammon at Monte Sacro (H2). This sequence strongly suggests a nice astronomical relationship between two selective sacred areas of the Punic, and later Roman, town: Arx Hasdrubalis (H1) and Mons Cronos (H2). Image series courtesy of Andrés Ros.

presumably located in the wall opposite the entrance. From this altar, an open view of the eastern horizon would have been obtained through the gate, ranging from the summer solstice sunrise (over Monte Sacro – H2 – and Baal Hammon temple, see Fig. 5.4) to sunrise at the Equinoxes. These relationships could have been significant for the rituals having place on the site. The building had a possible opening to the north (not documented) and another gate oriented to the south, perhaps to the area of the port although, within the errors, a stellar alignment cannot be ruled out (see Table 5.1).

On the other hand, the latter construction was a sanctuary devoted to an unknown deity (perhaps healing divinities), with a splendid view and an orientation that fully justified the new Roman dominion over the city and its port (see Fig. 5.2). The main axis of this temple was facing a spot in the southern slope on Cerro de la Concepción (*Mons Aesculapii*, H4) where the ruined church of Saint Mary (popularly known as the 'cathedral') is today located. If this was already a significant sacred spot in antiquity, it still needs to be confirmed by modern excavations. However, the secondary axis, the perpendicular one towards the east, offers a most interesting clue. It is facing sunrise at April 18 in the Gregorian proleptic calendar (with a margin of ±1 day) and hence would have been facing sunrise on 21 April in the first times of the Julian calendar (see Table 5.1). This is the traditional date of the foundation of Rome and it could have been deliberately incorporated in the temple design to highlight the Roman dominion over the city.[8] Actually the temple seems to be a little older than Caesar´s epoch but still may fit the same date within a certain year of the Republican calendar prior to the Julian reform.[9] In this regard, consideration should be given to the fact that the temple could be archaeologically dated mid-first century BCE.

At the end of the Republic, Cartagena was re-founded under the title of *Colonia Urbs Iulia Nova Carthago* and, especially under Caesar and Augustus, a new urban grid was applied to the city.[10] This was orthogonal in the western sector of the city and certainly followed a solstitial axis linking sunrise and sunset at summer and winter solstices, respectively (see Fig. 5.5). This axis could easily have been established by solar observations towards the eastern horizon, perhaps in a line of sight passing through the main eastern gate of the city, as stressed in the diagram. The western horizon was mostly obscured by relatively high nearby mountains presenting a large angular height and would have been useless for such a purpose. Later on, presumably via a process of *centuriatio* –perhaps using a *varatio* procedure – this solstitial alignment was extended north, to the area of the forum (see below), and to the south to the sector of the impressive Roman theater built under Augustus and dedicated to his adopted sons Lucius and Caius Caesar, children of Agrippa and his third wife Julia, Augustus' daughter.

This fact is well illustrated in Table 5.1 where it is shown that the *frons scaenae* and the *orchestra* of the theater followed the solstitial axis at 0° of angular height. Considering that the eastern horizon is obscured by *Mon Aesculapius* and the western one by the hills encircling the port, this alignment could only be produced if the orthogonal grid had been extended to the south. The theater showed a second conspicuous fact: it was designed and built in such a way that the winter solstice sun was rising on the axis of the grades and the *cavea*.[11] Antefixes with representations of Capricorn (see Fig. 5.3) have been found in the sector of the *Augusteum*, and it is possible that the importance given by Augustus to the zodiacal sign of Capricornus,[12] and hence to the winter solstice, was also known in Carthago Nova and reflected in its architectural design and symbolism.[13]

On the opposite side of the urban grid, to the area of the Forum, the solstitial arrangement was also extended, but it included an important peculiarity (see Fig. 5.5). Table 5.1 shows that the *cardines* and *decumani* located in the so-called 'Barrio de las Termas' did not follow the solstitial arrangement precisely.[14] Local possibly played a role on this. However, the area of the Forum was re-arranged so that

---

[8] Interestingly, the same date could be reflected in the architecture of the Pantheon in Rome: J. A. Belmonte and M. Hoskin, *Reflejo del Cosmos: atlas de arqueoastronomía del Mediterráneo occidental* (Madrid: Equipo Sirius, 2002), pp. 219-23. See also R. Hannah and G. Magli, 'The role of the sun in the Pantheon's design and meaning', *Numen - Archive for the History of Religion* 58, no. 4 (2011): pp. 486-513.

[9] J. Rüpke, *The Roman Calendar from Numa to Constantine: Time, History, and the Fasti* (New York: Wiley-Backwell, 2011).

[10] For a discussion on the urban planning of Roman cities, see J. Rykwert, *The idea of a town: the anthropology of urban form in Rome, Italy, and the ancient world* (Cambridge, MA: MIT Press, 1999); or more recently: R. Laurence, S. Esmonde-Cleary and G. Sears, *The city in the Roman West* (Cambridge: CUP, 2011).

[11] S. F. Ramallo and E. Ruiz, *El teatro romano de Cartagena* (Murcia: Editorial KR, 1998).

[12] See J. M. Noguera Celdrán, 'Un edificio del centro monumental de *Carthago Nova*: Análisis arquitectónico-decorativo e hipótesis interpretativas', *Journal of Roman Archaeology* 15 (2002): pp. 63-96.

[13] After the presentation of this work, and before its publication, a new most interesting work supporting the importance of the winter solstice in relationship to Augustus has been published, see: S. V. Bertarione and G. Magli, 'Augustus' power from the stars and the foundation of Augusta Pretoria Salassorum', *Cambridge Archaeological Journal* 25 (2015): 1-15.

[14] J. M. Noguera Celdrán and Mª. J. Madrid, eds., *Arx Hasdrubalis. La ciudad reencontrada. Arqueología en el Cerro del Molinete, Cartagena* (Murcia: Región de Murcia, 2009).

it could fit the pattern despite the orography. Our data of the temple of the Forum,[15] and the planimetry[16] provided for the excavations of the *Curia* and the *Augusteum*, show that sunrise at the summer solstice was produced on top of *Mons Aletes* (H3, see Table 5.1), the hill devoted to the legendary founder of the city to which Augustus was perhaps assimilated after the radical reform – nearly a second foundation – of the city in this period. Summarizing, Carthago Nova can be considered an astronomical and topographic materialization of sacred space, as is the case for many other cities in the ancient world strongly related to their founder such as, for example, Rome or even Alexandria.[17]

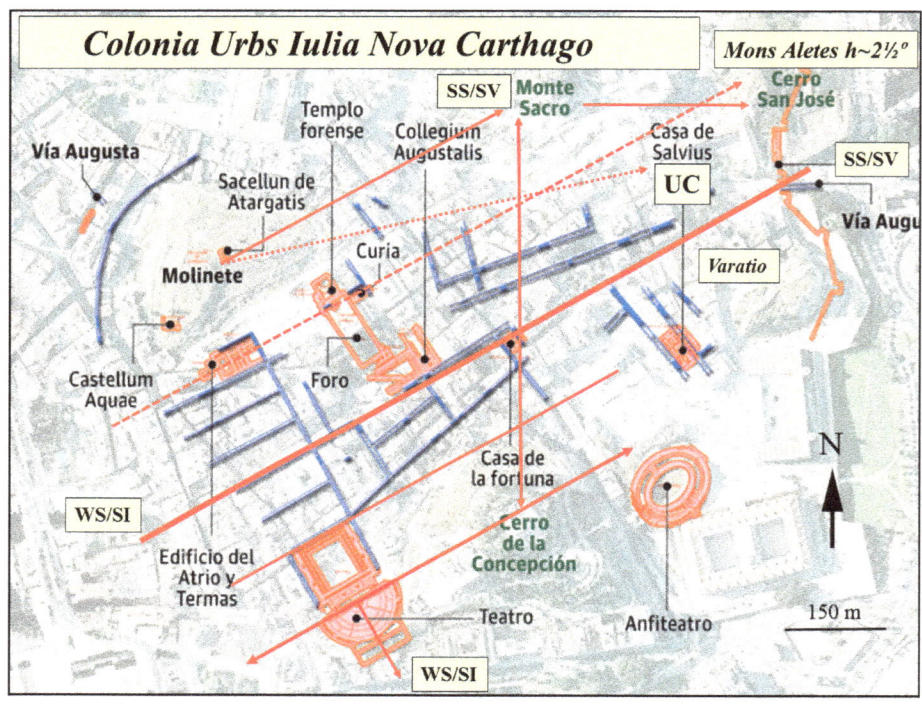

Fig. 5.5: The principal buildings and part of the recovered street grid of Roman Carthago Nova over imposed to a cardinally orientated modern city aerial view. A general solstitial arrangement relating important building and sacred spots can be ascertained. UC stands for urbe condita showing the secondary alignment of the Republican temple. See the text for a more detailed discussion. For a more detailed archaeological description and maps of the sites, see González-García et al. (2015).[18] Diagram by the authors, adapted from an image courtesy of 'Cartagena, Puerto de Culturas' Consortium.

---

[15] J. M. Noguera Celdrán, B. Soler, Mª. J. Madrid, and J. Vizcaíno, 'El foro de Carthago Nova: estado de la cuestión', in *Fora Hispaniae. Paisaje urbano, arquitectura, programas decorativos y culto imperial en los foros de las ciudades hispanorromanas*, ed. J. M. Noguera Celdrán (Murcia: Museo Arqueológico de Murcia, 2009), pp. 217-302.

[16] J. A. Antolinos, 'El trazado urbanístico y viario de la colonia romana', in *Arx Hasdrubalis. La ciudad reencontrada. Arqueología en el cerro del Molinete / Cartagena*, ed. J. M. Noguera Celdrán and Mª. J. Madrid (Murcia: Región de Murcia, 2009), pp. 59-67.

[17] See e.g., L. Ferro and G. Magli, 'The astronomical orientation of the urban plan of Alexandria', *Oxford Journal of Archaeology* 31 (2012): pp. 381-389.

[18] A. C. González-García, J. M. Noguera Celdrán, J. A. Belmonte, A. Rodríguez Antón, E. Ruiz Valderas, M. J. Madrid Balanza, E. Zamora, and J. y Bonnet, 'Orientatio ad sidera: astronomía y paisaje urbano en Qart Hadast / Carthago Nova', *Zephyrus* 75 (2015): pp. 139-60.

Regarding the two main orientations just described in Carthago Nova, it is worth noting that they appear to be two of the most common orientations used for Roman towns in Hispania, as our group is presently unveiling.[19]

Cartagena may still deserve surprises. After our campaign, we realized that we have skipped the analysis of a possible relevant site: the *sacellum* of *Iuppiter Stator* to the south east of the city.[20] This mostly ruined sanctuary was excavated in the early 90s and abandoned since then and today there is hardly any potential evidence on site.[21] However, its location can be fixed precisely at the slope of Cerro Gallufo, a small hill at the base of Cerro San Juan where the El Calvario chapel – an important Christian pilgrimage site – is located. Winter solstice sunrise occurred upon this distant spot as observed from the sacred area of *Arx Hasdrubalis* (H1) and may have indicated the relevance from a symbolic point of view of the area in the past. Most important, from the site of the *sacellum* of Jupiter, sunset at the summer solstice (and on Saint John's night) could be observed over the summit of Sierra Espuña in the distant western horizon (see Fig. 5.6). Sierra Espuña is the highest mountain range visible for the horizon of Carthago Nova, it was heavy forested in antiquity and it is often covered by snow. It could have been the perfect residence for a storm – *stator* – deity such as Baal (or Jupiter in Roman times), who was born precisely at the summer solstice according to earlier sources from the Levant.[22] This will need further research and analysis in the future.

Fig. 5.6: Sunset at summer solstice on the highest peak of Sierra Espuña as observed from the sacellum of Iuppiter Stator (Jupiter Tonante) to the southeast of the original urban area. Snow cupped (in winter) Espuña was the most imposing orographic feature in the area of Carthago Nova. Image courtesy of Andrés Ros.

---

[19] González-García, Rodríguez-Antón, and Belmonte, 'The orientation of Roman towns in Hispania: preliminary results'. New data have been obtained in the last couple of years that, once analysed, will hopefully confirm or deny this suspicion.

[20] S. F. Ramallo Asensio, *Carthago Nova: Puerto mediterráneo de Hispania* (Murcia: Fundación Cajamurcia, 2011).

[21] M. Amante Sánchez, M. Martín Camino, M. A. Pérez Bonet, R. González Fernández, and M. A. Martínez Villa, 'El sacellum dedicado a Iupitter Stator en Cartagena', *Antigüedad y Cristianismo* 12 (1995): pp. 533-62.

[22] A. Polcaro, A. C. González García, and J. A. Belmonte, 'Study of the orientation of the Bronze age temple of Pella, Jordan', *Anthropological Notebooks* 19 (2013): pp. 481-92.

Acknowledgements

We would like to thank Encarnación Zamora, José (Pepe) Bonnet and Andrés Ros for their assistance during fieldwork and after, during the imaging documentation process of the solstitial phenomena. Without their enthusiastic collaboration, this work would have not been possible. Elena Ruiz Valderas, Director of the Roman Theater Museum, and María José Madrid, Director of the archaeological excavations in the Molinete, acted as excellent hosts. 'Cartagena: Puerto de Culturas' Consortium and its supervisor Agustina Martínez Molina strongly supported the fieldwork in ancient Carthago Nova. This work has been partly financed under the framework of the projects P/310793 "Arqueoastronomía" of the IAC, and AYA2011-26759 "Orientatio ad Sidera III" and HAR2012-37405-C04-02 "Roma, las capitales provinciales y las capitales de Hispania: difusión de modelos en la arquitectura y el urbanismo. Paradigmas del conventus Carthaginiensis" of the Spanish MINECO.

Bibliography

Amante Sánchez, M., M. Martín Camino, M.A. Pérez Bonet, R. González Fernández and M.A. Martínez Villa. 'El sacellum dedicado a Iupitter Stator en Cartagena'. *Antigüedad y Cristianismo* 12 (1995): pp. 533-62.

Antolinos, J. A. 'El trazado urbanístico y viario de la colonia romana'. In *Arx Hasdrubalis. La ciudad reencontrada. Arqueología en el cerro del Molinete / Cartagena*, edited by J. M. Noguera Celdrán and Mª. J. Madrid, pp. 59-67. Murcia: Región de Murcia, 2009.

Belmonte, J. A. and M. Hoskin. *Reflejo del Cosmos: atlas de arqueoastronomía del Mediterráneo occidental*. Madrid: Equipo Sirius, 2002.

Belmonte, J. A. and A. C. González-García. 'The pillars of the Earth and the sky: capital cities, astronomy and landscape', *Journal of Skyscape Archaeology* 1 (2014): 8-37.

Bertarione, S. V. and G. Magli. 'Augustus' Power from the Stars and the Foundation of Augusta Pretoria Salassorum', *Cambridge Archaeological Journal* 25 (2015): 1-15

Esteban, C. and J. L. Escacena Carrasco. 'Archaeology of the sky. Astronomical orientations in Protohistoric buildings of the south of the Iberian Peninsula'. *Trabajos de Prehistoria* 70, no. 1 (2013): 114-139.

Ferro, L. and G. Magli. 'The astronomical orientation of the urban plan of Alexandria'. *Oxford Journal of Archaeology* 31 (2012): pp. 381-389.

González-García, A. C., L. Costa Ferrer, M. Zedda, and J. A. Belmonte. 'The orientation of the Punic tombs of Ibiza and Sardinia'. In *Light and Shadow in Cultural Astronomy*, edited by M. Zedda and J. A. Belmonte, pp. 47-56. Dolianova: Agora Nuraghica, 2007. 47-56.

González-García, A. C. and Costa-Ferrer, L. (2011) 'The diachronic study of orientations: Mérida, a case study'. In *Archaeoastronomy and Ethnoastronomy: Building Bridges Between Cultures*, edited by C. L. N. Ruggles. IAU Symposium 278: pp. 374-381.

González García, A. C. and J. A. Belmonte. 'Thinking Hattusha: astronomy and landscape in the Hittite lands'. *Journal for the History of Astronomy* 42 (2011): pp. 461-494.

González-García, A.C., A. Rodríguez-Antón, and J. A. Belmonte. 'The orientation of Roman towns in Hispania: preliminary results'. *Mediterranean Archaeology and Archaeometry* 14, no. 3 (2014): pp. 107-19.

González-García, A.C., J. M. Noguera Celdrán, J. A. Belmonte, A. Rodríguez Antón, E. Ruiz Valderas, M. J. Madrid Balanza, E. Zamora, and J. y Bonnet. 'Orientatio ad sidera: astronomía y paisaje urbano en Qart Hadast / Carthago Nova'. *Zephyrus* 75 (2015): pp. 139-60.

Hannah, R. and G. Magli. The role of the sun in the Pantheon's Design and Meaning. *Numen - Archive for the History of Religion* 58, no. 4 (2011): pp. 486-513.

Koch, M. 'Aletes, Mercurius und das prönikisch-punische Pantheon in Neukarthago'. *Madrider Mitteilungen* 23 (1982): pp. 101-13.

Laurence, R., S. Esmonde-Cleary, and G. Sears. *The city in the Roman West*. Cambridge: CUP, 2011.

Magli, G. 'On the orientation of Roman towns in Italy'. *Oxford Journal of Archaeology* 27 (2008): pp. 63–71.

Noguera Celdrán, J. M. 'Un edificio del centro monumental de Carthago Nova: Análisis arquitectónico-decorativo e hipótesis interpretativas'. *Journal of Roman Archaeology* 15 (2002): pp. 63-96.

Noguera Celdrán, J. M. and Mª. J. Madrid, eds. *Arx Hasdrubalis. La ciudad reencontrada. Arqueología en el Cerro del Molinete, Cartagena*. Murcia: Región de Murcia, 2009.

Noguera Celdrán, J. M., B. Soler, Mª. J. Madrid, and J. Vizcaíno. 'El foro de Carthago Nova: estado de la cuestión'.

In *Fora Hispaniae. Paisaje urbano, arquitectura, programas decorativos y culto imperial en los foros de las ciudades hispanorromanas*, edited by J. M. Noguera, pp. 217-302. Murcia, Museo Arqueológico de Murcia, 2009.

Noguera Celdrán, J. M. 'Carthago Nova: Urbs privilegiada del Mediterráneo occidental'. In *Hispaniae urbes. Investigaciones arqueológicas en ciudades históricas*, edited by J. Beltrán Fortes and O. Rodríguez Gutiérrez, pp. 121-190. Sevilla: Servicio de Publicaciones de la Universidad de Sevilla, 2012.

Noguera Celdrán, J. M. 'Qart Hadast, capital bárquida de Iberia'. In *Fragor Hannibalis. Anibal en Hispania*, edited by M. Bendala Galán, M.ª Pérez Ruiz, and I. Escobar, pp. 134-173. Madrid: Comunidad de Madrid, 2013.

Polcaro, A., A. C. González García, and J. A. Belmonte J.A. 'Study of the orientation of the Bronze age temple of Pella, Jordan'. *Antropological Notebooks* 19 (2013): pp. 481-92.

Ramallo, S. F. and E. Ruiz. *El teatro romano de Cartagena*. Murcia: Editorial KR, 1998.

Ramallo Asensio, S. F. *Carthago Nova; Puerto mediterráneo de Hispania*. Murcia: Fundación Cajamurcia and Darana Editorial, 2011.

Rykwert, J. *The idea of a town: the anthropology of urban form in Rome, Italy, and the ancient world*. Cambridge, MA: MIT Press, 1999.

Rüpke, J. *The Roman Calendar from Numa to Constantine: Time, History, and the Fasti*. New York: Wiley-Backwell, 2011.

*A STATUS REPORT: A REVIEW OF RESEARCH
ON THE ORIGINS AND DIFFUSION
OF THE BELIEF IN A SKY BEAR*

Roslyn M. Frank

ABSTRACT: This presentation provides an overview of the research that has been carried out on the constellation of Ursa Major in Europe, Eurasia and North America and attempts to explain some of the reasons behind the widespread association of the figure of a bear, a bear hunt or a hunter in the form of a bear-human with these stars and ones nearby. In doing so it takes into consideration a set of beliefs associated with 'bear ceremonialism' found among circumboreal hunter-gatherer peoples as well as among Europeans and the narrativization of these beliefs. Beginning with the assertion made by Gingerich that 'In the widespread mythological connections of the dipper stars with a Great Bear (Ursa Major) we have a hint that a few constellations may date back as far as the Ice Ages', special attention is given to the motif of the Cosmic Hunt and its diverse manifestations across the study area as well as the archaic cosmology that supports these celestial narratives and related motifs, most particularly the belief that humans descended from bears. In addition, the role played by the underlying hunter-gatherer mode of subsistence in shaping these cultural conceptualizations and enculturating the skyscape is addressed.

>   Everybody says, 'After you take a bear's coat off, it looks just like a human'.[1]
>   As Petiri Prébende, the last Basque bear hunter, said, in 1983: 'Lehenagoko eüskaldünek gizona hartzetik jiten zela sinhesten zizien' ('Basques used to believe that humans descended from bears').[2]

Introduction

Although we often speak of 'cultural astronomy', the exact scope of this term is still being defined. The current study is intended to contribute to this ongoing dialogue and at the same time demonstrate the importance of enculturating the skyscape so that what is viewed above is conceptualized more broadly. This will be accomplished by using a more integrated cross-disciplinary approach. While this chapter focuses attention on the constellation figure of Ursa Major, more specifically on the seven dipper stars of this constellation as well as other stars nearby, the purpose of the study is to bring into view the way that this constellation has been enculturated across the Northern hemisphere. Specifically, research on circumpolar bear ceremonialism will be utilized to show the ways that this constellation figure has been integrated into a pre-existing belief system; in the process, certain basic cosmological understandings that came to be linked to it across time and space are revealed. The emphasis will be on cognitive continuity rather than rupture and consequently on reconstructing the ethnocultural substratum that underlies the cross-cultural manifestations of the belief system itself, including its various representations in Europe.

For the most part, studies of constellation lore have been characterized by the tendency to see celestially coded stories and legends as folklore; that is, as quaint survivals from pre-modern mentalities, rather than as serious sources of information about what were once coherent and highly meaningful belief systems for the populations in question. This is also true for the vast majority of studies concerning the constellation of Ursa Major. Or in the best of cases, it is assumed that it was the remarkable characteristics of bears that led them to be projected skyward, an explanation that, indeed, does attempt to incorporate elements

---

[1] Gary Snyder, *The Practice of the Wild* (San Francisco: North Point, 1990), p. 164.
[2] Txomin Peillen, 'Le culte de l'ours chez les anciens basques', in *L'ours brun: Pyrénées, Agruzzes, Mts. Cantabriques, Alpes du Trenti*, ed. Claude Dendaletche (Pau: Acta Biologica Montana, 1986), pp. 171–3.

from the daily lives and experiences of hunter-gatherers into the cognitive process of enculturating the sky above.[3] Moreover, in line with that set of assumptions, in the past there have been many attempts to project, quite literally, the image of a bear onto the stars of Ursa Major, as if the resulting putative resemblance explained in some way how a bear originally came to be associated with these circumpolar stars, especially the dipper stars.

What has not been noticed is that this process was connected to a much more complex cosmology that linked humans to bears and specifically the belief that humans descended from bears. There are two basic perspectives concerning bear ceremonialism: one that emphasizes a diffusionist view and another that concentrates more on what is best understood as an environmental niche perspective in which the similarities in the cosmology of hunter-gatherer populations can be explained by the fact that the parallels evolve quite naturally from human interactions with the same type environmental resources, ones that are both cognitively and materially available to them.

At the same time, I would emphasize that when we are talking about bear ceremonialism and the cosmology that accompanies it, these two perspectives are not mutually exclusive. For instance, thirty years ago Owen Gingerich noted that the Big Dipper is seen by many different tribes of North America as a Great Bear, and although it is difficult to be sure how old their oral traditions are, it seems plausible that a very early tradition of a celestial bear crossed the Bering Straits with ancient migrants, especially since the same identification is found across Siberia. Gingerich states that '[s]uch an early tradition could well have diffused throughout the world from the ancient cave dwellers of Europe' and he concludes by saying that '[i]n the widespread mythological connections of the dipper stars with a Great Bear (Ursa Major) we have a hint that a few of the constellations may date back as far as the Ice Ages'.[4]

On the other hand, we have the environmental niche perspective which is the view espoused by the American environmentalist and nature writer Paul Shepard, whose extensive writings on bears and bear ceremonialism are well known. In his essay 'The Biological Bases of Bear Mythology and Ceremonialism', Shepard argued that in the absence of primates, in zones where humans and bears cohabited, it was natural for these forager populations to conclude that the animal anatomically most similar to them was the bear.[5] Indeed, where there have been bears and hunter-gatherers, one regularly finds evidence for bear ceremonialism along with the closely related belief in an ursine ancestor.

Consequently, the role played by bears and by humans interacting with bears is a key element in the enculturation of the skyscape. Clearly, this animal's intellectual endowments, its keen intelligence and great memory, along with its ability to walk upright, nurse its young as female humans do and leave footprints that resemble ever so much those of humans, all contributed to it being elevated to the status of the ancestor of humans.[6] Creating sacred narratives and projecting imagery linked to this ursine cosmovision skyward, e.g., onto the dipper stars, would have followed quite naturally without the need of a diffusionist model. Or, conversely, if and when such narratives and celestial imagery did diffuse and move into new territories, the receiving forager populations would have been predisposed to embrace them.

---

[3] Edwin C. Krupp, *Beyond the Blue Horizon: Myths and Legends of the Sun, Moon, Stars, and Planets* (New York: HarperCollins Publishers, 1991), p. 240.

[4] Owen Gingerich, 'Astronomical scrapbook. The origin of the zodiac', *Sky and Telescope* 67 (1984): p. 220.

[5] Paul Shepard, 'The biological bases of bear mythology and ceremonialism', *The Trumpeter: Journal of Ecosophy* 23, no. 2 (2007): pp. 74–9.

[6] While bears are perfectly capable of standing and even walking on their hind legs, they usually do so to get a better view of their surroundings or to walk only short distances. If, however, one of their front paws is injured they can choose to walk upright, as is clearly shown in these two recent videos: Anon., 'Bear walking on its legs', https://www.youtube.com/watch?v=m3_tch2HTEY [accessed 4 January 2015] and Anon. 'Just a bear walking upright like a man', https://www.youtube.com/watch?v=JuMw-tTaWCY [accessed 4 January 2015].

Enculturating the Skyscape: Back to Europe

No matter which model you choose, the dipper stars of the Great Bear constellation are best understood as belonging to the most archaic stratum of the star figures known to Europeans.[7] At the same time, the Greek myths intended to explain Ursa Major are best understand as recent overlays on a much older ursine-oriented cosmovision that was once alive and well in Europe.

Since Gingerich's comment in 1984, a tremendous amount of cross-cultural research has been published outlining the basic tenets of bear ceremonialism as well as charting the variety of celestial projections of this belief system onto the stars of the Big Dipper and/or the stars of constellations nearby (e.g., Arcturus in Boötes), both in North America and in northern Eurasia.[8] More recently the focus has shifted to Europe itself and to the recuperation of the ursine cosmovision indigenous to this zone, where remnants from this archaic hunter-gatherer oriented system are still abundantly present in European performance art and folk belief.[9] Indeed, the collection and analysis of these materials constitute one of the goals of a newly formed working group on bear ceremonialism, made up of researchers from across much of Europe.

In other words, to understand why Europeans see a bear associated with the dipper stars requires knowledge of the tenets of bear ceremonialism along with a careful investigation of the mechanisms that allowed this cognitive framework to be projected skyward and to perpetuate itself across time, passing from one generation to the next, albeit in an increasingly fragmented manner.

Types of Celestial Projection and Narrativization

There are two basic ways in which the image of a bear is projected skyward and narrativized, although both cultural conceptualizations are grounded in an ursine informed cosmovision. The first projection relates to the dwelling place of the primal Sky Bear in the heavens above. The second type of projection focuses on and narrativizes a hunting scene from one of two basic points of view.

First, we will examine the way the home of the Sky Bear is imagined. With respect to the first type of projection we are dealing with the celestial home of the Sky Bear. We find the belief in a bear ancestor, a kind of divinity whose dwelling place was in the heavens above, much in the same amorphous manner that Christian believers conceptualize God as residing in the sky above, but without specifying exactly where his abode is among the stars.

Across northern Eurasia evidence for the practice of bear ceremonialism is widespread along with the ritualized treatment of the slain bear's remains. Other ritual practices, such as the 'bear oath' ('swearing by the Bear'), reflect the belief in the supernatural powers of this creature. Similarly, one of the basic tenets of bear ceremonialism concerns the celestial home of the Sky Bear itself. The dipper stars and the Pole Star are frequently implicated in the 'sending home' ceremony. After the bear is hunted and killed,

---

[7] Roslyn M. Frank, 'The origins of "Western" constellations', in *The Handbook of Archaeoastronomy and Ethnoastronomy*, ed. C. L. N. Ruggles (Berlin: Springer Publishing Company, 2014), pp. 147–163; Roslyn M. Frank, 'The skylore of the indigenous peoples of northern Eurasia', in Ruggles, ed., *The Handbook of Archaeoastronomy and Ethnoastronomy*, pp. 1679–1686.

[8] Yuri Berezkin, 'The Cosmic Hunt: Variants of a Siberian-North-American myth', *Folklore: Electronic Journal of Folklore* 31 (2005): pp. 79–100.

[9] See for example: Maurizio Bertolotti, 'La fiaba del figlio dell'orso e le culture siberiane dell'orso', *Quaderni di Semantica*, XV, no. 1 (1994): pp. 39–56; Robert Bosch, *Fêtes de l'ours en Vallespir* (Perpignan: Trabucaire, 2013); Migel Mari Elosegi, *Hartz arrea Pirinioetan: Biologia, kultura eta kontsebazioa* (Donostia: Elkar, 2006); Jean-Dominique Lajoux, *L'homme et l'ours* (Grenoble: Glénat, 1996); Dominique Pauvert, 'Le rituel de l'ours des Pyrénées aux steppes', in *Traditions en devenir (coutumes et croyances d'Europe et d'Asie face au monde moderne)*, ed. Société des Études euro-asiatiques, EURASIE No. 2 (Paris: Hartmattan, 2014), pp. 17–51; Michel Pastoureau, *The Bear: History of a Fallen King* (Cambridge/London: The Belknap Press of Harvard University Press, 2011).

the soul of the slain bear is sent back up to heaven to give a report to the Sky Bear concerning whether humans treated the earthly bear properly. This is particularly true of the 'sending home' ceremony of the Algonquians.[10]

Moreover, the logic intrinsic to the ursine cosmovision can be detected in the celestial orientation and projection of the narrative. According to Ingold, 'Certain animals are always regarded as extraordinary, and are approached as one would approach other human beings, in the manner appropriate to particular persons. Chief among these is the bear, which for all circumboreal peoples—apparently without exception—is the object of special respect'.[11] From the hunter-gatherer point of view the killing and consumption of animals is an integral and necessary part of the creative cycle of renewal. These ritualized actions release the vital essence of the animals so that they can be reclothed with flesh. This requires the proper ritual killing of game animals and the propitiation of the 'spirit master' or 'owner' of each species, especially the bear who acts as the 'spirit master' of all beings. Doing this ensures the reproduction of the animals themselves and the food supply. This is a crucial point and one of the primary reasons for the 'sending home' ceremony of the slain bear and its report to the Sky Bear.

Cosmic Hunt Motif

A second common narrative motif found in North America and northern Eurasia is referred to as the Cosmic Hunt.[12] In North America the four stars of the bowl of the dipper are often associated with a bear while the three stars in the handle are portrayed as hunters. In Northern Eurasia, we often find an elk playing the role of prey, rather than a bear.[13] It is pursued by hunters, at times by a single hunter whose identity can sometimes be linked to a bear-human, known as Mangi, represented by Arcturus in Boötes. This figure may be identified in some fashion with the son/daughter of the Sky Bear, that is, the little bear cub who comes down from the sky to teach humans how to conduct themselves and venerate and ritually honour the slain bear and the rest of Nature.[14]

There are a number of circumpolar celestial parallels. Speaking of the ursine cosmology of the Khanty people of western Siberia, Schmidt gives this summary: through the institution of the 'bear oath' and the sacred narratives concerning the first-born bear son (or daughter) of Numi-Torum (the Sky Bear) and his/her descent to Earth, we see 'bears form, as it were, a controlling super society, sent down from the divine sphere, over human society'.[15] More on this last point shortly.

---

[10] Frank G. Speck, *When the Celestial Bear comes down to Earth: The Bear Ceremony of the Munsee-Mahican in Canada as related by Nekatcit, in collaboration with Jesse Moses. Delaware Nation. Ohsweken, Ontario* (Reading, PA: Reading Public Museum and Art Gallery, 1945), pp. 27-78. Cf. also Roslyn M. Frank, 'Hunting the European sky bears: When bears ruled the earth and guarded the gate of heaven', in *Astronomical Traditions in Past Cultures,* ed. Vesselina Koleva and Dmiter Kolev (Sofia: Institute of Astronomy, Bulgarian Academy of Sciences, and National Astronomical Observatory Rozhen, 1996), pp. 116–142; ''Evidence in Favor of the Palaeolithic Continuity Refugium Theory (PCRT): *Hamalau* and its linguistic and cultural relatives. Part 1', *Insula: Quaderno di Cultura* 4 (2008): pp. 91–131; 'Recovering European ritual bear hunts: A comparative study of Basque and Sardinian ursine carnival performances', *Insula: Quaderno di Cultura Sarda* 3 (2008): pp. 41–97. http://www.tinyurl.com/hamalau14. [accessed 3 June 2016]

[11] Tim Ingold, *The Appropriation of Nature: Essays on Human Ecology and Social Relations* (Manchester: Manchester University Press, 1986), p. 249.

[12] Berezkin, 'The Cosmic Hunt,' pp. 79–100.

[13] Alla Lushnikova, 'Early notions of Ursa Major in Eurasia', in *Astronomy of Ancient Societies,* ed. Tamila M. Potyomkina and Vladimir Obridko (Moscow: Nauka, 2002), pp. 254–261.

[14] Paul Shepard and Barry Sanders, *The Sacred Paw: The Bear in Nature, Myth and Literature* (New York, NY: Arkana, 1992), pp. 66–8.

[15] Éva Schmidt, 'Bear cult and mythology of the northern Ob-Ugrians,' in *Uralic Mythology and Folklore,* ed., Mihály

In other words, the landscape that bears originally inhabited was celestial and hence, the necessity of 'sending home' ceremonies which were conducted for the souls of individual earthly bears slain by hunters. Hence, flesh-and-blood bears were viewed as the earthly representatives of the celestial bears and elicited similar respect and veneration.

Another Interview and Enculturated Skyscape: Evidence for the Belief in the Celestial Bear in Europe

In the case of Europe, because of its physical appearance and great intelligence, the bear was, in fact, the animal that most closely replicated a human being. However, in contrast to its human relatives, the bear seemed to be capable of dying and being resurrected from a death-like sleep in the spring of each year. Moreover, evidence from many native peoples demonstrates that this ability has been perceived by them as one of supernatural, even mystic, proportions. The same appears to be true of Europeans in times past.

Among the Basques who speak a pre-Indo-European language, belief in the sacredness of bears as well as their role as ancestors of humans persisted into the latter part of the 20th century, as has been documented in an interview with the last Basque bear hunter which took place in 1983.[16] While scholars are familiar with the Greek tales concerning Callisto's transformation into a Great Bear, few are aware of the significance of the Basque belief or the truly remarkable information found in an interview carried out in Biarritz, nearly a hundred years earlier, in 1891, by the British folklorist, Thomas Hollingsworth.[17] Hollingsworth's informants were two Navarrese Basque bear-trainers, a man and his wife, accompanied by their bear. Their first language was Basque and they spoke Spanish with some difficulty. As we will see, in the interview the two informants speak of the special powers of bears in general, e.g., their ability to understand human speech and hear all that is said. They emphasize the relationship between their ward, an earthly bear, and another bear, conceptualized as a sort of ursine divinity who resides in heaven, but without indicating a specific location for it among the stars.

While Hollingsworth states that he is unfamiliar with the time frame associated with a reference that the pair makes to an earlier time when bears ruled the earth, the comment suggests that it could be related to the mindset that humans must have had long before the invention of firearms, at a point in time when humans were far out-numbered by bears, yet often ended up sharing mountain trails, salmon streams and berry patches with them.

In regard to the uncanny and often remarkable abilities that the pair attributed to their bear, similar beliefs are found among the Asiatic Eskimos who held that during the festival of the slain bear, the bear's shadow-soul could hear and understand the speech of humans, no matter where the people were,[18] while the North American Tlingit said, 'People must always speak carefully of bear people since bears [no matter how far away] have the power to hear human speech. Even though a person murmurs a few careless words, the bear will take revenge'.[19]

Other analogous beliefs are found among the Ket (Yenesei Ostyaks), an Ugric-speaking people of Siberia with a rich tradition of bear worship, who believe that the bear is chief among animals, that beneath its skin is a being in human shape, divine in wisdom. For them the bear was also invested with

---

Hoppál and Juha Pentikäinen (Budapest / Helsinki: Ethnographic Institute-Finnish Literary Society, 1989), p. 199.
[16] Peillen, 'Le culte de l'ours', pp. 171–3.
[17] Thomas Hollingsworth, 'A Basque superstition', *Folklore* II (1891): pp. 132–3.
[18] Shepard and Sanders, *The Sacred Paw*, p. 86.
[19] David Rockwell, *Giving Voice to Bear: North American Indian Myths, Rituals and Images of the Bear* (Niwot, Colorado: Roberts Rinehart Publishers, 1991), p. 64.

the capability of understanding the speech of all beasts as well as of man.[20] Among Finno-Ugric peoples and Native American groups, the bear is viewed as omnipotent and omnipresent. The animal has the power to hear all that is said. For this reason, hunters avoided mentioning the bear's real name, choosing rather to address him with circumlocutions. The Basque bear keepers' words echo a similar belief in the bear's ability to understand human speech. And, far from describing him as a cuddly pet, their comments depict their bear as a familiar yet awesome being.

That these might have been the qualities attributed to the European Sky Bear and his earthly representatives in times past appears to be evidenced linguistically by the avoidance taboo of Germanic and Slavic-speaking peoples who, in order not to mention the animal's real name, ended up employing semantically veiled expressions such as the 'brown one' in German and 'honey-eater' in Slavic.

Hollingsworth concludes his report with these pertinent revelations:

> I endeavored to learn when this sad state of affairs existed [when bears ruled humans], but could only ascertain that it was *antes*—before, in other times. 'El Orso', [sic] said his keepers, 'es el perro de Dios, el perro de San Pedro [the bear is the dog of God, the dog of Saint Peter]; he is very wise and thoughtful; he sits beside the blessed saint at the gate of Heaven, and if those who seek to enter have been cruel and unkind to bears in this world, the saint will turn them away, and they will have to go and live in hell […]'.[21]

Throughout the conversation the couple would constantly interrupt themselves to speak to the animal, assuring Hollingsworth that their bear perfectly understood all that was said. The hybrid nature of these last remarks shows the blending of two belief systems, that is, the pre-Christian ursine informed cosmology and Christianity. Their words end up positioning the bear as 'the dog of God' or as 'the dog of St Peter', sitting next to the Christian saint. However, the key to entering heaven is based on the answer the person gives to the question posed by St Peter, namely, whether the person has treated bears properly in this life.

From this perspective, St Peter is actually acting on behalf of the bear figure, seated silently beside him, 'very wise and thoughtful'. Stated differently, while St Peter is in charge of interrogating new arrivals concerning whether they have treated earthly bears with proper respect, the main character in this drama is the Celestial Bear. In this way the soul's entrance into to Heaven is conditioned by the way the person has interacted with bears on Earth.

In summary, this 19th century interview provides information concerning the role of the Celestial Bear as the guardian of the Gate of Heaven and resonates strongly with the tenets of northern bear ceremonialism.[22] Similarly, the report demanded by St. Peter concerning the conduct of humans vis-à-vis bears parallels the report central to the purpose of the sending-home ceremony in which the soul of slain bear, upon reaching heaven, would tell the Sky Bear whether humans had behaved properly. In other words, there was a moral component embedded in this aspect of the ritual.

Ethnographic Parallels
The following table summarizes the ethnographic parallels between the two data sets. The first set of characteristics is drawn from Basque, Pyrenean and European sources while the second data set comes from research on bear ceremonialism in Northern Eurasia and North America. These cross-cultural comparisons show the degree to which similarities can be identified and at the same time the way in which bear ceremonialism has played a role in enculturating the skyscape in these geographic locations.

---

[20] Evgeniia A. Alekseenko, 'The cult of the bear among the Ket (Yenisei Ostyaks),' in *Popular Beliefs and Folklore Tradition in Siberia*, ed. Vilmos Diószegi (The Hague, The Netherlands: Mouton & Co. 1968), p. 177.
[21] Hollingsworth, 'A Basque superstition', p. 133.
[22] Speck, *When the Celestial Bear comes down to Earth*.

Cross-cultural comparisons of the tenets of bear ceremonialism	Basque/Pyrenean/European	Northern Eurasia/ North America
Evidence of residual bear ceremonialism: bear as ancestor of humans, an ursine genealogy	Yes	Yes
Dipper stars associated conceptually with the figure of a bear[23]	Yes	Yes
Dipper stars associated with a narrative of a Cosmic Hunt	Yes, but only in Northern and North-Eastern Europe	Yes
Bears viewed as capable of hearing all that is said	Yes	Yes
The bear viewed generically as a guardian animal, protector of all other animals, humans and perhaps Nature itself	Yes	Yes
Residual evidence of a 'sending home' ceremony	Yes	Yes

Table 6.1

Concluding comments

Cultural astronomy has been defined as an interdisciplinary study bridging the sciences and humanities, whose goal is to investigate the relationships between human societies and their conceptualizations of sky resources. As is now well recognized, a full understanding of the human environment and culture must include the sky as well as the land, combining skyscape and landscape, for there is no cultural group which has not integrated sky resources into the construction of its belief system, a cognitive process that allows the populations in question to structure their cosmology, religion, politics and built environment.

Moreover, the interdisciplinary framework intrinsic to cultural astronomy helps us overcome drawbacks of the earlier more dichotomous mindset in which astronomical objects and phenomena received an independent ontological status. They were understood by researchers primarily as categories separated from the human mind. As such, they were perceived as *objects*, and as *phenomena* that had to be understood. They were observed, analysed, measured, and quantified according to standards of Western science.

In short, they were conceptualized in utilitarian and functional terms.[24] While they were conceived of as entities ready to be discovered and appropriated by humankind, on this view they remained separated from social practice and human culture, an orientation that reflects the ideas of modern astronomy. This conceptual framework also reflects a Cartesian dualism, separating nature from culture and hence astronomy from social life. However, this is not the epistemological premise on which the methodology and theory of cultural astronomy are being constructed as this chapter, hopefully, has demonstrated.

---

[23] It should be kept in mind that in Northern Eurasia, we find the bear-hunter Mangi chasing the figure of an elk while it is the latter animal who is linked to Ursa Major. See Berezkin, 'The Cosmic Hunt', pp. 79–84; Lushnikova, 'Early notions of Ursa Major in Eurasia', pp. 254–9.

[24] See Stanislaw Iwaniszewski, 'The erratic ways of studying astronomy in culture', in *Calendars, Symbols and Orientations: Legacies of Astronomy in Culture. Proceedings of the 9th Annual Meeting of the European Society for Astronomy in Culture (SEAC), Stockholm, 27–30 August 2001*, No. 59, ed. Mary Blomberg, Peter E. Blomberg and Göran Henriksson (Uppsala: Uppsala Astronomical Observatory, 2003), p. 10.

Keeping in mind the ursine genealogy and associated cosmovision that have informed the process of enculturation of the skyscape treated in this study, I believe the following comment by Griffin-Pierce is an especially appropriate way to conclude our discussion: 'Nearly every culture perceives the stars to be in groupings of constellations due to the uneven distribution of stars across the celestial sphere. Throughout time people have imposed order on the stars, as they perceive the heavens in terms of their own value systems'.[25]

Bibliography

Alekseenko, Evgeniia A. 'The cult of the bear among the Ket (Yenisei Ostyaks)'. In *Popular Beliefs and Folklore Tradition in Siberia*, edited by Vilmos Diòszegi, pp. 175–191. The Hague, The Netherlands: Mouton & Co. 1968.

Anon. 'Bear walking on its legs'. https://www.youtube.com/watch?v=m3_tch2HTEY.

Anon. 'Just a bear walking upright like a man'. https://www.youtube.com/watch?v=JuMw-tTaWCY.

Berezkin, Yuri. 'The Cosmic Hunt: Variants of a Siberian-North-American myth'. *Folklore: Electronic Journal of Folklore* 31 (2005): pp. 79–100.

Bertolotti, Maurizio. 'La fiaba del figlio dell'orso e le culture siberiane dell'orso'. *Quaderni di Semantica*, XV, no. 1 (1994) : pp. 39–56.

Bosch, Robert. *Fêtes de l'ours en Vallespir*. Perpignan: Trabucaire, 2013.

Elosegi, Migel Mari. *Hartz arrea Pirinioetan: Biologia, kultura eta kontsebazioa*. Donostia: Elkar, 2006.

Frank, Roslyn M. 'Evidence in Favor of the Palaeolithic Continuity Refugium Theory (PCRT): *Hamalau* and its linguistic and cultural relatives. Part 1'. *Insula: Quaderno di Cultura* 4 (2008): pp. 91–131. http://www.tinyurl.com/hamalau14.

Frank, Roslyn M. 'Hunting the European sky bears: When bears ruled the earth and guarded the gate of heaven'. In *Astronomical Traditions in Past Cultures*, edited by Vesselina Koleva and Dmiter Kolev, pp. 116–142. Sofia: Institute of Astronomy, Bulgarian Academy of Sciences, and National Astronomical Observatory Rozhen, 1996.

Frank, Roslyn M. 'Recovering European ritual bear hunts: A comparative study of Basque and Sardinian ursine carnival performances'. *Insula: Quaderno di Cultura Sarda* 3 (2008): 41–97. http://www.tinyurl.com/hamalau14.

Frank, Roslyn M. 'The origins of "Western" constellations'. In *The Handbook of Archaeoastronomy and Ethnoastronomy*, ed. C. L. N. Ruggles, pp. 147–163. Berlin: Springer Publishing Company, 2014.

Frank, Roslyn M. 'The skylore of the indigenous peoples of northern Eurasia'. In *The Handbook of Archaeoastronomy and Ethnoastronomy*, ed. C. L. N. Ruggles, pp. 1679–1686. Berlin: Springer Publishing Company, 2014.

Gingerich, Owen. 'Astronomical scrapbook. The origin of the zodiac'. *Sky and Telescope* 67 (1984): pp. 218–220.

Griffin-Pierce, Trudy. 'Ethnoastronomy in Navajo sand paintings of the heavens'. *Archaeoastronomy* 9, no. 1/4 (1986), pp. 62–9.

Hollingsworth, Thomas. 'A Basque superstition'. *Folklore* II (1891): pp. 132–3.

Ingold, Tim. *The Appropriation of Nature: Essays on Human Ecology and Social Relations*. Manchester: Manchester University Press, 1986.

Iwaniszewski, Stanislaw. 'The erratic ways of studying astronomy in culture'. In *Calendars, Symbols and Orientations: Legacies of Astronomy in Culture. Proceedings of the 9th Annual Meeting of the European Society for Astronomy in Culture (SEAC), Stockholm, 27–30 August 2001*, No. 59, edited by Mary Blomberg, Peter E. Blomberg and Göran Henriksson, pp. 7–10. Uppsala: Uppsala Astronomical Observatory, 2003.

Krupp, Edwin C. *Beyond the Blue Horizon: Myths and Legends of the Sun, Moon, Stars, and Planets*. New York: HarperCollins Publishers, 1991.

Lajoux, Jean-Dominique. *L'homme et l'ours*. Grenoble: Glénat, 1996.

Lushnikova, Alla. 'Early notions of Ursa Major in Eurasia'. In *Astronomy of Ancient Societies*, edited by Tamila M. Potyomkina and Vladimir Obridko, pp. 254–262. Moscow: Nauka, 2002.

---

[25] Trudy Griffin-Pierce, 'Ethnoastronomy in Navajo sand paintings of the heavens', *Archaeoastronomy* 9, no. 1/4 (1986): p. 62.

Pastoureau, Michel. *The Bear: History of a Fallen King*. Cambridge/London: The Belknap Press of Harvard University Press, 2011.

Pauvert, Dominique. 'Le rituel de l'ours des Pyrénées aux steppes'. In *Traditions en devenir (coutumes et croyances d'Europe et d'Asie face au monde moderne)*, edited by Société des Études euro-asiatiques, *EURASIE No. 2*, pp. 17–51. Paris: Hartmattan, 2014.

Peillen, Txomin. 'Le culte de l'ours chez les anciens basques'. In *L'ours brun: Pyrénées, Agruzzes, Mts. Cantabriques, Alpes du Trentin*, edited by Claude Dendaletche, pp. 171–3. Pau: Acta Biologica Montana, 1986.

Rockwell, David. *Giving Voice to Bear: North American Indian Myths, Rituals and Images of the Bear*. Niwot, Colorado: Roberts Rinehart Publishers, 1991.

Schmidt, Éva. 'Bear cult and mythology of the northern Ob-Ugrians'. In *Uralic Mythology and Folklore*, edited by Mihály Hoppál and Juha Pentikäinen, pp. 187–232. Budapest / Helsinki: Ethnographic Institute-Finnish Literary Society, 1989.

Shepard, Paul. 'The biological bases of bear mythology and ceremonialism'. *The Trumpeter: Journal of Ecosophy* 23, no. 2 (2007): pp. 74–79.

Shepard, Paul, and Barry Sanders. *The Sacred Paw: The Bear in Nature, Myth and Literature*. New York, NY: Arkana, 1992.

Snyder, Gary. *The Practice of the Wild*. San Francisco: North Point, 1990.

Speck, Frank C. *When the Celestial Bear comes down to Earth: The Bear Ceremony of the Munsee-Mahican in Canada as related by Nekatcit, in collaboration with Jesse Moses. Delaware Nation. Ohsweken, Ontario*. Reading, PA: Reading Public Museum and Art Gallery, 1945.

IN SEARCH OF PÄIVÄTÄR, THE FINNISH SOLAR GODDESS

Marianna Ridderstad

ABSTRACT: It is well known that the Iron Age solar goddess of the Eastern Finnic peoples, the largest groups of which are the Finns and the Estonians, was female. However, little is known about her role relative to other deities or the significance of solar rites in relation to other cultic practices in Finland in the late Iron Age and early historical times. In this study, these questions are addressed by an interdisciplinary approach, by combining the information on the solar deity contained in Finnish folklore and Kalevala metric poetry with evidence provided by other ethnographical, archaeological and archaeoastronomical sources of solar symbolism and solar rites. It is found that by the careful analysis of the existing data it is possible to create a partial reconstruction of the role of the solar deity and her rites in Southern Finland in the late Iron Age.

1. Introduction

Finland and the Baltic area in general was one of the last regions in Europe to become Christian; pre-Christian beliefs and customs were preserved locally for many centuries, e.g. in Eastern Finland as late as the early 20th century. Yet relatively little is known of the solar worship and the solar deity of the Baltic Finns in the Iron Age. The reason for this situation may be that in Northern Europe, where the sun deity was one of the most important deities due to the strongly varying seasonal climatic cycle, it also was one of the first to be assimilated with prominent Christian figures. For example, representations of Christ as the Sun or Virgin Mary as *Maria in Sole* are well known and may have effectively replaced the pre-Christian deities with similar epithets.

The late Iron Age solar goddess of the Eastern Finnic peoples, the largest groups of which are the peoples speaking Finnish and Estonian dialects, as well as the solar goddess of the Saami are known to have been female. Also the solar deity of the Balts is female, and there may have been Scandinavian and Germanic Iron Age female solar deities as well. Thus, there seems to have existed in the Baltic region a long tradition of female solar deities, one that can perhaps be traced as far back as to the Bronze Age.[1] In this paper, the worship and role of the Finnish solar goddess in Southern Finland, especially Tavastia, in the late Iron Age is investigated by tracing mentions of her in Finnish folk poetry and by examining the possible traces of solar worship in the imagery of artefacts, calendars, beliefs and customs that can be traced back to the pre-Christian Iron Age.

2. Traces of Finnish solar worship in literal sources and archaeological remains

In the 16th century, Bishop Mikael Agricola and Olaus Magnus both noted the practice of solar and lunar worship in the medieval Nordic countries.[2] In spite of their accounts, there are in the literature only few explicit mentions of pre-Christian beliefs and customs of historical times having been related to solar worship; those include, for example, the rite of saluting the rising sun and the belief of the sun protecting the churning of butter.[3] There are, however, many other rites that have a probable solar origin, including the burning of bonfires on certain festivals related to key solar days, drawing circled crosses,

---

[1] Kristian Kristiansen and Thomas B. Larsson, *The Rise of Bronze Age Society: Travels, Transmissions and Transformations* (Cambridge: Cambridge University Press, 2005), pp. 297–308.
[2] Mikael Agricola, the Prologue of *Dauidin Psaltari* (1551); Olaus Magnus, *History of the Northern Peoples* (1555).
[3] Uno Harva, *Suomalaisten muinaisusko* (Helsinki: WSOY, 1948), pp. 153–55.

i.e. *rota* on the door jambs on the feast day of St. Thomas (the winter solstice), and performing blood sacrifices to *ristinkannat* (literally, 'bases of the cross'), which were *rota* standing at the heads of upright poles standing on a wooden or stony 'altar'.[4]

Also the various dances and other games performed on the annual main solar days as well as on some Christian feast days of late spring and summer everywhere in Northern Europe have been seen as solar rites that had eventually turned into profane games after the arrival of Christianity. Especially the dances and games performed in labyrinths have been widely seen as solar, as the spiralling form of the labyrinth can be seen as a representation of the annual journey of the sun. Stone labyrinths (Finn. *jatulintarha*, 'Giant's Ring'; Swe. *jungfrudans*, 'Maiden's Dance') are found all over the Baltic area and in coastal Lapland. Most of them are probably medieval, but the oldest ones may date as early as to the early Iron Age or even the Bronze Age.[5] The labyrinth dances were usually performed by young girls and boys. The recorded labyrinth dances or games often included the rescuing of a young girl, possibly representing the solar maiden, from the centre of the labyrinth.[6]

From the earliest historical times (the 13th – 15th centuries) in Finland no written sources concerning solar worship have been preserved. However, the cremation burial practiced in Finland continuously from the Bronze Age until the late Iron Age and the arrival of Christianity has been seen as a solar rite. Also the placement of Iron Age cremation cemeteries on the southern or south-western slopes of small hills, where they would face the sun during the whole annual cycle and the setting winter and autumn sun, may have been related to beliefs concerning the role of the sun in the death cult.

More indirect evidence of solar worship in the late Iron Age and early historical times can be found in the symbolism of artefacts. The jewellery, household items and clothing of those periods contain symbols that are usually interpreted as solar and/or stellar.[7] The weaponry of the time, on the other hand, was often decorated with animal and mythological symbolism in the Germanic animal style, although also celestial imagery can be detected (cf. the weaponry of the Bronze Age, which was decorated which ample solar symbolism).[8]

Much of the metal jewellery in Iron Age Finland was imported from the Baltic countries and Scandinavia, while the weaponry was mainly from Germanic regions.[9] Especially the objects produced by the Estonians and the Balts often contained celestial symbolism.[10] Only from the 6th century on did a unique Finnish style develop in the making of jewellery; also many of these new types of objects then contained solar symbolism, in particular various types of crosses, wheels and stellar forms.[11] The symbols seen in the jewellery were also used in other artefacts: for example, clothing items were often decorated with crosses, circles and swastikas, and loom weights had crosses and concentric circles printed on

---

[4] Christfrid Ganander, *Mythologia Fennica* (1789), p. 78 ; Harva, *Suomalaisten muinaisusko*, pp. 153–5.
[5] John Kraft, *The Goddess in the Labyrinth* (Turku: Åbo Akademi, 1985), p. 12; Marianna Ridderstad, 'Knossoksesta Ultima Thuleen – jatulintarhat aurinkoriittien paikkoina', *Muinaistutkija* 2 (2013): pp. 11–24.
[6] Kraft, *The Goddess in the Labyrinth*, pp. 15–7; Ridderstad, 'Knossoksesta Ultima Thuleen', pp. 11–24.
[7] See, e.g., Ella Kivikoski, *Suomen Rautakauden kuvasto I* (Helsinki: WSOY, 1947), figs. 388-9, 399-403, 405, 413, 421, 443-4, 454-5, 554-6; and Ella Kivikoski, *Suomen Rautakauden kuvasto II* (Porvoo: WSOY, 1951), figures 615-625, 631-640, 652-9, 669-670, 708-714, 717-9, 702, 726, 731-2, 746-9, 971, 985, 1042-5, 1049-55.
[8] See, e.g., Kivikoski, *Suomen Rautakauden kuvasto I*, pp. 466–385; Kivikoski, *Suomen Rautakauden kuvasto II*, pp. 784–808; Kristiansen and Larsson, *The Rise of Bronze Age Society*.
[9] See Kivikoski, *Suomen Rautakauden kuvasto I-II*.
[10] See, e.g., Aivar Kriiska and Andres Tvauri, *Viron esihistoria* (Jyväskylä: The Finnish Literature Society, 2007), pp. 154–5, 167, 217–8; Marija Gimbutas, *The Balts* (London: Thames & Hudson, 1963), plates 19, 23, 29-34, 45, 56-59, 61, 70.
[11] Kivikoski, *Suomen Rautakauden kuvasto I*, 12.

them, perhaps as protective signs of the sun goddess who was the patroness of weaving.[12]

In general, solar and celestial symbolism is encountered in the artefacts of the Scandinavian, Baltic and Finnic cultures continuously from the Bronze Age to the end of the Iron Age.[13] It is probable that the beliefs related to the celestial symbolism of the jewellery among the Balts were transmitted between the Baltic and Finnic cultures during the hundreds of years of interaction, and that also the unique Finnish style developed therefore came to express many of the same astronomically related themes as the Baltic originals; an equally close connection can also be observed in the relations between Finnic mythology and the most central themes of the mythology of the Balts (see below). Naturally, similar reasoning also applies to the contacts with Scandinavia, which gradually became ever more important from the early Viking Age on.

Interestingly, the symbolism of the jewellery and other artefacts did not drastically change in the course of the arrival of Christianity in Finland: the various crosses, swastikas, pentagrams, etc. continued to be popular. It can therefore be assumed that it was mainly the interpretations of the symbols that changed, e.g. the solar and stellar crosses and other symbols were now seen as referring to the Christian belief system. This assumption is supported by the distinctly 'Byzantine' style of many of the artefacts of the 12th and 13th centuries that contain e.g. crosses, rota and other celestially connected symbols; some of the similarly stylized jewellery of the period also contains representations of Christ and the Virgin.[14]

The use of the runic staff calendars in the Nordic countries can be traced at least to the 13th century.[15] The calendar staffs of the general Nordic type that were based on the solar year and the 19-year lunar cycle were likely used also in Finland at least from the 13th century on, although the tradition of the use of wooden practical calendars itself probably predated Christianity also in Finland. In their simplest form, the calendar staffs contained runes and often simple crosses, rota and other simple symbols marking the most important feast days and the key solar days. Later in the Medieval, however, the calendars often also depicted more elaborate symbolism.[16]

In addition to jewellery and calendars, in the early Medieval period solar and celestial symbolism could also be found inside churches. Those symbols are found in the earliest church art all around the Baltic and in Finland from the 12th or 13th century on, and there is evidence that already the earlier wooden churches had contained paintings of similar kind.[17] The astronomical and other kinds of symbolic, markedly non-Christian imagery of the early church murals has been connected to the imagery in the church calendars and the wooden folk calendars (the Nordic runic staffs), as well as to the practice of ritual games played in the early spring and summer to ensure the fertility of the crops and, via those, to the course of the church year and the agricultural year, i.e. the annual ritual cycle related to agricultural fertility, the most central interest of the human society.[18] The similarity of

---

[12] See, e.g. Kivikoski, *Suomen Rautakauden kuvasto I*, figs. 363-4; Kivikoski, *Suomen Rautakauden kuvasto II*, pp. 927, 950, 1194–5, 1197, 1201–2; Jaana Riikonen, 'Hedelmällisyys- ja suojelumagiaa muinaissuomalaisessa naisenasussa', *Arkeologia NYT!* 4 (2004), pp. 5–7.

[13] See Kristiansen and Larsson, *The Rise of Bronze Age Society*; Gimbutas, *The Balts*; Kriiska and Tvauri, *Viron esihistoria*; Kivikoski, *Suomen Rautakauden kuvasto I-II*; L. Jaanits, S. Laul, V. Lõugas, and E. Tõnisson, *Eesti esiajalugu* (Tallinn: Eesti Raamat, 1982).

[14] See: Kivikoski, *Suomen Rautakauden kuvasto II*, pp. 1002–12.

[15] Elisabet Svärdström, *Nyköpingsstaven och de medeltida kalenderrunorna* (Stockholm: Almqvist & Wiksell, 1966).

[16] Anna-Lisa Stigell, *Kyrkans tecken och årets gång: tideräkningen och Finlands primitiva medeltidsmålningar* (Helsinki: The Finnish Literature Society, 1974), pp. 17–8.

[17] Stigell, *Kyrkans tecken och årets gång*; Markus Hiekkanen, *Suomen keskiajan kivikirkot* (Helsinki: The Finnish Literature Society, 2007), p. 31.

[18] Stigell, *Kyrkans tecken och årets gång*; Marianna Ridderstad, 'Solar and calendrical symbolism in the Early Medieval

the symbols in the church art to those on the jewellery and other artefacts indicates that also the latter were likely seen as celestial and that the onlookers would have interpreted them in a similar way, as solar or stellar symbols – albeit there must have been in the past a point from which the symbols had started to be seen in a Christian interpretative context instead of the pre-Christian one.

One popular theme of the earliest church murals was labyrinths and games. In the church of Sipoo, which is known for the ample decoration of its ceiling with solar wheels, there is a large mural depicting a maiden standing in the middle of a labyrinth. The painting probably refers to the spring and summer games where a girl was to be 'rescued' from the centre of a stone labyrinth.[19]

3. Sun goddess in Finnic poetry

On the Finnish wooden folk calendars the feast days of the Virgin were usually denoted with solar symbols, suggesting that it was her connection to solar symbolism that was seen as her most central character by the medieval Finns. Possibly she had replaced an earlier solar deity? Anna-Leena Siikala showed that this had indeed been the case: in Kalevala-metric poetry and Finnish folklore and beliefs of ethnic religion, mainly recorded from the late 18th to the early 20th century, the role, actions and epithets of Päivätär were similar to those of the Virgin Mary, called Maaria in the old Finnish, to the extent that the two were often indistinguishable.[20]

In the Finnish poetry, Päivätär is the mistress of bees, especially the honeybee, and wasps.[21] Päivätär and Kuutar, the lunar goddess, and their daughters weave threads of gold and silver, i.e. the rays of the sun and the moon.[22] In Estonian poems, the sun goddess was depicted combing her hair, and in Finland, the son of the sun washing his hair, both of which are allegories of the solar rays shining, perhaps especially when the sun is grazing low above the sea horizon.[23] Thus, Päivätär was the main protector of weaving and also combing. She also 'baked the [yellow] honey bread', whereas Kuutar baked 'the [white] lunar bread'.[24] In general, Päivätär was connected to all things yellow and golden, e.g. to the churning of butter, yellow flowers, golden jewellery, etc., as well as to fire: she was the mother or grandmother of the personified fire (Panu).[25]

In addition to Panu, the male deity most closely related to Päivätär in the mythology is Lemminkäinen. In the famous poem *The Wedding of Päivölä*, Lemminkäinen, who is a bold and amorous warrior hero, crashes the wedding in Päivölä, the heavenly habitat of celestial gods (cf. *päivä*, the sun), and is killed in a fight. His mother rakes his pieces from the River of Underworld and, with the help of a bee, an animal of the sun goddess, brings him back to life.[26] In one version of the tale of Lemminkäinen, his

---

Finnish church murals', in *Stars and Stones: Voyages in Archaeoastronomy and Cultural Astronomy, Proceedings of SEAC 2011*, ed. F. Pimenta, N. Ribeiro, F. Silva, N. Campion, A. Joaquinito, and L. Tirapicos (Oxford: Archaeopress/British Archaeological Reports, 2015), pp. 280–5.

[19] Kraft, *The Goddess in the Labyrinth*, pp. 16–17; Ridderstad, 'Solar and calendrical symbolism', pp. 280–5.

[20] Anna-Leena Siikala, 'What Myths Tell about Past Finno-Ugric Modes of Thinking', in *Myth and Mentality, Studies in Folklore and Popular Thought*, ed. Anna-Leena Siikala (Helsinki: Finnish Literature Society, 2002), pp. 15–32.

[21] *Suomen Kansan Vanhat Runot I-XV* [hereafter SKVR], I4: 349, 350, XII2: 4788, 7291, XV: 260 (The Finnish Literature Society), available at: http://dbgw.finlit.fi/skvr/ [accessed 30 January 2015]; Anna-Leena Siikala, *Itämerensuomalaisten mytologia* (Helsinki: The Finnish Literature Society, 2012), p. 288.

[22] E.g., SKVR I1: 467, 480, I3: 1560, 1660, 1678, 1701, 1732a, 1739, III2: 1416, VII1: 219, VII4: 2224, VII5: 3674, 4197.

[23] Aado Lintrop, 'The great oak, the weaving maidens and the red boat, not to mention a lost brush', *Electronic Journal of Folklore*, no. 11 (1999): pp. 7–30.

[24] SKVR VII5: 3253.

[25] SKVR XII1: 4535, 4544, 4546, 4585, 4596.

[26] E.g., SKVR I2: 816, 834, 849, VII1: 777.

mother is identified as Päivätär.[27] The wedding described in *The Wedding of Päivölä* is based on the Baltic mythological motif of the celestial wedding of the daughter of the sun and the son of the moon; in Finland and Karelia, though, the pair is the son of the sun and the lunar maiden.[28] In some poems, Lemminkäinen is called *Pätöinen poika*, the Sunshiny Son (archaic *pätö* – sunshine, the gleam of the sun).[29] Thus, Lemminkäinen is *Päivän poika*, the son of the sun, who is wedded in the wedding of Päivölä and who in some versions of *The Great Oak* prepares for the wedding by washing his hair, an act belonging to solar imagery.[30]

The poem of *The Origin of Beer* was often sung as a prelude for *The Wedding of Päivölä*, while *The Wedding of Päivölä* itself could be sung as a prelude for *The Great Oak*.[31] These three poems together form a complex of songs related to the annual cycle and agricultural fertility.[32] In *The Origin of Beer*, a bee, 'the bird of the heavens' brings honey to be used as a fermenting agent.[33] In some versions, the goddess preparing the beer for the wedding of Päivölä, who usually is called Osmotar, is identified as Päivätär.[34] Thus, Päivätär is both the mistress of Päivölä and the mother of Lemminkäinen.

Confusingly, it thus seems that Lemminkäinen is both the solar son, the bridegroom of the wedding of Päivölä, *and* the wedding crasher who dies in the wedding. This combination immediately brings to mind the myths about the great Mother Goddess and her companion the 'dying god', a mythological theme common to the beliefs of many ancient cultures. Kari Sallamaa has shown that via his 'dying god' character, Lemminkäinen is closely connected to the dying vegetation and grain god Sampsa Pellervoinen.[35] Päivätär resurrects her son Lemminkäinen in a way similar to how the warmth of the sun awakens Sampsa and revives the vegetation in the spring. It seems that the tale of Lemminkäinen is a mix of at least three different stories or myths: a Viking Age tale of a bold warrior who conquers many difficulties on his journey, an older Baltic myth of the wedding of the celestial lights, and a version of the myth of the dying god, which is a myth related to the annual fertility and agricultural cycle.

Via the tale of Lemminkäinen, Finnish mythology connects Päivätär with a resurrecting god, an international fertility motif; the production of beer, i.e., an agricultural motif; and the idea of bringing together the sun and the moon, i.e., a calendric motif. We may thus conclude that Finnish solar worship was mostly related to the calendric cycle and rites of the agricultural year.

Perhaps the most interesting feature of the story of Lemminkäinen is its implication that the sun goddess was able to wake the dead. It brings to mind the belief system behind Iron Age cremation burials and the placement of graves to face the setting sun of Kekri, the ancient Finnic New Year's festival during which the dead could visit their former homes.

In some versions of *The Wedding of Päivölä* the master of Päivölä is called 'Ilman Ukko', i.e. Ukko

---

[27] SKVR I2: 823.
[28] Matti Kuusi, *Suomen kirjallisuus I: Kirjoittamaton kirjallisuus* (Helsinki: The Finnish Literature Society, 1963), pp. 142–146.
[29] Siikala, *Suomalainen šamanismi*, p. 268.
[30] E.g., SKVR VII4: 2642; Lintrop, 'The great oak', pp. 7–30.
[31] E.g., SKVR I2: 816, 850, VII1: 777.
[32] Marianna Ridderstad, 'The Great Oak: an annual calendric and agricultural fertility myth of the Baltic Finns', in *Astronomy: Mother of Civilization and Guide to the Future, Proceedings of SEAC 2013*, ed. K. Malville and M. Rappenglück. *Mediterranean Archaeology and Archaeometry* 14.3 (2014), pp. 319–330.
[33] SKVR I2: 1019.
[34] Anna-Leena Siikala, *Itämerensuomalaisten mytologia* (Helsinki: The Finnish Literature Society, 2012), p. 289.
[35] Kari Sallamaa, 'Maan poika – kolme suursuomalaista pellonjumalaa', in *Ei kiveäkään kääntämättä – Juhlakirja Pentti Koivuselle*, ed. J. Ikäheimo and S. Lipponen (Oulu: Pentti Koivusen juhlakirjatoimikunta, Tornion kirjapaino, 2009), pp. 47–57.

Ilmarinen, the sky creator god, whom Unto Salo has identified as the same as Ukko the god of thunder, also called Perkele.[36] As the master of Päivölä, the sky and thunder god can also be perceived as a companion of Päivätär. The pair is similar to the sun goddess Saule and the thunder god Perkunas of the Balts.[37] In the 18th century Karelia people were able to tell that their ancestors had worshiped 'Ilmarinen and Virgin Mary' or 'old Väinämöinen and Virgin Mary, the great mother'.[38] The first of these pairs seems to have been formed by the heavenly pair Ukko Ilmarinen and the sun goddess Päivätär-Maaria; again, the pair is similar to Saule and Perkunas.

The remark from Karelia also suggests that in many rituals and festivals there had been both a head male deity and a head female one. We may thus suggest that Ukko the sky god, as well as Lemminkäinen 'the dying god' and Sampsa the vegetation god, who all were related to the annual vegetation cult, could with Päivätär have formed the main pair in the rituals of the Finnish fertility cult.

4. Rituals of the Tavastian Helkajuhla festival as examples of solar fertility rites

The Helkajuhla festival celebrated on the Pentecost in Ritvala of Sääksmäki in Tavastia in inland Southern Finland is the sole living remnant of the ancient Finnish agricultural fertility spring and early summer rituals (cf., e.g. the labyrinth dances). It was believed that if the ritual ever ceased to be celebrated, the fields of Ritvala would never bear fruit again.[39] Nowadays Helkajuhla is celebrated on the Pentecost, but in the 19th century, the Helka songs were sung on several evenings between the Pentecost and the feast day of St Peter at the end of June. The feast includes a group of maidens from Ritvala walking a cross-shaped route among the fields singing, ascending a local hill, and dancing in a circle. After that, bonfires are lit in the evening.

The contents of the present three medieval Helka songs are markedly Christian: 1) Mary Magdalene repenting, 2) a faithful bride gets finally wed, and 3) a sinful woman kills her deceitful lover by magical means in medieval Turku.[40] Only in the End Song are there a few strange verses about a drooling elk and a mythical tree that grows out of the saliva. At the very end of the End Song there are Christian verses in which the Virgin Mary and Christ are thanked and welcomed again next year – a detail that also reveals the two main characters worshiped in the ritual. However, if we remember that the contents and verses of the songs have been changed and censored many times due to the demands of Christianity especially since the 17th century, a more complex picture emerges. What actually happens in the songs are the following acts: 1) a female saint character meets a male god beside a well, 2) a wedding happens, and 3) a woman raises a storm and kills her lover (note that the woman is not necessarily a witch, as in the Medieval period it was widely believed that all Finns and Lapps could alter the weather). From the last act it suddenly and unexpectedly follows that an elk runs, drools and causes the tree of love and life grow. I have shown elsewhere that these verses were originally part of the poem of the Great Oak, which is a myth related to the annual solar and agricultural cycle.[41] The detail of the meeting that happens beside a well in the first Helka song is important: as shown by Anna-Leena Siikala, there is also a well in the centre of the Finnic cosmos, beside which a Virgin Mother goddess, often identified as

---

[36] SKVR III: 728a; Unto Salo, *The God of Thunder of the Ancient Finns and His Indo-European Family* (Washington, DC: Institute for the Study of Man, 2006).
[37] Vytautas Straizys and Libertas Klimka, 'The cosmology of the ancient Balts', *Archaeoastronomy* 22 (1997): pp. S57–S81.
[38] Kaarle Krohn, *Suomalaisten runojen uskonto* (Helsinki: WSOY, 1914), pp. 325–6.
[39] See, e.g., Elsa Enäjärvi-Haavio, *Ritvalan Helkajuhla* (Helsinki: WSOY, 1953), p. 13.
[40] E.g., SKVR IX 1: 76-79.
[41] Ridderstad, 'The Great Oak', pp. 319–330.

Maaria-Päivätär and her son are sitting under an oak.[42] We therefore see that the Helka songs actually describe a series of events very similar to two ancient myths with the central theme of *hieros gamos*: the celestial wedding of deities and the myth of the dying god.

As seen above, the Finnish dying god characters were Lemminkäinen and Sampsa Pellervoinen. It is difficult to conclude which one of these deities was the original primary male fertility deity of the Ritvala ritual. It is possible also that the rites of Ukko the thunder and weather deity were performed during the festival; it is even possible that he would have been seen as the companion of Päivätär in some rites (cf. Saule and Perkunas of the Balts, and the rites of Perkunas related to an oak tree).[43]

Based on the solar rites of the dances and bonfires, the main female deity in the predecessor of the present Helkajuhla in the late Iron Age seems to have been Päivätär, who was probably replaced by the Virgin from early on. The only other female candidate would be Mother Earth, the (presumed) companion of Sampsa Pellervoinen, but she is non-personified in the Finnic mythology and has no connection to the sun or solar rites.

It is of course possible that originally there were *two* divine pairs in the festival and two different kinds of rituals: sowing rituals for Sampsa and Mother Earth, and other rituals (e.g. the bonfires and the dances) for the sun deity and her companion. It is also possible that the present Helkajuhla contains remnants of two festivals: a sowing ritual and the midsummer festival of the late Iron Age – the celestial wedding and the solar rites would well suit to the latter. In some fertility myths of the Balts related to the summer solstice festival the main divine couple of the celestial wedding was Saule or Saules Meita, the solar maiden, and the sky god Dievs, who also was a vegetation god resembling Sampsa.[44] The Finnish summer solstice festival was also known as the festival of Ukko, where the honour name Ukko may refer either to the sky god or Ukko Sampsa. Thus, while there definitely were solar rites for Päivätär in the pre-Christian Ritvala fertility ritual festival(s), we end up with several possibilities when it comes to the male deities worshiped in those rituals. It can be concluded that while we are perhaps one step closer to solving the 'enigma' of Helkajuhla, there are still many details in need of further investigation.

5. Conclusions

The Finnish solar worship and the significance of the sun goddess can be traced by examining the existing literal, archaeological and ethnographical sources. In this study, solar symbolism has been detected in jewellery and other artefacts of the late Iron Age and early historical times. In the literature, explicit mentions of solar worship are scarce, but other evidence of possible solar rites has been found. Many of the solar rituals seem to have had the nature of fertility rituals and have been connected to the festivals celebrated in the spring and summer time, which can be understood on the basis that in the harsh natural conditions of the Nordic countries, the sun was ever important not only as the provider of light and warmth but also in her role producing annual agricultural fertility.

The character of the Finnish solar goddess Päivätär can be traced mainly through ethnographical sources and Finnic Kalevala-metric poetry. The interpretations made are complicated by the fact that in mythologies, there really are no 'original' or 'true' versions of the myths, but each singer interpreted, created and presented his/her own versions of the common mythical themes, often enriched with

---

[42] Siikala, *Itämerensuomalaisten mytologia*, pp. 178–199.

[43] See, e.g. Jonas Vaiškunas, 'On the possible astronomical significance of Lithuanian autumn goat ceremony', in *Lights and Shadows in Cultural Astronomy, Proceedings of SEAC 2005*, ed. M. P. Zedda and J. A. Belmonte (Isili: Associazione Archeofila Sarda, 2007), pp. 344–54.

[44] Siikala, *Itämerensuomalaisten mytologia*, p. 268.

themes of contemporary popular songs.[45] Therefore, what we can deduce with certainty are often only the central mythical themes and their relations to the contemporary society, i.e. what meaning they could have had to the peoples of a certain period and region. Based on this type of reasoning, we can deduce that the significance of the sun goddess and the solar worship in the Iron Age Finland was most probably connected to the demands of sustaining life and livelihood in the cold climate. Via her role as the central provider of light and warmth and in ensuring the growth of vegetation and the agricultural yield, she may have become associated with good luck and fertility in general. She may have been important for the women as a giver of fertility, a protector of certain tasks (e.g. weaving) and perhaps as a provider of general protection as well. Although there is no direct proof of the central role of the sun goddess in the mortuary cult of the late Iron Age, the cremation burial and the placement of burial grounds to face the sun, as well as the resurrecting character of the sun in the tale of Lemminkäinen, certainly point towards this conclusion.

Bibliography
Agricola, Mikael. *Dauidin Psaltari*. 1551.
Enäjärvi-Haavio, Elsa. *Ritvalan Helkajuhla*. Helsinki: WSOY, 1953.
Ganander, Christfrid. *Mythologia Fennica*. 1789.
Gimbutas, Marija. *The Balts*. London: Thames & Hudson, 1963.
Harva, Uno. *Suomalaisten muinaisusko*. Helsinki: WSOY, 1948.
Hiekkanen, Markus. *Suomen keskiajan kivikirkot*. Helsinki: The Finnish Literature Society, 2007.
Jaanits, L., S. Laul, V. Lõugas, and E. Tõnisson. *Eesti esiajalugu*. Tallinn: Eesti Raamat, 1982.
Kivikoski, Ella. Suomen rautakauden kuvasto I. Helsinki: WSOY, 1947.
Kivikoski, Ella. Suomen rautakauden kuvasto II. Porvoo: WSOY, 1951.
Kraft, John. *The Goddess in the Labyrinth*. Religionsvetenskapliga skrifter 11. Turku: Åbo Akademi, 1985.
Kriiska, Aivar and Andres Tvauri. *Viron esihistoria*. Jyväskylä: The Finnish Literature Society, 2007.
Kristiansen, Kristian and Thomas B. Larsson. *The Rise of Bronze Age Society: Travels, Transmissions and Transformations*. Cambridge: Cambridge University Press, 2005.
Krohn, Kaarle. *Suomalaisten runojen uskonto*. Helsinki: WSOY, 1914.
Kuusi, Matti. *Suomen kirjallisuus I: Kirjoittamaton kirjallisuus*. Helsinki: The Finnish Literature Society, 1963.
Lintrop, Aado. 'The great oak, the weaving maidens and the red boat, not to mention a lost brush'. *Electronic Journal of Folklore* 11 (1999): pp. 7–30.
Magnus, Olaus. *History of the Northern Peoples*. 1555.
Ridderstad, Marianna. 'Knossoksesta Ultima Thuleen – jatulintarhat aurinkoriittien paikkoina'. *Muinaistutkija* 2 (2013): pp. 11–24.
Ridderstad, Marianna. 'The Great Oak: an annual calendric and agricultural fertility myth of the Baltic Finns'. In K. Malville and M. Rappenglück, eds., *Astronomy: Mother of Civilization and Guide to the Future, Proceedings of SEAC 2013*, pp. 319-330. *Mediterranean Archaeology and Archaeometry* 14.3, 2014.
Ridderstad, Marianna. 'Solar and calendrical symbolism in the Early Medieval Finnish church murals'. In *Stars and Stones: Voyages in Archaeoastronomy and Cultural Astronomy, Proceedings of SEAC 2011*. Edited by F. Pimenta, N. Ribeiro, F. Silva, N. Campion, A. Joaquinito and L. Tirapicos, pp. 280–5. Oxford: Archaeopress/British Archaeological Reports International Series 2720, 2015.
Riikonen, Jaana. 'Hedelmällisyys- ja suojelumagiaa muinaissuomalaisessa naisenasussa'. *Arkeologia NYT!* 4 (2004), pp. 5–7.
Sallamaa, Kari. 'Maan poika – kolme suursuomalaista pellonjumalaa'. In *Ei kiveäkään kääntämättä – Juhlakirja Pentti Koivuselle*. Edited by J. Ikäheimo and S. Lipponen, pp. 47–57. Oulu: Pentti Koivusen juhlakirjatoimikunta, Tornion kirjapaino, 2009.
Salo, Unto. *The God of Thunder of the Ancient Finns and His Indo-European Family*. Journal of Indo-European Studies

---

[45] Siikala, *Itämerensuomalaisten mytologia*, 452.

Monograph 51. Washington, DC: Institute for the Study of Man, 2006.
Siikala, Anna-Leena. *Suomalainen šamanismi*. Suomalaisen Kirjallisuuden Seuran Toimituksia 565. Hämeenlinna: Finnish Literature Society, 1992.
Siikala, Anna-Leena. 'What Myths Tell about Past Finno-Ugric Modes of Thinking'. In *Myth and Mentality. Studies in Folklore and Popular Thought*. Edited by Anna-Leena Siikala, pp. 15–32. Studia Fennica Folkloristica 8. Helsinki: Finnish Literature Society, 2002.
Siikala, Anna-Leena. *Itämerensuomalaisten mytologia*. Helsinki: The Finnish Literature Society, 2012.
Suomen kansan vanhat runot I–XV (SKVR). The Finnish Literature Society. http://dbgw.finlit.fi/skvr.
Stigell, Anna-Lisa. *Kyrkans tecken och årets gång: tideräkningen och Finlands primitiva medeltidsmålningar*. Helsinki: The Finnish Literature Society, 1974.
Straizys, Vytautas, and Libertas Klimka. 'The cosmology of the ancient Balts'. *Archaeoastronomy* 22 (1997): pp. S57–S81.
Svärdström, Elisabet. *Nyköpingsstaven och de medeltida kalenderrunorna*. Stockholm: Almqvist & Wiksell, 1966.
Vaiškunas, Jonas. 'On the possible astronomical significance of Lithuanian autumn goat ceremony'. In *Lights and Shadows in Cultural Astronomy, Proceedings of SEAC 2005*. Edited by M. P. Zedda and J. A. Belmonte, pp. 344–54. Ilisi: Associazione Archeofila Sarda, 2007.

THE NORDIC CALENDAR AND THE GREAT MIDWINTER SACRIFICE AT OLD UPPSALA

Göran Henriksson

ABSTRACT: The pagan Great Midwinter Sacrifice at Old Uppsala took place every eighth year. The starting date was defined by the full moon that occurred between 28 January and 26 February in the Gregorian calendar. By combining historical data and calculations of the dates of full moons, the author has established the exact years of the eight-year cycle. One such year was 852 CE. According to the rule of semi-legendary King Aun, the phases of the moon fell one day earlier after 304 years. Such displacements in the eight-year cycle took place in 1692, 1388, 1084, 780, and 476. King Aun is considered to have reigned about 450–500 CE and to have been buried at Old Uppsala. The three 'royal' burial mounds have been dated to 450–550 CE. They are oriented in such a way that they could have been used to regulate the sacrificial calendar. During excavations in 2013, two long rows of postholes were discovered. An 850 m long row was oriented towards the northern limit for the rising midwinter full moon on the earliest date, and the approximately 600m long row was oriented towards the rising sun on 6–8 February. If it was a full moon on that day, the Great Midwinter Sacrifice the following year should begin on its earliest date.

The Great Midwinter Sacrifice at Old Uppsala

According to archbishop Adam from Bremen, ca. 1075, every ninth year all the Swedish provinces had to send representatives to Old Uppsala for a common celebration.[1] However, when the early Swedes said *every ninth year* this corresponds to *every eighth year*, as they had no zero and counted the beginning of the first year as year one and reached year nine when only eight years had elapsed. In fact this celebration took place every eighth year according to an eight-year cycle determined by the phases of the moon.

The eight-year cycle is the shortest period after which the same lunar phase is repeated approximately on the same date, as eight tropical years = 2921.934 days and 99 synodic months = 2923.528 days. This means that the same phase of the moon will appear delayed by one and a half days after eight years. After 18 or 19 eight-year cycles, 144 or 152 years, the cycle is shifted by a whole month, something already mentioned by the Greek astronomer Geminos ca. 70 BCE.[2] He also explained why this cycle was called both the nine-year and the eight-year cycle in antiquity.

For a period of nine days, one man and seven male domestic animals were sacrificed. When the sacrifice was completed, one could see altogether 72 bodies hanging in the holy tree.

The years for the Great Midwinter Sacrifice

The historical records give three conditions for identifying the sacrificial years of the eight-year cycle:

1) There was a Great Sacrifice 851–853, just before the second visit of bishop Ansgar at Birka, according to the Chronicle by Rimbert, written before 876.[3]
2) The Great Sacrifices took place around the time of the vernal equinox just before 1075, when Adam from Bremen wrote: 'Hoc sacrificium fit circa aequinoctium vernale', which means this sacrifice took place

---

[1] Adam of Bremen, *History of the Archbishops of Hamburg-Bremen*, trans. F. Tschan (New York: Columbia University Press, 1959).
[2] Geminos, *Introduction aux Phènoméns*, trans. G. Aujac (Paris: Les Belles Lettres, 1975).
[3] H. Rieper, *Ansgar und Rimbert: die beiden ersten Erzbischöfe von Hamburg/Bremen und Nordalbingen* (Hamburg: EB-Verlag, 1995).

around the time of the vernal equinox, i.e., at the end of February and beginning of March.

3) The Great Sacrifices at Lejre, the Danish counterpart to Old Uppsala, took place in January before 934, when they were forbidden, according to Thietmar of Merseburg, writing about 1000.[4]

To solve this problem I computed all the full moons between 28 January and 26 February in the Gregorian calendar for the period 200-1200 CE.[5] The only possible solution is an eight-year cycle including the year 852 CE as the year for the second visit of Ansgar in Birka. This means that the exact year for all the Great Midwinter Sacrifices can be computed as multiples of eight years counted from the year 852. The original dates in the Julian calendar were 21 January and 19 February.

Year	Important events in early Swedish History in accordance with the eight-year cycle
852	Second visit of St. Ansgar at Birka.
980	Foundation of Sigtuna, the royal city succeeding Birka, according to dendrochronological dating.
1060	The first bishop in Lund, province of Scania.
1076	King Anund had to abdicate because he refused to lead the Great Sacrifice at Old Uppsala.
1084	The last Great Midwinter Sacrifice in Uppsala. The heathen temple was burnt down in 1087.
1124	The bishop at Old Uppsala had to leave the country after a few months.
1156	King Sverker was murdered. According to Nyberg, this may have been an act of revenge from the last heathens in Sweden.[6]
1164	The first archbishop at Old Uppsala.

Table 8.1

Very few important years from early Swedish history are known. However, all important decisions, both secular and ecclesiastical, were taken during years in accordance with the eight-year cycle, see Table 8.1. By establishing the bishopric at Lund, on a major pagan cult site, and the archbishopric at Old Uppsala, at the most important pagan cult centre in Scandinavia, during the Great Midwinter Sacrificial years, they defeated the pagan religion both in space and time.

The importance of the Great Midwinter Sacrifice, later called the Disting
The original meaning of the Disting was threefold; there should be:[7]

1) a great sacrifice for peace and victory for the king,
2) a general meeting with representatives from all the Swedish provinces,
3) a major market.

At the general meeting important common political decisions were taken, such as election of a new king

---

[4] Thietmar von Merseburg, *Chronik*, trans. W. Trillmich (Darmstadt: Rütten & Loening, 1957).
[5] G. Henriksson, 'Riksbloten och Uppsala högar'. *Tor* 27, no. 1, (1995): pp. 337-394.
[6] T. Nyberg, *Monasticism in North-Western Europe, 800-1200* (Aldershot: Ashgate, 2000).
[7] J. Granlund, *Kulturhistoriskt lexikon för nordisk medeltid*, Vol. 3, s. v. 'Disting' (Malmö: Allhem, 1958).

or solution of judicial questions that not could be solved at local courts. The participation of the representatives was compulsory, and Christian representatives who refused to come because of the human sacrifice had to pay a great fine. The dates for the Disting were linked to the phases of the moon according to an ancient rule preserved in medieval texts.

Already Tacitus had pointed out that important meetings among the Germanic peoples had to take place at the new or full moon.[8] In his *Historia de gentibus septentrionalibus,* written in 1555 during his exile in Rome, Olaus Magnus, the last Roman Catholic archbishop in Sweden, explained that the Disting was started at the full moon because the light from the moon facilitated travel to Uppsala during the short days at midwinter.[9]

The precise definition as to when the Disting should take place
The exact rule for determining the starting date of the Disting was given by Olof Rudbeck, professor in medicine and astronomy at the University of Uppsala, a scholar with broad scientific interest:

> The moon that shines in the sky on Twelfth Day (6/1) is the Christmas moon and after this follows the Disting's moon.[10]

The full moon appeared on the 14th day of the month and the earliest date for the beginning of the Disting was 21 January (7/1+14 days) and the latest date was 19 February (7/1+29+14 days) according to the Julian calendar used in Sweden at that time. In our modern calendar, the corresponding interval for the Disting is 28 January and 26 February.

However, it may seem strange that this originally heathen rule was related to Twelfth Day, or the Epiphany, as the rule for the start of the Disting in Magnus. The explanation is that the rule for the dates of the Disting was related to the Christian calendar in the twelfth century. At that time there was a shift by seven days between the Julian calendar and our Gregorian calendar that is closely related to the solstices and equinoxes.

This fact also explains why, according to the Swedish tradition, the night of St. Lucia, 13 December, is the longest and darkest night of the year. If seven days are added to this date, we get the date of the winter solstice in our Gregorian calendar.

The Julian calendar was delayed by seven days
The important pre-Christian days on the runic calendar staff fall 7 days after the dates that give a natural division of the solar year into 12 months with 30 days reckoned from the winter solstice.

> Midwinter day: 13 January + 7 = 20 January
> The first summer day: 14 April + 7 = 21 April
> The first winter day: 14 October + 7 = 21 October
> St. Lucia, the longest night: 13 December + 7 = 20 December
> The winter solstice took place during prehistoric times on 20-21 December.

The Julian calendar was already delayed by 7 days when the important pre-Christian festival days were included in the written calendar during the twelfth century. The reason was that the ecclesiastical Julian

---

[8] Tacitus, *Agricola. Germania. Dialogue on Oratory,* trans. M. Hutton (Cambridge, MA and London: Loeb Classical Library 35, 1970).
[9] Olaus Magnus, *A description of the northern peoples, Rome 1555*, Vol. 1. trans. P. G. Foote (London: Hakluyt Society, 1996).
[10] O. Rudbeck, *Atlantica,* Vol. 1 (1679; reprint Uppsala och Stockholm: Almquist och Wiksells, 1937), pp. 68.

calendar, calibrated in 325, should have excluded 1 day every 128th year.

Determination of the phases of the moon one year in advance according to the farmer's rule

During the Disting market in 1689, Rudbeck talked to a 90-year-old farmer, Anders from Röklunda in the province of Uppland, who demonstrated his old runic calendar staff (*runstav*). The farmer taught him the following rule for the shifts of the phases of the moon in the same month of the next year:

> The phases of the moon will be shifted either 12 days backwards or 20 days forwards in the same month with 30 days the next year. [11]

According to modern arithmetic we subtract 11 or add 19 days to the actual date instead of 12 and 20, as in the old way of calculating without zero. This rule is easy to understand because one solar year has 365 days and 12 lunar months correspond to 354 days.

This rule could have been especially useful in the determination, one year in advance, of the day of the Great Midwinter Sacrifice. For instance, if it was a full moon on 8 February the Great Sacrifice the next year would be shifted from the latest date, 26 February, to the earliest date, 28 January = 8 February – 11 days. The 8 February could be directly determined in Old Uppsala as the day when the sunset took place along the three big burial mounds (see Fig. 8.1). Before the existence of these mounds this important day was determined as the day when the sun was rising along the more than 600m long row of posts put up in a field some hundred metres to the south of the big mounds, see below.

Fig. 8.1: Sunset along the big mounds on 8 February 1993. If it was a full moon 8 February, the great eight-year sacrifice would to be held the following year on its earliest date of 28 January. This photograph was taken from the observation mound at the southeastern corner of the churchyard wall. The computer has removed the trees.

The important sacrificial dates can be determined by solar observations at Old Uppsala[12]

The dating limits for the beginning of the Great Midwinter Sacrifice (later the Disting) could be established by observing sunset from the menhir on the Tingshög (Fig. 8.2).

---

[11] O. Rudbeck, *Atlantica*, Vol. 2. (1689; reprint Uppsala och Stockholm: Almquist och Wiksells, 1939), pp. 652.
[12] G. Henriksson, 'The Pagan Great Midwinter Sacrifice and the "royal" mounds at Old Uppsala', in *Proceedings of the SEAC2001 meeting in Stockholm*, ed. M. Blomberg, P. E. Blomberg, and G. Henriksson (Uppsala: Uppsala University Printers, 2003), pp. 15–25.

In 468 CE, there was a full moon on 26 February, when the sun set where the burial cairn in the middle mound stands. In 476 CE, it was time to re-start the cycle of sacrifices on 28 January at sunset over the highest point of Tunåsen. Even as late as the seventeenth century, the ancient King Aun was regarded as the inventor of the rule that predicted the displacement of the moon's phases by one day after 304 years, see below. The semi-legendary King Aun may have reigned in Uppsala about 450–500 CE and been buried there.[13]

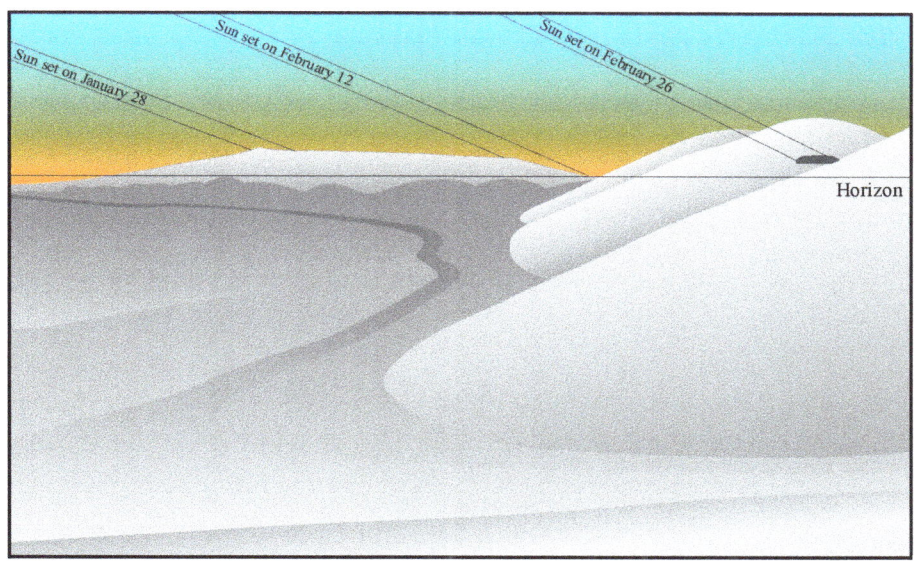

Fig. 8.2: On the earliest date for the Great Midwinter Sacrifice, 28 January, the sun set straight over the highest point on Tunåsen and on the last date, 26 February, at the position of the burial cairn in the middle mound (Aun's?), as observed from the menhir at Tingshögen (Thing Mound). Every eighth year a Great Midwinter Sacrifice was made at the full moon, which fell between these dates.

The farmers' method to determine the date of the full moon

The Uppland farmers determined the date of the next full moon by measuring the moon's distance at sunset from the 'pale yellowish-red' night-ring.

The distance between the thumb and the forefinger was called a 'span' and corresponds to the daily motion of the moon.

The phases of the moon fall one day earlier after 304 years in the Julian calendar. The older people of Uppsala knew that in 1689 the Disting's full moon would fall one day earlier than it had during the last 300 years. When Rudbeck asked the farmer Anders why, he answered that 'there exists an old rule according to which the Disting's full moon had now completed the cycle of Aun and that it now should be shifted by one day every nineteenth year during 300 and some years'.[14]

He explained that the rule was called after the ancient King Aun and that 1689 was the correct year to adjust the date of the Disting, as one of his forefathers had engraved a half moon on his rune staff on

---

[13] Snorre Sturlason, *Heimskringla or the lives of the Norse kings* (Iceland 1230), trans. E. Monsen (Cambridge: Dover Publications, 1932).
[14] Rudbeck, *Atlantica*, Vol. 2. . pp. 652–653.

the day of the Disting, the year, and the golden number of that year. I realized that the '300 and some years' was 304 years = 16 x 19 years, and that a marking of the Disting's full moon as a half moon in 1385 (1689–304 years) could be interpreted as the result of a lunar eclipse. This hypothesis could easily be verified because in 1385 the Disting's full moon was totally eclipsed after midnight and could be seen as a half moon about 7 o'clock in the morning.

Shifts in the date for the Great Midwinter Sacrifice according to the 304-year rule

The first year after 1689 to be shifted in the old eight-year cycle for the pagan Great Midwinter Sacrifice was 1692.

This means that the earlier shifts had taken place in 1388 with no sacrifice, in 1084 with the last sacrifice, and in 780 and 476 with sacrifices. The year 476 falls within the estimated time of rule for the semi-legendary King Aun, 450–500, mentioned in *Heimskringla or the lives of the Norse kings* by Snorre Sturlason.[15]

This is very interesting as King Aun, according to the tradition preserved by the old farmer Anders, established this 304-year rule. This implies that the Julian calendar was introduced in Uppsala by at least 476, see below, or that the ancient Swedes had independently invented a calendar with an intercalation of one day every fourth year. The latest shift in the 304-year cycle occurred in 1996 (1692 + 304 years), on 3 February, which corresponds to 21 January in the Julian calendar.

Re-start of the Great Midwinter Sacrifice from its earliest date

The date for the Great Midwinter full moon was delayed every eights year by somewhat more than 1.5 days, which means that after 152 years = 8 x 19 years the total shift was a complete month and the next Great Midwinter Sacrifice had to start from the earliest date, 28 January.

If one wanted the sacrificial dates to fall within fixed dates in the Julian calendar, the re-start should take place after alternating 152 years and 160 years with the total period 152 + 160 years = 312 years.

If one instead wanted the sacrificial dates to fall within fixed dates in the solar year, for instance reckoned from the date when the sun set on the peak of Tunåsen as observed from the menhir on the Tingshög (see Fig. 8.2), the re-start should take place after alternating 144 years and 152 years with the total period 144 + 152 years = 296 years.

The later case is supported by the description from Snorre Sturlason's visit to the province of Västergötland in 1219 when he reported that the Disting that year was started on 2 February, and that the Disting the next year should take place in January for the first time since the Christianisation of Sweden. This is an independent proof for the correlation with the year 476 CE, already considered as an important year in my first papers: 1220 – (2 x 296 + 152) years = 476 CE.[16]

Two long rows of postholes discovered at Old Uppsala in 2013

A row of large stone fundaments, about 850m long, has been discovered beneath an old road that was the main entrance to Old Uppsala from the north (Fig. 8.3).

---

[15] Sturlason, *Heimskringla or the lives of the Norse kings*.
[16] G. Henriksson. 'Riksbloten i Uppsala', *Gimle* 20 (1992): pp. 14–23; G. Henriksson, *Arkeoastronomi i Sverige*. Literature for course in Archaeoastronomy at Uppsala University (Uppsala: Uppsala University Press, 1994).

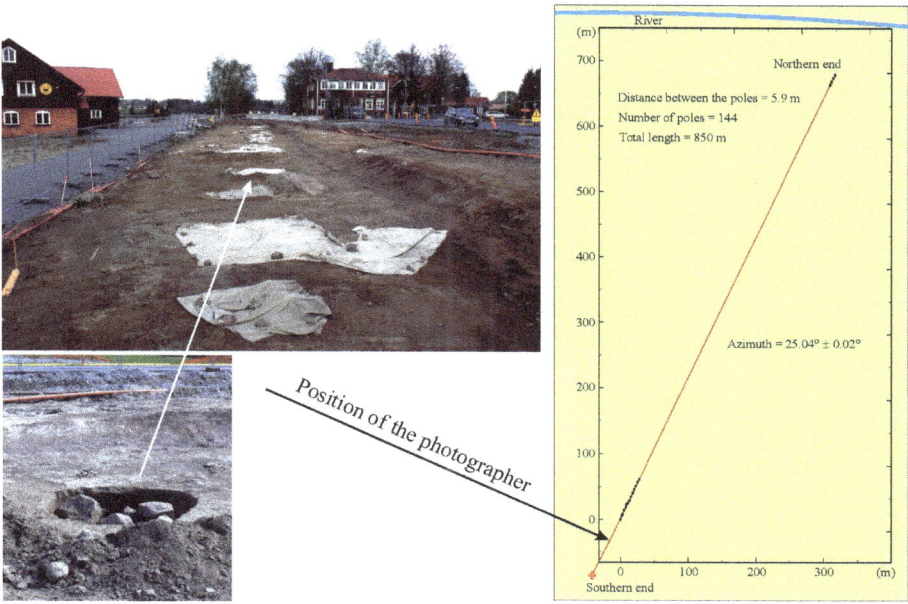

Fig. 8.3: The southern part, first excavated ca. 70 m, of the about 850m long row of postholes discovered in May 2013 beneath the old main road leading out from Old Uppsala toward the north is seen in the upper left photo. Today this road is called Vattholmavägen, and continues in the straight birch avenue in the background. One of the postholes in the northern row is seen in the lower left photo. The diameter of the pit is more than one metre. They contain up to 1500kg of rocks, necessary to stabilize 8m high vertical posts. The original ground level is 1.3m below the asphalt surface in the background. The plan to the right shows the measured postholes in the northern row. The northern most posthole is situated on the flat ground just before the depression beside the river. Photos by G. Henriksson taken in May 2013.

Another more than 600m long similar row of stone fundaments has been found to the south of the famous big 'royal' burial mounds.

Each one of the stone fundaments was used to hold wooden posts up to 8m high (Fig. 8.3). They were preliminarily thought to be from the fifth or sixth century CE and related to the building period of the three monumental burial mounds; they add new information to a site that has been considered as one of the most important in Swedish prehistory. It was a major centre for royal power, trade, religion, handicraft and justice administration.

This unexpected important find was made during archaeological surveys made before a new rail tunnel for a railway was built, linking northern Sweden to Stockholm. The excavations started in May 2013. The postholes in both lines were made about 6m apart, and each hole contained up to 1500kg of rocks to keep the huge vertical posts of pine trees in place (see Fig. 8.3). In some of the holes animal bones from horses, cows and pigs were found, which indicates animal sacrifices. They are thought to have been sacrificed during the construction process. In about 20 of the stone fundaments remains of the wooden posts were found. Most of them have now been C^{14}-dated to get an idea of their age, see Table 8.2.

Archaeologists are not sure what the rows of posts were meant for, but they believe that it may have been a way to demonstrate the local chieftain's power. However, there are no records of huge lines of posts in any written sources. The archaeologists have found a house, dated to 705 ±65, built above one of the postholes in the northern row. This means that the posts must have been removed before about 700.

Some of the posts have been lifted up from their fundaments, others have been cut at ground level and others have been burnt down above ground level.

The larger northern row, with 144 posts, is now completely excavated. It is situated about 200m from an Iron Age burial site. A final report from these excavations will be published in 2017.

Orientation of the alignments

The excavation of the southern end of the northern postholes was finished in the beginning of June 2013. On 5 June, when the fences around the excavation were removed, it became possible for me to measure the position and orientation of 11 postholes with a total station and solar orientation of the coordinate system. On 29 June the four visible postholes at the northern end were measured and on 7 July both ends of the northern row were connected in the same system. The mean distance between the posts was 5.9m = 20 feet (0.295m) and the total length can be estimated to about 144 x 5.9m = 850m. A linear least square fit with all measurements in the northern row was performed and its final azimuth was determined as 25.04° ±0.02° (see Fig. 8.3).

On 16 May 2014 the altitude of the northern horizon was determined as 0.65° with trees and 0.47° without trees. The total station was placed on the surface of the road 1.4m above the original ground level and after correction the original horizon altitude became 0.67° with trees and 0.49° without trees.

Identification of the alignments

The measured combination of azimuth and horizon altitude for the northern row of postholes corresponds to the northern limit for the rising midwinter full moon, on 26–27 January during prehistoric times, the earliest possible second full moon after the winter solstice. In this case we are only interested in full moons in the established eighth-year cycle.

The calculations were performed for the interval 2000 BCE–1000 CE. There exists only one such solution, 26 January 565 BCE. The altitude of the upper limb of the moon was 0.73° when the azimuth was 25.04°, which means that the upper limb of the moon was 0.06° above the horizon. This was 1/5 of the radius of the rising moon that day (see Fig. 8.4).

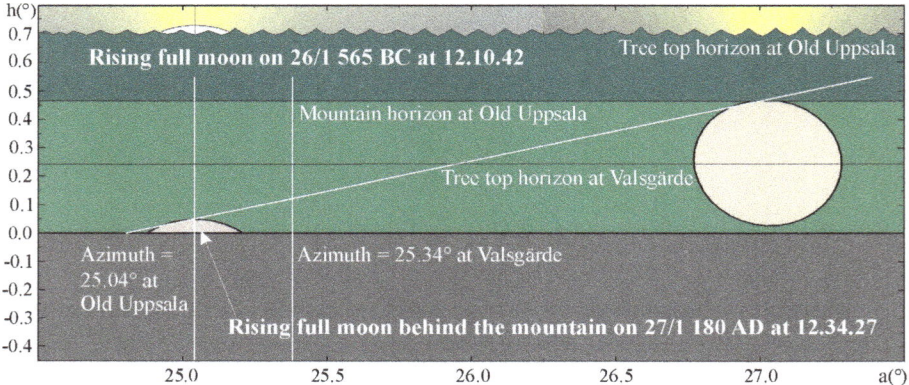

Fig. 8.4: If the northern row of postholes was oriented towards the upper limb of the rising Midwinter full moon, in the established eighth-year cycle, there exists only one solution, 26 January 565 BCE. It is assumed that there were trees on the distant hill as there are today. The depiction in the lower part of the figure shows the position of the upper limb of the rising Midwinter full moon, on 27 January 180 CE, for a distant zero horizon beyond the hill. Under favourable atmospheric circumstances the first vertical rays from the rising full moon may have been visible above the hill in the same azimuth as the posts. This may explain the correlation with the orientation of a ship setting with 14 postholes on the nearby important burial ground Valsgärde.

Only 19 postholes in the eastern end of the southern row have been excavated. This corresponds to 112.7m, but the total length can be estimated to about 600m from geo-radar investigations.

The postholes in the southern row were measured on 1 July 2013. The orientation was determined by the least square method to be 120.66° ±0.45°. The precision was not so good as for the northern row because it is much shorter and it was only possible to determine the centre of 6 postholes because all postholes were filled up and the surface smoothed. A shallow stone, used for grinding corn, was found in a fundament in the southern row. This may indicate that there has been a settlement in the vicinity.

The interpretation of the orientation of the southern row of posts is simple because the measured azimuth corresponds to the rising sun on 6–7 February, a fundamental date for the definition of the re-start of the sacrificial cycle at its earliest date, 26–27 January. This date was later defined by the sunset along the northern side of the royal burial mounds (see Fig. 8.1).

The rule was simple: *when it was full moon on 6–7 February, the Great Midwinter sacrifice should take place the next year at its earliest date.*

A ship setting at Valsgärde burial ground

One of the most interesting prehistoric burial grounds in Sweden is situated at Valsgärde, 2.5 kilometres NNW of the northern row of posts at Old Uppsala. The burials are from about 500 BCE to about 1100 CE. The most remarkable type of grave contains a warrior with full equipment, including a horse, buried in a boat. Fifteen such boat-graves from 550–1100 CE have been investigated.

A preliminary investigation shows that the boats are oriented towards the rising full moon of the Great Midwinter Sacrifice in the eight-year cycle. The skeleton of the horse is well preserved, but the skeleton of the dead warrior chieftain is very eroded. The horseshoes have frost-nails, indicating that the burial took place in wintertime.

From these observed facts it seems reasonable to assume that the burial of the dead chieftain was performed at the next Great Midwinter Sacrifice when all of the important chieftains from the country were gathered in Old Uppsala. The body of the dead chieftain may meanwhile have been put on a high platform to prevent animals from eating it but allowing birds to eat the flesh and the weather to erode the bones.

The unique ship setting in the northern part of Valsgärde, consisting of 14 vertical posts, may have been the fundament for such a platform. No remains of the posts have been found, but the relation to the surrounding graves indicates that they probably were raised before 400 CE.

It is remarkable that the orientation of this ship setting, 25.34°, differs only from the orientation of the northern row of posts at Old Uppsala by 0.32°, a little more than the radius of the moon, ca. 0.25°.

Results for the oldest C^{14}-dated post

The oldest remains of a C^{14}-dated post have the calibrated date 1241 ±45 BCE. The nearest year after that in the earlier defined series of 144 and 152 years between the re-start of the sacrificial cycle was 1165 BCE. Because the southern row of posts was used for prediction one year in advance, the posts in this row were oriented on 6 February 1166 BCE and erected the same year. That morning the sun was rising in azimuth 120.55° for a horizon altitude of 0.17°, at 08.06.25 local time. At that moment the full moon was setting in azimuth 306.22° and altitude –0.67°, and its phase was 174.18°. This was a perfect year for defining the fundamental date of the rule for prediction of the re-start of the Great Midwinter Sacrifice at its earliest date.

The post, from which the oldest preserved C^{14} sample was taken, had an age of 75 years when it became a post. The present day ground level for the eastern, lowest, end of the southern row of postholes, 20.3m above mean sea level, was determined on 16 December 2014. This means that the original ground level for

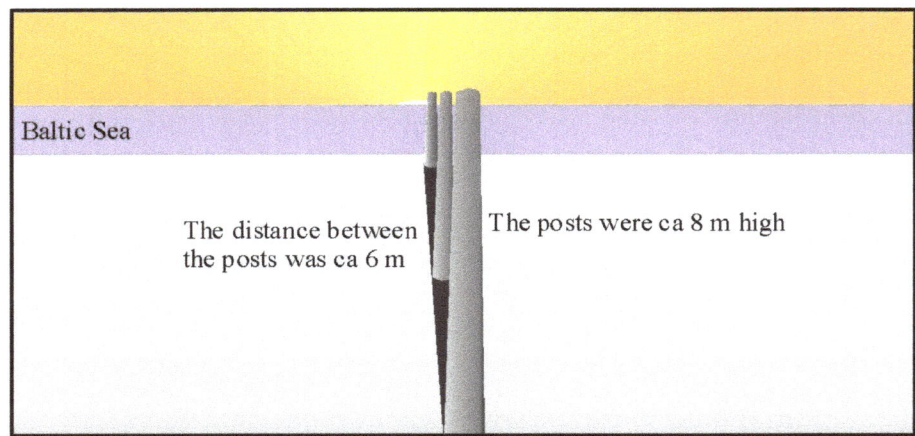

Fig. 8.5: Sunrise at Old Uppsala on 6-7 February, the day in the middle of the winter when the sun rose along the more than 600 m long southern row of posts. When these posts were raised, during the twelfth century BCE, the eastern most post was standing directly at the shore of the Baltic Sea. If it was full moon that day, the Great Midwinter Sacrifice would start the next year on its earliest date 26–27 January.

the lowest post is 19.9m above the present mean sea level, after reduction with 0.4m for the later layers. In 1166 BCE, when this C^{14}-dated post was erected, the mean sea level can be estimated to have been ca. 20.1 m above the present mean sea level.

This means that the eastern end of the southern row of posts was placed directly at the shore of the Baltic Sea. The western end was situated 3-4m higher, and when the important sunrise on 6 or 7 February took place, the rays of the sun were reflected in the sea on both sides of the posts and the shadows from the posts were aligned with the row (see Fig. 8.5).

This was a magnificent and very precise way to determine the correct date for the re-start of the Great Midwinter Sacrifice the next year.

The age of the posts

After calibration of all the 14 C^{14}-dated posts it seems possible to draw the conclusion that five of the posts in the northern row were erected during the important years of re-start of the 144/152 year cycle from its earliest date, and two posts in the southern row were erected one year before these years because this row of posts was used for prediction one year in advance, see Table 8.2.[17] The years in the southern row are 1165 and 421 BCE and in the northern row 180, 476 (with 3 posts) and 628 CE. It is interesting that 3 posts have been preserved from the very important year 476 CE when the semi-legendary King Aun formulated the rule for changes of the phases of the moon 'during 300 and some years' (304 years), see above.

A linear least square fit to the deviations from the expected years gives the very high coefficient of correlation 0.999997. The calibrated C^{14}-scale is 1.0027 ±0.0011 times the calendar years determined between 1241 BCE and 556 CE. The C^{14}-samples were taken from the well-preserved core of the pine and they represent approximately the year of birth of the tree.

It is therefore possible to determine the mean age of the trees as 72.0 ±1.8 years from the constant term in the least square fit. It is easier to understand this extremely low standard deviation if the trees were

---

[17] M. Stuiver, P. J. Reimer, and R. Reimer, p.j.reimer@qub.ac.uk, *Calib* 7.1, available at http://calib.qub.ac.uk/calib/calib.html [accessed 18 December 2014].

Sample[18] number	Radio-carbon age BP	Calibrated year BC/AD = mean of 1 σ limits	Relative probability within 1 σ	Re-start in the 144/152 year cycle	Age of the post (Year)	Deviation in the linear least square fit	Sample description
Ua-29932	3003 ±30	-1240 ±45	0.96	-1165	75*	0.2	Post remains
Ua-31482	2407 ±112	-494 ±97	0.67	-421	73*	0.2	Post remains
Ua-31686	1882 ±30	106 ±34	0.92	180	74	-2.5	Post remains
Ua-31574	1637 ±30	405 ±24	0.85	476	71	-0.3	Post remains
Ua-31213	1632 ±31	407 ±23	0.76	476	69	1.7	Post remains
Ua-31207	1630 ±33	408 ±23	0.69	476	68	2.3	Post remains
Ua-31480	1528 ±30	556 ±23	0.60	628	72	-1.7	Post remains

* Posts from the southern row have been erected one year before the re-start of the 144/152 year cycle

Table 8.2

deliberately cut when they were 72 years old. There may have been a certain sacred plantation for holy trees with new plants every 72 years. This means that the age of these trees is equal to the number of individuals, 72, which were sacrificed during the Great Midwinter Sacrifice.

One unburnt animal bone has been found in a posthole from 476 and 10 remains of food and unburnt animal bones have been found in postholes from 628. The mean value for the deviations from these years is 2.8 ±16 years, which further supports the idea that the posts were erected these years.

The last year during which old posts were replaced by new ones was in 628. In two of the fundaments for the postholes in the southern row datable material from 689 ±14 have been found. These finds can probably be related to the destruction of all the posts. The destruction of the northern row must have taken place before 705 ±65 because a house had been built above one of the postholes.

Acknowledgements

I would like to thank Professor Göran Possnert, Department of Physics and Astronomy, Uppsala University, for useful discussions, and Associate Professor Mary Blomberg, Department of Classical Archaeology and Ancient History, Uppsala University, for correction of my English.

Bibliography

Adam of Bremen. *History of the Archbishops of Hamburg-Bremen*. Translated by F. Tschan. New York: Columbia University Press, 1959.
Geminos. *Introduction aux Phènomèns*. Translated by G. Aujac. Paris: Les Belles Lettres, 1975.
Granlund, J. *Kulturhistoriskt lexikon för nordisk medeltid*, Vol. 3, s. v. 'Disting'. Malmö: Allhem, 1958.
Henriksson, G. *Arkeoastronomi i Sverige*. Uppsala: Uppsala University Press, 1994.
Henriksson, G. 'Riksbloten i Uppsala'. *Gimle* 20 (1992): pp. 14–23.
Henriksson, G. 'Riksbloten och Uppsala högar'. *Tor* 27, no. 1 (1995): pp. 337–394.
Henriksson, G. 'The Pagan Great Midwinter Sacrifice and the "royal" mounds at Old Uppsala'. in *Proceedings of the SEAC2001 meeting in Stockholm*, edited by M. Blomberg, P. E. Blomberg, and G. Henriksson, pp. 15–25. Uppsala:

---

[18] Jonas Wikborg, Societas Archaeologica Upsaliensis, presented these C14-data at a public lecture in Uppsala, 17 November 2014. The analysis of the carbon samples have been performed by professor Göran Possnert, Department of Physics and Astronomy, Uppsala University, Box 516, SE-751 20 Uppsala, Sweden.

Uppsala University Press, 2003.

Magnus, Olaus. *A description of the northern peoples, Rome 1555*, Vol. 1. Translated by P. G. Foote. London: Hakluyt Society, 1996.

Nyberg, T. *Monasticism in North-Western Europe, 800-1200*. Aldershot: Ashgate, 2000.

Rieper, H. *Ansgar und Rimbert: die beiden ersten Erzbischöfe von Hamburg/Bremen und Nordalbingen*. Hamburg: EB-Verlag, 1995.

Rudbeck, O. *Atlantica*, Vol. 1. 1679. Reprint Uppsala och Stockholm: Almquist och Wiksells, 1937.

Rudbeck, O. *Atlantica*, Vol. 2. 1689. Reprint Uppsala och Stockholm: Almquist och Wiksells, 1939.

Stuiver, M., P. J. Reimer, and R. Reimer. p.j.reimer@qub.ac.uk. *Calib* 7.1, http://calib.qub.ac.uk/calib/calib.html.

Sturlason, Snorre. *Heimskringla or the lives of the Norse kings* (Iceland 1230). Translated by E. Monsen. Cambridge: Dover Publications, 1932.

Tacitus. *Agricola. Germania. Dialogue on Oratory*. Translated by M. Hutton. Cambridge, MA and London: Loeb Classical Library 35, 1970.

Thietmar von Merseburg. *Chronik*. Translated by W. Trillmich. Darmstadt: Rütten & Loening, 1957.

FIRE FROM THE HEAVENS: THE IDEA OF COSMIC FIRE ACROSS ARCHAIC CULTURES

Michael A. Rappenglück

ABSTRACT: Fire is an essential part of archaic cosmovisions worldwide. People thought that the cosmic generation of fire resulted in a main fiery celestial object, the sun, radiating light and heat, but also in other celestial luminaries, e.g. the moon, the planet Venus, the stars, the Milky Way, the meteors, lightning, or auroras. Primeval cosmic fire, antagonistic and complementary to cosmic water, was essential for creation, transformation, destruction and regeneration of the world's entities. People often linked cosmic fire to the domestic hearth or the cultic fire and delivered rituals of fire kindling at the turning points of time cycles. Later philosophical ideas about fire as an element and about the central fire are a kind of aftermath of the mythical cosmic fire. The paper is a literary review of the idea of cosmic fire across archaic cultures.

Producing and taming fire was distinctive for humanization. Hitherto, the first indubitable evidence (microstratigraphic) for *in situ* use of fire in the early Acheulean comes from Wonderwork Cave (Northern Cape Province, South Africa) about 1 million years ago.[1] Hence, since time of early Homo erectus mankind has dealt with the essence and origin of fire. The invention of tools for generating fire as well as technologies for changing matter by fire went parallel with the cognitive and symbolic abilities of humans.[2]

Fire in the heavens, in the Earth, between both, and in living beings
Ancient cultures all over the world reflected much upon the origin and production of fire: fire consumes ('eats'), transforms ('cooking', 'transmutation'), and destroys matter (into smoke and ash).[3] Examples handed down by people worldwide show that producing, taming, and using fire means to have great power and to be able to control acts of production, transformation and destruction. People distinguished different kinds of fire. In Vedic tradition fire in heaven, in the atmosphere, and on earth is a manifestation of the primordial cosmic fire.[4] Fire was located in the hidden deeper interior of the Earth, too (e.g. Polynesian, Hawaiian, or Greek concepts), which, e.g., could be seen in a volcano's red-hot lava, gases, and the smoke column lit by fiery bolts, all escaping from underground spaces.[5] The natives of Northern Califor-

---

[1] Francesco Berna et al., 'Microstratigraphic evidence of in situ fire in the Acheulean strata of Wonderwerk Cave, Northern Cape province, South Africa', *Proceedings of the National Academy of Sciences* 10, no. 20 (2012): pp. E1215–E1220.

[2] Kyle S. Brown et al., 'Fire as an Engineering Tool of Early Modern Humans', *Science New Series* 325, no. 5942 (2009): pp. 859–862.

[3] Somewhat outdated, but partially useful: James George Frazer, *Myths of the Origin of Fire* (New York: Barnes and Noble, 1996). Certain overviews are given by: Johannes Maringer, 'Das Feuer in Kult und Glauben der vorgeschichtlichen Menschen', *Anthropos* 69, nos. 1 and 2 (1974): pp. 68–112; David M. Knipe, 'Fire', in *Encyclopedia of Religions, Second Edition*, ed. Lindsay Jones (New York: McMillan, 2005), pp. 3116–3121; Anatilde Idoyaga Molina, 'Fuego, purificación y metamorfosis. Significación y simbolismo ígneo entre los Pilagá (Chaco Central, Argentina)', *Anthropos* 104, no. 1 (2009): pp. 113–129; and Gregory Schrempp, 'Catching Wrangham: On the Mythology and the Science of Fire, Cooking, and Becoming Human', *Journal of Folklore Research* 48, no. 2 (2011): pp. 109–132.

[4] David M. Knipe, 'One Fire, Three Fires, Five Fires: Vedic Symbols in Transition', *History of Religions* 12, no. 1 (1972) pp. 28–41.

[5] Otto Gilbert, *Die Meteorologischen Theorien des griechischen Altertums* (Leipzig: B. G. Teubner, 1907); Tamra Andrews, *Legends of the Earth, Sea, and Sky. An Encyclopedia of Nature Myths* (Santa Barbara, CA: ABC-CLIO, 1998), pp. 81–82.

nia (USA) thought that the world was originally a globe of fire or molten matter. This fiery substance was absorbed by the trees from where it could be extracted by friction of wood, especially using a fire drill.[6] Fire was thought to be present in the air and the sky, by people who don't make any conceptual difference regarding astronomical and meteorological events or proceedings as it is done today. The ancient Greek term μετέωρος (metéōros) – 'lofty and high (in the sky)' – includes atmospheric and celestial phenomena as well.[7] The Inuit concept of a vivid sphere (*Sila/Silarjuaq*) around the Earth, below the celestial realm, is comparable.[8] This category contains the stars (especially red sparkling stars, like Antares (α Sco, 0.91-1.07 mag[9]), the Milky Way, the sun (especially showing the corona during an eclipse), the moon, the planets (especially red Mars), but also the red sky and sunset glow, the rainbow, the aurora borealis and australis, the crepuscular rays, the iridescent clouds, halos, parhelia ('sundogs'), and St. Elmo's fire. Fire coming down from heaven to earth was recognized, for example, in light pillars (sun, moon, and Venus pillars), lightning and meteors (especially bolides). Moreover people related fire to a kind of internal power entity moving and altering living organisms.[10] Fire, in particular the celestial type, is linked to the creatures souls' substance, alchemistic teaching (transmutation of matter), and religious traditions about altered states of mind, ecstasy, afterlife, and transcendence.

People considered fire responsible for producing light and heat.[11] Among the celestial bodies only the sun embodies 'hot' fire, because her heat can be felt.[12] All other celestial bodies and phenomena typify 'cold' fire.

Fire and water: The cosmic water cycle
People created myths about fire and water as the main agents of the earth's water cycle, thought to be driven by the fiery sun.[13] Fire was associated with sky-water – male power, rain, celestial blood (sperm), saltwater, dryness, heat (fire), the sun and light – antagonistic but also complementary to the earth-water, which was linked to female power, terrestrial and underground aquatic reservoirs, earthly blood (menstrual blood), freshwater, wetness, cold and the moon.[14] According to the ancient Indians there is an invisible, primordial fiery power (*Apam Napat*) beyond space and time, hidden, but nevertheless perpetually active in the infinite and dark undulating cosmic waters owning the potential creativity of the void.[15] *Agni*, in his threefold manifestation as heaven, air and earth, originated from Apam Napat. People frequently said that the first light, the sun, appeared in the greatest depth of the aquatic domain beyond the boundary, produced by some kind of initial fiery breath.[16] The tools for fire making by friction are also

---

[6] Frazer, *Myths of the Origin of Fire*, Vol. II, Part VII, p. 295.
[7] Otto Gilbert, *Die Meteorologischen Theorien des griechischen Altertums*.
[8] John MacDonald, *The Arctic Sky. Inuit Astronomy, Star Lore, and Legend* (Ontario: Royal Ontario Museum and Nunavut Research Institute, 1998), pp. 35–37.
[9] Stephen Field, 'Cosmos, Cosmograph, and the Inquiring Poet: New Answers to the "Heaven Questions"', *Early China* 17 (1992): pp. 83-110 (p. 102) and fn. 55.
[10] Knipe, 'Fire', p. 3116.
[11] Jean Chevalier and Alain Gheerbrant, *Dictionary of Symbols* (London: Penguin, 1996), pp. 379–382; Knipe, 'Fire', pp. 3116–3121.
[12] Ed C. Krupp, 'Rayed Disks'. *Griffith Observer* 53, nos. 1 and 2 (1989): p. 2.
[13] For primary and secondary sources see Michael A. Rappenglück, 'The Cosmic Deep Blue: The Significance of the Celestial Water World Sphere Across Cultures', *Mediterranean Archaeology and Archaeometry* 14, no. 3 (2014): pp. 293–305.
[14] For primary and secondary sources as well as for examples see Rappenglück, 'The Cosmic Deep Blue', p. 298.
[15] Stella Kramrisch, 'The Triple Structure of Creation in the Ṛg Veda'. *History of Religions* 2, no. 22 (1963): pp. 256–285; Rappenglück, 'The Cosmic Deep Blue', pp. 293–305.
[16] For primary and secondary sources as well as for example see Rappenglück, 'The Cosmic Deep Blue', pp. 295, 298.

located there.[17] Later manifestations of the fire are the astral objects, lightning, rainbows, and the Milky Way. The primordial fire embodied in the sun, or in the moon (the sun's night time manifestation), was thought to be engulfed in a giant water animal's dark belly like an embryo in a womb, for example in Indian tradition and elsewhere.

The worldly water cycle driven by the sun seems to be addressed in the myths of the combat between a solar hero (an avian, often a raptor) and an aquatic hero (a sea monster, usually a reptile, a dragon, or a giant fish).[18] They signify the antagonism and polarity of the upper world with the highest point in the sky (the celestial pole or zenith) and the lower world with the deepest point in the subterranean water realm, along the (polar or zenithal) world axis. One hemisphere is supernal, filled with primordial fire. It is related to light and day, creation, order, consciousness, life and omniscience. Frequently this realm is dominated by a primeval male avian world being, ruling the sun and shaping the world.[19] The other hemisphere is a chthonic one, containing primordial, chaotic water, that is associated with dark and night, destruction, disorder, unconsciousness, and death, but also possessing creative potency. Often this realm is dominated by a primeval female aquatic world being, controlling the moon, essential waters and vital juices, such as plant saps, (menstrual) blood, and sperm.[20] The raptor sits at the crown of the world tree (polar world axis), growing out of the cosmic waters; the serpent coils at the tree root.

Both celestial fire and water are impelling the cosmos and the processes within. Graeco-Roman authors wrote about fire as one of the four classical elements responsible for the origin and shapeshifting of objects in the world.[21] They identified the sphere of fire either with aether far above the sphere of the fixed stars or with our own realm located below the Moon's orbit. Primal elements including fire (with the antagonistic water) were also part of the cosmovisions of other cultures.[22]

Based on the continuity compared to the average of human life people distinguished *persistent* and *transient* fire phenomena.[23] The first type includes the sun, the moon, the stars and the Milky Way. The second contains all other ephemeral fiery objects, phenomena and processes.

Transient cosmic fire phenomena

People related the red Aurora borealis and Aurora australis to sky fires, bloody wars in the heavens and, on earth, celestial signals of catastrophes or even the apocalyptic end of the world.[24] The Karelians (North-

---

[17] Following the idea of Giorgio de Santillana and Hertha von Dechend, *Die Mühle des Hamlet* (Berlin: Kammerer & Unverzagt, 1993), pp. 196–205, 290–291, 294.
[18] For primary and secondary sources as well as for examples see Rappenglück, 'The Cosmic Deep Blue', pp. 299–300.
[19] Rappenglück, 'The Cosmic Deep Blue', pp. 299–300; e.g. David M. Knipe, 'The Heroic Theft: Myths from Ṛigveda IV and the Ancient Near East', *History of Religions* 6, no. 4 (1967): pp. 328–360, 337–338, 351–358; John Irwin, '"Aśokan' Pillars": A reassessment of the evidence - IV: Symbolism', *The Burlington Magazine* 118 (1976): pp. 884, 734–736, 751, 753.
[20] For primary and secondary sources see as well as for examples Rappenglück, 'The Cosmic Deep Blue', pp. 299–300; e.g. Peter G. Roe, *The Cosmic Zygote. Cosmology in the Amazon Basin* (New Brunswick, NJ: Rutgers University Press, 1982), pp. 138–139, 215–217.
[21] M. R. Wright, *Cosmology in Antiquity* (London, New York: Routledge, 1995), pp. 62–63, 97, 104, 109–125, 141, 152, 168–169.
[22] Andrews, *Legends of the Earth, Sea, and Sky*, p. 75; Willibald Kirfel, *Die fünf Elemente, insbesondere Wasser und Feuer: ihre Bedeutung für den Ursprung altindischer und altmediterraner Heilkunde. Eine medizingeschichtliche Studie* (Walldorf, Hessen: Verlag für Orientkunde, 1950); Marcel Granet, *Das Chinesische Denken. Inhalt, Form, Charakter* (München: dtv, 1980), pp. 122–127, 230–233.
[23] David H. Kelley and Eugene F. Milone, *Exploring Ancient Skies. An Encyclopedic Survey of Archaeoastronomy* (New York: Springer, 2005), pp. 109–153.
[24] D. Justin Schove, 'Sunspots, Aurorae and Blood Rain: The Spectrum of Time', *Isis* 42, no. 2 (1951): pp. 137; Richard

ern Europe) thought that the Aurora borealis was caused by the movements of a giant celestial red fire fox.[25] Some cultures identified lightning with powerful fire from heaven, giving life and creation as well as causing destruction and death.[26] It is regarded as a kind of gigantic phallus (of the sky god or the sun), which inseminates the Earth's womb by mixing fire and water together. The rays of the sun, stars and planets caused lightning. Fiery matter, shaped as bolts, is streaming out of the celestial bodies, rapidly falling down, and striking objects on earth or finally the ground.[27] Graeco-Roman authors explain that the planets Jupiter, Saturn and Mars, especially during conjunctions, are responsible for fiery lightning bolts linked to meteors and comets.[28] The Etruscan *Libri Fulgurales* contained a complete doctrine about the mantic and astrological interpretation of lightning corresponding to certain locations in the cosmos and their generation by stars, planets and zodiacal signs.[29] Roman writers thought that during the stormy season certain stars, e.g. Castor and Pollux, descended from their constellation Gemini and alighted as St. Elmo's fire on the masts and sails, helping the mariners during dangerous situations.[30]

People identified single meteors, especially bolides, meteor streams, with pieces of cosmic fire falling down from the sky to earth.[31] North American tribes (Huron-Wynadot, Seneca) identified bright comets and meteors appearing out of the watery depths of the northern polar sky with sparks of cosmic fire, which comes out of their bodies.[32] Similar Mesoamerican people linked meteors with the process of cosmic fire drilling.[33] Comets had been a sign of celestial fire according to the traditions of some cultures.[34] The Greek myth of Phaeton for example describes a fiery comet like object, which once caused immense devastation on earth and chaos in the sky.[35] According to Aristoteles (384–322 BCE) celestial fire produces comets as well as meteors, haloes, parhelia and auroras. People also associated parhelia ('sundogs') and certain pillars with fire in the sky.[36]

Stothers, 'Ancient Aurorae', *Isis* 70, no. 1 (1979): pp. 85–95; Roslynn D. Haynes, 'Astronomy and the Dreaming', in *Astronomy across Cultures. The History of Non-Western Astronomy*, ed. Helaine Salin (Dordrecht: Kluwer, 2000), p. 86; Birgit Schlegel and Kristian Schlegel, *Polarlichter zwischen Wunder und Wirklichkeit: Kulturgeschichte und Physik einer Himmelserscheinung* (Heidelberg: Spektrum, 2011), pp. 3–8, 31, 36–41, 42, 44, 46–52, 61–64.

[25] Schlegel and Schlegel, *Polarlichter*, pp. 28–29.

[26] Chevalier and Gheerbrant, *The Penguin Dictionary of Symbols*, pp. 606–607; Andrews, *Legends of the Earth, Sea, and Sky*, pp. 128–130.

[27] Eugene S. McCartney, 'Classical Weather Lore of Thunder and Lightning', *The Classical Weekly* 25, no. 23 (1932): p. 186; Wilhelm Gundel, *Sterne und Sternbilder im Glauben des Altertums und der Neuzeit* (Hildesheim: Georg Olms, 1981), pp. 218.

[28] Van der Sluijs, Marinus Anthony, 'Phaethon and the Great Year', *Apeiron: A Journal for Ancient Philosophy and Science* 39, no.1 (2006): pp. 80–84.

[29] Stefan Weinstock, 'Libri Fvlgvrales'. *Papers of the British School at Rome* 19 (1951): pp. 122–153.

[30] Marek Hermann, 'Zur Astrometeorologie bei römischen Autoren', *Rheinisches Museum für Philologie* NF 148, nos. 3 and 4 (2005): p. 288.

[31] M. E. Chauvin, 'Useful and Conceptual Astronomy in Ancient Hawaii', in *Astronomy across Cultures. The History of Non-Western Astronomy*, ed. Helaine Salin (Dordrecht: Kluwer, 2000), pp. 98, 100.

[32] George R. Hamell, 'Long-Tail: The panther in Huron-Wynadot and Seneca myth, ritual, and material culture', in *Icons of Power. Feline Symbolism in the Americas*, ed. Nicholas J. Saunders (London and New York: Routledge, 1998), pp. 264–265.

[33] Karen Bassie, *Maya Creator Gods*, 2002, available at http://www.mesoweb.com/features/bassie/CreatorGods/CreatorGods.pdf, pp. 1–60 [accessed?].

[34] Anton Scherer, *Gestirnnamen bei den indogermanischen Völkern* (Heidelberg: Carl Winter, 1953), pp. 103–105, 107.

[35] Van der Sluijs, 'Phaethon and the Great Year', pp. 77–78, 80–83.

[36] Peter Metevelis, 'The Dog Star and the Multiple Suns Motif: An Asian Contribution to European Mythology', *Asian Folklore Studies* 64, no. 1 (2005): pp. 135–136; Richard Stothers, 'Ancient Meteorological Optics', *The Classical Journal* 105,

Manifestations of cosmic fire: living beings and gods

People visualized their perceptions of astronomical and meteorological phenomena by identifying them with certain animals or parts of their bodies, like avian animals (most notably high flying raptors), certain mammals (deer, bull, horse, lion etc.), reptiles and others. Feather crowns, beaks, plumages, tails, horns and antlers, manes, hoofs and paws have been associated with light phenomena in the sky. Perceptions of what today are called *atmospheric optics* (halos, including 22° circular halos, sundogs, light pillars and others; crepuscular and anti-crepuscular rays) together with events, which today are categorized as astronomical (the corona during a total eclipse; a red moon at total lunar eclipse; bright meteors, bolides, comets), linked to the concept of cosmic fire seem to be delivered in ancient traditions.[37] Plants, too, e.g. the lotus, have been associated with cosmic fire and especially the power of the sun.[38] Finally, ancient artwork worldwide show the sun as a human face with a fiery wreath of rays or as a rayed disk.[39]

Catching cosmic fire: myths and tools

Catching celestial (cosmic) fire, bringing it down to earth and leaving it in man's custody was, on one hand, taken by the people as a brilliant heroic act, but on the other hand as a theft of originally divine cosmic power.[40] Traditions worldwide were about the theft of fire from the heavens by a trickster or a hero and the donation of this cultural gift to humankind. Certain animals, mostly birds, but also mammals, insects, fish, or even plants, were active in the production or the theft of cosmic fire. Either way, the fire is stolen from the sun or from the celestial pole, being the essential centre of cosmic power.

People noticed that certain tools for fire making could provide explanations for the natural process generating cosmic fire.[41] People in Mesoamerica thought that striking sparks by clashing the rocks of the sky vault produced meteors.[42] Cultures in the ancient Near East thought the fire sawing method was used by the sun (god) for producing light and heat.[43] The Skidi-Pawnee (North America) link the male Morning Star (red shining Mars) with the fire-place, and the fire drill (together with the female Evening

---

no. 1 (2009): pp. 27–42; Richard Stothers, 'Ancient Aurorae', p. 89.

[37] Les Cowley, *Atmospheric Optics*, available at http://www.atoptics.co.uk/opod.htm [accessed 30 January 2015]. According to Janet McCrickard, *Eclipse of the Sun. An Investigation into the Sun and Moon Myths* (Glastonbury: Gothic Image Publications, 1990), pp. 90–91, 128, 131, 136, 157, 243; Michael A. Rappenglück, 'Tracing the Celestial Deer – An Ancient Motif and its Astronomical Interpretation Across Cultures', in *Astronomy and Cosmology in Folk Traditions and Cultural Heritage*, ed. J. Vaiškūnas (Klaipeda: Klaipeda University Press, 2008), pp. 62–65; Michael A. Rappenglück, 'Heavenly Messengers: The Role of Birds in the Cosmographies and the Cosmovisions of Ancient Cultures,' in *Cosmology across cultures*, ed. J. A. Rubiño-Martín; J. A. Belmonte, F. Prada and A. Alberdi (Orem, UT: ASP, 2009). pp. 145–150.

[38] For primary and secondary sources as well as for examples see Rappenglück, 'The Cosmic Deep Blue', p. 301.

[39] Ed C. Krupp, 'Rayed Disks'; Ed C. Krupp, 'Facing the Sun', *Griffith Observer* 54, no. 1 (1990): pp. 2–13.

[40] Adalbert Kuhn, *Die Herabkunft des Feuers und des Göttertranks. Ein Beitrag zur vergleichenden Mythologie der Indogermanen* (Berlin: Ferdinand Dümmler, 1859); David M. Knipe, 'The Heroic Theft: Myths from Ṛigveda IV and the Ancient Near East', pp. 328–360; Beate Seiffert, *Die Herkunft des Feuers in den Mythen der nordamerikanischen Indianer* (Bonn: Holos, 1990); Rappenglück, 'Cosmic Spinning and Weaving: Making the Texture of the World', pp. 162–163; Rappenglück, 'Tracing the Celestial Deer', p. 64; Rappenglück, 'Heavenly Messengers', 2009, pp. 146–147.

[41] Winifred S. Blackman, 'The Magical and Ceremonial Uses of Fire', *Folklore* 27, no. 4 (1916): pp. 352–377; John Loewenthal and Bruno Mattlatzki, 'Die europäischen Feuerbohrer,' *Zeitschrift für Ethnologie* 48, no. 6 (1916): pp. 349–369.

[42] Bassie, *Maya Creator Gods*.

[43] Ernst E. Ettisch, 'Die Säge als Sonnensymbol im Alten Orient und ihre Darstellung in der jüdischen Mystik', *Paideuma* 7, no. 7 (1961): pp. 345–351.

star, Venus) with flint, meteors and creation.[44] The Aztec identified parts of Orion (the Belt Stars) with a fire-drill asterism.[45] There exists, however, some types of myths about cosmic fire drilling, which try to comprehensively explain the creation and spreading of cosmic fire.

The fiery celestial objects turning around the celestial pole(s), implied the existence of a certain rotational power located there. People had the impression of a gigantic cosmic being located at the centre of the sky, having only one single leg (the polar world axis) and being in charge for the rotation of the world.[46] Such chimeras related to the kindling of cosmic fire are well known from ancient European, Asian, Mesoamerican and even South American traditions. During certain epochs, stars and asterisms had been close or just at the celestial pole, giving the one-legged polar being a specific appearance such as a one-footed bull, ram, goat, bird, human, etc.[47]

Some people realized that the polar world axis, identified as a giant cosmic fire drill, correspond to the style of a polar aligned shadow stick. The sun, a fiery disc, pivoting on the style illustrates rotational power, too. Somehow, the sun and the celestial pole (or polestar/pole asterism) seemed to be invisibly coupled by the world axis, indicating that both are of the same impelling cosmic fire.[48] Therefore in some traditions they were symbolically and mythically linked to each other. The power of all celestial bodies essentially originates from the celestial pole.

The shadow stick, aligned to the celestial pole at night, is pointing to the very centre of cosmic fire. The Vedic *Ajá ekapâda* ('the he-goat with only one foot') is a very good example for the symbolism of cosmic fire drilling and its relation to the shadow stick. It was associated with the ray of the sun, the thunderbolt (fire and light), the column of fire, and the smoke of the sacrificial altar at the world's navel. It illustrates the fire descending and ascending between the cosmic spheres. They all correspond to the threefold manifestation of the fire (*Agni*) and are related to the three hierarchical ordered spheres (*triloka*) around the (polar) world axis.[49] *Ajá ekapâda*, the sun, and *Dhruva*, the polestar, are both coupled by an invisible net of power threads. The rotational energy at the celestial pole and the sun's force belong to the same fiery cosmic entity.

The sun is culminating on the shadow stick every day at noon and climbs up to its top end at noon on the summer solstice, assuming the power of movement, heat and light from the world axis. In the traditions of some cultures the style could be a stick, a column, a cross, a mountain, a nail, a phallus, a pillar, a pole, a plug, a sceptre, a temple, a tree or a vertebral column.[50] The cosmic fire, produced at the celestial pole, where creation started, was thought to be stored predominantly in the sun, as the very important source of heat and light, but also in other celestial bodies. The fiery bird at the celestial pole had a derivative localized in the ardent solar bird, circling around the cosmic fire drill and related to it. This leads to the globally widespread motif of the sun-bird perched on a tree. A pole, crowned by a bird,

---

[44] Von Del Chamberlain, *When Stars Came Down to Earth. Cosmology of the Skidi Pawnee Indians of North America* (Los Altos, CA, and Maryland: Ballena Press/Center for Archaeoastronomy Cooperative Publication, 1982), pp. 55–59, 142, 158–161; Kelley and Milone, *Exploring ancient skies*, pp. 422-423, 492-493.

[45] Anthony F. Aveni, *Skywatchers of Ancient Mexico* (Austin: University of Texas Press, 1980): pp. 35.

[46] Michael A. Rappenglück, 'The Whole Cosmos Turns around the Polar Point: One Legged Polar Beings and their Meaning', in Astronomy and Cultural Diversity, ed. César Esteban and Juan Belmonte Aviles (La Laguna, Tenerife: OACIMC, 1999), pp. 170–174.

[47] Rappenglück, ‚The Whole Cosmos Turns around the Polar Point', pp. 170–174.

[48] Michael A. Rappenglück, 'The Pivot of the Cosmos. The Concept of the World Axis Across Cultures', in *Cosmic catastrophes. A collection of articles*, ed. M. Kõiva, I. Pustylnik, and L. Vesik (Tartu, 2005), pp. 157–165.

[49] Alessandro Grossato, '"Shining Legs" The One-footed Type in Hindu Myth and Iconography', *East and West* 37, no. 1/4 (1987): pp. 262–263, 267, 269, 272; Kramrisch, 'The Triple Structure of Creation in the Rg Veda', pp. 256–285.

[50] Rappenglück, 'Heavenly Messengers', pp. 146–147.

was often part of midsummer fire rituals.[51] Frequently the solar bird and the bird of the celestial pole have been intertwined and mingled.

According to an old German custom at Midsummer, a ton of burning pitch, symbolizing the sun, was chained to the top of a high pole. The barrel was passed around the post speedily up to the highest point, which could be reached. Finally, it came to rest near the foot of the post. Meanwhile the chain was wound around the mast. This ritual symbolises the movement of the sun helically up and down around the polar world axis.[52]

Cultures sometimes identified the sun's disc with a wheel, because of its ability to roll along the sky vault. There is however the concept of a spoked solar wheel visualizing an abstraction of cyclical time given by the orbits of the sun, the moon, and the planets. It could have 7 (sun, moon, five planets), 12 (synodic months), 28 (integer of days in the sidereal month) or even 365 (days) spokes. That concept is about the distribution of the fiery cosmic power by rotation down into the belt of the Zodiac.[53] Ajá ekapáda is linked to the later motif of ekacâkra, the 'one wheel', on which all creatures stand.[54] Examples of such wheels around a vertical axis are well known from the ancient Egyptians, Germans, Greeks, and Hittites. Indian and Iranian traditions tell of the chords of winds, which bind the planets, stars, and asterisms to the celestial pole (Dhruva), turning around him and causing him to revolve. They embody a fire-wheel, which is driven by the wind-wheel.[55] The sun occupies the tip of a main spoke of the fire-wind-wheel: thus, it seems to be one-legged.[56] Seasonal fire-festivals, especially on Midsummer (Germany, Hungary) often are related to producing fire by a rotating wheel round a wooden axle.[57]

The idea of cosmic fire drilling could go back 13,600 years: then a fire-drilling turtle asterism with the bright star Vega (α Lyr; 0.03 mag) was 3.4° close to the celestial North Pole.[58] Myths concerning a cosmic drilling of fire related to a giant world turtle, generating and owning the fire as well as sometimes producing the fiery ring of the Milky Way, are known from Europe, North America, Mesoamerica and Asia.

Following the fire-drilling process people identified the driller ('the spindle') with the (zenithal or polar) world axis, and the wooden board (the so-called 'hearth') with the solid sky vault. By analogy they associated the spindle with a fertilizing phallus (and father), the board with a fertile vulva (and mother), the process of drilling with sexual intercourse, and the generating of the fire flame with creation.[59]

---

[51] Frazer, *Myths of the Origin of Fire*, Vol. II, Part VII, p. 111; Michael A. Rappenglück, *Eine Himmelskarte aus der Eiszeit?* (Frankfurt a. Main: Peter Lang, 1999), pp. 171–177, 184–185.

[52] Frazer, *Myths of the Origin of Fire*, Vol. II, Part VII, p. 169; Michael Rappenglück, 'Cosmic Spinning and Weaving: Making the Texture of the World'. *BAR International Series* 1647 (2007): p. 165.

[53] Abraham Seidenberg, 'The Ritual Origin of the Circle and Square'. *Archive for History of Exact Sciences* 25, no. 4 (1981): pp. 269–327.

[54] Paul Horsch and F. B. J. Kuiper: 'Aja ekapād und die Sonne', *Indo-Iranian Journal*, no. 1 (1965): pp. 9, 11, fn. 42; Rappenglück, 'The Whole Cosmos Turns around the Polar Point', p. 171.

[55] Antonio Panaino, 'Uranographia Iranica II: Avestan hapta.srū- and mərəzu-: Ursa Minor and the North Pole?', *Archiv für Orientforschung* 42/43 (1995/1996): pp. 190–207 (pp. 196–198).

[56] P. E. Dumont, 'The Indic God Aja Ekapad, the One-Legged Goat.' *Journal of the American Oriental Society* 53, no. 4 (1933): pp. 332–334.

[57] Frazer, *Myths of the Origin of Fire*, Vol. II, Part VII, pp. 116–117, 201, 334–336.

[58] Michael A Rappenglück, 'The Whole World Put between two Shells: The Cosmic Symbolism of Tortoises and Turtles', *Mediterranean Archaeology and Archaeometry* 3 (2006): pp. 225–227.

[59] W. Schwarz, 'Der (rothe) Sonnenphallos der Urzeit. Eine mythologisch-anthropologische Untersuchung', *Zeitschrift für Ethnologie* 6 (1874): pp. 167–188; Horsch and Kuiper, 'Aja ekapād und die Sonne', pp. 14–15; Frazer, *Myths of the Origin of Fire*, Vol. II, pp. 208–209, 235, 238, 239, 248-249, Vol. VIII, p. 65, Vol. IX, p. 391; Hermann Baumann, *Das doppelte*

The fireplace: centre of domestic world and the cosmos

People put the fireplace, the hearth (the same term as the wooden board used for fire drilling), the cremation place, and later the (alchemical) oven but also the female uterus on one level with the creative vessel of the cosmos.[60] The fireplace and the hearth, as the centre of a settlement, of a structure, or of a cult building, symbolizes and repeats the primordial and also perpetual act of creation, the process of transformation, going through phases of formation and destruction. The cosmos is considered to be a giant womb, a kind of hearth, in which female and male powers (water and fire) come together generating the world's entities by transforming matter in different shapes. The spindle (phallus) serves as a whisk for propelling the process. The fire-place or hearth was identified with both – the navel of the human body and the centre of the world – being the origin of transformational physical and spiritual power. One of the oldest examples for that idea can be found in Lepenski Vir (Romania), 7000-6000 BCE. Thus the place where a fire is kindled offers the possibility to come into contact with the primordial origin of the cosmos and to communicate with the ancestors who are thought to be present there. The fireplace and hearth anchored a family or a clan at a place, giving life, protection, stability, unity, and ensuring linkage and communication with the ancestral domain or even the powerful centre of the world itself.

Fire is one of the main agents impelling the world destructively and constructively. Therefore, it was considered the most important power in sacrifices and ritual cremations. These are linked to certain cosmovisions which are handed down in symbols, myths and rituals but which also manifest in archaeological records, as for example in Scandinavian prehistory in the light of Indo-European cosmological and cosmogonical concepts.[61]

Ancient cultures in Africa, Asia, Mesoamerica and North America often intentionally placed certain objects (bones, clay balls, animal or human figurines) near or into the fire, indicating the relation to the chain of ancestors, bound to the origin of the cosmos.[62] A similar ritual act might already have existed 7400–6200 BCE in Çatalhöyük (Anatolia, Turkey).

The position of fire-altars in the dwellings, close to the centre and the main supporter, if one existed, was important. The smoke rising up from the fire-place/altar was considered supporting the relationship between the cosmic spheres. In some cases, for example the Harlan-Style Charnell House (North America), the smoke column resulting from the ritual burning of a complete house was identified with the axis mundi and the path the souls take to the other world. The fireplace in a Hooghan, the dwelling of the Navajo (North America) is thought to be directly linked to Polaris (α UMi).[63] The asterisms 'Male Revolver' (parts of Ursa Major) and 'Female Revolver' (parts of Cassiopeia) are thought to be a male / female couple. The Lodge of the Skidi Pawnee (North America) has a smoke hole, which allows viewing the asterism of 'The Chiefs in Council' (Corona Borealis).[64] The smoke hole of a Mongolian yurt was considered to connect the terrestrial realm (fireplace) with the celestial realm.[65] Finally, Graeco-Roman tradition identified the constellation Altar (*Ara*) as the producer of the Milky Way.[66]

Geschlecht. Studien zur Bisexualität in Ritus und Mythus (Berlin: Dietrich Reimer, 1986), pp. 310–314.

[60] For primary and secondary sources as well as for examples see Michael A. Rappenglück, 'The Housing of the World: The Significance of Cosmographic Concepts for Habitation', *Nexus Network Journal* 15, no. 3 (2013): pp. 387–422.

[61] Anders Kaliff, *Fire, Water, Heaven and Earth. Ritual Practice and Cosmology in Ancient Scandinavia: An Indo-European Perspective* (Stockholm: Riksantikvarieämbetet, 2007).

[62] Rappenglück, 'The Housing of the World', pp. 400–401.

[63] Trudy Griffin-Pierce, 'The Hooghan and the stars', in *Foundations of New World Cultural Astronomy*, ed. Anthony Aveni (Boulder, CO: University Press of Colorado, 2008), pp. 441–448.

[64] Anthony F. Aveni, *People and the Sky. Our Ancestors and the Cosmos* (London: Thames and Hudson, 2008), pp. 114–116.

[65] Rolf A. Stein, *Le monde en petit* (Paris: Flammarion, 2001), pp. 150–154.

[66] Julius D. W. Staal, *The New Pattern in the Sky. Myths and Legends of the Stars* (Blacksburg VA: McDonald and Woodward

Catching cosmic fire with advanced tools

Finally, some people used special tools for catching and subduing fire from the sky. In Late Han-Dynasty (23/25–220 CE) China, during a special ceremony called the 'Director of Sun Fire', a concave bronze mirror facing the sun at noon served to focus solar fire.[67] In addition it was used to collect dew, being exposed to the full moon's light. Thus the mirrors were associated to both fire and water, corresponding to sun and moon. The Taoist bronze mirrors showing the so-called TLV design are related to sundials decorated in a similar way. Minoan people used a sophisticated procedure for attracting divine fire from the sky: bronze pillars may have served as lightning-rods in ancient Crete peak sanctuaries.[68] During Antiquity also lenses seem to be used for generating ritual pure fire from the sky.[69]

New fire ceremonies / bonfire rituals

The concept of cosmic fire figured prominently in rituals of destruction, purification, cremation and renewal. Astronomically motivated fire rituals (New Fire, Bonfire etc.) were linked to the turning points in time cycles, for example main as well as intermediate calendric dates. Whirling fires, fire wheels, fire drilling, burning of straw dolls, candles on trees, etc. are important elements of such ceremonies. New fire rituals are widespread all over the world and throughout time.[70] A very long cycle is ritualized in the Aztec New Fire ceremony, which is held every 52 years: all fires were quenched and kindled anew using the fire drill.[71] The myth of the Phoenix (Egypt, India, China, and Greek) links together the sun, death and rebirth, the destructive and constructive power of fire, and time cycles related to the sun and longer periods.[72]

Ekpyrosis: destruction of the world by fire

Finally, the Babylonians, Indians, Chinese, Iranians, Greeks (Pythagoreans, Stoics) and people of Medieval world, but also Mesoamerican or North American, thought about the conflagration of the world by fire.[73] The initial and the end point given by the conjunction of all wandering stars is accompanied by recurrent cosmic disasters, a catastrophe by a global flood (deluge, *kataklysmos*) alternating with one by a global conflagration (*Ekpyrosis*). The Great Year is divided into seasons: the Great Year's winter is related to the deluge (conjunction of all wandering stars in Capricorn) and the Great Year's summer to the *Ekpyrosis* (conjunction of all wandering stars in Cancer). It is an extension of the primeval combat of fire and water.

Publishing, 1988), pp. 229–230.

[67] Joseph Needham, *Science and Civilisation in China*, Vol. IV, 1 (Cambridge: Cambridge University Press, 1996), p. 87; Anna Seidel and Marc Kalinowski, 'Tokens of Immortality in Han Graves', *Numen* 2, no. 1 (1982): pp. 97–98.

[68] Doro Levi, 'Features and Continuity of Cretan Peak Cults', in *Temples and High Places in Biblical Times*, ed. A. Biran (Jerusalem: Nelson Glueck School of Biblical Archaeology, 1981), pp. 40.

[69] Robert Temple, *The Crystal Sun. Recovering a Lost Technology of the Ancient World* (London: Century, 2000), pp. 92–120.

[70] Frazer, *Myths of the Origin of Fire*.

[71] Kay Almere Read, *Time and sacrifice in the Aztec cosmos* (Bloomington and Indianapolis, IN: Indiana University Press, 1998), pp. 11, 26–28, 45, 47, 89, 101–103, 105, 118, 123–127, 156–169, 170–175, 199, 223, 226, 231, 238, 241; Davíd Carrasco, 'Star Gatherers and Wobbling Suns: Astral Symbolism in the Aztec Tradition', *History of Religions* 26, no. 3 (1987).

[72] Chevalier and Gheerbrant, *The Penguin Dictionary of Symbols*, pp. 752–753.

[73] Frank Waters, *The Book of the Hopi* (New York: Penguin, 1963), pp. 13–14; Åke V. Ström, 'Indogermanisches in der Völuspá', *Numen* 14, no. 3 (1967): pp. 167–208 (pp. 192–193); Wright, *Cosmology in Antiquity*, pp. 32, 90, 141–144; Bartel Leendert Van der Waerden, 'Das Grosse Jahr und die Ewige Wiederkehr', *Hermes* 80, no. 2 (1952): pp. 129–155; Wayne Elzey, 'The Nahua Myth of the Suns: History and Cosmology in Pre-Hispanic Mexican Religions', *Numen* 23, no. 2 (1976): pp. 114–135.

Conclusion

The concept of cosmic fire (together with cosmic water) was an important part of archaic world views. The ideas of ancient people ranged from simple analogical identifications to very elaborate and profound cosmovisions.

Bibliography

Andrews, Tamra. *Legends of the Earth, Sea, and Sky. An Encyclopedia of Nature Myths.* Santa Barbara, CA: ABC-Clio, 1998.
Aveni, Anthony F. *People and the Sky. Our Ancestors and the Cosmos.* London: Thames and Hudson, 2008.
Aveni, Anthony F. *Skywatchers of Ancient Mexico.* Austin: University of Texas Press, 1980.
Bassie, K. 'Maya Creator God', 2002, http://www.mesoweb.com/features/bassie/CreatorGods/CreatorGods.pdf.
Baumann, Hermann. *Das doppelte Geschlecht. Studien zur Bisexualität in Ritus und Mythos.* Berlin: Dietrich Reimer, 1986.
Berna, Francesco, et al. 'Microstratigraphic evidence of in situ fire in the Acheulean strata of Wonderwerk Cave, Northern Cape province, South Africa'. *Proceedings of the National Academy of Sciences* 10, no. 20 (2012): pp. E1215–E1220.
Blackman, Winifred S. 'The Magical and Ceremonial Uses of Fire'. *Folklore* 2, no. 4 (1916): pp. 352–377.
Brown, Kyle S., et al. 'Fire as an Engineering Tool of Early Modern Humans'. *Science New Series* 325, no. 5942 (2009): pp. 859–862.
Carrasco, Davíd. 'Star Gatherers and Wobbling Suns: Astral Symbolism in the Aztec Tradition'. *History of Religions* 26, no. 3 (1987): pp. 279–294.
Chamberlain, Con Del. *When Stars came down to Earth. Cosmology of the Skidi Pawnee Indians of North America.* Los Altos, CA, and Maryland: Ballena Press/Center for Archaeoastronomy Cooperative Publication, 1982.
Chauvin, Michael E. 'Useful and Conceptual Astronomy in Ancient Hawaii'. In *Astronomy across Cultures. The History of Non-Western Astronomy.* Edited by Helaine Salin, pp. 91–126. Dordrecht: Kluwer Academic Publishers, 2000.
Chevalier, Jean, and Alain Gheerbrant. *The Penguin Dictionary of Symbols.* London: Penguin, 1996.
Cowley, Les. 'Atmospheric Optics'. http://www.atoptics.co.uk/opod.htm.
de Santillana, Giorgio, and Hertha von Dechend, *Die Mühle des Hamlet.* Berlin: Kammerer & Unverzagt, 1993.
Dumont, P. E. 'The Indic God Aja Ekapad, the One-Legged Goat.' *Journal of the American Oriental Society* 53, no. 4 (1933): pp. 326–334.
Elzey, Wayne. 'The Nahua Myth of the Suns: History and Cosmology in Pre-Hispanic Mexican Religions'. *Numen* 23, no. 2 (1976): pp. 114–135.
Ettisch, Ernst E. 'Die Säge als Sonnensymbol im Alten Orient und ihre Darstellung in der jüdischen Mystik'. *Paideuma* 7, no. 7 (1961): pp. 345–351.
Field, Stephen. 'Cosmos, Cosmograph, and the inquiring poet: New Answers to the "Heaven Questions"'. *Early China* 17 (1992): pp. 83–110 (p. 102) and fn. 55.
Frazer, James George. *Myths of the Origin of Fire (12 Volumes).* New York: Barnes and Noble, 1996.
Gilbert, Otto. *Die Meteorologischen Theorien des griechischen Altertums.* Leipzig: Teubner, 1907.
Granet, Marcel. *Das Chinesische Denken. Inhalt, Form, Charakter.* München: dtv, 1980.
Griffin-Pierce, Trudy 'The Hooghan and the stars'. In *Foundations of New World Cultural Astronomy.* Edited by Anthony Aveni, pp. 439–455. Boulder, CO: University Press of Colorado, 2008.
Grossato, Alessandro. '"Shining Legs"s The One-footed Type in Hindu Myth and Iconography'. *East and West* 37, nos. 1/4 (1987): pp. 247–282.
Gundel, Wilhelm. *Sterne und Sternbilder im Glauben des Altertums und der Neuzeit.* Hildesheim: Olms, 1981.
Hamell, George R. 'Long-Tail: The panther in Huron-Wynadot and Seneca myth, ritual, and material culture'. In *Icons of Power. Feline Symbolism in the Americas.* Edited by Nicholas J. Saunders, pp. 258–291. London and New York: Routledge, 1998.
Haynes, Roslynn D. 'Astronomy and the Dreaming'. In *Astronomy across Cultures. The History of Non-Western Astronomy.* Edited by Helaine Salin, pp. 53–90. Dordrecht: Kluwer, 2000.
Hermann, Marek. 'Zur Astrometeorologie bei römischen Autoren'. *Rheinisches Museum für Philologie* NF 148, nos. 3

and 4 (2005): pp. 272–292.

Horsch, Paul, and F. B. J. Kuiper. 'Aja ekapād und die Sonne.' *Indo-Iranian Journal* 9, no. 1 (1965): pp. 1–31.

Irwin, John. '"Aśokan' Pillars": A reassessment of the evidence - IV: Symbolism'. *The Burlington Magazine* 118 (1976): pp. 884, 734, 736–751, 753.

Kaliff, Anders. *Fire, Water, Heaven and Earth. Ritual Practice and Cosmology in Ancient Scandinavia: An Indo-European perspective*. Stockholm: Riksantikvarieämbetet, 2007.

Kelley, David H., and Milone, Eugene F. *Exploring Ancient Skies. An Encyclopedic Survey of Archaeoastronomy*. New York: Springer, 2005.

Kirfel, Willibald. *Die fünf Elemente insbesondere Wasser und Feuer: ihre Bedeutung für den Ursprung altindischer und altmediterraner Heilkunde. Eine medizingeschichtliche*. Walldorf: Verlag für Orientkunde, 1950.

Knipe, David M. 'Fire'. In *Encyclopedia of Religions, Second Edition*. Edited by Lindsay Jones, pp. 3116-3121. New York: McMillan, 2005.

Knipe, David M. 'One Fire, Three Fires, Five Fires: Vedic Symbols in Transition'. *History of Religions* 12, no. 1 (1972): pp. 28–41.

Knipe, David M. 'The Heroic Theft: Myths from Ṛigveda IV and the Ancient Near East'. *History of Religions* , no. 4 (1967): pp. 328–360.

Kramrisch, Stella. The Triple Structure of Creation in the Ṛg Veda. *History of Religions* 2, no. 2 (1963): pp. 256–285.

Krupp, Ed C. 'Facing the Sun'. *Griffith Observer* 54, no. 1 (1990): pp. 2–13.

Krupp, Ed C. 'Rayed Disks'. *Griffith Observer* 53, nos. 1 and 2 (1989): p. 2.

Kuhn, Adalbert. *Die Herabkunft des Feuers und des Göttertranks. Ein Beitrag zur vergleichenden Mythologie der Indogermanen*. Berlin: Ferdinand Dümmler, 1859.

Levi, Doro. 'Features and Continuity of Cretan Peak Cults'. In *Temples and High Places in Biblical Times*. Edited by A. Biran, pp. 38–46. Jerusalem: Nelson Glueck School of Biblical Archaeology, 1981.

Loewenthal, John, and Bruno Mattlatzki. 'Die europäischen Feuerbohrer'. *Zeitschrift für Ethnologie* 48, no. 6 (1916): pp. 349–369.

MacDonald, John. *The Arctic Sky. Inuit Astronomy, Star Lore, and Legend*. Ontario: Royal Ontario Museum and Nunavut Research Institute, 1998.

Maringer, Johannes. Das Feuer in Kult und Glauben der vorgeschichtlichen Menschen. *Anthropos* 69, nos. 1 and 2 (1974): pp. 68–112.

McCartney, Eugene S. 'Classical Weather Lore of Thunder and Lightning.' *The Classical Weekly* 25, no. 23 (1932): pp. 183–192.

McCrickard, Janet. *Eclipse of the Sun. An Investigation into the Sun and Moon Myths*. Glastonbury: Gothic Image Publications, 1990.

Metevelis, Peter. 'The Dog Star and the Multiple Suns Motif: An Asian Contribution to European Mythology.' *Asian Folklore Studies* 64, no. 1 (2005): pp. 133–137.

Molina, Anatilde Idoyaga. 'Fuego, purificación y metamorfosis. Significación y simbolismo ígneo entre los Pilagá (Chaco Central, Argentina)'. *Anthropos* 104, no. 1 (2009): pp. 113–129.

Needham, Joseph. *Science and Civilisation in China*, Vol. IV, 1. Cambridge: Cambridge University Press, 1996.

Panaino, Antonio. 'Uranographia Iranica II: Avestan hapta.srū- and mərəzu-: Ursa Minor and the North Pole?' *Archiv für Orientforschung* 42/43 (1995/1996): pp. 190-207.

Rappenglück, Michael. 'Cosmic Spinning and Weaving: Making the Texture of the World'. *BAR International Series* 1647 (2007): pp. 161–171.

Rappenglück, Michael A. *Eine Himmelskarte aus der Eiszeit?* Frankfurt a. M.: Peter Lang, 1999.

Rappenglück, Michael. 'Heavenly Messengers: The Role of Birds in the Cosmographies and the Cosmovisions of Ancient Cultures'. In *Cosmology across cultures*. Edited by J. A. Rubiño-Martín, J. A. Belmonte, F. Prada and A. Alberdi, pp. 145–150. Orem, UT: ASP, 2009.

Rappenglück, Michael. 'The Cosmic Deep Blue: The Significance of the Celestial Water World Sphere across Cultures'. *Mediterranean Archaeology and Archaeometry* 14, no. 3 (2014): pp. 293–305.

Rappenglück, Michael. 'The Housing of the World: The Significance of Cosmographic Concepts for Habitation'. *Nexus Network Journal* 15, no. 3 (2013): pp. 387–422.

Rappenglück, Michael. 'The pivot of the cosmos. The Concept of the world axis across cultures'. In *Cosmic catastro-*

phes. A collection of articles. Edited by Mare Kõiva, Izold Pustylnik, and Liisa Vesik, pp. 157–165. Tartu, 2005.

Rappenglück, Michael A. 'The Whole Cosmos Turns around the Polar Point: One Legged Polar Beings and their Meaning'. In *Astronomy and Cultural Diversity*, ed. César Esteban and Juan Belmonte Aviles, pp. 169-176. La Laguna, Tenerife: OACIMC, 1999.

Rappenglück, Michael A. 'The whole world put between to shells: The cosmic symbolism of tortoises and turtles'. *Mediterranean Archaeology and Archaeometry* 3 (2006): pp. 225–227.

Rappenglück, Michael. 'Tracing the Celestial Deer – An Ancient Motif and its Astronomical Interpretation Across Cultures'. In *Astronomy and Cosmology in Folk Traditions and Cultural Heritage*. Edited by Jonas Vaiškūnas, pp. 62–65, Klaipeda: Klaipeda University Press, 2008.

Read, Kay Almere. *Time and sacrifice in the Aztec cosmos*. Bloomington and Indianapolis, IN: Indiana University Press, 1998.

Roe, Peter G. *The Cosmic Zygote. Cosmology in the Amazon Basin*. New Brunswick, NJ: Rutgers University Press, 1982.

Scherer, Anton. *Gestirnnamen bei den indogermanischen Völkern*. Heidelberg: Carl Winter, 1953.

Schlegel, Birgit, and Kristian Schlegel. *Polarlichter zwischen Wunder und Wirklichkeit: Kulturgeschichte und Physik einer Himmelserscheinung*. Heidelberg: Spektrum, 2011.

Schove, D. Justin. 'Sunspots, Aurorae and Blood Rain: The Spectrum of Time'. *Isis* 42, no. 2 (1951): pp. 133–138.

Schrempp, Gregory. 'Catching Wrangham: On the Mythology and the Science of Fire, Cooking, and Becoming Human'. *Journal of Folklore Research* 48, no. 2 (2011): pp. 109–132.

Schwartz, W. 'Der (rothe) Sonnenphallos der Urzeit. Eine mythologisch-anthropologische Untersuchung'. *Zeitschrift für Ethnologie* 6 (1874): pp. 167–188.

Seidel, Anna, and Mar Kalinowski. 'Tokens of Immortality in Han Graves (With an Appendix)'. *Numen* 29, no. 1 (1982): pp. 79–122.

Seidenberg, Abraham. 'The Ritual Origin of the Circle and Square'. *Archive for History of Exact Sciences* 25, no. 4 (1981): pp. 269–327.

Seiffert, Beate. *Die Herkunft des Feuers in den Mythen der nordamerikanischen Indianer*. Bonn: Holos, 1990.

Staal, Julius D. W. *The New Pattern in the Sky. Myths and Legends of the Stars*. Blacksburg VA: McDonald and Woodward Publishing, 1988.

Stein, Rolf A. *Le monde en petit*. Paris: Flammarion, 2001.

Stothers, Richard. 'Ancient Aurorae'. *Isis* 70, no. 1 (1979): pp. 85–95.

Stothers, Richard. 'Ancient Meteorological Optics'. *The Classical Journal* 105, no. 1 (2009): pp. 27–42.

Ström, Åke V. 'Indogermanisches in der Völuspá'. *Numen* 14, no. 3 (1967): pp. 167–208.

Temple, Robert. *The Crystal Sun. Recovering a Lost Technology of the Ancient World*. London: Century, 2000.

Van der Sluijs, Marinus Anthony. 'Phaethon and the Great Year'. *Apeiron: A Journal for Ancient Philosophy and Science* 39, no. 1 (2006): pp. 57–90.

Van der Waerden, Bartel Leendert. 'Das Grosse Jahr und die Ewige Wiederkehr'. *Hermes* 80, no. 2 (1952): pp. 129–155.

Waters, Frank. *The book of the Hopi*. New York: Penguin, 1963.

Weinstock, Stefan. 'Libri Fulgurales'. *Papers of the British School at Rome (New Series Volume 6)* 19 (1951): pp. 122-153.

Wright, M. R. *Cosmology in Antiquity*. London and New York: Routledge, 1995.

ASTRONOMICAL ORIENTATIONS

ON THE ORIENTATION OF THE HISTORIC CHURCHES OF LANZAROTE: WHEN HUMAN NECESSITY DOMINATES OVER CANONICAL PRESCRIPTIONS

Alejandro Gangui, A. César González García, Mª Antonia Perera Betancort, and Juan Antonio Belmonte

ABSTRACT: The orientation of Christian churches is a well-known distinctive feature of their architecture. There is a general tendency to align their apses within the solar range of rising directions on the horizon, favouring orientations close to the east (astronomical equinox), although the alignments in the opposite direction, namely, with the apse towards the west, are not unusual. The case of the churches built in northwest Africa before the arrival of Islam is representative in this regard, and may reflect earlier traditions. As the Canary Islands are the western end of this North African common culture, it is relevant to study compact sets of churches located in the different islands of the archipelago. We started our project with Lanzarote, seeking for pre-European traditions – including astronomical ones – being merged with new ones as in other aspects of Canarian culture. We have measured the orientation of 30 churches built prior to 1810, as well as a few buildings of later times, nearly a complete sample of all the island's Christian sanctuaries. The analysis here indicates that a definite orientation pattern was followed on the island but, unlike what is often found in most of the Christian world, it has two interpretations. On the one hand, the representative orientation to the east (or west) is present. However, the sample has also a marked orientation towards north-northeast which is, as far as we know, a pattern exclusive to Lanzarote. We analyse the reasons for this pattern and suggest that one possible explanation could be a rather prosaic one, namely, that sometimes needs of everyday life are more relevant than – and push individuals to make decisions at odds with – religious beliefs. This work is the beginning of the first systematic archaeoastronomical study ever conducted with old churches in the Canary Islands.

Introduction

The study of the orientation of Christian churches has been of interest for long time and has recently received new impetus in the literature, as it was recognized that it represents a key feature of their architecture.[1] According to the texts of the early Christian writers and apologists, churches' apses should lie along a particular direction, that is, the priest had to stand facing eastward during services. This is recognized by Origen, Clement of Alexandria and Tertullian, and the first Council of Nicaea (325 CE) decreed that it should be that way. Also in the fourth century, St. Athanasius of Alexandria declared that the priest and participants should face east, where Christ, the Sun of Justice, would shine at the end of time (*'ecclesiarum situs plerumque talis erat, ut fideles facie altare versa orientem solem, symbolum Christi qui est sol iustitia et lux mundi interentur…'*). A thorough analysis of the early sources and methods of orientation can be found in the Bibliography.[2]

However, these requirements are ambiguous, making it possible to choose between different interpretations: should the church be oriented towards sunrise on the precise day its foundations were

---

[1] A. C. González-García and J. A. Belmonte, 'The orientation of pre-Romanesque churches in the Iberian Peninsula', *Nexus Network Journal* 17 (2015): pp. 353–377.

[2] C. Vogel, 'Sol aequinoctialis. Problèmes et technique de l'orientation dans le culte chretien', *Revue des Sciences Religieuses* 36 (1962): pp. 175–211; S. C. McCluskey, *Astronomies and cultures in early Medieval Europe* (Cambridge: Cambridge University Press, 1998).

prepared? Or at the sunrise of another day that may have been relevant, such as the feast day of the saint to whom the church was originally dedicated? Or was the orientation to be strictly towards the east? Were the churches oriented to sunrise at the equinox? In this case, toward which equinox? There are several possibilities:[3] the Roman vernal equinox occurred on March 25, while the Greek equinox was on March 21, as was reflected in the Council of Nicaea; but other definitions might also have been in use, such as the entrance of the Sun in the sign of Aries (which might be different from the canonical equinox as established by the church) or the autumnal equinox. From each of these definitions we would obtain dates, and therefore orientations, that are slightly different.[4]

North Africa, in spite of the Roman dominance, is an exception to the rule. In many regions of Africa, such as Proconsularis and Tripolitania, a number of churches with orientations towards the west – which is a usual custom in the early times of Christianity – are found.[5] These regions are relevant for our study, because they are possible homelands of the Canarian aboriginal population. Note also that most of these churches are oriented within the solar range (with orientations between the winter and summer solstices), with clustering around the equinoxes and solstices.

We started a large-scale project in the Iberian Peninsula and the Canary Islands. In the latter area, this work is the first such systematic study. Our interest is to check the orientation of the churches of Lanzarote, as it can provide us with a compact set of old churches where we can search for pre-European or canonical religious traditions, including astronomical ones, or a mix of both. This could provide a broader understanding of one key aspect of Canarian culture.[6]

Churches and chapels of Lanzarote – a description

Religious architecture on the island of Lanzarote began with the building of modest chapels endowed with a single room. In some, over time, small shrines or altars were added in their headers, together with vestries on their sides and other elements of practical use, such as low walls bordering the atrium (barbicans) and calvaries (Fig. 10.1). In general, these constructions were not subjected to strict building rules; thus, their structure was erected according to the needs of the moment. Nowadays, small chapels are in general located far from the cities, while some of the churches eventually achieved some monumental dimensions, like those in Teguise and Arrecife.

Given the large number of historical monuments, close to thirty and therefore suitable for a statistical analysis, Lanzarote was chosen as the first test area to study the orientation of the Canarian churches in the immediate post-conquest centuries (fifteenth century onwards). The research aim was to analyse whether in this location the orientation of the monuments was influenced by factors such as the presence of aboriginal people in the islands, who had worship habits and calendric systems completely different

---

[3] S. C. McCluskey, 'Astronomy, Time, and Churches in the Early Middle Ages', in *Villard's legacy: Studies in Medieval Technology, Science and Art in Memory of Jean Gimpel*, ed. M.-T. Zenner (Aldershot: Ashgate, 2004), pp. 197–210.

[4] C. L. N. Ruggles, 'Whose equinox?', *Archaeoastronomy, The Journal of Astronomy in Culture*, 22 (1999): pp. S45–50; González-García, A.C. and J.A. Belmonte, 'Which Equinox?', *Archaeoastronomy, The Journal of Astronomy in Culture* 20 (2006): pp. 97–107.

[5] C. Esteban, J. A. Belmonte, M. A. Perera Betancort, R. Marrero, and J. J. Jiménez González, 'Orientations of pre-Islamic temples in North-West Africa', *Archaeoastronomy, The Journal of Astronomy in Culture* 26 (2001): pp. S65–84; J. A. Belmonte, A. Tejera, M. A. Perera, and R. Marrero. 'On the orientation of pre-Islamic temples of North-west Africa: a reappraisal. New data in Africa Proconsularis', *Mediterranean Archaeology and Archaeometry* 6, no. 3 (2007): pp. 77–85; J. A. Belmonte, M. A. Perera Betancort, and A. C. González-García, 'Análisis estadístico y estudio genético de la escritura líbico-bereber de Canarias y el norte de África', in the Proceedings of the *VII Congreso de patrimonio histórico: inscripciones rupestres y poblamiento del Archipiélago Canario*, held in Arrecife, Lanzarote, 6–8 October 2010), in press.

[6] J. A. Belmonte and M. Sanz de Lara, *El Cielo de los Magos* (Islas Canarias: La Marea, 2001).

Fig. 10.1: Two churches in Lanzarote with unique features: (a) the church of Nuestra Señora de las Mercedes (Our Lady of Mercy) in Mala is the only one on the island with a precise equinoctial orientation (nearly 90° of azimuth), and that may also be oriented to the solar sunrise on the day of its Marian devotion (September 24), a not very frequent feature in the island environment; (b) the chapel of San Rafael, located alone and isolated on a plateau overlooking El Jable on the outskirts of the village of Teguise; it dates from the nineteenth century but it is already cited in 1661. Its orientation is eastward (72°.5) and has the peculiarity that it possesses a large 'L'-shaped barbican that protects the entrance from the prevailing winds, as we can infer from the sand that is seen deposited on it.

from the newly arrived people.[7]

Table 10.1 shows the data obtained in our campaign of fieldwork. The identification of the churches and their coordinates are presented along with their orientation (archaeoastronomical data): the measured azimuth (rounded to 1/2° approximation) and the angular height of the point of the horizon towards which the apse of the church is facing, as well as the corresponding computed declination. Regarding the construction dates provided for the churches, there is some ambiguity: in general, the year is only estimated and it corresponds to the first reported reference of the building (and for one church it has not been clarified). This dating can be useful for our later analysis of the data.

We obtained our measurements using a pair of tandem instruments which incorporate a clinometer and a compass with a precision of half a degree, and also by analysing the landscape setting of each of the buildings (see Fig. 10.2). We then corrected the data according to the local magnetic declination (geomag.nrcan.gc.ca/calc/mdcal-en.php). Our values of the magnetic variation for different sites on the island range from 4°38′ to 4°46′ W. As the precision of the measured magnetic azimuths is about 1/2°, we have used the same magnetic declination (4°42′ W) for all 32 values of orientation (this may be considered a limitation in precision but it simplifies the calculation). The obtained values are the average of two and sometimes even three measurements, and we must emphasize that, with few exceptions, the various measurements differed by less than 1/2°. In any case, given the magnetic disturbances recorded at various locations on the island (especially in the vicinity of the volcanic eruptions of the Timanfaya), some measures (one fifth of the total) have been verified with photo satellite images, and we found few differences. Therefore, we estimate the error of our measurements to be around ±3/4° (upper bound), and thus the data is suitable for a statistical study of the monuments' orientation.

---

[7] J. A. Belmonte, C. Estéban, A. Aparicio, A. Tejera Gaspar, and O. Gónzalez, 'Canarian Astronomy before the conquest: the pre-hispanic calendar', *Revista de la Academia Canaria de Ciencias* 6, nos. 2–3–4 (1994): pp. 133–156.

LOCATION	NAME / DATE	L (°/′) NORTH	l (°/′) WEST	a (°)	h (°)	δ (°)	PATRON SAINT DATE / Orientation
(1) Femés	San Marcial de Limoges (1630)	28/54	13/46	52½	1¼	32¾	16 Abril - 8 Julio / ----
(2) Yaiza	Ntra. Sra. de los Remedios (1699)	28/57	13/46	77½	2¾	12	vv.ff. 9 Sept / 21 Abr – 22 Ago
(3) Uga	San Isidro (1956)	28/57	13/44	15	3½	60	15 Mayo / -----
(4) Masdache	La Magdalena (s. XX)	28/59	13/39	339½	-0½	54¼	22 Julio/ -----
(5) La Geria	Ntra. Sra. de la Caridad (1706)	28/58	13/43	44½	0½	38¾	8 Sept R 15 Agos / -----
(6) Mancha Blanca	Ntra. Sra. de los Dolores (1782)	29/02	13/41	92	1¾	-1¼	15 Sept / 17 Mar – 26 Sept
(7) Tinajo	San Roque (1669)	29/04	13/40	34½	-0¾	45¼	16 Agos R/ -----
(8) Yuco	Ntra. Sra. de Regla (1663)	29/03	13/39	356½	-1¼	59	vv.ff. / -----
(9) Tiagua	Ntra. Sra. Del Socorro (1625)	29/03	13/38	12	-1¼	57¼	8 Sept / -----
(10) Sóo	San Juan Evangelista (1749)	29/05	13/37	142	0¾	-43½	27 Dic / -----
(11) Tao	San Andrés (1627)	29/02	13/37	6½	-1	58¾	30 Nov / -----
(12) Mozaga	Ntra. Sra. de la Peña (1785)	29/01	13/36	278	1¼	7½	R 8-13 Agos / 8 Abr – 4 Sept
(13) San Bartolomé	San Bartolomé (1661)	29/00	13/36	26½	0 B	51	24 Agos R / -----
(14) Nazaret	Ntra. Sra. de Nazaret (1648)	29/02	13/34	105	1¾	-12¼	R 26 Agos / 17 Feb – 25 Oct
(15) Teguise	San Rafael (1661)	29/04	13/34	72½	1¾	15¾	29 Sept / 4 May – 9 Ago
(16) Teguise	El Cristo de la Vera Cruz (1625)	29/04	13/33	82	0 B	6¾	3 Mayo / 7 Abr – 5 Sept
(17) Teguise	San Juan de Dios y San Fco. de Paula (Sto. Domingo) (1698)	29/03	13/34	254½	0½	-13¾	8 Mar – 2 Abr / 11 Feb – 1 Nov
(18) Teguise	Ntra. Sra. de Guadalupe (1680)	29/04	13/34	128½	5½	-30¼	6 Sept / -----
(19) Teguise	Ntra. Sra. Miraflores. Convento de San Francisco (1588)	29/03	13/33	84	6½	8	? – 4 Oct / 10 Abr – 2 Sept
(20) Teseguite	San Leandro (1674)	29/03	13/32	71	0¾	16½	13 Nov / 6 May – 6 Ago
(21) El Mojón	San Sebastián (1661)	29/04	13/31	42½	0¾	40¼	20 Ene / -----
(22) Los Valles	Santa Catalina (1749)	29/05	13/31	339½	14¼	65¼	25 Nov – 20 Abr / ----
(23) La Montaña (Teguise)	Ntra. Sra. de las Nieves (1661)	29/06	13/32	15½	-1	56	5 Agos / -----
(24) Haria	San Juan (1625)	29/09	13/30	98½	-0½	-7¾	24 Jun / 1 Mar – 12 Oct
(25) Haria*	Ntra. Sra. de Encarnación (1631)	29/09	13/30	74	4½	15¾	25 Mar / 4 May – 9 Ago
(26) Mala	Ntra. Sra. de las Mercedes (1809)	29/06	13/28	89½	-0½	0	24 Sept / 20 Mar – 23 Sep Equinoccio
(27) Guatiza	El Cristo de las Aguas (1915)	29/04	13/29	107½	1	-15	R 13 Sept / 8 Feb – 3 Nov
(28) Arrecife	San Ginés (1570)	28/57	13/33	53½	0 B	31¼	R 25 Agos / ----
(29) Tahiche	Santiago Apóstol (1779)	29/01	13/33	70	12½	23	25 Jul / 11 Jun – 2 Jul
(30) Tías	Ntra. Sra. de la Candelaria (1795)	28/58	13/39	323½	8¾	49¾	2 Feb / -----
(31) Conil	María Magdalena (1794)	28/59	13/40	118½	12	-18½	22 Jul / 27 Ene – 15 Nov
(32) Tegoyo	Sagrado Corazón de Jesús (1863)	28/58	13/41	52½	9¼	36¾	Móvil. Junio / ----

Table 10.1: Table showing orientations for the chapels and churches of Lanzarote. For each building, we show the location, identification (name and most likely date of construction), the geographical latitude and longitude (L and l), the astronomical azimuth (a) taken along the axis of the building towards the apse (rounded to 1/2° approximation), the horizon angular height (h) in that direction (0 B means horizon is blocked; we take h = 0°) and the corresponding resultant declination (δ). Some of the churches were surrounded by nearby hills; this justifies the high value of angular height (h) for them.

In Figure 10.2(a), we show the orientation diagram for the churches and chapels. The diagonal lines on the graph indicate, in the eastern quadrant, the extreme values of the corresponding azimuth for the sun (azimuths of 62°.5 and 116°.7 –continuous lines – equivalent to the northern hemisphere summer and winter solstices, respectively) and for the moon (azimuths of 56°.4 and 123°.8 – dotted lines – equivalent to the position of the major lunistices).

Of the 32 chapels and churches we measured, 12 are oriented in the northern quadrant (between 315° and 45°), 2 in the western quadrant, 17 are oriented in the eastern quadrant (with 13 of them in the solar range), and only 1 in the southern quadrant; see Figure 10.2(a). The sample is representative of the island of Lanzarote (although it is not of all the Canary Archipelago), and here two distinct orientations are distinguished: (i) to the North, with 'entrance' on the leeward/downwind side, avoiding perhaps the dominant winds of the place, and (ii) eastward, with the apse of the chapel pointing toward the eastern quadrant. The 13 monuments facing *ad orientem* fall within the logic observed in other studies of orientations of churches, but what is remarkable here is the large number of monuments oriented to the northern quadrant, falling outside of the solar range. It seems to be a case singular of Lanzarote where practical and prosaic issues (the orientation against the trade winds from the NNE, see Figure 10.2(b)) appear side by side with cultic and canonical traditions (i.e., the orientations within the solar range).

While there might be different underlying causal factors for this church orientation pattern, the idea that they may be astronomically oriented is suggestive. Regarding the solar range, there are two particularities. On the one hand we have the mother church of the historic capital city of the island, Teguise, oriented

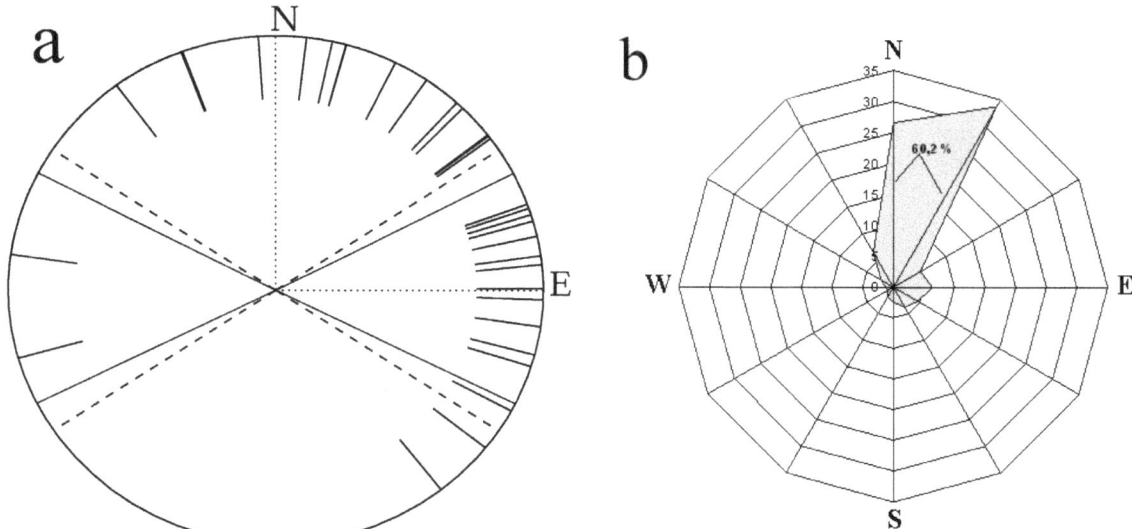

Fig. 10.2: (a) Orientation diagram for the churches and chapels of Lanzarote, obtained from the data in Table 10.1. Although a significant number of monuments follows the canonical orientation pattern in the solar range, a non-negligible number of churches qre oriented to the north-northeast. (b) Diagram of winds for Arrecife Airport in Lanzarote, illustrative of the prevailing winds on the island. Note the enormous concentration in the range N-NE (azimuths between 345° and 45°), similar to the exceptional orientations of several churches on the island.

with an azimuth 128°.5, roughly five degrees from the direction (123°.8) of the southernmost rising point of the moon (an orientation already found, in the aboriginal world, in some paradigmatic cases like in the Roque Bentayga, in the island of Gran Canaria, or in the Mount Tindaya, in the island of Fuerteventura).[8] Secondly, the church of Nuestra Señora de las Mercedes (Our Lady of Mercy) in Mala (Fig. 10.1), is the only one on the island precisely oriented towards the equinox, which also seems to follow a uncommon rule on the island, namely to be oriented to the sunrise on the date of its devotion, as we see in Table 10.1. (The church Nuestra Señora de los Dolores – Our Lady of Sorrows – in Mancha Blanca, also seems to be equinoctial, but with more discrepancy.) In this sense, it is also to be noted that most of the churches of Teguise – where most of the initial settlers established – are oriented (with the notable exception of the mother church of Guadalupe, with its 'anomalous' orientation, as stated above) to declinations included in the canonical range. While the equinox was also important in the aboriginal world,[9] it seems that in the site which the majority of the European population selected for settlement the rules/habits from their places of origin were respected.

To better understand the above discussion, in Figure 10.3 we present the declination histogram, which is independent of the geographical location and the local topography. This shows the astronomical declination versus the normalized relative frequency, which enables a clear and more accurate determination of the structure of the peaks. Again, the peak associated with the orientations to the north-northeast, absolutely outstanding, dominates the chart.

---

[8] J. A. Belmonte and M. Hoskin, *Reflejo del Cosmos: atlas de arqueoastronomía en el Mediterráneo occidental* (Madrid: Equipo Sirius, 2002).

[9] J. A. Belmonte., C. Estéban, R. Schlueter, M.A. Perera Betancort, and O. González, 'Marcadores equinocciales en la prehistoria de Canarias', *IAC Noticias (Instituto de Astrofísica de Canarias)* 4 (1995): pp. 8–12.

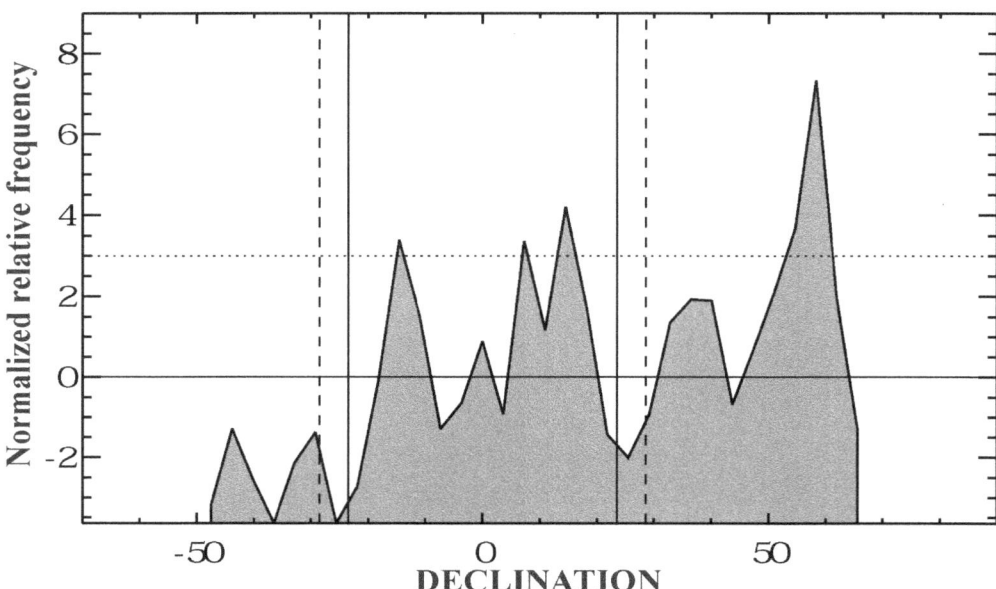

Fig. 10.3: Declination histogram for the chapels and churches of Lanzarote. Only a few statistically significant peaks are found above the 3σ level (dotted horizontal line). Continuous vertical lines represent declinations corresponding to the extreme positions of the sun at the solstices, while the dashed vertical lines represent the same for the moon in major lunistices. Three statistically significant minor peaks are found in the solar range (canonical orientation). However, the highest peak, located around 58° is exceptional and is certainly associated with a accumulation peak due to orientations near the meridian line.

When we encountered this phenomenon in the data, we pondered over the possible causes, establishing different interpretations: could it be the result of the influence of previous aboriginal traditions?[10] Was this related to the settlement on the island of Moorish slaves imported from the nearby African coast and of Islamic tradition?[11] However, it turned out to be that the most prosaic explanation was the one that hinted at the most plausible solution, as we will see in the next section.[12]

The orientation of churches and landscape

The particular features of the churches we measured in Lanzarote have little correlation with other studies already mentioned in the previous sections. In the Iberian Peninsula and overall in the Mediterranean,

---

[10] Belmonte and Hoskin, *Reflejo del Cosmos*, 2002.
[11] L. A. Anaya Hernández 'El Corso Magrebí y Canarias. El último ataque berberisco a las islas: la incursión a Lanzarote de 1749', in *Actas X Jornadas de Estudios sobre Lanzarote y Fuerteventura* (Santa Cruz de Tenerife: Servicio de Publicaciones del Excmo. Cabildo Insular de Lanzarote, Litografía Romero, 2004), Vol. 1, pp. 13–29; L. A. Anaya Hernández, 'La liberación de cautivos de Lanzarote y Fuerteventura por las Órdenes Redentoras', in *Actas XII Jornadas de Estudios sobre Lanzarote y Fuerteventura* , Vol. 1, no. 1, pp. 65–93.
[12] J. de León Hernández, M. A. Robayna Fernández, and M. A. Perera Betancort, 'Aspectos arqueológicos y etnográficos de la comarca del Jable', in *Actas II Jornadas de Historia de Lanzarote y Fuerteventura* (Santa Cruz de Tenerife: Servicio de Publicaciones del Excmo. Cabildo Insular de Lanzarote, 1990), Vol. 2, pp. 283–319.

the orientation ranges are predominantly within the solar limits.[13] In particular, the important proportion of churches oriented roughly northward we found is new: examples of this are the church Nuestra Señora de las Nieves (Our Lady of the Snows) in La Montaña, patron saint of the island, and also many others, very old ones, such as those located at Tiagua or Tao. It is noteworthy that a significant proportion of churches oriented in this way belong to the northwestern and central sectors of the island, as shown in Figure 10.4.

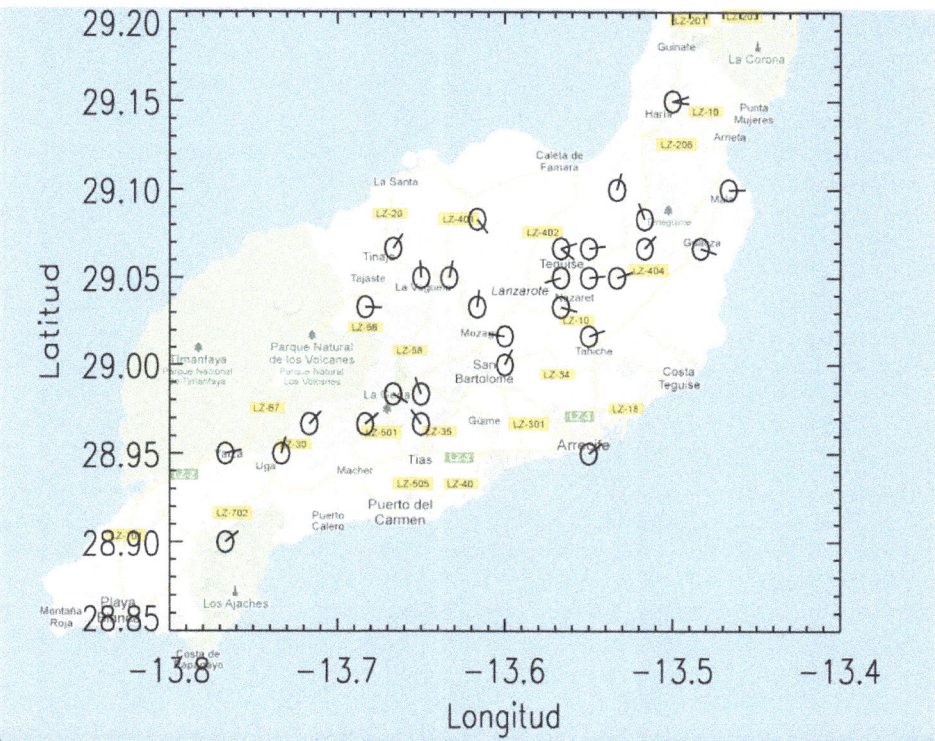

Fig. 10.4: Map showing the geographical location of all the measured churches (indicated by ellipses), together with the orientation of the axis of the buildings towards the apse (oriented according to the azimuths recorded in Table 10.1). In the town of Haría, two geographically very close churches have different orientations, which explains the presence there of a single ellipse with two stripes. The same happens with a couple of churches in the town of Teguise. Image based on a map courtesy of Google Maps.

The difference between our results for Lanzarote, and those of other studies that have been completed elsewhere,[14] leads us to look for alternatives in trying to understand the pattern of orientations of the churches of this island. If these monuments, in general, are not oriented according to the sun, could it be due to such prosaic reasons as the need to orient the porch of the constructions contrary to the dominant winds coming from the NE direction onto the island (as in Fig. 10.1) or otherwise protect it? Or else, could it be due to the topography (perhaps changing over time) of different regions of the island? In any case, it

---

[13] González-García and Belmonte, 'The orientation of pre-Romanesque churches in the Iberian Peninsula'.
[14] S. Čaval, 'Church orientations in Slovenia', in *Handbook of Archaeoastronomy and Ethnoastronomy*, ed. C. L. N. Ruggles, (New York: Springer, 2014), pp. 1719–1726.

seems clear by looking at the graphics and images that the environmental issue is relevant.

Regarding the winds, the areas where more churches facing north-northeast have been built (with their entrance oriented towards the southern quadrant) is on the verge of El Jable (north and centre of the island), where it becomes imperative to avoid the sand driven by the wind, sometimes in raging storms as that of 1824 that buried several villages, and that even today, despite the changes in the landscape, shows its lasting effects.[15] Interestingly, the highest number of canonical orientations (i.e., eastward) is found in buildings located in the northeast of the island, in the lee of the wind, in areas protected of the sand by the cliffs of Famara.

In relation to the changing topography of the island, the volcanic eruptions of the Timanfaya that occurred between 1730 and 1736, perhaps also may have played a role. For example, the chapel of St. Catherine in Los Valles, shows a construction of the XVIII century with a peculiar orientation, which replaced a previous one dedicated to the same cult, which had been destroyed during the Timanfaya lavas. The current chapel is oriented with azimuth 339°.5, i.e. in the NNW direction, but is protected by the mountains that surround it, so it may be considered anomalous from any point of view (it would be interesting to know the orientation of the early church). In any case, these ideas seem attractive, as they suggest that human needs can sometimes override obligations of the cult.

To check any noticeable evolution in the characteristics of the buildings over time, we conducted a study on the evolution of the orientation with respect to building age. In Figure 10.5 we show the values of azimuth and declination versus the probable dates of construction for the churches: 6(a) includes all of the oldest churches for which we know these dates (28 of the 32 in the sample); 6(b) selects those churches oriented within the solar range, i.e., whose azimuths are located between the two horizontal lines (azimuths 62°.5 and 116°.7).

The third panel (c) shows the declination values as a function of date (with horizontal lines at declinations -23°.5 and +23°.5). We note that building orientations stay in the canonical range throughout the entire period; but starting in the second quarter of the seventeenth century, large-scale construction of north-facing chapels began, perhaps in a time when human needs and associated environmental factors far exceeded those of worship. The chart also reflects eruptions on Timanfaya, since the construction of chapels almost ceases in the first half of the eighteenth century, reactivating soon after, but again showing the characteristic pattern of two representative orientations.

Conclusions

After the conquest and colonization of Lanzarote by the European population in the early fifteenth century, in the decades immediately following it began the large-scale colonization of the island with the establishment of small farms and villages, together with some larger towns like Teguise or Femés. This was accompanied by the construction of a non-negligible number of Christian churches reflecting the new social and religious situation.

It is possible that in a few places, the orientation of sacred buildings followed the pattern of the aboriginal cult, which was centred in the celestial sphere, notably on the sun and the moon. In others, the canonical tradition of aligning the temples eastward was respected (with some exceptions to the western quadrant), although with a much greater degree of tolerance than was usual. In this regard, only one church in Lanzarote, the one in Mala, appears to have an orientation consistent with the solar sunrise on the day of the (Marian) dedication (Fig. 10.1).

---

[15] M. A. Perera Betancort, 'Aportación al problema de El Jable a principios del siglo XIX', in *Actas X Jornadas de Estudios sobre Lanzarote y Fuerteventura* (Santa Cruz de Tenerife: Servicio de Publicaciones del Excmo. Cabildo Insular de Lanzarote, Litografía Romero, 2004), Vol. 1, pp. 205–212.

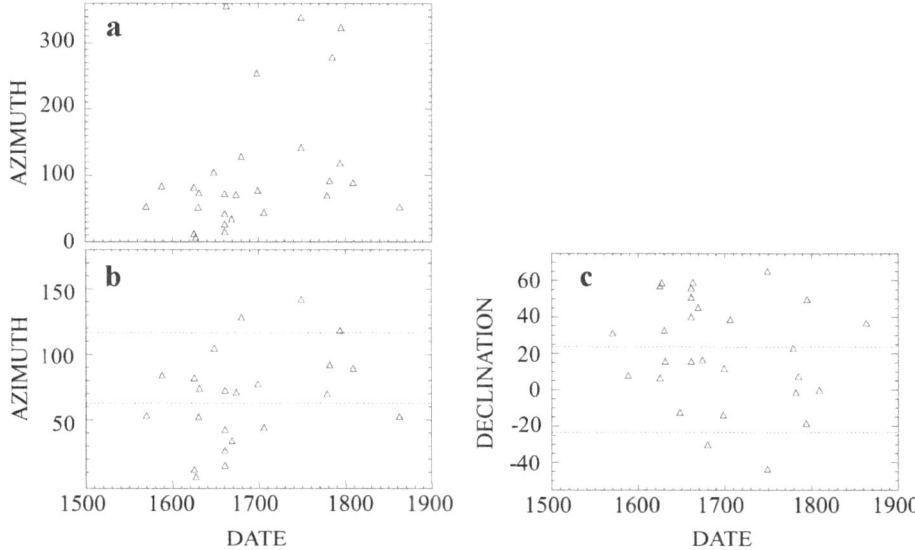

Fig. 10.5: (a) Diagram showing the azimuth of the churches and chapels versus the most probable date of construction (or its first mention in the sources). We include 28 out of the 32 buildings studied. (b) Same as the above diagram, but with extended area close to the solar azimuth range. Except for those buildings with a clear orientation towards the N or NE, a good proportion of churches fall within the solar range, as indicated by the two parallel lines. (c) Declination diagram versus the date. The horizontal lines indicate the extreme declinations where the sun is located at solstices. Note the gap in the construction of chapels possibly associated with eruptions of Timanfaya in the first half of the eighteenth century.

Finally, we find a relatively large number of north-northeast oriented churches in Lanzarote, which is a notable exception to the rule. We analysed different possibilities to explain this anomaly, and concluded that this pattern of orientation appears to reflect the desire to avoid the strong winds prevailing on the island, which come precisely from that direction, and in particular to minimize the discomfort caused by the sand displaced by the wind on buildings located close or bordering with El Jable.

This is the first stage of a project which will be developed in the coming years, namely measuring the orientation of the oldest Christian churches and chapels in other islands of the Canary Archipelago. For example, it would be interesting to study the orientation of old churches in the island of Fuerteventura, subjected to the same wind flow – even more intense – than that blowing on the neighbouring island of Lanzarote. Will the churches of Fuerteventura also show the same pattern of orientations? Have their builders dared to violate the canonical precept to impose the human needs over the cult? Future work will help us to answer this question.

Bibliography

Anaya Hernández, L. A. 'El Corso Magrebí y Canarias. El último ataque berberisco a las islas: la incursión a Lanzarote de 1749'. In *Actas X Jornadas de Estudios sobre Lanzarote y Fuerteventur*, Vol. 1, pp. 13–29. Santa Cruz de Tenerife: Servicio de Publicaciones del Excmo. Cabildo Insular de Lanzarote, Litografía Romero, 2004.

Anaya Hernández, L. A. 'La liberación de cautivos de Lanzarote y Fuerteventura por las Órdenes Redentoras'. In *Actas XII Jornadas de Estudios sobre Lanzarote y Fuerteventura* (Santa Cruz de Tenerife: Servicio de Publicaciones del

Excmo. Cabildo Insular de Lanzarote, Litografía Romero, 2008), Vol. 1, no. 1, pp. 65–93.
Belmonte, J. A., and M. Hoskin. *Reflejo del Cosmos: atlas de arqueoastronomía en el Mediterráneo occidental*. Madrid: Equipo Sirius, 2002.
Belmonte, J. A., C. Estéban, A. Aparicio, A. Tejera Gaspar, and O. Gónzalez. 'Canarian Astronomy before the conquest: the pre-hispanic calendar'. *Revista de la Academia Canaria de Ciencias* 6, nos. 2–3–4 (1994): pp. 133–156.
Belmonte, J. A., C. Estéban, R. Schlueter, M.A. Perera Betancort, and O. González. 'Marcadores equinocciales en la prehistoria de Canarias'. *IAC Noticias (Instituto de Astrofísica de Canarias)* 4 (1995): pp. 8–12.
Belmonte, J. A., M. A. Perera Betancort, and A. C. González-García. 'Análisis estadístico y estudio genético de la escritura líbico-bereber de Canarias y el norte de África'. In the Proceedings of the *VII Congreso de patrimonio histórico: inscripciones rupestres y poblamiento del Archipiélago Canario*, held in Arrecife, Lanzarote, 6–8 October 2010, in press.
Belmonte, J. A., A. Tejera, M. A. Perera, and R. Marrero. 'On the orientation of pre-Islamic temples of North-west Africa: a reappraisal. New data in Africa Proconsularis'. *Mediterranean Archaeology and Archaeometry* 6, no. 3 (2007): pp. 77–85.
Belmonte, J. A., and M. Sanz de Lara. *El Cielo de los Magos*. Islas Canarias: La Marea, 2001.
Čaval, S. 'Church orientations in Slovenia'. In *Handbook of Archaeoastronomy and Ethnoastronomy*, edited by C. L. N. Ruggles, pp. 1719–1726. New York: Springer, 2014.
Esteban, C., J. A. Belmonte, M. A. Perera Betancort, R. Marrero, and J. J. Jiménez González. 'Orientations of pre-Islamic temples in North-West Africa'. *Archaeoastronomy, The Journal of Astronomy in Culture* 26 (2001): pp. S65–84.
González-García, A. C., and J. A. Belmonte. 'Which Equinox?'. *Archaeo-astronomy, The Journal of Astronomy in Culture*. 20 (2006),:pp. 97–107.
González-García, A. C., and J.A. Belmonte. 'The orientation of pre-Romanesque churches in the Iberian Peninsula'. *Nexus Network Journal* 17 (2015): pp. 353–377.
León Hernández, J. de, M. A. Robayna Fernández, and M. A. Perera Betancort. 'Aspectos arqueológicos y etnográficos de la comarca del Jable'. In *Actas II Jornadas de Historia de Lanzarote y Fuerteventur*, Vol. 2, pp. 283–319. Santa Cruz de Tenerife: Servicio de Publicaciones del Excmo. Cabildo Insular de Lanzarote, 1990.
McCluskey, S. C. *Astronomies and cultures in early Medieval Europe*. Cambridge: Cambridge University Press, 1998.
McCluskey, S. C. 'Astronomy, Time, and Churches in the Early Middle Ages'. In *Villard's legacy: Studies in Medieval Technology, Science and Art in Memory of Jean Gimpel*, edited by M.-T. Zenner. pp. 197–210 (Aldershot: Ashgate, 2004).
Perera Betancort, M. A. 'Aportación al problema de El Jable a principios del siglo XIX'. In *Actas X Jornadas de Estudios sobre Lanzarote y Fuerteventura*, Vol. 1, pp. 205–212. Santa Cruz de Tenerife: Servicio de Publicaciones del Excmo. Cabildo Insular de Lanzarote, Litografía Romero, 2004.
Ruggles, C. L. N. 'Whose equinox?'. *Archaeoastronomy, The Journal of Astronomy in Culture* 22 (1999): pp. S45–50.
Vogel, C. 'Sol aequinoctialis. Problèmes et technique de l'orientation dans le culte chretien'. *Revue des Sciences Religieuses* 36 (1962): pp. 175–211.

ORIENTATION OF ROMAN CAMPS AND FORTS IN BRITANNIA: RECONSIDERING ALAN RICHARDSON'S WORK

Andrea Rodríguez-Antón,[1] Antonio César González-García[2] and Juan Antonio Belmonte[3]

ABSTRACT: At the time of establishing a new settlement Romans needed to decide the right location and layout previously to its building, which normally required the presence of a surveyor.[4] There are several ancient treatises about how Roman cities and military settlements had to be designed, some of them taking into consideration possible astronomical orientations.[5], Alan Richardson presented a number of orientations, measured by a protractor, on published archaeological plans of Roman military settlements in Britain, Masada and few in Spain.[6] He proposed that, since some angles are more common than others and more than could be expected from a random distribution, the null hypothesis (no tendency to cluster significantly in any particular direction) must be rejected. This result and the statistical method followed by him have been disputed by other authors.[7] In the present work, we will discuss the results of our analysis of Richardson's data. We have studied the orientation of 93 Roman camps and forts in modern-day Britain. We have also complemented the data acquisition by estimating the angular altitudes of the horizon at each site by using a reconstruction of the horizon given by a digital elevation model, *Hey What's That?*, thus also considering the local topography.[8]

1. Introduction

In his article published in 2005, Alan Richardson presented a statistical study of the orientations of a great number of Roman temporary camps and forts in Britain, Spain and the Judea Desert, measured on archaeological plans. This work generated great controversy, both for its results and the method that he employed, leading other authors, such as Peterson, to review his work and publish an article where he questioned the procedure followed by Richardson. Britannia was a province with an important and constant presence of the Roman army due to the continuous struggles against the British tribes, so it is possible to find a wide sample of military settlements. From this point, it is interesting to reconsider these previous works and try to shed light on this topic in an attempt to discern if the election of the orientations was deliberate and to determine what reason(s), possibly cultural or political, may lie behind this. Remains of warfare activity are quite visible, mainly in northern Britain, where two main frontiers were established: Hadrian's Wall and the Antonine Wall. In general, legionary fortresses were shaped like

---

[1] Instituto de Astrofísica de Canarias, La Laguna, Tenerife, Spain; Departamento de Astrofísica, Universidad de La Laguna, Tenerife, Spain.
[2] Instituto de Ciencias del Patrimonio, Incipit, CSIC, Santiago de Compostela, Spain.
[3] Instituto de Astrofísica de Canarias, La Laguna, Tenerife, Spain; Departamento de Astrofísica, Universidad de La Laguna, Tenerife, Spain.
[4] C. M. Gilliver, *The Roman art of war* (Stroud: Tempus, 2001), pp. 68–88.
[5] Carl Thulin, *Corpus agrimensorum romanorum. Opuscula Agrimensorum Veterum* (Lipsia: Teubner, 1913), Vol. I, Fasc. I; Vegetius, *Epitome of military Science*, N.P. Pilmer (Liverpòol: Liverpool University Press, 1996), Book I, p. 23.
[6] Alan Richardson, 'The Orientation of Roman Camps and Forts', *Oxford Journal of Archaeology* 24, no.4 (2005): pp. 415–26.
[7] J. M. W. Peterson, 'Random Orientation of Roman Camps', *Oxford Journal of Archaeology* 26, no.1 (2007): pp. 103–8; http://alunsalt.com/2007/02/17/vidi-7/ [Accessed 10 December 2015].
[8] http://www.heywhatsthat.com/ [Accessed 10 December 2015].

a playing card, rectangular with rounded angles, surrounded by a rampart and one or more defensive ditches.[1] There were two main streets: the *Vía Principalis,* which connected the two *portae principalis,* and the *Via Praetoria,* which met in a T-junction at the *groma* and passed through *Porta Praetoria* (the main front gate). The *Via Decumana,* a secondary road, entered through the rear gate *Porta Decumana*.[9] These main streets would be equivalent to the *cardus* and *decumanus maximus* in a Roman city as being the main axes. At the intersection of these two streets was where the headquarters (*Principia*) and the house of the commanding officer (*Praetorium*) were located, surrounded by public buildings. Camps in Britain were generally based on a similar model to the permanent forts, though their temporary character made them less complex.[10] For instance, they would have had leather tents instead of buildings, but the street plan would be rectangular rather than square (a very common proportion being 3:2)[1]. Surveyors would tailor the design according to the size of the unit using whole number proportions, but there are also examples of square and irregular marching camps since terrain conditions were a substantial factor to deal with at the time of the construction.[11]

In the present work, we attempt to complete and implement Richardson's and subsequent works by taking into account the local topography by means of the measurement of the angular altitude in the horizon with a digital elevation model. We do not overlook the limited precision acquired with these tools and acknowledge that the data obtained with them will never be as accurate as those obtained *in situ*. That is the reason why we must consider these results as preliminary in advance of a more comprehensive one, with a fieldwork campaign in order to broaden the sample and to obtain more accurate data. We must bear in mind that even if a new military settlement had an offensive or a defensive aim, we should not underestimate the role of the religion and the liturgical activities in the daily life of Roman soldiers, even with political purposes, and the fact that, in most cases, these military settlements represented the power of Rome in conquered lands as much as the colonies did.[12]

2. Religion in the Roman army

As we have said before, Britain provides a body of information on military architecture, so it is therefore natural to turn to the evidence from Britannia to better understand the religion practiced by Roman soldiers. Apart from the reverence of the Roman army for ensigns or standards, known as *signa militaria*, reliefs and altars dedicated to different deities have been found in most of the military settlements, reminding us that the Roman army kept the most important ceremonies of the Roman religious year as long as the pagan Empire lasted.[13] The existence of an official calendar such as *Feriale Duranum,* where the days of the festivals for the different deities were dictated, could be seen as a liturgical link between the army, spread along the Empire, and the *Urbs* by the simultaneous celebration[14]. There also existed other non-official cults, due to the tendency the Romans had for importing divinities from settled lands or assimilating them with Roman ones. In Lancaster, for example, the Celtic god *Cocidius* was endowed

---

[9] M. C. Bishop, *Handbook to Roman Legionary Fortresses* (Barnsley: Pen & Sword, 2012).
[10] R. H. Jones, *Roman camps in Britain* (Stroud: Amberley Publishing, 2012)
[11] Humphrey Welfare and Vivien Swan, *Roman camps in England: The Field Archaeology,* (London Royal Commission on Historic Monuments, 1995).
[12] R. Laurence, C. Simon Esmonde, and G. Sears, *The City in the Roman West c. 250 BC- c. AD 250* (Cambridge: Cambridge University Press, 2011).
[13] Tertullian. *Ad nationes,* Q.Howe (www.tertullian.org, 2007), 1.12.
[14] S. O. Flink, S. A. Hoey, and W. F. Snyder, *The Feriale Duranum,* Yales Classical Studies 7 (Princeton, NJ: Yale University Press, 1940); Pat Southern, *The Roman Army: A Social and Institutional History* (Santa Barbara, CA: ABC-CLIO, 2006).

with the attributes of Mars, becoming Mars *Cocidius*.[15] Some British deities assumed by the Romans included *Belatucadrus*, *Conventia* or *Antenociticus*, who had a Roman temple in his honour at Benwell, whose remains are still visible.[16]

Another factor to bear in mind is that, at the time of establishing or re-founded a new settlement, Romans performed a number of sacred rituals to consecrate the space and prepare it for use.[17] In the most common rite, we can find some elements performed by Romulus in the foundation of Rome.[18] During its development, one of the first steps was done by the augur, who looked for auspicious signs in the sky to determine if the project was appropriate or not (*spectio*) and defined the *templum*, a terrestrial image of the heavens where two main perpendicular axes dividing the celestial sphere were traced. It was in that place where the ceremony took place and where he could have a proper view of the landscape to establish the limits. It was followed by the ploughing of the *pomerium*, or sacred enclosure where the town or camp would be erected. A pit named *mundus* where offerings were thrown would be dug and, supposedly, near it the *agrimensores* placed the *groma*, an instrument that these ancient topographers employed to determine perpendicular directions. It could be used to define the direction of the *decumanus*, perhaps according to the way of the sun, and the *cardus*, perpendicular to it.[19]

At Pen Llystyn, a hole dug near the *Principia* has been found containing carbonized materials.[20]. This is the typical ritual followed in the foundation of a city and we cannot assert whether it was performed in the army, but the religious factor was also present at the time of establishing a new military settlement, as it can be observed in some scenes of the Trajan Column in Rome. This information provides us with an idea of how important the celebration of cults and festivities were to official regimental life and for the stability of the Empire. It is because of this symbolic aspect of the Roman army that we should not underestimate the role of the astronomy at the time in determining the exact date of those festivities, something that could be incorporated in the design of the military settlements, adding a religious meaning to them.

3. Textual evidence

Textual evidence from ancient writers constitute one of the starting points of our research into the orientation of Roman buildings. A number of texts from the *agrimensores*, the ancient topographers that we have mentioned before, are collected in the *Corpus Agrimensorum*[2], where they proposed a number of steps to follow in land surveying and demarcation. Apart from measuring land, some of them referred also to the task of orientation, such as Frontinus and Hyginus Gromaticus, who tell us that the path of the sun should be followed by the time of laying out the *decumanus*.[21] Meanwhile, Vitruvius supported the idea

---

[15] Thomas Baines, *Lancashire and Chesire: Past and Present, Volume I* (1868; repr. Heritage Publications, 2012), p. 281.
[16] Martin Henig, *Religion in Roman Britain*, (London: B T Batsford Ltd, 1984).
[17] Dominique Briquel, 'L'espace consacré chez les Étrusques: reflexions sur le rituel étrusco-romain de fondation des cites', in *Saturnia Tellus. Definizioni dello spazio consacrato in ambiente etruco, italico, fenicio-punico, iberico e céltico*, ed. X. Dupré Raventós, S. Ribichini, and S. Verger (Rome: Consiglio Nazionale delle Ricerche, 2008), pp. 27–47.
[18] Livy, *Ab Urbe Condita*, B.O. Foster (London: Harvard University Press, 1919), I, 7; and Plutarch, *Romulus*, B. Perrin (Cambridge: Loeb Classical Library, 1923), XI.
[19] Joseph Ryckwert, *The Idea of a Town. The Anthropology of Urban Form in Rome, Italy and the Ancient World* (Princeton, NJ: Princeton University Press, 1988).
[20] Gloria Andrés Hurtado, 'Los lugares sagrados. Los campamentos militares' (Tesis para la obtención del Certificado Diploma de Estudios Avanzados, Instituto de Estudios Riojanos, Vol. 96, 2002) pp. 137–160.
[21] Frontinus. *De Agrimensura*, C.Thulin (Lipsiae: Teubner, 1913), p. 27: 'The limits and the origin, just as described by Varro, came from the Etruscan Discipline; the soothsayers [aruspices] divided the world into two parts, the right hand towards the north, which they called Septentrion, to the left would be the meridian of the earth, from east to west, where you can see the paths of the sun and the moon... '; Hyginus Gromaticus, *Constitutio*, C. Thulin (Lipsiae: Teubner,

that the main factor to take into account regarding the time to orientate, and also to locate, a city was to avoid the direction of the principal winds because of health reasons.[22] He explained a method for determining the cardinal points, and therefore the main direction of the winds, based on the use of a *gnomon* and, consequently, on the observation of the solar way along one day. Furthermore, in the same book[23], he highlighted the importance for an architect to manage astronomy and the theory of the heavens. Both pieces of evidence reveal the necessity of astronomical observations to determine the direction of the main axes of a town, even in the case of Vitruvius. Regarding the military camps, we cannot avoid the texts of Hyginus and Vegetius.[24] For the former, the *Porta Praetoria* should face the enemy while the *Porta Decumana* should be fixed at the highest point, and for the latter the *Porta Praetoria* should either face east or towards the enemy, except when the army is on the march. In this case, it should face in the direction in which they are to proceed. In these two texts, there are no references to astronomy, and if the Romans strictly followed them we should find a totally random distribution of orientations, which does not occur. So, were these treatises followed by Roman surveyors? And which of those precepts would be the most common in practice?

4. Orientation of Roman camps and forts in Britannia
4.1 Data sample
The data sample that we have used in this work contains the azimuth data extracted from Richardson's article, the angular altitude of the horizon for a value of azimuth and the latitude of the different sites. There are in total 93 azimuth values of Roman camps and forts in Britain, measured by Richardson on published archaeological plans, supposedly bearing from true north. To ensure the validity of the data, we have measured some orientations on Google Earth and compared them with those obtained by Richardson; we have found that, in those cases, both measures usually do not differ by more than 1.5°. Although Richardson only measured the azimuth of one of the axes, due to the square or rectangular shape of the camps, we have used the four perpendicular azimuths with their corresponding angular altitudes of the horizon. They were extracted from a reconstruction of the horizon from radar satellite images, available on the internet (http://www.heywhatsthat.com). For that, we have previously located the settlements to obtain their geographic coordinates. Through this program it is possible to measure the angular altitude for an azimuth with respect to the true north, so corrections of the magnetic declination were not necessary.

4.2 Orientation of the whole sample
Maybe the major contrast between our and other author's dealing with the issue is the consideration of the local horizon in the present work.[25] The first step was to calculate the declination towards the west and east directions and to represent them in histograms (Fig. 11.1). Richardson did not clarify what streets, *Praetoria* or *Decumana*, he had measured so in every case we have considered as *Via Decumana* those whose azimuth values are nearer to the east or west cardinal points, assuming a parallelism with *decumanus* in cities. The azimuth ranges considered are from 30° to 150° towards east and from 210° to

---

1913), p. 1: 'The limits are set out not without a reason, but direct the decumani in accordance to the course of the sun, and the cardines towards the polar axis'.
[22] Vitruvius, *De Architectura*, J. L. Olivier (Madrid: Alianza Editorial, 1995), Book I, 6.
[23] Vitruvius, *De Architectura*, Book I, 1.
[24] Hyginus Gromaticus, *De Munitionibus Castrorum*, G.Gemoll (Lipsiae: Teubner, 1879), p. 56; Vegetius, *Epitome of military Science*, Book I, 23.
[25] Richardson, 'The Orientation of Roman Camps and Forts', pp. 415–26; Peterson 'Random Orientation of Roman Camps', pp. 103–8; http://alunsalt.com/2007/02/17/vidi-7/.

330° towards west in order to compare with the solar arc, and also taking into account the lunar azimuth extremes in their major standstill for British latitudes.

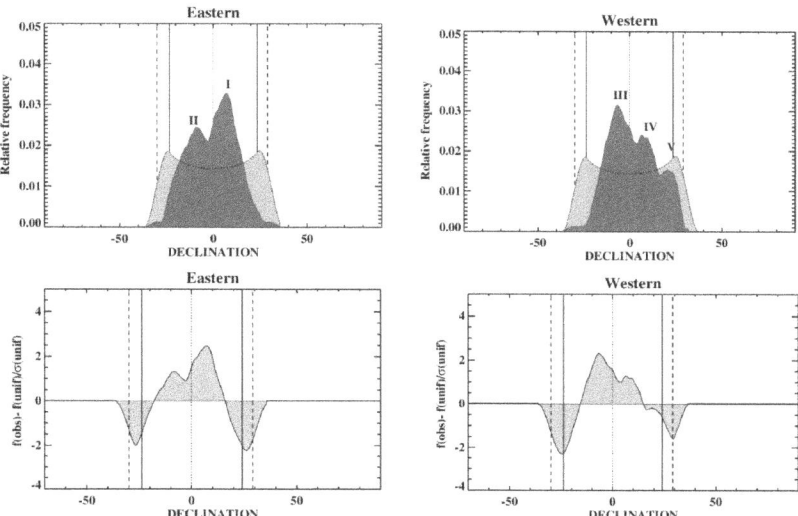

Fig. 11.1: Declination histograms towards west (top right) and east (top left). In the top row the dark grey area is the sample declination distribution and the light grey area is given by a distribution set of data of the same size of our sample, homogeneously distributed in the decumanus azimuth range. In the bottom row, we have subtracted the homogeneous distribution to the sample. The corresponding dates for the different peaks, identified by Roman numerals, are expressed in Table 11.1. Vertical continuous lines indicate the declination extremes for the sun, and the dashed vertical lines for the moon.

Since the amplitude of these azimuth ranges is wider than 90° in Britain, two perpendicular azimuths of the same camp could fall in the same sector. In order to consider just one azimuth per camp in a sector, we have chosen as eastern or western azimuths those values nearest to the east or west cardinal points and calculated the declination with an error of 1.5° by comparing Richardson measures from Google Earth ones. Once we had all the data, we calculated the declination histograms towards east and west, using a density distribution with an Epanechnikov kernel with a pass band of 3°. Furthermore, we have compared those histograms with a homogeneous distribution (see Fig. 11.1). It represents the one we might expect by populating homogeneously a sector between 30° and 150°, or 210° and 330°, in azimuth with a sample of 93 independent measures, assuming a flat horizon and then calculating the declination considering an average latitude of 54.83°. The second step was to subtract the homogeneous distribution to the sample histogram and normalize it with the standard deviation of the former (Fig. 11.1, bottom). By this comparative exercise we tried to estimate how much our results differ from a random distribution and if some directions are more common. We took into consideration a number of peaks cantered on a limited number of declination values, which would be at odds with the expectations of Vegetius and Hyginus's rules of orientating towards the enemy or marching direction. It could also provide further information about Roman army *modus operandi* to that extracted from previous studies.

In the histograms towards the east, we find a maximum corresponding with dates around 9 April (I) or the end of February (II), and towards the west there is one which corresponds with dates around the beginning of March (III), with this one and that of 9 April (I) being above 2σ level. This means a degree of confidence higher than 95%. The beginning of March has been considered the traditional date on which agricultural and war rituals were performed. It was so at least in earlier times, when war had an annual regularity, with the goal of preparing troops at the start of the campaigning season.[26] Since we have considered a bin width of 3°, which means 11 days approximately, the peak around the end of February towards the east (II) could also refer to this date. In any event, the western result would seem more plausible, at least in the case of marching camps, if we consider that Roman troops arrived at a new place at the end of the journey, and it was at that time when they would be built. In case that the solar position was the direction that guided the surveyors to orientate the settlement's main axes, it would be that of the sunset.

Direction	Peak	Declination	Date
Eastern	I	7.25°	9 Apr / 4 Sep
	II	-9°	17 Oct / 25 Feb
	III	-6.5°	4 Mar / 10 Oct
Western	IV	8°	10 Apr / 1 Sep
	V	23.5°	Summer Solstice

Table 11.1: Declination values and corresponding dates of the peaks obtained in eastern and western directions for the whole sample.

Regarding peaks I and IV, related to 9 and 10 April, we cannot give any certain explanation from the point of view of the Roman ritual year but it could shed light about the season when those settlements were built, in case that solar orientations were followed, as well as about the performing of a certain activity or ritual related to the army during that period. Finally, there appears to be a solstitial peak towards the west (V). Although this is not significant, it would be consistent with similar orientations found in Hispania, which would lead us to think about the importance of the solstices in Roman religious life.[27]

4.3 Hadrian's Wall as the main frontier

Due to the presence of such a formidable barrier as Hadrian's Wall, we decided it convenient to divide the sample and to analyse the northern and southern settlements separately with respect to this frontier. The building of Hadrian's Wall started in 122 CE, during the rule of Emperor Hadrian, and extended west from Wallsend on the River Tyne to the Solway Firth. Twenty years later a northern frontier was erected, the Antonine Wall, but it was finally abandoned around 164–9 CE, maybe due to troop withdrawal for service in other campaigns.[28] Then, Hadrian's Wall again became the permanent northern limit of Roman territory in Britannia. There are 43 sites represented to the north of the Wall and 50 to the south (see Fig. 11.2).

We have thus calculated the declination histograms separating those camps to the north and south of this barrier. Figure 11.3 includes all the corresponding histograms towards eastern and western directions, with camps and forts to the north and south of Hadrian's Wall listed separately. Some of them, around 17 in our sample, were attached to it. In relation to the histograms of the southern sites (bottom, Fig.

---

[26] John Rich, 'Roman Rituals of War', in *The Oxford Handbook of Warfare in the Classical World*, ed. B Campbell and L. A. Tritle (Oxford: Oxford University Press, 2013), pp. 542–51.
[27] A. C. González-García et al., 'The orientation of Roman towns in Hispania: Preliminary results', *Mediterranean Archaeology and Archaeometry* 14, no. 3 (2014): pp. 107–119.
[28] David J. Breeze, *Roman frontiers in Britain*, Classical world series (London: Bloomsbury, 2007).

11.3), we have satisfactory obtained a single peak towards both eastern and western directions (XI and XII). The eastern peak (XI) corresponds with dates around 7 April, for which we have mentioned that we do not have any certain explanation, while the western peak (XII) corresponds with the beginning of March. As we have discussed before, it was an important date in the Roman army and it would agree with the hypothesis of the evening establishing of settlements. Regarding the northern results, in these two histograms (top, Fig. 11.3) we have more than one peak, with two towards the east and three to the west. The main one towards the eastern direction (VI) corresponds with dates around 1 March, as we had previously obtained, while the secondary peak (VII) and the main peak in the western histogram (VIII) match with dates of the middle of April and the end of August. There is also a solstitial peak towards the sunset on the summer solstice (X), which would agree with the developing of campaigns during the summer (in case that the sun direction was followed in the trace of the axes). The declination values of the different peaks can be seen in Table 11.2.

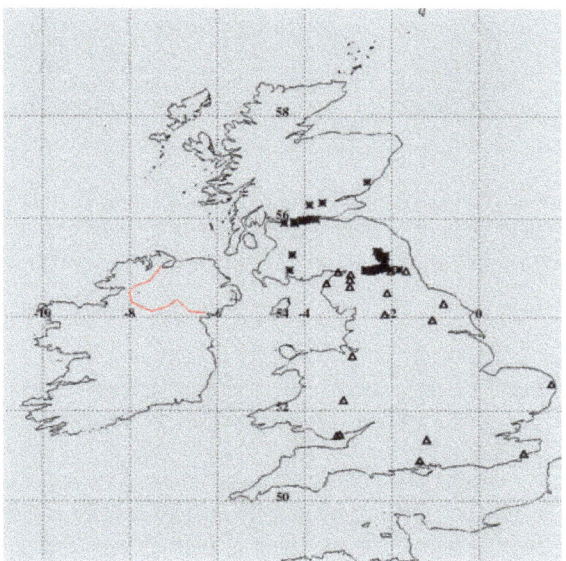

Fig. 11.2: Map of Britannia with the 93 camps and forts in the sample. Triangles indicate settlements to the south of Hadrian's Wall, and asterisks those to the north of Hadrian's Wall.

Region	Direction	Peak	Declination	Date
North of Hadrian's Wall	Eastern	VI	-10.75°	28th Feb/22th Oct
		VII	~9°	14th Apr/31th Aug
	Western	VIII	9.5°	15th Apr/30th Aug
		IX	~-10°	28th Feb/30th Oct
		X	~23.5°	Summer Solstice
South of Hadrian's Wall	Eastern	XI	6.5°	7th Apr/6th Sep
	Western	XII	-6.25°	5th Mar/10th Oct

Table 11.2: Declination values and corresponding dates of the peaks obtained at the south and the north of Hadrian's Wall towards eastern and western direction, respectively.

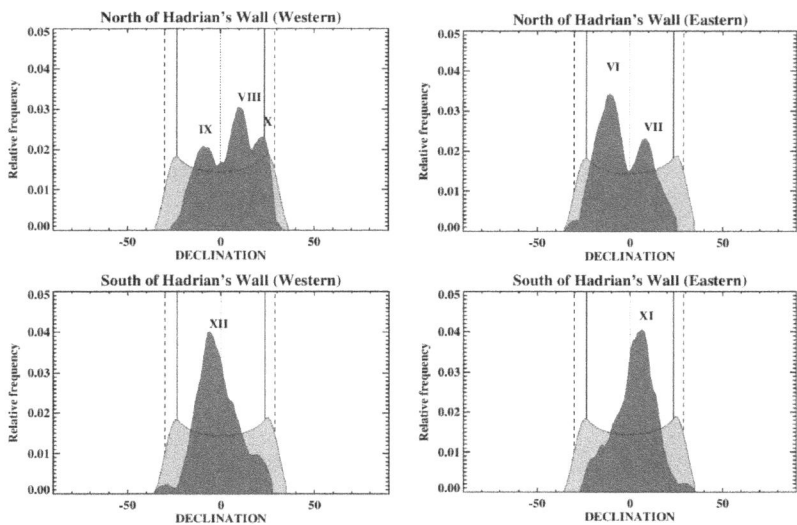

Fig. 11.3: Declination histograms towards west (left) and east (right) of the data set of Roman camps and forts to the north (top) and south (bottom) of Hadrian's Wall. The corresponding dates for the different peaks, identified by Roman numerals, are expressed in Table 11.2. The vertical continuous lines indicate the declination extremes for the sun, and the dashed vertical lines for the moon.

5. Conclusions

We have analysed a large sample of data in Britain, from which we could say that the orientations of Roman military settlements are not randomly distributed. It seems that there exist some preferred directions that, compared with the solar declination values, can be related to the sunrise or sunset on specific dates during a year. From the general analysis two orientations should be highlighted: that of 1 March towards the sunset (II, III, IX, XII) or sunrise (VI), and that of the middle of April or the end of August (I, IV, VII, VIII, XI), some of them with a degree of confidence higher than 95% (I, III, VI, VIII) and 99% (XI, XII). For 1 March we cannot ignore that, at least in early Roman times, a feast in honour of Mars, a Roman god of war, was celebrated; also, that date was linked with the beginning of the campaigning season even in the Republican era.[29] There are also some inscriptions and later writings referring to that feast, for example in the *Feriale Duranum* and Ovid's *Fasti*.[30] For the date of April we have not found any direct reference to a relevant festivity, apart from possibly emperor's birthdays or anniversaries of their glorious acts. Nonetheless, further study of ancient sources and comparison with more Roman military sites could clarify what is behind these results, and we are opened to the possible existence of relevant dates and Roman religious practices that we may have overlooked.

The existence of a solstitial peak, although discrete, would be consistent with the results obtained in other parts of the Empire – Hispania and Italy.[31] Concerning the settlements to the south of Hadrian's Wall, the existence of only one and significant peak would agree with the fact that this region was more

---

[29] Breeze, *Roman frontiers in Britain*.
[30] Henig, *Religion in Roman Britain*, p. 12; Duncan Fishwick, 'Dated inscriptions and the Feriale Duranum', *Syria. Archéologie, Art et Histoire* 65, no. 3 (1988): pp. 349–61.
[31] G. Magli, 'On the orientations of Roman Towns in Italy', *Oxford Journal of Archaeology* 27, no. 1 (2008): pp. 63–71.

stable and peaceful than the northern one, which was under continuous threat, and it would have been easier to perform an intended layout. If we consider that the camps were established in the evening and that a solar position was followed, the surveyors would observe the western one, at dusk.

Nevertheless, whether military campaigns were developed in a particular season or not, or if a feast in honour of Mars was celebrated or not, what remains clear from these results is that there exists a tendency of orientation in the military settlements in Britain and that can be related to those dates. And it might be conceivable that the election of those presumable preferred orientations was linked to certain religious or political traditions in order to add a special meaning to the dates in which the Roman army celebrated important events. Based on all these results, which we can consider as preliminary because of the way the data were obtained, it would be desirable to carry out fieldwork on the terrain in order to broaden the sample and obtain more accurate data. It would also be interesting to develop more specific studies in specific regions in Britannia and maybe discard those camps and forts attached to walls, near main roads or where topographical limitations exists, as Salt and Peterson suggested and we agree on. All the above results are attractive since we cannot underestimate the possibility of an astronomical nature in the orientations of Roman military settlements in Britannia, considering what has been outlined in this work.

Bibliography

Andrés Hurtado, Gloria. 'Los lugares sagrados: Los campamentos militares'. Tesis para la obtención del Certificado Diploma de Estudios Avanzados, Instituto de Estudios Riojanos Militar, Vol. 96, 2002.
Baines, Thomas. *Lancashire and Cheshire: Past and Present*, Vol. I. Heritage Publications, 2012.
Beard, Mary, John North, and Simon Price. *Religions of Rome. Volume I. A History*. Cambridge: Cambridge University Press, 1998.
Bishop, M. C. *Handbook to Roman Legionary Fortresses*. Barnsley: Pen & Sword, 2012.
Breeze, David J. *Roman frontiers in Britain*. Classical World series. London: Bloomsbury, 2007.
Brodersen, Kai. '"The grand old lady still has plenty of surprises left": Hadrian's Wall'. In *Walls, Ramparts, and Lines of Demarcation*. Edited by Natalye Fride and Dirk Reitz, pp. 5–11. LIT Verlag (2009).
Briquel, Dominique. 'L'espace consacré chez les Étrusques: réflexions sur le rituel étrusco-romain de foundation des cites'. In *Saturnia Tellus. Definizioni dello spazio consacrato in ambiente etruco, italico, fenicio-punico, iberico e céltico*. Edited by X. Dupré Raventós, S. Ribichini, and S. Verger, pp. 27–47. Rome: Consiglio Nazionale delle Ricerche, 2008.
Fishwick, D. 'Dated inscriptions and the Feriale Duranum'. *Syria. Archéologie, Art et Histoire* 65, no. 3 (1899): pp. 349–61.
Flink, S. O., S. A. Hoey, and W. F. Snyder. *The Feriale Duranum*. Yale Classical Studies VII. Princeton: Yale University Press, 1940.
Frontinus. *De Agrimensura*. Translated by C. Thulin, Lipsiae, Teubner, 1913.
Gilliver, CM. *The Roman art of war*. Stroud: Tempus, 2001.
Goldsworthy, Adrian Keith. *The Roman Army at War, 100 BC-AD 200*. Oxford: Oxford University Press, 1998.
González-García, A. C., and L. Costa-Ferrer. 'The diachronic study of orientations: Mérida, a case study'. In *Archaeoastroastronomy and Ethnoastronomy: Building Bridges Between Cultures*. Edited by C. L. N. Ruggles, pp. 374–381. IAU Symposium 278 (2011).
González-García, A. C., A. Rodríguez-Antón, and J. A. Belmonte. 'The orientation of Roman towns in Hispania: Preliminarily results'. *Mediterranean Archaeology and Archaeometry* 14, no. 3 (2014): pp. 107–119.
Henig, Martin. *Religion in Roman Britain*. London: B T Batsford Ltd, 1984.
Hicky Morgan, M. *Vitruvius. The ten books on architecture*. Cambridge, MA: Harvard University Press, 1914.
Hyginus Gromaticus. *Constitutio*. Translated by C. Thulin, Lipsiae, Teubner, 1913.
Hyginus Gromaticus. *De Munitionibus Castrorum*. Translated by G. Gemoll. Lipsiae,Tubner, 1879.
Irby-Massie, Georgia L. *Military Religion in Roman Britain*. Leiden: Brill, 1999.
Jones, R. H. *Roman camps in Britain*. Stroud: Amberley Publishing, 2012.
Laurence, R., C. Simon Esmonde, and G. Sears. *The City in the Roman West c. 250 BC- c. AD 250*. Cambridge: Cam-

bridge University Press, 2011.
Livy, *Ab Urbe Condita*. Translated by B. O. Foster, London: Harvard University Press, William Heinemann, Ltd. 1919.
Magli, Giulio. 'On the orientations of Roman Towns in Italy'. *Oxford Journal of Archaeology* 27, no. 1 (2008): pp. 63–71.
Peterson, J. W.M. 'Random Orientation of Roman Camps'. *Oxford Journal of Archaeology* 26, no. 1 (2007) pp. 103–108.
Plutarch. *Romulus*. Translated by B. Perrin, Loeb Classical Library edition (Cambridge, MA and London), 1923.
Rich, John. 'Roman Rituals of War'. In *The Oxford Handbook of Warfare in the Classical World*. Edited by B. Campbell and L. A. Tritle, pp. 542–51. Oxford: Oxford University Press, 2013.
Richardson, Alan. 'The Orientation of Roman Camps and Forts'. *Oxford Journal of Archaeology* 24, no. 4 (2005): pp. 415–426.
Ryckwert, J. *The Idea of a Town. The Anthropology of Urban Form in Rome, Italy and the Ancient World*. Princeton, NJ: Princeton University Press, 1988
Ryckwert, J. *La idea de ciudad. Antropología de la forma urbana en Roma, Italia y el mundo antiguo*. Salamanca: Ediciones Sígueme, 2002.
Salvador Oyonate, J.A. *Higinio el Agrimensor. El establecimiento de los límites. Higinius Gromaticus. Constitutio Limitum*. Documentos Arqueológicos Históricos (D.A.H) nº1, Ediciones ACEAB, 2015.
Southern, Pat. *The Roman Army: A Social and Institutional History*. Santa Barbara, CA: ABC-CLIO, 2006.
Tertullian. *Ad nationes*, 1.12, Q. Howe, www.tertullian.org, 2007.
Thulin, C. *Corpus agrimensorum romanorum*. Opuscula Agrimensorum Veterum. Lipsia: Teubner, 1913.
Erdkamp, Paul, ed. *A companion to the Roman army*. Chichester: Wiley-Blackwell, 2011.
Vegetius. *Epitome of military Science*. Translated by N.P. Pilmer. Liverpool: Liverpool University Press, 1996.
Vitruvius. *De Architectura*. Translated by J. L. Oliver. Madrid: Alianza Editorial, 1995.

EVIDENCE FOR THE EXISTENCE OF SOLAR AND LUNAR ALIGNMENTS IN WESTERN SCOTLAND: THE CONTRASTING NATURE OF BACKSIGHTS, FORESIGHTS AND ALIGNMENTS

Thomas Gough

ABSTRACT: Alexander Thom determined the declinations at many standing stone sites. Since many of them gave significant declinations for the sun or moon he concluded that some of them may have been intentional alignments.[1,2] Clive Ruggles and Douglas Heggie examined Thom's results, in particular the high precision lunar lines, finding some errors and inconsistencies.[3,4] Partly because Thom's chosen sites were geographically widely spread, Ruggles concluded that there was insufficient evidence to support the claim that the putative alignments had been intentionally set up, and argued that they could be explained as due to chance.

This paper is a report on research begun in 2007. In a suitable region in western Scotland, all standing stone sites (but not circles) with clear horizons were assessed for the declinations of foresight features in the direction indicated by the freestanding menir/ short row. There were 92 sites of which 88 were visited. Trees, or fallen stones leaving no indicated direction, prevented assessment at 43 sites. The remainder were assessed. The declinations found were compared with the declinations of key horizon positions to be expected for the sun and moon in the late Neolithic/ Early Bronze Age. Substantial agreement between the two was found. This was especially evident for the lunar declinations, which is of particular significance as the close agreement, if not due to chance, has potentially important implications. A separate investigation gave evidence that chance alignments were not common and were therefore unlikely to be able to explain the results found.

Later examination of the nature of the backsights, foresights and alignments which had been identified as possibly planned showed that the features of the alignments for the sun and for the moon were very different from each other. This non-random property is seen as supporting evidence for the intentionality of the alignments concerned.

Keywords: alignments, sun, moon, Scotland

Introduction

Many Neolithic structures show an interest in the sun or moon. Ruggles, for example, has shown that the Newgrange passage grave entrance was aligned on the winter solstice sunrise and that many recumbent stone circles were orientated on the extreme southern moon.[5] Any direction indicated by the above examples depends upon the orientation of the structure alone. Thus while Stonehenge may be capable of indicating moderately accurate directions to within a degree or so but lacking, as it does, any horizon features for use as foresights precise alignments are not possible. However when a distant skyline feature is used the precision achievable is much greater. The principle is analogous to that of a rifle with a backsight and a foresight. This distinction between orientations and alignments using an indicated foresight is important.

Thom found many such alignments, mostly in Scotland, which he believed had been intentionally set up. Many of Thom's claimed alignments were supposedly accurate to a minute of arc. However the subsequent re-evaluation of these claims by archaeoastronomers has rejected this level of accuracy. In

---

[1] Alexander Thom, *Megalithic Sites in Britain* (Oxford: Oxford University Press, 1967).
[2] Alexander Thom, *Megalithic Lunar Observatories* (Oxford: Oxford University Press, 1971).
[3] C. N. L. Ruggles, *Astronomy in Prehistoric Britain and Ireland* (Princeton: Yale University Press, 1999), pp. 49–67
[4] D. Heggie, *Megalithic Science* (London: Thames and Hudson, 1981), pp. 170–79.
[5] Ruggles, *Astronomy in Prehistoric Britain and Ireland*, pp. 17–19, 91–101.

particular, Clive Ruggles criticised Thom's claimed alignments, emphasising that widely scattered sites and foresights that were open to subjective choice nullified these claims. In response to this, the present investigation adopted the method of:

- Assess one region at a time.
- Assess all sites in that region.
- A backsight is a flat-sided stone, pair or short row with an indicated direction; and,
- Within this group, NO selection.

The region assessed was in western Scotland: Argyll and the nearby islands of Mull and Islay.

Following Ruggles' criteria, no consideration is given to possible alignments where the horizon is nearer than 1km.[6] There follows later in Sections A (moon) and B (sun) a summary of earlier investigations regarding alignments by the present author.[7,8] There after the different nature of the putative alignments is discussed.

It is important to have an understanding of the movements of the sun and moon and the resulting appearance of the moon – the 'standstills', the 'wobble' and their timescale – if the significance of the results, particularly the lunar results, presented later are to be appreciated.

The declination of the sun varies each year between $\pm\varepsilon$ (the obliquity of the ecliptic), and the moon between $\pm(\varepsilon \pm i)$ where i is the inclination of the moon's orbit to the plane of the earth's orbit round the sun.[9] The value of ε has been decreasing at the rate of about 0'.7 of arc per century for several thousand years. Thom deduced a possible value for ε of about 23°54',[10] which implies a date of approximately 1700 BCE. The value of i is unchanging and is about 5°8'.7. Thus the moon's maximum declination is about $\pm 29°03'$, $\pm(\varepsilon+i)$, and the minimum $\pm 18°45'$, $\pm(\varepsilon-i)$. Although the moon is seen to move from north to south and back again each month, the maximum declination reached at each lunation changes very slowly, taking 18.6 years for a complete cycle.

A detailed understanding of the reasons for the moon's movements as seen from Earth may be useful but is not necessary.

In the region chosen (Argyll, between latitudes 56° 0'N and 56°45'N, together with the nearby islands of Mull and Islay) there are 92 sites. All of these sites were visited, excepting only 4 remote sites. Of the remaining 88 sites:

- 43 – stones fallen, trees, no indicated direction etc.
- 17 – an indicated direction but horizon less than 1km or no foresight. Some of these are likely to be waymarkers. E.g., 5 stones in western Mull are probably medieval guides to the important island of Iona.
- 14 – indicated foresight region, declinations suggesting calendrical (solar) alignments. (See later)
- 11 – indicated foresight region, lunar declination.
- 3 – other plausible explanation. E.g., remains of a stone circle.

---

[6] C. L. N. Ruggles, 'The stone alignments of Argyll and Mull', in *Records in Stone: Papers in Memory of Alexander Thom*, ed. C. L. N. Ruggles (Cambridge: Cambridge University Press, 1985), p. 235.
[7] T. T. Gough, 'New Evidence for Precise Lunar Alignments in Argyll, Scotland in the Early Bronze Age', in the *Proceedings of the 20th Conference of the SEAC, supplement*, ed. I. Sprajc and P. Pehani (Ljubljana: Slovene Anthropological Society, 2013), pp. 157–175, available at http://www.academia.edu.
[8] T. T. Gough, 'Further Evidence for the Existence of Prehistoric Celestial Alignments in Western Scotland: Calendrical Alignments on the Island of Mull', in *the Proceedings of the SEAC XXIII Conference*, 2–6 September 2013, Athens, Greece. In press, available at http://maajournal.com/Issues/2014/Vol14-3/Full23.pdf [accessed 23 July 2016].
[9] Thom, *Megalithic Lunar Observatories*, pp. 15–19
[10] Thom, *Megalithic Lunar Observatories*, pp. 36–43

A. Lunar Alignments: The Lunar Band

The moon's orbit and its apparent movement as seen from earth are not simple. A detailed explanation is available elsewhere.[11] As described above, the declination of the moon at a lunation maximum varies in a regular manner between ±29°03′ and ±18° 45′ ca. 1700 BCE over an 18.6 year period. The moon does not come smoothly to the maximum ('major standstill') and minimum ('minor standstill'), but in a series of small but regular variations in declination with a period of about 173 days and amplitude of about 9′ arc, termed the lunar perturbation or 'wobble', symbol Δ *(delta)*.

Figure 12.1 shows one of the lunar bands which define the extremes of the moon's monthly movements across the sky – that for the major standstill north of the celestial equator. Note that the only directly measurable positions are for one of the moon's limbs at the upper or lower limit of the 9′ of arc 'wobble'. There are 8 lunar bands: rising/setting, north/south, major and minor standstills.

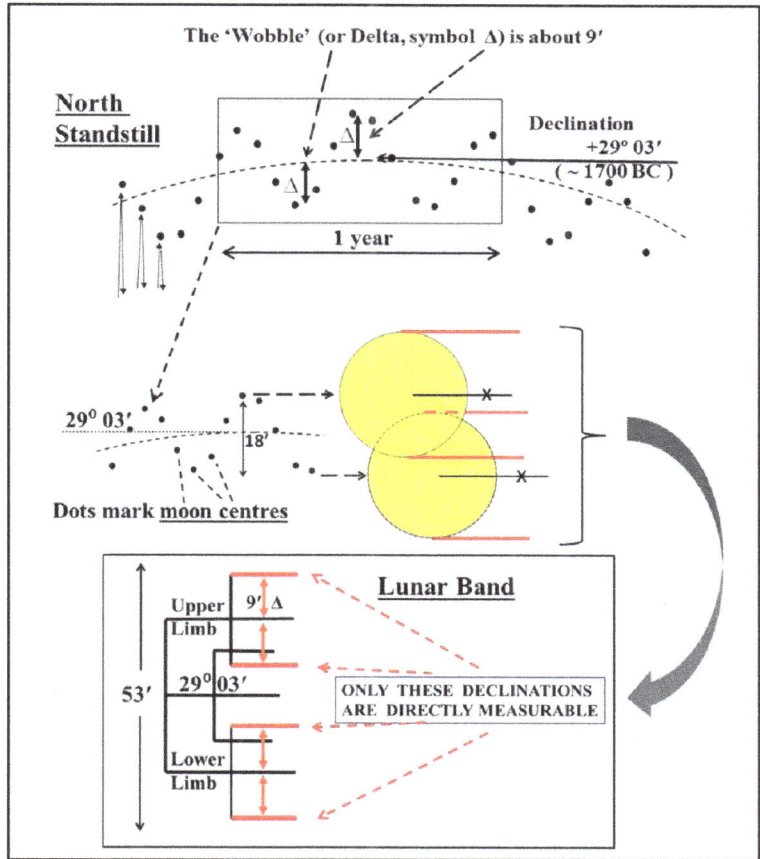

Fig. 12.1: Explanation of a Lunar Band for the 'standstill' north showing the relevant declinations

---

[11] Thom, *Megalithic Lunar Observatories*, pp. 15–27; J. E. Wood, *Sun, Moon and Standing Stones* (Oxford: Oxford University Press, 1978), pp. 66–70; A. S. Thom, 'Megalithic lunar observatories; an assessment of 42 lines', in *Astronomy and Society in Britain during the period 4000-1500BC*, ed. C. L. N. Ruggles and A. W. R. Whittle (Oxford: British Archaeological Reports, British Series 88, 1981), pp. 13–61; J. D. North, *Stonehenge: Neolithic Man and the Cosmos* (London: Harper Collins, 1996), pp. 553–68.

Examples of Lunar Alignments

A stone row, pair or flat side of a stone will typically indicate a direction to only about ±2° at best. This is fully sufficient to identify the major standstill lunar bands in both the north and south, since at about latitude 56°N they are separated by about 10° azimuth from the nearest other possible target: namely the solstices (about ±24°). In every case where the horizon region of a major lunar band was indicated, there was a suitable foresight in that lunar band with a declination within about 2' of arc of one of the four theoretical values (Fig. 12.1). The minor standstill, the declination of which is about ±18°45', lies within the sun's range and is problematical. However no alignments for the minor standstill were found.

Two illustrations are given below. One is an example of an alignment (Fig. 12.2); the other is for the combined lunar bands in the north with alignments shown (Fig. 12.3). All of the results, together with tables and discussion, can be found on the website www.lunarsites-scotland.net.

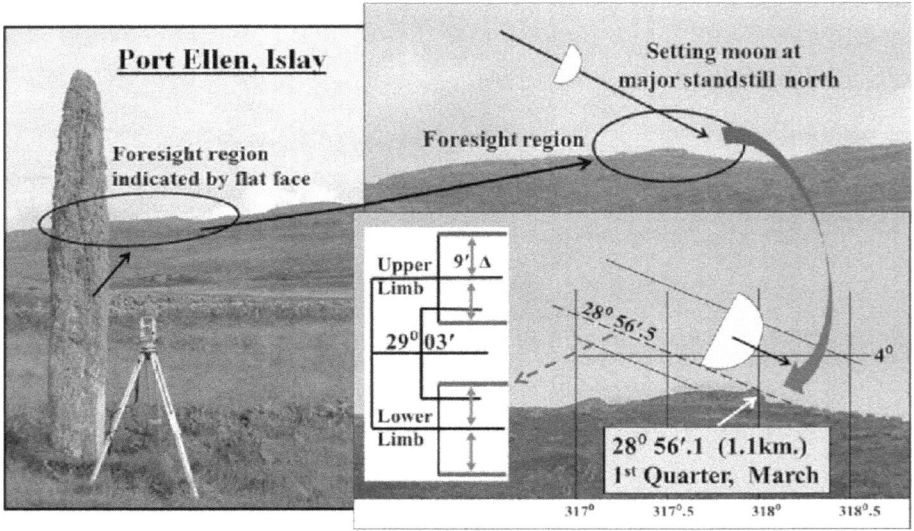

Fig. 12.2: Lunar alignment, Port Ellen, Islay (Map ref. NR 3715 4559).

The alignment here is for the first quarter moon at the spring equinox. Note that the moon is at its extreme northerly declination; it could never have been further north. The position of the backsight has apparently been arranged so that the lower limb is used, which would have ensured that the moon was clearly seen as it sets. The precision required for the placement of the stone depends on the alignment length. In this case (1.1km), if the stone were more than 2m to the left or right, there would be no alignment of the required precision.

An alignment (i.e., for a limb ±9' of arc 'wobble') is only accepted if it is within ±~2' of arc of one of the 4 key theoretical declinations. Of the 7 alignments shown, 6 are for a limb with +'wobble' (in red) and 1 for a limb with −'wobble' (in orange). The declination values shown are for the maximum north of the moon around 1700 BCE. The alignments with 'wobble' to the north for one or other limb (6 out of 7) would be the easiest to observe because the moon is then in a unique position. We have empirical data, and a test of statistical significance can be provided by a binomial distribution calculator.[12] We needed to start by

---

[12] http://www.calcul.com/show/calculator/binomial-distribution [accessed 10 July 2016].

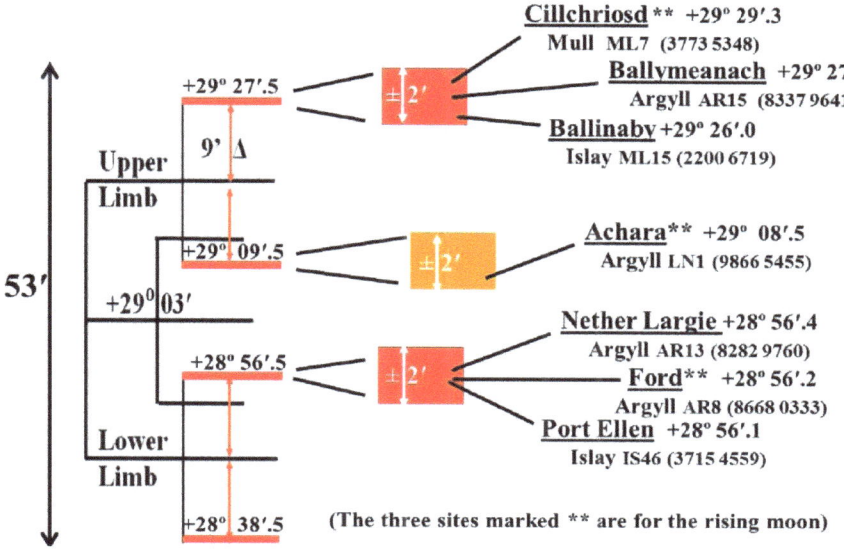

Fig. 12.3: The combined lunar bands in the north with alignments.

ensuring that all possible alignments were 'hits' somewhere in the lunar band. That is true since there were no other suitable foresights anywhere either within each lunar band or near to it.

Each acceptable alignment range is 4' wide. We are concerned with the two shown in red, so the 'target' is 8' of arc wide. Therefore the chance of a declination value being in one of the red squares is 8/53. In the formula, k (number of successes) = 6, n (number of trials) = 7, probability is 8/53 or 0.151. We entered these values and found p=0.00007. This told us that the probability of the result being due to chance is negligible. All of the declinations were checked during at least one return visit, giving confidence that the values given are unlikely to be in error by more than ±0'.5.

Note 1. Using all 7 results: 7 trials, 7 successes, probability is 16/53. p=0.0002
Note 2. There were also 4 alignments for the moon at its southern major standstill each of which was within ±2' of arc of a key declination. In no case was there any alternative usable foresight in or near to a lunar band.

Chance alignments

It has been claimed that the precise lunar alignments, which Thom believed he had found, could be explained as being due to chance.[13] This claim is based mainly upon the facts that the moon can have a number of significant declinations in the region of the standstills (although not the 9 originally assumed); that in hilly country potential foresights are likely to be frequent; and that Thom's alignments were selected from a wide geographical area. In view of these claims field work was undertaken to assess the likely frequency of chance alignments.

In Argyll a total of 1500 degrees of random skylines in suitable hilly country from identifiable random viewing positions was undertaken. This contained 30 lunar bands. Three chance alignments were found: one for a major standstill and two for minor standstills. This is insufficient to explain the alignments that occur.

---

[13] Ruggles, *Astronomy in Prehistoric Britain and Ireland*, pp. 59–67; Heggie, *Megalithic Science*, p. 234.

B. Possible Solar Alignments

The sun has no unique positions except at the solstices: i.e., at all other times its rising/setting azimuth changes. The daily movement increases during the 3 months after a solstice to a maximum at the equinox before decreasing over the next 3 months to zero at the next solstice. Thom's proposed prehistoric solar calendar was based on the idea that the year was divided into 16 periods of about 23 days each.[14] He calculated probable declinations for the start of each of these 'months' or epochs. These declinations cannot be precise – except at the solstices – as they must serve for a range of declinations.[15] Thom's data was obtained from a large number of sites. Although the results included many declinations that seemed to fit with the expected theoretical ones, they had not been obtained in a systematic manner, but were from geographically widely spread sites. As a consequence – both of the spread of sites and that a precise declination was not a possibility – the evidence supporting the possible existence of a prehistoric solar calendar was not conclusive.[16] During the present project, in which all possible sites in a defined region were assessed, a significant number gave declinations that did fit with those predicted (See Table 12.1). These were all on the islands of Mull and Islay.

Sunrise	Map ref.	Dec.	#	Epoch (Month) + Present dates	Dec.	Sunset	Map ref.
Uluvalt C (ML 25)	5468 2996	+0°.4	0	Mar 21st	+0°.6	Tenga (ML 13) Baile Tharbhach (IS7)	5040 4632 3636 6762
		+9°.0	1	Apr.13th	+9°.2	Rossal (ML 27) Ardalanish (ML 33)	5434 2820 3784 1888
Knocklearach (IS11) Cnoc Ard (IS49)	3989 4830 3264 4600	+16°.6	2	May 5th	+16°.7		
Gruline (ML16) Scanistle (IS5)	5456 3960 4108 6724	+22°.0	3	May 28th	+22°.1	Druim nan Madagan (IS –)	3824 4595
Gruline (ML 16) Scanistle (IS5)	5456 3960 4108 6724	+23°.9	4	June 21st	+23°.9		
Gruline (ML 16) Scanistle (IS5)	5456 3960 4108 6724	+22°.1	5	July 15th	+22°.0	Druim nan Madagan (IS –)	3824 4595
Knocklearach (IS11) Cnoc Ard (IS49)	3989 4830 3264 4600	+16°.8	6	Aug 5th	+16°.6		
		+9°.3	7	Aug 29th	+9°.1	Rossal (ML 27) Ardalanish (ML 33)	5434 2820 3784 1888
Uluvalt C (ML 25)	5468 2996	+0°.5	8	Sept 21st	+0°.3	Tenga (ML 13) Baile Tharbhach (IS7)	5040 4632 3636 6762
Tenga (ML 13)	5040 4632	-8°.4	9	Oct 14th	-8°.6	Lochbuie A (ML 28)	6163 2543
		-16°.2	10	Nov 5th	-16°.4	Lag (ML 6)	3626 5331
Uluvalt B (ML 25)	5463 3002	-21°.9	11	Nov 29th	-22°.0	Tenga (ML 13)	5040 4632
		-23°.9	12	Dec 21st	-23°.9	Ardnacross (ML 12)	5422 4915
Uluvalt B (ML 25)	5463 3002	-21°.8	13	Jan 13th	-21°.7	Tenga (ML 13)	5040 4632
		-16°.3	14	Feb 6th	-16°.2	Lag (ML 6)	3626 5331
Tenga (ML 13)	5040 4632	-8°.5	15	Feb 27th	-8.4	Lochbuie A (ML28)	6163 2543

Table 12.1: Possible calendrical sites on Mull and Islay

Mull and Islay sites with indicated foresight features in the region of theoretical 16 'month' (epoch) declinations (Islay in italics). The reference numbers are to Ruggles' *A new archaeological and statistical study of 300 western Scottish sites*.[17]

There is a discussion of 4 of the above sites in an earlier paper by the present author.[18]

---

[14] Thom, *Megalithic Sites in Britain*, pp. 107–117.
[15] Thom, *Megalithic Sites in Britain*, Table 9.1, 110.
[16] That there is positive evidence for such a calendar is discussed in the final conclusions.
[17] C. L. N. Ruggles, *A new archaeological and statistical study of 300 western Scottish sites* (Oxford: British Archaeological Reports, British Series 123, 1984).
[18] Gough, 'Prehistoric Celestial Alignments', Figures 3–9, pp. 250–252.

Evidence for the Existence of Solar and Lunar Alignments in Western Scotland

The different nature of alignment features for the sun and moon

1. Backsights
   (a) The Sun. The backsights are not usually large. They are typically slabby blocks with at least one fairly flat side, about 1.0–1.6 m high.
   (b) The Moon. The backsights are usually large and substantial. Single stones are typically 3–4 m high with flat or nearly flat sides. Stones in a row are usually smaller, about 2 – 2.5m. high, and may be flat-sided blocks rather than slabs. E.g. Dervaig B (ML 10)[19]

2. Foresights
(a) The Sun. At or near the solstices the sun's rising/setting position changes slowly, and a clear bump or hilltop tends to be used. At all other times the sun's movement on consecutive days is noticeable and so for these epochs each foresight must serve for a small range of declinations.[20]

Four types of foresight were found for the 14 putative solar alignments in the region (Fig. 12.4).

A. The sun's movement is parallel to a ridge or slope. The sun is hidden on the day of the epoch. It is only visible above the ridge one day earlier or later depending on its direction of movement. The most common type.
B. Into or out of a hollow
C. Into or out of a hill intersection
D. Use of an horizon bump

Note: Only type D is found for lunar alignments. The others give insufficient precision.

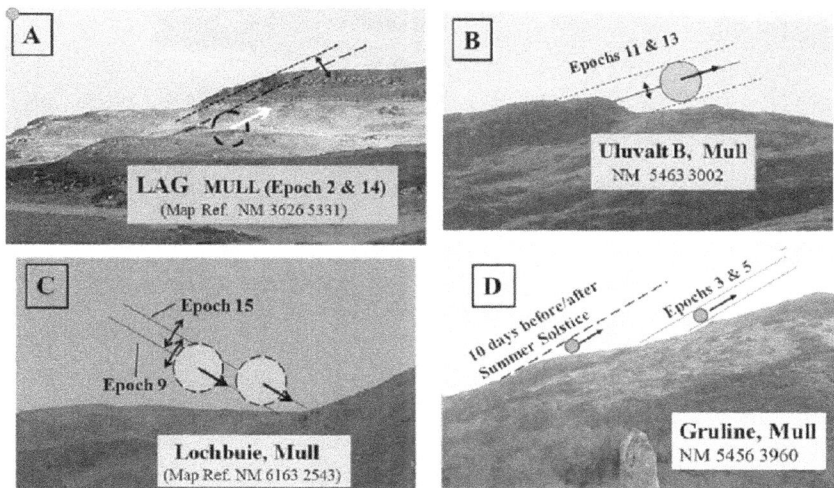

Fig. 12.4: Solar foresights.

(b) The Moon
The foresights must be suitable for observation of the moon's very small changes in declination at the standstills. It is probably for this reason that foresights for the moon are almost always rocky thus avoiding the uncertainties that would result from vegetation

---

[19] Ruggles, *300 western Scottish sites*, p. 127.
[20] Gough, 'Prehistoric Celestial Alignments', p. 249.

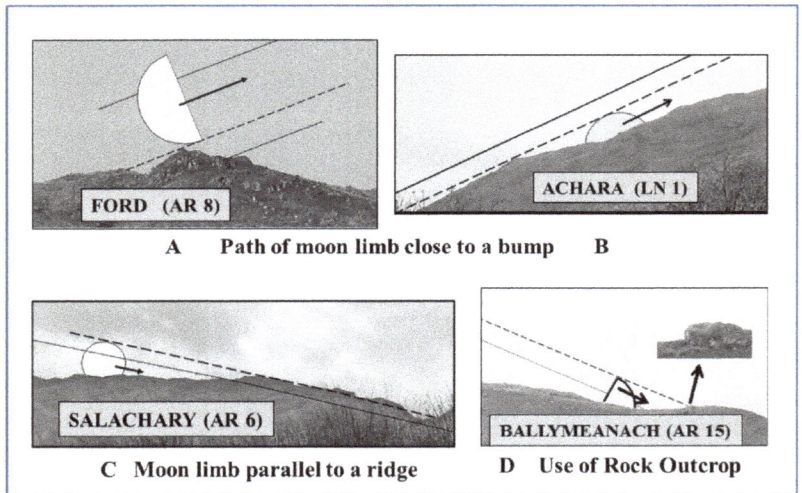

Fig. 12.5: Examples of lunar foresights found. (Dotted lines are for one of the four key lunar declinations at a major standstill, Figs. 12.1 and 12.3)

Discussion and Conclusion

The evidence for the existence of planned precise lunar alignments, as presented here, is strong. The evidence for solar alignments is weaker. This arises because there exists the possibility of precise alignments for the moon but not, except at the solstices, for the sun. Possible solar alignments at or near the solstices, being more precise, carry more weight. Other solar alignments, which must of necessity be for a significant range of declinations, individually carry less weight. However it should be recognised that a significant proportion of assessable sites in Mull and Islay have alignments matching theoretical calendrical declinations (Table 12.1); that the nature of the foresights of these putative alignments is suitable for the purpose; and that, as shown here, the nature of the alignments (backsights and foresights) form a group which is distinct from that for lunar alignments.

The survey in 2009 in Argyll, when a total of 1500 degrees of random hilly horizon was measured, assessed the likelihood of chance alignments. Three chance lunar alignments were found (30 lunar bands) but only one solar (calendrical) alignment. This suggests that the putative calendrical alignments are more sophisticated than they appear.

Bibliography

Anon. http://www.calcul.com/show/calculator/binomial-distribution.
Gough, T. T. ' New Evidence for Precise Lunar Alignments in Argyll, Scotland in the Early Bronze Age'. In *Ancient Cosmologies and Modern Prophets, Proceedings of the 20th Conference of the European Society for Astronomy in Culture, Anthropological Notebooks, supplement*, edited by I. Sprajc and P. Pehani, pp. 157–175. Ljubljana: Slovene Anthropological Society, 2013. http://www.academia.edu.
Gough, T. T. 'Further evidence for the existence of Prehistoric Celestial Alignments in western Scotland: Calendrical Alignments on the Island of Mull'. In *the Proceedings of the SEAC XXIII Conference*, 2–6 September 2013, Athens, Greece. In press. http://maajournal.com/Issues/2014/Vol14-3/Full23.pdf.
Heggie, D. *Megalithic Science*. London: Thames and Hudson, 1981.
Morrison, L. V. 'On the analysis of Megalithic Lunar Sightlines in Scotland'. *Supplement to the Journal for the History of Astronomy* 11, no. 2 (1980): p. xi.

North, J. D. *Stonehenge: Neolithic Man and the Cosmos*. London: Harper Collins, 1996.
Ruggles, C. L. N. *A new archaeological and statistical study of 300 western Scottish sites*. Oxford: British Archaeological Reports, British Series 123, 1984.
Ruggles, C. L. N. *Astronomy in Prehistoric Britain and Ireland*. Princeton: Yale University Press, 1999.
Ruggles, C. L. N. 'The stone alignments of Argyll and Mull'. In *Records in Stone: Papers in Memory of Alexander Thom*, edited by C. L. N. Ruggles. Cambridge: Cambridge University Press, 1988.
Thom, A. *Megalithic Lunar Observatories*. Oxford: Oxford University Press, 1971.
Thom, A. 'Megalithic lunar observatories; an assessment of 42 lines'. In *Astronomy and Society in Britain during the period 4000-1500BC*, edited by C. L. N. Ruggles and A. W. R. Whittle. Oxford: British Archaeological Reports, British Series 88, 1981.
Thom, A. *Megalithic Sites in Britain*. Oxford: Oxford University Press, 1967.
Wood, J. E. *Sun, Moon and Standing Stones*. Oxford: Oxford University Press, 1978.

ARCHITECTURE, ILLUMINATION AND COSMOLOGY: THE ARLES-FONTVIEILLE MONUMENTS, ARCHAEOASTRONOMY AND MEGALITHIC STUDIES

Morgan Saletta

ABSTRACT: In light of my discovery and documentation[1] of seasonal light and shadow hierophanies produced by the setting sun within the Arles-Fontvieille monuments this article considers the interpretive implications of these monuments for archaeoastronomers. While horizon astronomy was very likely used for time-reckoning in the Neolithic, I argue that the 'seasonal illumination hypothesis' is more appropriate than the 'celestial targeting paradigm' not only for the interpretation of the role of astronomy and cosmological symbolism in the construction of the Arles-Fontvieille monuments but in many, if not the majority, of European late prehistoric tombs with chambers and passages as well. Multiple lines of evidence suggest a cosmologically symbolic link between houses of the living and houses of the dead in late prehistoric Europe. But is there a functional link? It has been suggested that Neolithic dwellings were, in many cases, oriented with respect to the sun not only for symbolic reasons, but also out of functional considerations – for both heating and illumination. These dwellings were an intimate part of their inhabitants' life-world; the heating power of the sun and the changing seasonal patterns of solar illumination in these dwellings very likely acquired cosmological importance and as such were incorporated into funerary and ancestral monuments. The fundamental similarity between the seasonal illumination events at the Arles-Fontvieille hypogées with the illumination events at monuments such as New Grange and Maeshowe is also a crucial piece of evidence, convergent with other archaeological evidence, suggesting cosmological principals spread by contact and diffusion between major centres of megalithism in the Middle and Late Neolithic.

The archaeoastronomy of the Arles-Fontvieille monuments, seasonal illumination and the statistical orientation pattern of European passage graves

The Arles-Fontvieille monuments, or *hypogées*, have a special place both in local history as well as in European archaeology and megalithic studies more generally. Their unique architecture, which blends elements of 'Atlantic' megalithic construction with rock cut architecture more commonly found in the western Mediterranean, and their size, especially that of the Grotte or Hypogée de Cordes, place them among the most important monuments in France and Europe, as emphasized by Glyn Daniel, Jean Guilaine and an ongoing collaborative research project under the aegis of the CNRS of which I am part.[2] There are five monuments generally considered to be part of the Arles-Fontvieille group. The largest of the group is the Hypogée or Grotte de Cordes, located on the Montagne des Cordes. The other four monuments are located on the neighbouring but much lower hill named Castelet. These are: the Hypogée du Castelet, the Hypogée (or Grotte) de Bounias, the Hypogée (or Grotte) de la Source and the Dolmen de Coutignargues (in ruins). The monuments are long, trapezoidal chambers, cut directly in the local bedrock (with the exception of the Dolmen de Coutignargues), roofed with megalithic slabs and accessed via an entry

---

[1] Morgan Saletta, 'The Arles-Fontvieille Megalithic Monuments: astronomy and cosmology in the European Neolithic' (PhD thesis, University of Melbourne, 2014).

[2] Glyn Edmund Daniel, *The prehistoric chamber tombs of France : a geographical, morphological and chronological survey* (London: Thames and Hudson, 1960); Jean Guilaine, *Au temps des dolmens: mégalithes et vie quotidienne en France méditerranéenne il y a 5000 ans* (Toulouse: Ed. Privat, 1998); Xavier Margarit, *Les monument mégalithiques d'Arles-Fontvieille, état des connaissances, contexts et nouvelles données* (Projet collectif de recherche 2014), ed. Xavier Margarit (Aix-en-Provence: CNRS/LAMPEA, 2014).

ramp. The chambers of the smaller monuments range in overall length from 14 to 19 meters. The Grotte de Cordes has an interior chamber of some 24m in length and an overall length of almost 43m. Figure 13.1 shows the cut and plan of the Grotte de Cordes.³

Fig. 13.1: Sunlight striking the rear wall of the Hypogée de la Source, 23 September 2011. A full time-lapse video sequence is available on the author's academia.edu page. Photo by author.

As shown in Table 13.1, the monuments all open in a westerly or south-westerly direction, which is unusual within the larger European context, a fact already noted by Prosper Mérimée during his visit to the monuments as *Inspecteur des Monuments Historiques* in 1834.⁴ While the chronology of the Arles-Fontvieille monuments remains a topic of some debate, the general consensus is that they were built between 3300 BCE and 2900 BCE.⁵ This would make them roughly contemporary with centres of megalithic construction elsewhere in Atlantic and Northern Europe, including Newgrange and Maeshowe.

The analysis of orientation patterns on a supra-regional western European scale shows a custom of broadly easterly orientation, toward the sun as it rises or climbs, extending over a vast geographic area of Western Europe ranging from Denmark to Portugal and spanning several millennia of construction. On the French Mediterranean coastal region, part of nearby Catalonia and the Balearic islands, however, the orientation pattern is very different – monuments face toward the sunset or the sun as it descends in the

---

³ Paul Cazalis de Fondouce, *Les Temps préhistoriques dans le Sud-Est de la France: Allées couvertes de la Provence* (C. Coulet, 1873).
⁴ P. Mérimée, *Notes D'Un Voyage Dans Le Midi De La France* (Paris: Librarie de Fournier, 1835).
⁵ For discussion of the chronologie of the Arles-Fontvieille monuments see Jean Philippe Sargiano et al., 'Les Arnajons (Le-Puy-Sainte-Réparade, Bouches-du-Rhône) : un nouveau dolmen dans le Sud-Est de la France', *Préhistoires Méditerranéennes* (2010): pp. 119–53; Jean Guilaine, *Méditerranée mégalithique: dolmens, hypogées, sanctuaires* (Lacapelle: Éd. Archéologie Nouvelle, 2011), p. 116.

sky.⁶ While these statistical surveys are invaluable, their interpretation is not unproblematic. While often interpreted as evidence of solar orientation, it has been suggested that they may be lunar, and this 'solarist vs. lunatics' debate has included the Arles-Fontvieille hypogées and surrounding monuments.⁷

Monument	Azimuthal Orientation
Hypogée de la Source	259°
Hypogée de Bounias	267°
Hypogée de Castellet	262°
Hypogée de Cordes ⁸	240°?
Dolmen de Coutignargue	266°

Table 13.1: Azimuthal orientations of the Arles-Fontvieille monuments.

My own field research at the Arles-Fontvieille monuments, whose orientations are shown in Table 13.1, has shown that three of the intact monuments are seasonally illuminated by the rays of the setting sun, which penetrate the length of the monuments and strike their back walls for several weeks on and around the equinoxes (Fig. 13.2 shows this event in the Grotte de la Source). ⁹

Exploratory 3D simulation shows that the Grotte de Cordes (to which access is more limited) is illuminated during the winter months though the accuracy of my model is not sufficient to pinpoint conclusively the exact period(s) of time. Thus, while the regional westerly tradition of orientation centred on the Arles-Fontvieille monuments is indeed relatively unique in the European context,¹⁰ I argue that it is a variation of a larger tradition of solar orientation for the purpose of seasonal illumination which

---

⁶ Yves Chevalier, 'Orientations of 935 dolmens of southern France', *Journal for the History of Astronomy Supplement* 30 (1999): p. 47; Gérard Sauzade, 'Des Dolmens en Provence', in *Mégalithismes : de l'Atlantique à l'Ethiopie*, ed. Jean Guilaine (Paris : Errance, 1999), pp. 125–40; Michael Hoskin, *Tombs, Temples and their Orientations: a new perspective on Mediterranean prehistory* (Bognor Regis: Ocarina Books, 2001). Clive L. N. Ruggles, 'Later Prehistoric Europe', in *Heritage Sites of Astronomy and Archaeoastronomy in the context of the UNESCO World Heritage Convention*, ed. C. L. N. Ruggles and Michel Cotte (Paris: IAU, 2010).

⁷ Hoskin, *Tombs, Temples and their Orientations : a new perspective on Mediterranean prehistory*, p. 216; A. César González-García, Lourdes Costa Ferrer, and Juan A Belmonte, 'Solarists vs. Lunatics: modelling patterns in megalithic astronomy', in *Lights and Shadows in Cultural Astronomy Proceedings of the SEAC 2005*, ed. Mauro Peppino Zedda and Juan Antonio Belmonte (Comune di Isili: Instituto de Astrofisica de Canarias, 2007).

⁸ It should be noted that the orientation of the Grotte or Hypogée de Cordes was incorrectly listed as 270° in Hoskin's *Tombs, Temples and their Orientations*. My own measure, taken with magnetic compass, is subject to some uncertainty due to the presence of steel or iron I-beams reinforcing the roof of the monument. Satellite imagery, though limited by resolution and possible distortion, suggests an orientation between 240-250° (the latter figure being close to Cazali de Fondouce's (1873:9) measure of N 71 E – though clearly he was giving the orientation facing the entrance. It is, in any case, certainly a south-westerly to west south-westerly orientation. The as yet unpublished survey of the Grotte de Cordes by Xavier Margarit (personal communication) may help clarify this.

⁹ Morgan Saletta, 'The archaeoastronomy of the megalithic monuments of Arles–Fontvieille: the equinox, the Pleiades and Orion', in *Archaeoastronomy and Ethnoastronomy: Building Bridges between Cultures, Proceedings of the International Astronomical Union* ed. Clive L.N. Ruggles (Cambridge: Cambridge University Press, 2011); Saletta, 'The Arles-Fontvieille Megalithic Monuments: astronomy and cosmology in the European Neolithic'.

¹⁰ Hoskin, *Tombs, Temples and their Orientations: a new perspective on Mediterranean prehistory*; Juan Antonio Belmonte and Edmundo Edwards, 'Archaeoastronomy: archaeology, topography and celestial landscape—From the Nile to Rapa Nui' (paper presented at the Highlights of Spanish Astrophysics VI, 2011).

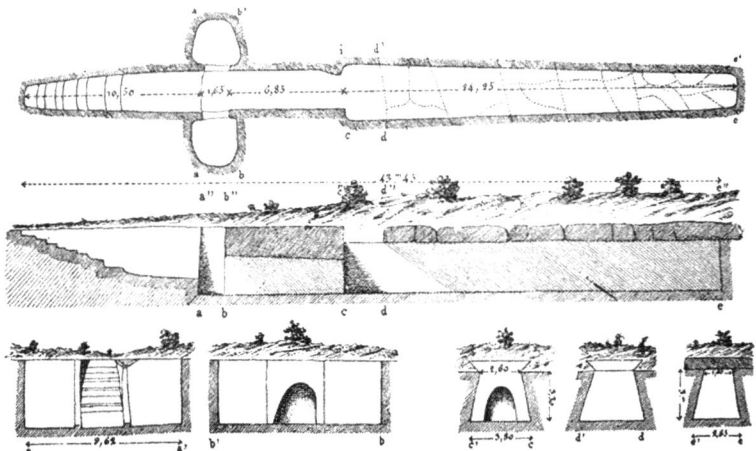

Fig. 13.2: Cut and plan of the Grotte de Cordes (de Fondouce, 1873).

extended throughout much of Neolithic Europe.[11] While perhaps not enough to conclusively resolve the 'solarist vs. lunatics' debate concerning these monuments, the light and shadow hierophany is certainly strong phenomenological evidence for a solar rather than a lunar orientation.

Moving beyond the celestial targeting paradigm: seasonal illumination and ritual time

Despite a relatively extensive literature concerning light and shadow hierophanies within monuments, most archaeoastronomers investigating European megaliths continue to think in terms of what I term the 'celestial targeting hypothesis' which assumes that a monument's orientation is targeting or aligned to the position of a celestial body at some point in the sky, generally on or near the horizon. This paradigm is implicit in the interpretation of regional orientation or azimuthal signatures by Hoskin and others.[12]

As a result of an over emphasis on the celestial targeting paradigm, as well as continued emphasis on celestial events considered significant from a modern astronomical perspective, the possibility that the light and shadow hierophanies might help to explain the statistical orientation pattern of passage graves and similar monuments has largely been ignored with the exception of a few well-known examples including Maeshowe and Newgrange. Clearly, given the many morphological differences in monumental structures, not all monuments would be deeply illuminated by the sun, but many, perhaps even the majority, would be.[13] Clusters of monuments may also be oriented with respect to the moon or even to the heliacal risings of asterisms.[14] However, while horizon astronomy may explain the orientation of some

---

[11] Richard Bradley, 'Darkness and Light in the Design of Megalithic Tombs', *Oxford Journal of Archaeology* 8, no. 3 (1989): pp. 251–59.

[12] e.g. Hoskin, *Tombs, Temples and their Orientations: a new perspective on Mediterranean prehistory*.

[13] For a discussion of angled passage graves and possible shadow and light hierophanies in Brittany see Bradley, 'Darkness and Light in the Design of Megalithic Tombs'.

[14] Clive L. N. Ruggles, 'The stone rows of south-west Ireland: a first reconnaissance', *Journal for the History of Astronomy Supplement* 25 (1994): S1–S20; Fabio Silva, 'A Tomb with a View: New Methods for Bridging the Gap between Land and Sky in Megalithic Archaeology', *Advances in Archaeological Practice: A Journal of the Society for American Archaeology* 2, no. 1 (2014): pp. 24–37; A. César González-García and Lourdes Costa Ferrer, 'Possible astronomical orientation of the Dutch hunebedden', in *Calendars, Symbols, and Orientations: Legacies of Astronomy in Culture. Proceedings of 9th*

monuments, it is insufficient from an interpretive perspective if we wish to fully understand the role of astronomy in the construction of late prehistoric European monuments.

In the context of the Neolithic landscape of Orkney, Jane Downes and Colin Richards suggest that the broad range of monument orientations was creating an 'ancestral time' associated with specific places in the landscape through the successive illumination of monuments at different times of the year.[15] A more general anthropological term for this kind of time is 'ceremonial time' or 'ritual time'.[16] Perhaps, in Orkney and elsewhere, the wide variation in orientations served to distribute celebrations across a period of time, or ritual/festive season(s), thereby allowing various groups and clans to participate in their own and other's celebrations. Such distributions may very well have produced what Peter Dwyer has termed social synchrony, helping to organize ritual and seasonal ecological activities of various social groups in time and space.[17]

Houses of the Living, Houses of the Dead: solar orientation and cosmological symbolism in domestic and monumental architecture

Archaeological evidence and ethnographic analogy suggest that chambered megalithic monuments were cosmologically related to domestic structures and conceived of by those who built and used them as houses of the dead, spirit houses and shrines to the ancestors.[18] What has not previously been suggested, to my knowledge, is that the orientation of late prehistoric tombs and light and shadow hierophanies might have an origin in the *functional* design and orientation of domestic structures. Peter Topping argues European longhouses were oriented roughly south and south-easterly to maximize solar heating.[19] Functional solar orientation has also been suggested for dwellings at Barnhouse in Orkney.[20] Thus, seasonal light and shadow hierophanies in passage graves and similar monuments may have evolved from the seasonal experience of light and illumination within Neolithic domestic structures oriented deliberately to take advantage both of the sun's warmth and illumination. In the larger Western European context this might also help to explain the predominance of funerary

---

annual meeting of European Society for Astronomy in Culture (SEAC), ed. Mary Blomberg, Peter E. Blomberg, and Göran Henriksson (Stockholm: Uppsala University University Press, 2003), pp. 111–18.

[15] J. Downes and C. Richards, 'The Dwellings at Barnhouse', in *Dwelling among the monuments: the Neolithic village of Barnhouse, Maeshowe passage grave and surrounding monuments at Stenness, Orkney*, ed. Colin Richards (Cambridge: McDonald Institute for Archaeological Research, 2005), pp. 57–126.

[16] Barbara Bender, 'Time and landscape', *Current Anthropology* 43, no. S4 (2002): S103–S112; Maurice Bloch, 'The past and the present in the present', *Man* (1977): pp. 278–92; Richard Bradley, 'Ritual, time and history', *World Archaeology* 23, no. 2 (1991): pp. 209–19.

[17] P. D. Dwyer, *The pigs that ate the garden: a human ecology from Papua New Guinea* (Ann Arbor: University of Michigan Press, 1990); David W. Wheatley et al., 'Approaching the landscape dimension of the megalithic phenomenon in Southern Spain', *Oxford Journal of Archaeology* 29, no. 4 (2010): pp. 387–405; Fabio Silva, 'Landscape and Astronomy in Megalithic Portugal: the Carregal do Sal Nucleus and Star Mountain Range', *Papers from the Institute of Archaeology* 22 (2013): pp. 99–114; Silva, 'A Tomb with a View: New Methods for Bridging the Gap between Land and Sky in Megalithic Archaeology'.

[18] Ruggles, 'Later Prehistoric Europe', pp. 34–35; Christine Hugh-Jones, *From the Milk River: spatial and temporal processes in northwest Amazonia* (Cambridge: Cambridge University Press, 1979).

[19] Peter Topping, 'Structure and Ritual in the Neolithic house: Some examples from Britain and Ireland (paper presented at the Neolithic houses in northwest Europe and beyond, 1996), p. 162. Such orientation is fairly flexible, a long axis due south is ideal, but variations up to 20 degree have minimal impact, 'Passive Solar Design', California Energy Commission, available at http://www.consumerenergycenter.org/residential/construction/solar_design/orientation.html [accessed 29 July 2016].

[20] Downes and Richards, 'The Dwellings at Barnhouse'.

monument orientations toward the rising or setting sun in the winter months.

Light and shadow hierophanies were, of course, also imbued with cosmological meaning. Duncan Garrow et al., in emphasizing the relationship between Maeshowe, the village of Barnhouse and cosmological considerations, employ Arnold van Gannep's notion of a rite of passage, suggesting that after months of cold darkness, symbolically related to death, the monument is flooded with light and warmth from the sun 'allowing a rite of passage involving union with the ancestors'.[21] The illumination of monuments by the sun may also have been related to conceptions of the movement of the soul from one cosmological realm to another, possibly along the path of the sun as suggested by David Lewis-Williams and David Pearce, a notion which echoes Alasdair Whittle's suggestion that 'Passages, for example, may have been as much for the exits of souls and spirits as for the entrances of the living'.[22] Here too, the work of van Gannep on the importance of celestial bodies in rites of passage is particularly relevant.[23]

Conclusion

The seasonal illumination of the Arles-Fontvieille monuments, which is substantially similar to the illumination events at Newgrange in Ireland and Maeshowe in Scotland, is an important piece of evidence suggesting that astronomical alignment and light and shadow hierophanies were part of a shared core of cosmological conceptions and practices involving monumental construction in late prehistoric Europe during the fourth and early third millennium BCE.[24] Seasonal light and shadow hierophanies whose purpose, while cosmologically symbolic, may have had their origin in the functional orientation of Neolithic dwellings for the purpose of solar heating and illumination. In future, archaeoastronomers should consider to what extent and in which cases, the seasonal illumination hypothesis may offer fuller interpretive possibilities than the celestial targeting paradigm. One way to do this would be a collaborative research project, possibly enlisting enthusiastic amateurs in a 'citizen science' project designed to document solar illumination within these monuments when and where it occurs.

Bibliography

Belmonte, Juan Antonio, and Edmundo Edwards. 'Archaeoastronomy: Archaeology, Topography and Celestial Landscape—from the Nile to Rapa Nui'. Paper presented at the Highlights of Spanish Astrophysics VI, 2011.
Bender, Barbara. 'Time and Landscape'. *Current Anthropology* 43, no. S4 (2002): S103–S112.
Bloch, Maurice. 'The Past and the Present in the Present'. *Man* (1977): pp. 278–92.
Bradley, Richard. 'Darkness and Light in the Design of Megalithic Tombs'. *Oxford Journal of Archaeology* 8, no. 3 (1989): pp. 251–59.
Bradley, Richard. 'Ritual, Time and History'. *World Archaeology* 23, no. 2 (1991): pp. 209–19.
Chevalier, Yves. 'Orientations of 935 Dolmens of Southern France'. *Journal for the History of Astronomy Supplement* 30 (1999): p. 47.
Cunliffe, B. *Facing the Ocean: The Atlantic and Its Peoples, 8000 BC to AD 1500*. Oxford: Oxford University Press, 2001.
Daniel, Glyn Edmund. *The Prehistoric Chamber Tombs of France: A Geographical, Morphological and Chronological Survey*. London: Thames and Hudson, 1960.

---

[21] Duncan Garrow, John Raven, and Colin Richards, 'The Anatomy of a Megalithic Landscape', in *Dwelling among the monuments: the Neolithic village of Barnhouse, Maeshowe passage grave and surrounding monuments at Stenness, Orkney*, ed. Colin Richards (Cambridge: McDonald Institute for Archaeological Research, 2005), p. 253.
[22] David Lewis-Williams and David Pearce, *Inside the Neolithic Mind : consciousness, cosmos and the realm of the gods* (London: Thames & Hudson, 2005), p. 79; Alistair Whittle, *Europe in the Neolithic: the creation of new worlds* (Cambridge: Cambridge University Press, 1996), pp. 248.
[23] Arnold Van Gennep, *The rites of passage* (Chicago: University of Chicago Press, 1960), pp. 5, 84, 155–60
[24] B. Cunliffe, *Facing the Ocean: The Atlantic and Its Peoples, 8000 BC to AD 1500* (Oxford: Oxford University Press, 2001), pp. 211–12; Magdalena S. Midgley, *The Megaliths of Northern Europe* (London, New York: Routledge, 2008), p. 192.

de Fondouce, Paul Cazalis. *Les Temps Préhistoriques Dans Le Sud-Est De La France: Allées Couvertes De La Provence*. C. Coulet, 1873.
Downes, J., and C. Richards. 'The Dwellings at Barnhouse'. In *Dwelling among the Monuments : The Neolithic Village of Barnhouse, Maeshowe Passage Grave and Surrounding Monuments at Stenness, Orkney*, edited by Colin Richards, pp. 57–126. Cambridge: McDonald Institute for Archaeological Research, 2005.
Dwyer, P. D. *The Pigs That Ate the Garden: A Human Ecology from Papua New Guinea*. Ann Arbor: University of Michigan Press, 1990.
Garrow, Duncan, John Raven, and Colin Richards. 'The Anatomy of a Megalithic Landscape'. In *Dwelling among the Monuments: The Neolithic Village of Barnhouse, Maeshowe Passage Grave and Surrounding Monuments at Stenness, Orkney*, edited by Colin Richards, pp. 249–60. Cambridge: McDonald Institute for Archaeological Research, 2005.
González-García, A César, and Lourdes Costa Ferrer. 'Possible Astronomical Orientation of the Dutch Hunebedden'. In *Calendars, Symbols, and Orientations: Legacies of Astronomy in Culture. Proceedings of 9th Annual Meeting of European Society for Astronomy in Culture (SEAC)*, edited by Mary Blomberg, Peter E. Blomberg and Göran Henriksson, pp. 111–18. Stockholm: Uppsala University Press, 2003.
González-García, A. César, Lourdes Costa Ferrer, and Juan A Belmonte. 'Solarists Vs. Lunatics: Modelling Patterns in Megalithic Astronomy'. In *Lights and Shadows in Cultural Astronomy Proceedings of the SEAC 2005*, edited by Mauro Peppino Zedda and Juan Antonio Belmonte, pp. 23–30. Comune di Isili: Instituto de Astrofisica de Canarias, 2007.
Guilaine, Jean. *Au Temps Des Dolmens: Mégalithes Et Vie Quotidienne En France Méditerranéenne Il Y a 5000 Ans*. Toulouse: Ed. Privat, 1998.
Guilaine, Jean. *Méditerranée Mégalithique: Dolmens, Hypogées, Sanctuaires*. Lacapelle : Éd. Archéologie Nouvelle, 2011.
Hoskin, Michael. *Tombs, Temples and Their Orientations: A New Perspective on Mediterranean Prehistory*. Bognor Regis: Ocarina Books, 2001.
Hugh-Jones, Christine. *From the Milk River: Spatial and Temporal Processes in Northwest Amazonia*. Cambridge: Cambridge University Press, 1979.
Lewis-Williams, David, and David Pearce. *Inside the Neolithic Mind: Consciousness, Cosmos and the Realm of the Gods*. London: Thames & Hudson, 2005.
Margarit, Xavier. *Les Monument Mégalithiques D'arles-Fontvieille, État Des Connaissances, Contexts Et Nouvelles Données (Projet Collectif De Recherche 2014)*, ddited by Xavier Margarit. Aix-en-Provence: CNRS/LAMPEA, 2014.
Mérimée, P. *Notes D'un Voyage Dans Le Midi De La France*. Paris: Librarie de Fournier, 1835.
Midgley, Magdalena S. *The Megaliths of Northern Europe*. London, New York: Routledge, 2008.
'Passive Solar Design'. California Energy Commission, www.consumerenergycenter.org/residential/construction/solar_design/orientation.html.
Ruggles, Clive L. N. 'Later Prehistoric Europe'. In *Heritage Sites of Astronomy and Archaeoastronomy in the Context of the UNESCO World Heritage Convention*, edited by C. L. N. Ruggles and Michel Cotte, pp. 28–35. Paris: IAU, 2010.
Ruggles, Clive L. N. 'The Stone Rows of South-West Ireland: A First Reconnaissance'. *Journal for the History of Astronomy Supplement* 25 (1994): S1-S20.
Saletta, Morgan. 'The Archaeoastronomy of the Megalithic Monuments of Arles–Fontvieille: The Equinox, the Pleiades and Orion'. In *Archaeoastronomy and Ethnoastronomy: Building Bridges between Cultures. Proceedings of the International Astronomical Union*, edited by Clive L. N. Ruggles, pp. 364–73. Cambridge: Cambridge University Press, 2011.
Saletta, Morgan. 'The Arles-Fontvieille Megalithic Monuments: Astronomy and Cosmology in the European Neolithic'. PhD thesis, University of Melbourne, 2014.
Sargiano, Jean Philippe, André D'Anna, Céline Bressy, Jessie Cauliez, Muriel Pellissier, Hugues Plisson, Stéphane Renault, et al. 'Les Arnajons (Le-Puy-Sainte-Réparade, Bouches-Du-Rhône): Un Nouveau Dolmen Dans Le Sud-Est De La France'. *Préhistoires Méditerranéennes* (2010): pp. 119–53.
Sauzade, Gérard. 'Des Dolmens En Provence'. In *Mégalithismes: De L'atlantique À L'ethiopie*, edited by Jean Guilaine, pp. 125–40. Paris: Errance, 1999.
Silva, Fabio. 'Landscape and Astronomy in Megalithic Portugal: The Carregal Do Sal Nucleus and Star Mountain Range'. *Papers from the Institute of Archaeology* 22 (2013): pp. 99–114.
Silva, Fabio. 'A Tomb with a View: New Methods for Bridging the Gap between Land and Sky in Megalithic

Archaeology'. *Advances in Archaeological Practice: A Journal of the Society for American Archaeology* 2, no. 1 (2014): pp. 24–37.

Topping, Peter. 'Structure and Ritual in the Neolithic House: Some Examples from Britain and Ireland'. Paper presented at the Neolithic houses in northwest Europe and beyond, 1996.

Van Gennep, Arnold. *The Rites of Passage*. Chicago: University of Chicago Press, 1960.

Wheatley, David W., Leonardo Garcia Sanjuan, Patricia A. Murrieta Flores, and Joaquin Marquez Perez. 'Approaching the Landscape Dimension of the Megalithic Phenomenon in Southern Spain'. *Oxford Journal of Archaeology* 29, no. 4 (2010): pp. 387–405.

Whittle, Alistair. *Europe in the Neolithic: The Creation of New Worlds*. Cambridge: Cambridge University Press, 1996.

AN ETHNOASTRONOMY STUDY ON THE ASTRONOMICAL ORIENTATION AND ASTRAL DECORATION OF THE STONE GRANARIES (HÓRREOS) OF VILABOA (GALICIA, SPAIN)

Fátima Braña Rey and Ana Ulla Miguel[1]

ABSTRACT: Galicia is the Spanish region located on the northwest area of the Iberian Peninsula, where maize crops were grown extensively at the end of the seventeenth century and the beginning of the eighteenth.[2] The specific granaries called *hórreos* are a characteristic feature in the northern most humid regions of Spain and Portugal, and were built to store the grain in the best possible conditions. These singular buildings decorate the Galician landscape with different shapes and sizes. They are also located in privileged places near important buildings such as monasteries, rectory houses and dwellings. These *hórreos* gave prestige to any house, in the sense that they showed a healthy economy by being a visible sign of the main grain crops harvest in Galicia.[3] This privileged location seems to indicate that their orientation should, in theory, be associated with their efficiency as maize granaries[4] while the decorations displayed by the *hórreos* must be significant and related to the economic and social organization.

In the Peninsula do Morrazo (province of Pontevedra, southwestern Galicia), we found in the structure of these stone granaries a decoration that simulates astral motifs worth studying. In particular, decorations in the shape of moons, suns or stars, along with the presence of sundials in some cases, have claimed our attention and a sample of these *hórreos* has been considered for an initial study. Our research aims to increase the knowledge and insight analysis of the orientation of *hórreos* containing astral decorations in Galicia, together with that of the astral motifs themselves. The work has started with a sample of granaries found in the parishes of Santa Cristina and San Adrian de Cobres, belonging to the council of Vilaboa, where a great number of astral-related elements were found. In no other region of Galicia we have seen, so far, a similar number of astral decorative elements in the granaries.

Despite the difficulties of multidisciplinary work in cultural astronomy,[5] this paper focuses on the team work by astronomers and anthropologists, using a methodology based on a twofold foundation: on the one hand, the ethnography which helps us to elaborate an inventory and give meaning to all the decorations found, and, on the other hand, the archaeoastronomy accounting for the measuring of orientations of the *hórreos* in our sample. We want to test three main hypotheses: 1. whether the orientation of *hórreos* with astral motifs is similar to or different from that of *hórreos* with other types of decorations; 2. whether the orientations found may carry astronomical bearings of any sort; and 3. what the meanings of the astral motifs are, if any, in the context of the cosmic vision of the area inhabitants and of the *hórreos*' owners in particular.

Given that this initial analysis is yet limited, only preliminary results can be presented here for the moment.

---

[1] Universidade de Vigo

[2] Enrique Bande Rodríguez and Carlos Tain Carril, 'El mundo simbólico del hórreo gallego', in *Hórreos. Actas del primer Congreso europeo del Hórreo* (Santiago de Compostela: Compostela, 1990), pp. 53-60.

[3] Jozef Van Linthoudt, *Bueu: os seus canteiros e os seus reloxos de sol* (Vigo: Autoedición, 1997), at http://descubrobueu.wix.com/bueu#!bficha033/clov [accessed 26 April 2016].

[4] O. A. Perez-Garcia, X. C. Carreira, E. Carral, M. E. Fernandez, and R. A. Mariño, 'Evaluation of traditional grain store buildings (hórreos) in Galicia (NW Spain): analysis of outdoor/indoor temperature and humidity relationships', *Spanish Journal of Agricultural Research* 8, no. 4 (2010): pp. 925-935; César Saá, J. L. Míguez, J. C. Morán, J. A. Vilán, M. L. Lago, R. Comesaña, J. Collazo, 'The influence of slotted floors on the bioclimatic traditional Galician agricultural dry-store structure (hórreo)', *Energy and Buildings* 43, no. 12 (2011): pp. 3491-3496.

[5] C. Esteban, 'Is cultural astronomy an interdisciplinary science? An astrophysicist's reflections', *Complutum* 20, no. 2 (2009): pp. 69-77.

Introduction

The aim of this work is to perform a collaboration between two areas of knowledge (astronomy by Ana Ulla and anthropology by Fátima Braña) in order to understand orientations and decorations with astral bearings, in one of the most representative buildings from our rural Galician heritage. With this paper we intend to report our first findings into the study of the stone granaries in Galicia (northwest of Spain) called *hórreos* from a cultural astronomy approach. Since we are at the beginning of the research we opened our quest in many ways. This inductive perspective has been applied in a small territory, called Cobres parishes, which are part of the municipality of Vilaboa, where a large number of stone granaries with decorations that simulate astral motifs have been found that are worth studying.

Our work will be presented here in four parts: firstly we cover important general points to be taken into account when discussing Galician *hórreos*. Secondly, it is necessary to know about some of the *hórreos'* particular features. Next, we will present our data and sample on orientations and decorations of *hórreos* in the Cobres parishes of Vilaboa, and we finish with a summary of conclusions.

General points

Galicia is the Spanish region located in the northwest area of the Iberian Peninsula. The specific granaries called *hórreos* are characteristic of the northern and northwest areas of the Peninsula; that is to say, the north of Portugal, Asturias, Cantabria, the northern parts of Castilla and León, Navarra and the Basque Country are the territories where the *hórreos* were built in order to maintain the best conditions to hold crops in these humid regions.

These singular buildings decorate the Galician landscape in different shapes and sizes. They are also located in privileged places near important buildings such as monasteries (e.g., Poio, the largest in Galicia), rectory houses (e.g., Cobres rectory house has two *hórreos*: one with fourteen feet, or columns, and the other one is a *paneira*, which is the largest type of *hórreo* that can be found), as well as noble dwellings, called *Pazos* in Galicia (e.g., Pazo Lourizan has a big *hórreo* with sixteen feet or columns).

The first representation of an *hórreo* was made in the thirteenth century.[6] It shows monks praying next to small buildings that resemble granaries which are full of a cereal very similar to maize. They look similar to actual Galician *hórreos*: an elevated rectangular structure with a roof used in rural spaces as granaries. Thanks to this drawing we know that *hórreos* were in existence before maize was grown as a crop, but the origin of hórreos is not clear: different authors talk about references from the eighth century[7] or earlier.[8] This cereal came to Galicia in the seventeenth century from America. In Galicia, maize crops were grown extensively in the eighteenth century when the actual *hórreos* began to spread all over Galicia.[9]

Since the nineteenth century, as part of the romantic ideology, *hórreos* became a sign of Galician culture and in 1973 they were recognized as part of our cultural heritage. From that date on, all *hórreos* that are over one hundred years old are protected as Galician heritage by law, art.1 *of Decreto 449/1973*.[10]

Galician *hórreos* have been extensively studied by a significant number of researchers from

---

[6] Museo de Pontevedra. *Cantigas de Santa María*, at http://www.museo.depo.es/noticias/notas.de.prensa/es.02010152.html [accessed 9 May 2014].

[7] Perfecto Rodríguez Fernández, 'El hórreo en la diplomacia medieval asturiana en latín (siglos VIII-XIII)', *Aula abierta* 41-42 (1984): pp. 97-114.

[8] Eugeniusz Frankowski, *Hórreos y palafitos de la Península Ibérica* (Madrid: Istmo, 1986).

[9] Bande Rodríguez and Tain Carril, 'El mundo simbólico del hórreo gallego', pp. 53-60.

[10] Decreto 449/1973, February 22th, por el que se colocan bajo la protección del Estado los 'hórreos' o 'cabazos' antiguos existentes en Galicia y Asturias. B.O.E no. 62, 13 March.

different fields of study, including linguists,[11] historicists,[12] geographers,[13] engineers,[14] ethnographers[15] and architects.[16] A big part of these studies has focused on describing *hórreos* as logical or pragmatic architectural elements. Most of them show how the Galician popular culture includes a pragmatic use of environmental knowledge to work with natural resources. In this sense, architectural features are essential to understand *hórreos* as a functional and popular building.

Of the different kinds of *hórreos* that are found all over Galicia, we focus on two types: stone granaries (called the Morrazo type) and mixed ones, made of wood and stone (called the Pontevedra type).[17]

Hórreo features

As mentioned earlier, the most common shape of an *hórreo* in Galicia is the rectangular one. Width or *penal* – in Galician language – is the frontal structure, usually made of stone, and length or *costal* is the lateral side, made of stone and wood in the mixed type (typical of the town of Pontevedra and surrounding areas), or only of stone. *Hórreos* with lateral sides made only of stone can be found all around the Peninsula do Morrazo, in the province of Pontevedra. Vilaboa, where we are sourcing data from, is a municipality also within the province of Pontevedra and borders with the Peninsula do Morrazo. Thus, our area of study has strong influences from both regions.

Hórreos sit on a varied number of 'feet': columns and circles, or straddle stones, which support the building. They can be found sitting either on four, six, eight or even more feet. In Cobres parishes these feet usually adopt a special shape, which looks like a real boot (leg and foot). The role of the feet is to sustain the balance and weight of the building, and this particular leg-type shape is also used in other parts of Galicia like the Ourense region.

Between the main structure and any foot, there is a round piece of stone – called *tornaratos* – which provides a wider area for support and makes it difficult for rodents to get into the storage room. The main part of an *hórreo* are the inner walls where the corn, grains and other vegetables are kept.

They are preserved thanks to slits that allow the air to flow inside. These slits make it easier to store and dry the maize. As people say from interviews, these are the main functions of *hórreos*, to keep and dry maize, which is confirmed by scientific studies[18].

On top, *hórreos* have gable roofs, and in the Galician region these are made with tiles. *Pinche* is the

---

[11] Alberto Balil, 'Equivalentes y significados del término hórreo en el latín', in *Hórreos. Actas del primer Congreso europeo del Hórreo* (Santiago de Compostela: Compostela, 1990), pp. 43-51; Eligio Rivas Quintas, *Millo e hórreo, legumia e cestos* (Santiago de Compostela: Ediciόns Laiovento, 1996).

[12] María Teresa Rivera Rodríguez, *Los Pazos orensanos: arquitectura popular del siglo XVIII en la provincia de Orense* (Orense: Caja de Ahorros Provincial de Orense, 1981); Frankowski, *Hórreos y palafitos de la Península Ibérica*.

[13] Ignacio Martínez-Rodríguez, *El hórreo gallego*, 2nd edition (A Coruña: Fundación Pedro Barrié de la Maza, 1999).

[14] Perez-Garcia et al., 'Evaluation of traditional grain store buildings (hórreos) in Galicia', pp. 925-935; Saá et al., The influence of slotted floors', pp. 3491-3496.

[15] Xaquín Lorenzo Fernández, *A Casa* (Vigo: Galaxia, 1995; Begoña Bas López, 'Arquitectura para a produción campesiña. O millo: os hórreos en Galicia', in *Antropoloxía*, ed. Xosé Manuel Gónzalez Reboredo (A Coruña: Hércules,1997), pp. 238-282.

[16] Manuel Caamaño Suárez, *As construccións da arquitectura popular da arquitectura popular: Patrimonio etnográfico de Galicia* (A Coruña: Hércules, 2006); Manuel Caamaño Suárez, 'O hórreo galego na encrucillada' in *Hórreos. Actas del primer Congreso europeo del Hórreo* (Santiago de Compostela: Editorial Compostela, 1990), pp. 67-76; Pedro De Llano, *Arquitectura popular en Galicia* (Santiago de Compostela: Colegio Oficial de Arquitectos de Galicia, 1983).

[17] In this cataloguing we followed Martínez-Rodríguez, *El hórreo gallego*.

[18] Perez-Garcia et al., 'Evaluation of traditional grain store buildings (hórreos) in Galicia', pp. 925-935; Bas López, 'Arquitectura para a producción campesiña', pp. 192-535.

gable in the *hórreos* and it is usually the part where the decorations are drawn (astral or otherwise). On top of the roof, at both front sides, there are other two forms of decoration, which all authors who have studied *hórreos* mention: a cross and a pinnacle. We do not analyse these decoration elements in our work but they are interesting apotropaic elements. It has been suggested that they can contribute to keep away the evil influences – such as the devil or witches – from harvests[19].

Locations, orientations and decorations of *hórreos* in Vilaboa

Our first issue was to obtain the position of the largest number of *hórreos* possible that contained astral elements, in the context of the positions of all *hórreos* in general. We obtained the kind assistance of the Concello de Vilaboa (town council of Vilaboa) to find out how many *hórreos* there are in the municipality. Thanks to this data we could draw the map in Figure 14.1, showing the distribution of *hórreos* in the two Cobres parishes of Santa Cristina and San Adrián. All the rectangular spots in the picture are *hórreos*: they represent the exact location of the *hórreos*, with their real orientation, in original ED50 UTM (X, Y) system coordinates. This information, as explained below, was useful in computing a first approach to their azimuth orientations from the north (Y axis). Santa Cristina has 137 *hórreos* and San Adrián 108 *hórreos*. So both parishes account for 245 *hórreos* in total. It is to be noted that these two parishes are on coastline of the Ría de Vigo (bay) and correspond to inhabited areas, close to the sea, and are surrounded by forest (Fig. 14.2). In the Cobres parishes, we have observed that *hórreos* are always in a plot of houses. It is important to take this into account in order to plan our fieldwork, since not all owners welcome researchers into their houses!

For the 245 *hórreos* in Figure 14.1, we obtained preliminary statistical results on their azimuths in the following way: a) north is up and east is right on the plane, corresponding to +Y (0° N) and +X (90° E), respectively; b) for each rectangle (*hórreo*) we considered as the longitudinal direction that of its largest side; c) using the AutoCAD software we measured the angle of each longitudinal direction from north (+Y), towards south (coord. -Y, 180°), plus/minus 1 degree of uncertainty; d) as, on the plane, it is not possible to find any peculiar elements (the side on which the *hórreo's* door is located, for example) that could help us to discriminate between an E or W direction for the rectangles, only azimuths between 0° and 180° were considered. The azimuths' distribution collected this way is shown in Figures 14.3 and 14.4, from north (coord. +Y, 0°) to south (coord. -Y, 180°). For Figure 14.3 an average three degree west magnetic correction has been subtracted from the measurements for the whole area of the two parishes, making use of the data provided at the WEB facility of the National Geophysical Data Center in the USA (NOAA – see footnote 25). In the future, a detailed magnetic correction is intended to be applied for the position of each of the 245 *hórreos* individually, for which the ED50 UTM (X, Y) coordinates information for each rectangle will be needed.

As we can observe from the data in Figures 14.3 and 14.4, the 245 *hórreos* do not show a similar or homogenous orientation, nor a clearly preferred tendency towards a certain one but, perhaps, a slight indication towards northeast or south. In order to analyse these preliminary results we have to take into account that many *hórreos* were moved or destroyed. There are also private reasons for their current locations: an earlier expropriation of land to build the nearby motorway altered the location of some houses and their *hórreos*, and some of them were also lost or moved from their original positions for various other reasons. Other *hórreos* are new, they may been built simply to embellish new houses. When carrying out fieldwork in these conditions, it becomes hard to ascertain the true positional value of each *hórreo*.

As stated earlier, out of these 245 *hórreos* in Vilaboa we have selected a subsample for a more detailed

---

[19] Alfonso Castelao, *As cruces de pedra na Galiza* (Vigo: Galaxia, 1998).

An Ethnoastronomy Study on the Astronomical Orientation and Astral Decoration of the Stone Granaries (Hórreos) of Vilaboa (Galicia, Spain)

Fig. 14.1: Horreos in Santa Cristina and Santo Cobres, 2014.

Fig. 14.2: Santa Cristina and Santo Adrian of Cobres, from Googlemaps, 2014.

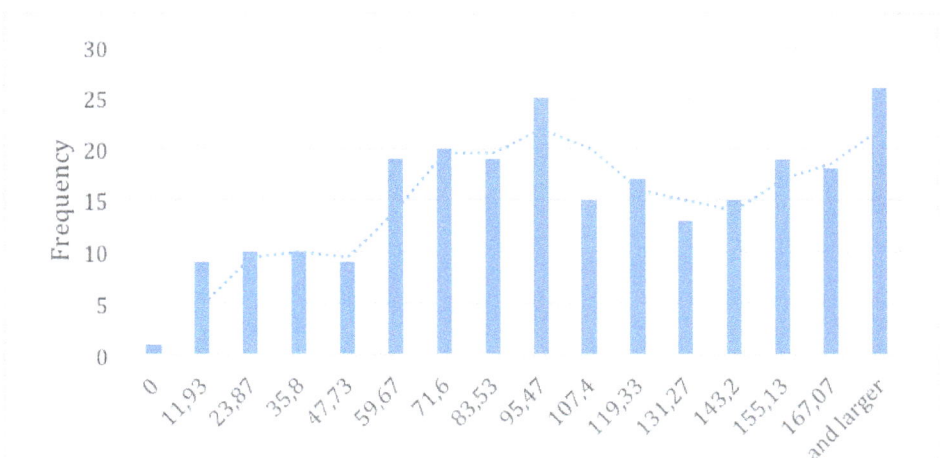

Fig. 14.3: Hórreos of Santa Cristina and Santo Adrián (Vilaboa, Spain). Magnetically uncorrected Azimuths distribution of their largest longitudinal direction; from 0° N to 180° S, (in degrees, plus/minus 1° of uncertainty).

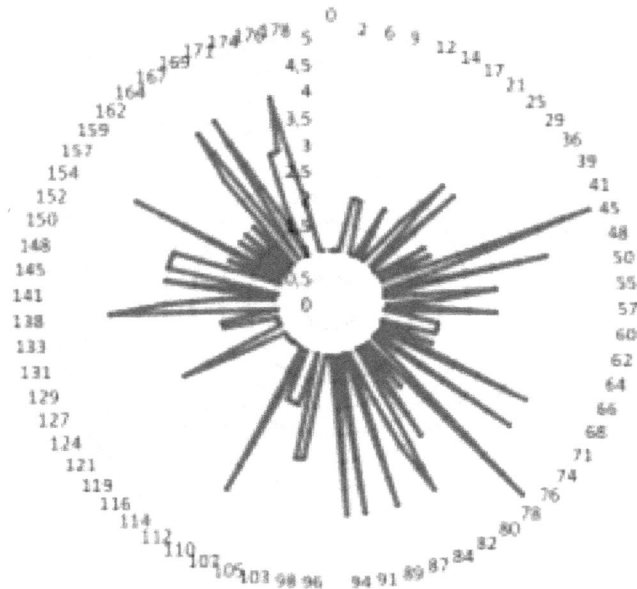

Fig. 14.4: Number of hórreos of Santa Cristina and Santo Adrián (Vilaboa, Spain) per Azimuth (from 0° N to 180° S) with an average magnetic correction of 3° for the whole (245) sample. In degrees, plus/minus 1° of uncertainty.

initial, although preliminary, study. In Cobres parishes we have found so far about 20 *hórreos* displaying decorative shapes such as stars or moons that we have called astral decoration.[20] Nine of them have so far been addressed for the study and Table 14.1 shows the summary of our findings (see also the text below). We must say that it was in fact a surprise for us to find such elements, and even for many of their owners, who had not paid attention to this type of decoration on their own *hórreos*.

However, more precise observations are required as, from an astronomical point of view, it is certain that we have not enough data and analysis yet in order to know whether *hórreos* have an orientation linked to any celestial body – the Sun and the Moon in particular. Orientations towards landscape, horizon, and meteorological elements must also be considered more thoroughly, all issues that we intend to address in the future.

In addition, the roaming method of approximation to the territory with people was the first ethnographical work made. We designed an outline and a survey with two objectives: firstly, locating informants and *hórreos* with astral decoration, and secondly identifying the cultural position of the *hórreo*. The uses and importance of *hórreos* have changed throughout history. Most of them were built in the eighteenth and nineteenth centuries. There are just a few people left that remember the process of building an *hórreo*. Besides, the traditional way to build since the 1980s in the rural spaces has experienced rapid changes and most of the *hórreos* ended up losing their original uses. These circumstances made it necessary to work with both the social memory and their present meanings.

Up to now we have carried out eleven interviews: nine men and three women, all of them between ages 39 to 88 years old, and all of them are from the Cobres parishes. Three interviewees were stonemasons.

---

[20] María Paz García-Gelabert Pérez, 'Los motivos decorativos de los hórreos y paneras de Riocastelo (Tineo, Asturias): un testimonio de la pervivencia de la iconografía astral prerromana', *Hispania antiqua*, no. 35 (2011), pp. 7-42.

We consider the data obtained just enough to do a first assessment of customs in relation to *hórreos* in the Cobres parishes.

Through the fieldwork we discovered different types of *hórreos* according to their materials, shapes, uses and owners, too. The largest *hórreos* are from church properties; there are *hórreos* that work as a barbecue on the inside, others can function as a kennel; some have dormers in the middle of the gable roof, and others are made just with wood over stone columns. So new uses and forms work along with old ones. And although they can be made with different materials and they can have different orientations, regardless, all of the *hórreos* are close to the houses.

When asking about the best place to build an *hórreo*, people said that the most important thing is the orientation of the storehouse taking into account the inclination in relation to the wind, so it keeps the goods safe from the rain. People took into account the Sun (identified as south) and dry wind (they said north/northeast).

Usually *hórreos* are placed in higher parts of the plot and separated from trees and walls. The reason is to keep animals out of the inner walls. The stone is usually clearly visible from old paths and commonly the decoration is visible too. So visibility could be other element to take into account when building *hórreos*.

In Cobres parishes not all people interviewed have an *hórreo* but all of them admitted wanting to have one. Some people mentioned that their ancestors were poor so they could not build an *hórreo*. Nowadays they are so expensive that they are considered a luxury, perhaps because maize crops declined by the end of twentieth century. *Hórreos* made only with wood have disappeared in high numbers but we were able to find some examples in the Cobres parishes. Unfortunately, we did not find astral decorations on them so we left the wood *hórreos* out of the research, at least for the moment.

In addition, the people interviewed consider *hórreos* as pieces of identity; they told about them as 'something of ours', 'something that represents us'. Land, house and *hórreos* connect one generation to the previous one, and interviewees claim that one of the most important characteristics is their adherence to the land.

In Galicia, family and house can be the same. The house is more than a building, it is a unit of production and reproduction, and the *hórreo* works as an important part of both tasks. The *hórreo* was the larder where the harvest of one year will be kept for the house. Furthermore, *hórreos* were the places where the church stored the harvest gathered by common labour and with tithe. According to other researches, it seems that the position, structure and decoration of *hórreos* in Cobres appears to be related to power and social position of the house as a productive unit in rural spaces.[21]

In summary, our data so far points to the orientation of *hórreos* having a relation to the local predominant winds and to social organization.

From a holistic perspective, we then obtained data from the meteorological station just in front of Cobres parishes, on the other side of the Ría de Vigo (bay). The station belongs to the Galician Meteorological Service (Meteogalicia) and it was with their kind help and assistance that the wind distributions for the area displayed in Figures 14.5 to 14.8, for the years 2010 to 2013, were provided to us.

As we can see in the four figures, the strong and dominant winds in the area are south-southwest and north- northeast. Winds from the south are humid and from the north dry.

At present, the data does not allow us to reach conclusions. As we mentioned, this is a work in progress so we are refining measures as well as having in mind to look for historical series of meteorological magnitudes for further analysis.

---

[21] Xosé María Lema, 'Os hórreos do extremo occidental de Galicia', *Gallaecia* 5 (1979): pp. 197-289. Bas López, , 'Arquitectura para a producción campesiña', pp. 238-282.

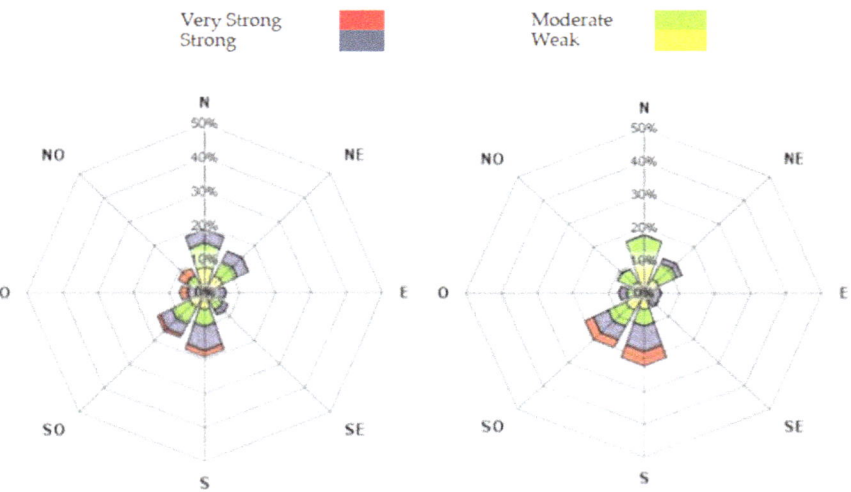

Fig. 5: Force weighted on direction. Viso 2010. Calm 18%.

Fig. 6: Force weighted on direction. Viso 2011. Calm 19%.

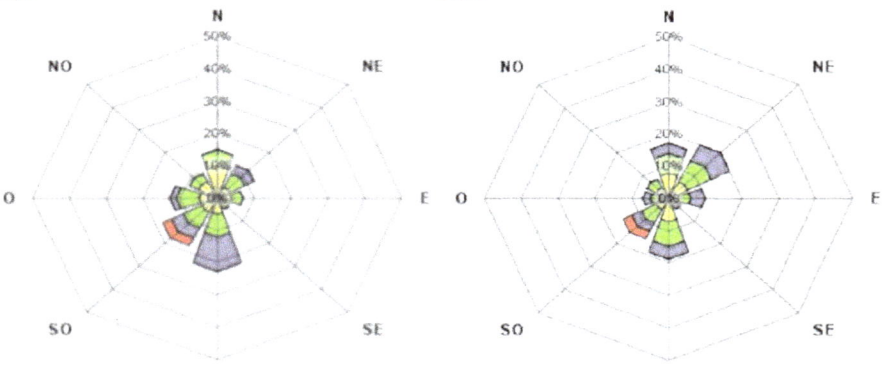

Fig. 7: Force weighted on direction. Viso, 2012. Calm 20%.

Fig. 8: Force weighted on direction. Viso, 2013. Calm 16%.

Figures 14.5-8 show one example of the kinds of stars we found in the hórreos of Cobres parishes. We imagine that they are stars as people said that but perhaps some of them at least could be suns. Stars usually were carved on the stone but they do not have openings to let air flow. They are just ornaments.

An Ethnoastronomy Study on the Astronomical Orientation and Astral 171
Decoration of the Stone Granaries (Hórreos) of Vilaboa (Galicia, Spain)

Fig. 14.9: Gamble with star, moon and circle.

Other slits in semi-circle (moon-like type) shapes made in the gable could have been used to keep pigeons (see Fig. 14.10).

Fig. 14.10: Gamble with moon and hexa-petal rosette.

When looking for explanations people said that they did not know the meaning and the reasons for these shapes in the gables of *hórreos*. It could have been the stone-carver's style or that the owners have ordered them. The people interviewed believed in both possibilities. In our opinion the second option has more meaning due to the value of *hórreos* as heritage and a sign of power of the house. These shapes are made without electrical machines so the time and effort to do them could be expensive.

In addition to stars or suns we can find other decorations, including the Moon or an hexa-petal rosette (Fig. 14.10). We also know of some cases of *hórreos* with sundials on, but those are left for future analysis.

Moon decorations are usually slits to help de air flow into the storeroom. As it is well known, the Moon has an important role in the whole plantation process and harvesting, because some people think that the decreasing moon is the time to harvest.[22]

And finally, we found many hexa-petal rosettes in *hórreos'* gables: different researchers understand this shape as a solar symbol and it would have an antropopaic function.[23] We have examples in Galicia that go back to the Iron Age in the Santa Tegra Castro, which is an archaeological site not very far from Vilaboa. We therefore can't establish a relationship between archaeological representations and these *hórreo* shapes since this decoration is very ordinary all over the world from Romanesque churches to rural tools. In fact, we can find hexa-petal rosette shapes anywhere: mills, churches, megalithic monuments, pottery[24] – but at the moment we don't know if nowadays they have any meaning or function for Cobres farmers.

Based on the astral-like shapes found, and after taking different aspects into consideration, Table 14.1 includes our field measurements for a sample of nine stone *hórreos* still situated in their original places according with their owners. For this initial approach to our study, those *hórreos* removed from their initial location (about six) have not been considered. In order to obtain the data we used a GPS, rule of five meters, SILVA tandem with compass and clinometer, and a camera. To calculate declinations we used Clive Ruggles WEB facility[25] and all azimuths have been magnetically (ß) corrected with the NOAA WEB facility,[26] according to coordinates and dates.

The sourcing of the data dates from June 2013 to September 2014. Table 14.1 shows that dimensions are not very different from one *hórreo* to another and most of them have eight feet. The most common displayed shape in the sample is the moon and we have noticed that it is common for *hórreos* (in general) to have more than one decorative shape. So far, azimuths, altitudes over horizon and apparent declinations in this small sample do not show any clear relationship in their orientations to solstices or equinoxes (in particular). We must however complete our acquisition of data for a much larger sample first, including at least the rest of *hórreos* with astral motif decoration already identified in the area (about ten more), before trying to draw conclusions. A comparative analysis between this group and the rest of *hórreos* with other decorations will also be attempted.

Summary of conclusions

To conclude, we would like to emphasize that all data collected so far correspond only to our first appraisal of the ethnoastronomy, having selected the topic of astronomical orientation and astral decoration of *hórreos* (granaries) in Galicia (NW, Spain) for study. The data here presented, a sample of objects in the

---

[22] Benito Vilas Estévez, "A review of the cosmological beliefs and traditions that have influenced farmers in Bueu, a rural village in Galicia", *SPICA* 2, no. 1 (2014): pp. 4-19.
[23] García-Gelabert Pérez "Los motivos decorativos de los hórreos', pp. 7-42; Navarro Caballero, 'Las estelas en la brecha de Santo Adriao: observaciones tipológico-cronológicas', *Boletín del Seminario de Estudios de Arte y Arqueología: BSAA* 64 (1998): pp. 175-206.
[24] Caballero "Las estelas en la brecha de Santo Adriao', pp. 175-206; Ramón Rodríguez Pérez, 'Acerca de algunos símbolos y "signos mágicos" representados en amuletos monetiformes andalusíes', *Revista Numismática OMNI* 1 (2014): pp. 65-78.; H. T. Bossert, *El Arte popular en Europa* (Barcelona: Gustavo Gili, 1955).
[25] Clive Ruggles, http://www.cliveruggles.net [accessed 24 July 2013].
[26] NOAA Magnetic Field Calculators, at http://www.ngdc.noaa.gov/geomag-web [accessed 3 September 2014].

An Ethnoastronomy Study on the Astronomical Orientation and Astral Decoration of the Stone Granaries (Hórreos) of Vilaboa (Galicia, Spain)

Item Nr.	Date (d/m/yr)	Latitude Longitude (° ′ ″)	Floor Dimensions (m)	Nr. of Feet	Conservation Status and Age
1	24/07/13	42 18 01 N 08 36 21 W	1,47 x 6,65	8	Good, but not in use anymore
2	31/07/13	42 19 45,59 N 08 40 07,91 W	1,35 x 6,50	8	Good, always in use (> 100 yrs)
3	19/09/13	42 19 47,9 N 08 40 12,3 W	1,57 x 6,74	8	Almost destroyed (1905)
4	19/09/13	42 20 51,6 N 08 38 36,96 W	1,64 x 4,78	6	Good, still in use (corn) / 'whole life'
5	19/09/13	42 19 46,82 N 08 40 05,5 W	1,46 x 6,99	8	Good, still in use (> 64 yrs)
6	27/09/13	42 20 55,56 N 08 38 38,49 W	1,57 x 5,08	6	Good, still in use (corn, potatoes...)
7	02/09/14	42 21 27,35 N 08 38 54,21 W	1,65 x 5,18	6	Good (> 48 yrs)
8	02/09/14	42 21 47,25 N 08 38 31 W	1,47 x 5,16	8	Good, still in use 'whole life'
9	02/09/14	42 21 45,25 N 08 38 31,09 W	1,47 x 4,88	3 and 4	Good 'whole life'

Table 14.1, part 1: Data collected in Cobres parishes for our initial sample of 9 hórreos (still at their original locations) with 'astral decoration'. Latitude and Longitude refer to the ITRF92 (Earth Rotation Service Terrestrial Reference Frame 01 1992) Geographical System. Right Ascension and Declination refer to Epoch 2000.

Item Nr.	Decoration	Azimuth ß corrctd. (° ′ ″)	Alt. over Horizon (°)	Apparent Declination (° ′ ″)
1	Big star/Sun and small Moon	50 57 00	+2	+29 03 31
2	Small Moon	13 25 48	+0	+45 24 48
3	Big star/Sun	79 57 00	+3,25	+09 27 40
4	One Moon on each side	113 57 36	+1,75	-16 27 21
5	Big star/Sun	126 57 00	+2,5	-24 41 45
6	One Moon on each side	62 57 36	+5	+23 03 13
7	One Moon on each side	99 05 24	+2	-05 33 03
8	One (odd) Moon	175 05 24	+0	-47 59 12
9	One Moon	102 05 24	+8	-03 29 04

Table 14.1, part 2: Data collected in Cobres parishes for our initial sample of 9 hórreos (still at their original locations) with 'astral decoration'. Latitude and Longitude refer to the ITRF92 (Earth Rotation Service Terrestrial Reference Frame 01 1992) Geographical System. Right Ascension and Declination refer to Epoch 2000.

two parishes of Cobres, of the council of Vilaboa, allows us to confirm that functional aspects seem to be decisive when it comes to the location of *hórreos,* as other authors have mentioned.[27] The *hórreo* is a functional architecture element and it is half way between the house (reproduction) and land/growing area (production) of the plot of land around the house to which it belongs. So we have found out that there are elements of social relevance that also help to make a decision regarding the place where the *hórreo* is finally built. In order to understand the orientation of *hórreos* we must take into account cultural patterns in rural areas. Analysing these, we can find out about elements of production, social organization and ideological aspects. We could take grooves, signs or symbols found in the gables as part of the decoration of *hórreos,* however we need to do more fieldwork to try to fully understand the worldview in the Cobres parishes. So in this sense, maybe, the astral decorations found so far (in the form of moons, stars or suns on some of the *hórreos* measured) could have some relation to the social position of *hórreos* and this research might contribute to understand the importance of *hórreos* in Galician culture. Orientations towards landscape, horizon and meteorological elements will also be considered thoroughly in the future.

The effort made to carry out this research shows that it would be worth continuing this line of work. We understand the need of gathering a larger, as well as more comprehensive, amount of data and we intend to do so it in next phases of the research.

Acknowledgments: This work has been possible thanks to the generosity from all the following people and organizations: Subdirección Xeral de Conservación e Restauracións de Bens Culturais, Xunta de Galicia, who allowed us to check the ethnological archives where we found a first representation of astral decoration in hórreos; Santiago Salsón Casado (Meteogalicia), who sent us the winds' measurements; Ornela Fernández Salgado and Concello de Vilaboa, who facilitated the access to both data and territory; Alberte Reboreda Carreira and Consultora Galega S.L., for letting us use their data about hórreos' locations; Benigno González Castro, for his help with the fieldwork; Pedro Mateo Lago, for his assistance with the usage of AutoCAD; O Barqueiro Association, which was our significant intermediary to get to know more about Cobres' culture; and, last but not least, the Vilaboa neighbours for their collaboration. Partial funding support was provided by the Vice-Chancellor's Office for Research of the University of Vigo.

Bibliography

Balil, A. 'Equivalentes y significados del término hórreo en el latín'. In *Hórreos. Actas del primer Congreso europeo del Hórreo*, pp. 43-51. Santiago de Compostela: Compostela, 1990.

Bande Rodríguez, E., Tain Carril, C. 'El mundo simbólico del hórreo gallego'. In *Hórreos. Actas del primer Congreso europeo del Hórreo*, pp. 53-60. Santiago de Compostela: Compostela, 1990.

Bas López, B. 'Arquitectura para a produción campesiña. O millo: os hórreos en Galicia'. In *Antropoloxía*, edited by Xosé Manuel Gónzalez Reboredo, pp. 238-282. A Coruña: Hércules,1997.

Bossert, H.T. *El Arte popular en Europa*. Barcelona: Gustavo Gili, 1955.

Caamaño Suárez, Manuel. *As construccións da arquitectura popular da arquitectura popular: Patrimonio etnográfico de Galicia*. A Coruña: Hércules, 2006.

Caamaño Suárez, Manuel. 'O hórreo galego na encrucillada'. In *Hórreos. Actas del primer Congreso europeo del Hórreo*, pp. 67-76. Santiago de Compostela: Editorial Compostela, 1990.

Castelao, Alfonso R. *As cruces de pedra na Galiza*. Vigo: Galaxia, 1998.

De Llano, Pedro. *Arquitectura popular en Galicia*. Santiago de Compostela: Colegio Oficial de Arquitectos de Galicia, 1983.

Decreto 449/1973, February 22th, por el que se colocan bajo la protección del Estado los 'hórreos' o 'cabazos' antiguos existentes en Galicia y Asturias. B.O.E no. 62, 13 March.

Esteban, C. 'Is cultural astronomy an interdisciplinary science? An astrophysicist's reflections'. *Complutum* 20, no. 2 (2009): pp. 69-77.

---

[27] Bas López, 'Arquitectura para a producción campesiña', pp. 238-282; De Llano, *Arquitectura popular en Galicia*; Lema, 'Os hórreos do extremo occidental de Galicia', pp. 197-289; Castelao, *As cruces de pedra na Galiza*.

Frankowski, Eugeniusz. *Hórreos y palafitos de la Península Ibérica*. Madrid: Istmo, 1986.

García-Gelabert Pérez, María Paz. 'Los motivos decorativos de los hórreos y paneras de Riocastelo (Tineo, Asturias):un testimonio de la pervivencia de la iconografía astral prerromana'. *Hispania antiqua*, no. 35 (2011): pp. 7-42

Lema Suárez, Xose María. 'Os hórreos do extremo occidental de Galicia'. *Gallaecia* 5 (1979): pp. 197–289.

Lorenzo Fernández, Xaquín. *A Casa*. Vigo: Galaxia, 1995.

Martínez-Rodríguez, Ignacio. *El hórreo gallego*, 2nd edition. A Coruña: Fundación Pedro Barrié de la Maza, 1999.

Museo de Pontevedra. *Cantigas de Santa María*. 2009. http://www.museo.depo.es/noticias/notas.de.prensa/es.02010152.html.

Navarro Caballero, Milagros. 'Las estelas en la brecha de Santo Adriao: observaciones tipológico-cronológicas'. *Boletín del Seminario de Estudios de Arte y Arqueología: BSAA* 64 (1998), pp. 175-206.

Perez-Garcia, O. A., Carreira, X. C., Carral, E., Fernandez, M. E., and Mariño, R. A. 'Evaluation of traditional grain store buildings (hórreos) in Galicia (NW Spain): analysis of outdoor/indoor temperature and humidity relationships'. *Spanish Journal of Agricultural Research* 8, no. 4 (2010): pp. 925-935.

Rivas Quintas, Eligio. *Millo e hórreo, legumia e cestos*. Santiago de Compostela: Ediciòns Laiovento, 1996.

Rivera Rodríguez, María Teresa. *Los Pazos orensanos: arquitectura popular del siglo XVIII en la provincia de Orense*. Orense: Caja de Ahorros Provincial de Orense, 1981.

Rodríguez Fernández, Perfecto. 'El hórreo en la diplomacia medieval asturiana en latín (siglos VIII-XIII)'. *Aula abierta*, no. 41-42 (1984): pp. 97-114.

Rodríguez Pérez, Ramón. 'Acerca de algunos símbolos y "signos mágicos" representados en amuletos monetiformes andalusíes'. *Revista Numismática OMNI* 1 (2014), pp. 65-78.

Ruggles, Clive. http://www.cliveruggles.net/.

Saá, César, Míguez, J. L., Morán, J. C., Vilán, J. A., Lago, M. L., Comesaña, R., Collazo, J. 'The influence of slotted floors on the bioclimatic traditional Galician agricultural dry-store structure (hórreo)'. *Energy and Buildings* 43, no. 12 (2011): pp. 3491-3496.

Van Linthoudt, Jozef. *Bueu: os seus canteiros e os seus reloxos de sol*. (Vigo: Autoedición, 1997). https://dl.dropboxusercontent.com/u/76189300/Libros/reloxos/Reloxos_todo.pdf accessed 10 May 2013]

Vilas Estévez, Benito. 'A review of the cosmological beliefs and traditions that have influenced farmers in Bueu, a rural village in Galicia'. *SPICA* 2, no. 1 (2014): pp 4-19.

CONNECTIONS: THE RELATIONSHIPS BETWEEN NEOLITHIC AND BRONZE AGE MEGALITHIC ASTRONOMY IN BRITAIN

Gail Higginbottom and Roger Clay

ABSTRACT: It has already been empirically verified that for many Bronze Age monuments erected in the Late Bronze Age of Scotland, there was a concerted effort on behalf of the builders to align their monuments to astronomical bodies on the horizon. It has also been found that there are two common sets of complex landscape and astronomical patternings, combining specific horizon qualities (like distance and elevation) with the rising and setting points of particular astronomical phenomena. However, it has only been very recently demonstrated by us that that the visible astronomical-landscape variables found at standing stone Bronze Age sites on the inner isles and mainland of western Scotland were likely first established nearly two millennia earlier, with the erection of possibly the first standing-stone 'great circles' in Britain: Callanish and Stenness of Scotland. In this paper we demonstrate our preliminary assessment of the connection between the monuments examined by us to date and the large Late Neolithic circles south of Scotland, namely those of Castlerigg and Swinside in Cumbria, England, and the Druids Circle in Gwenydd, Wales.

The standing stones of Scotland

The chronology, archaeological associations and various possible functions of free-standing stone (F–SS) monuments, discussed at length in Higginbottom et al. 2015, reveal a number of informative points on the archaeology of F–SS in Scotland.[1] This archaeological information tells us that these monuments seem to appear suddenly during the Late Neolithic (LN) approximately 3000–2900 cal BCE and their building continued until the end of the Bronze Age (BA).[2] The first F-SS built included the great circles Callanish and Stenness, using thin, tall slabs. It is possible that they were the first of such monuments.[3] Fascinatingly, linear F-SS sites in Scotland are so far scientifically dated from the end of the mid to the late Bronze Age (LBA) and single stones are associated with, or part of, monuments dated through the LN to the LBA. However, scientifically dated single standing stones that are not part of a stone circle (SC), but associated with other F-SS monuments, have only been dated to the Bronze Age (BA).[4]

---

[1] G. Higginbottom, A. G. K. Smith, and P. Tonner, 'A Re-creation of Visual Engagement and the Revelation of World Views in Bronze Age Scotland ', *Journal for Archaeological Theory and Method* 22, no. 2 (2015): pp. 584–645; first released online 12 December 2013, http://link.springer.com/article/10.1007/s10816-013-9182-7 [accessed 4 August 2016].

[2] P. J. Ashmore, *Calanais Survey and Excavation 1979-88* with contributions by T. Ballin, S. Bohncke, A. Fairweather, A. Henshall, M. Johnson, I. Maté, A. Sheridan, R. Tipping, and M. Wade Evans (in press); J. N. G. Ritchie, 'The Stones of Stenness, Orkney', *Proceedings of the Society of Antiquities of Scotland* 107 (1976): pp. 1–60; J. Barber, 'The excavation of the holed-stone at Ballymeanoch, Kilmartin, Argyll', *Proceedings of the Society of Antiquities of Scotland* 109 (1977–78): pp. 104–11; R. Schulting, A. Sheridan, R. Crozier, and E. Murphy, 'Revisiting Quanterness: New AMS dates and stable isotope data from an Orcadian chamber tomb', *Proceedings of the Society of Antiquities Scotland* 140 (2010): pp. 1–50; J. A. Sheridan, 'The National Museums' Scotland radiocarbon dating programmes: results obtained during 2005/6', *Discovery and Excavation in Scotland* 6 (2006): pp. 204–206; R. D. Martlew and C. L. N. Ruggles, 'Ritual and landscape on the West Coast of Scotland: An investigation of the stone rows of Northern Mull', *Proceedings of the Prehistoric Society* 62 (1996): pp. 1256–129.

[3] P. J Ashmore, *Calanais Survey and Excavation 1979-88*; P. Ashmore, 'Radiocarbon dating: Avoiding errors by avoiding mixed samples', *Antiquity* 73, no. 279 (1999): pp. 124–130; C. Richards and S. Griffiths, 'A time for stone circles, a time for new people', in *Building the Great Stone Circles of the North*, ed. C. Richards (Oxford: Windgather Press, 2013), pp. 281–91; R. Schulting et al., 'Revisiting Quanternes', pp. 35–36.

[4] P. Duffy, 'Excavations at Dunure Road, Ayrshire: A Bronze Age cist cemetery and standing stone', *Proceedings of the Society of Antiquaries of Scotland* 137 (2007): p. 53; J. A. Sheridan, 'Towards a fuller, more nuanced narrative of Chalcolithic

Interestingly, both Neolithic and BA circles, along with the simpler BA F-SS monuments, are usually directly associated with death, fire, burial, body transformations (cremation and possible separation of cremation remains) and pale/white/shiny stones or pebbles. Cremated remains are often placed in the stone pit and/or next to the F-SS.[5] Some interpretation of the monuments' role(s) could proceed at this point, but this kind of archaeological information cannot tell us why people chose to erect these monuments where they did, and does not explain fully why they erected them at all.[6]

In this paper, we will limit the discussion of our new work to the dominant context, the surrounding 360-degree landscape, in particular the intersection of the land and the sky – the horizon. This visual boundary, at the furthest point a person can see, defines and contains what is to be observed from a megalithic site at least during the time of construction.[7] Through the examination of this context, we can demonstrate the connection of place (or places) and continuity of cosmology over two millennia more firmly than can the accompanying archaeological evidence, though the latter is essential for any full interpretation and comprehension of these generally enigmatic sites. Specifically, our earlier work was carried out in western Scotland moving outwards to other parts of Scotland and then, much more recently, we began to investigate other regions in Britain with strong megalithic traditions, including Cumbria and Wales.

We will present an overview of the early work that contains some fundamental results and provides background to our most recent research. Note that GH has visited many sites on Mull in Argyll and in the northern Outer Hebrides, as well as the stone circles discussed in this paper. These visits were either initial site visits or confirmation visits to check the patterns discovered in the GIS research.

Earlier published work
Orientation studies – testing the distribution of observed horizon declinations indicated by monument alignments
Ruggles' 1984 aim was to test for the likelihood of highly accurate astronomical alignments and so he had many *a priori* rules set out for site choices. All our earliest work was carried out to test our new methodological approaches for archaeoastronomy and to reassess Ruggles' 1984 results; therefore we had to ensure that our database was consistent with Ruggles.[8] The best way to do this, then, was to use

---

and Early Bronze Age Britain 2500–1500 BC', *Bronze Age Review: The international journal of research into the archaeology of the British and European Bronze Age* 1 (2008): p. 61; J. A. Sheridan, 'The National Museums' Scotland radiocarbon dating programmes', p. 183.

[5] R. Bradley, ed., *The Good Stones: A new investigation of the Clava Cairns* (Edinburgh: Society of Antiquaries of Scotland, 2000); Duffy, 'Excavations at Dunure Road, Ayrshire', p. 53; C. Richards, 'Interpreting Stone Circles', in *Building the Great Stone Circles of the North*, ed. C. Richards (Oxford: Windgather Press, 2013), pp. 2–30; C. Richards, 'Wrapping the hearth: constructing house societies and the tall Stones of Stenness, Orkney', in C. Richards, ed., *Building the Great Stone Circles of the North*, pp. 64–89; J. N. G. Ritchie ,'The Stones of Stenness, Orkney', pp. 1–60; R. Schulting et al., 'Revisiting Quanternes', pp. 35–36.

[6] G. Higginbottom, A. G. K. Smith, and P. Tonner, 'A Re-creation of Visual Engagement and the Revelation of World Views in Bronze Age Scotland ', pp. 584–645.

[7] Higginbottom, G., A. Smith, K. Simpson, and R. Clay. 'Incorporating the natural environment: investigating landscape and monument as sacred space'. In Martin Gojda and Timothy Darvill, *Landscape archaeology: new approaches to field methodology and analysis*, British Archaeological Reports S987 (Oxford: Archaeopress, 2001); Higginbottom, G. 'Perception creates worlds: meaning and experience in the erection of the standing stones of western Scotland'. In *Yachay Wasi: a collection of papers in honour of Ian S. Farrington*, edited by Lisa Solling and Tom Knight. Oxford: Archaeopress, in preparation. Draft available at https://www.academia.edu/6367035/Perception_creates_worlds_-_the_phenomenological_origin_of_the_spatiality_of_Nature_Husserl_-_Draft [accessed 4 August 2016].

[8] C. Ruggles, *Megalithic astronomy: A new archaeological and statistical study of 300 Western Scottish Sites* (Oxford: British Archaeological Reports British Series 123, 1984).

exactly the same sites as did he, along with the raw data he gathered in the field, like orientations of alignments and the coincident declinations he produced from these, as found in the appendices of his 1984 volume. Section 2 of Ruggles' 1984 volume is devoted to detailing the reasoning behind and methodology of site choices. Sections 3.3 and 3.4 outline the *a priori* decisions for defining and gathering orientations in the field and the types of orientations. The latter were either the internal alignments of monument elements, like the axis of the standing stones of a stone row or an axis created by the width of a thin, wide slab, or external alignments created by the line drawn between two monuments, such as a small SC and a standing stone (how this was done is defined by Ruggles in these sections).

In this way, meaningful sets of data and statistical analyses were devised for sites with a single or a very small number of orientations (like single slab or stone row), to analyse the astronomical potential of the sites contained within western Scotland as a whole and then within separate geographical regions. The fundamental problem for observational astronomy is that the horizon elevation function (the relationship between azimuth and elevation, and therefore too the associated declination) is real and fixed to a specific site, it is not a probabilistic distribution. Therefore, we had to ensure that we were using real declination data for both our expected and observed data sets. To overcome such limits in the study of archaeoastronomy early on in the project, Smith developed the Horizon software, which allowed us to extract declinations for entire horizons at any location in our case-study regions and which we used in our previously published statistical investigations.

This process allowed us to discover that, for the islands of Mull, Coll and Tiree together with 2 sites from North Argyll, as well as Argyll with Lorn and Islay with Jura, the distribution of observed horizon declinations indicated by monument alignments was unlikely to be due to chance factors (Kolmogorov-Smirnov test, rejection of the null hypothesis: Mull $p=0.00817$ (n1=24 sites; n2= 40 declinations), Argyll $p=0.00593$ (n1=21, n2= 44) and Islay $p=0.00105$ (n1=23, n2= 41).[9] We therefore interpreted this outcome to mean that the regional monument alignments indicating particular patterns of declinations along the horizon were deliberately chosen by the builders of these monuments. As a graduated step in our declination assessment, this was a test for the distribution pattern only, and tested the likelihood that groups of declinations were sought for in the first place (like testing for 'clustering only' in azimuth); it was not testing the likelihood of any specific declination(s).

So, at this point, it wasn't clear what the preferred declinations were and whether they had any astronomical significance. In order to investigate this, we had to see where in the declination profile the differences occurred, and then determine if these declinations aligned with any astronomical phenomena. Binning the observed and expected (random) data into 5-degree bins for each region (e.g. 0^0–5^0, 5^0–10^0), we applied a simple probability test (p), to test whether or not the number within each observed bin differed from that found in the expected (null) bins. This tests the likelihood of any difference in number occurring by chance. Once this test was done, the statistically supported bins were studied to see if they overlapped with declinations of astronomical bodies or phenomena. Importantly, the statistically supported ranges could indicate an avoidance of, or clustering within, a declination range.[10] The declinations used for the moon and the sun were the same as those used by Ruggles in

---

[9] G. Higginbottom et al., 'Gazing at the horizon: sub-cultural differences in western Scotland?', in César Esteban and Juan A. Belmonte, *The Oxford VI International Conference on Archaeoastronomy and Astronomy and Culture* (Tenerife: Organismo Autónomo de Museos del Cabildo de Tenerife, 2001), pp. 43–49; G. Higginbottom et al. 'Gazing at the horizon', pp. 43-50; G. Higginbottom, A. G. K. Smith, K. Simpson, and R. Clay, 'More than orientation: placing monuments to view the cosmic order', in *Ad Astra per Aspera et per Ludum*, ed. Amanda-Alice Maravelia, BAR International Series S1154 (Oxford: Archaeopress, 2003): pp. 39-52. With typographical corrections for results' table published in the previous reference.

[10] G. Higginbottom et al., 'Gazing at the horizon'; G. Higginbottom et al., 'More than orientation: placing monuments

1984: solstices -23.9° & +23.9°; the moon at the major standstill (MajLS) -30° & +28.2°; the moon at the minor standstill (MinLS) -19.7° & +17.9°; all for the year 2000 BCE).[11]

We found statistical support from the orientations for an interest in the moon's rising and setting points most close to the MajLS and MinLS, both in the southerly and northerly directions, as well as the sun at the winter solstice (WSol) and areas that flank the midpoint between the solstices.[12] These statistical tests were carried out on groups of sites across the chronological range from the Neolithic to the end of the BA. Whilst no statistical support was found for the sun at the summer solstice (SSol) by region, a small number of sites were oriented in this direction within 2° (approximately nine orientations out of 276). However, it will be demonstrated by the 3D landscapes below that the SSol in the BA was important in ways similar to the great circles of the LN.

3D-landscape reconstruction with astronomical phenomena layer
The variables considered in the creation of the software of 3D landscapes are discussed in Higginbottom et al.'s 2015 paper and detailed in the manual by A. G. K. Smith (software designer) which can be found at http://www.agksmith.net/horizon/.[13] The manual can be accessed by downloading the entire software and is part of a .zip file. The testing of an early version of the 3D landscapes is discussed Chapter 8 of Gail Higginbottom's thesis.[14] Here a series of panoramic photos was taken by a local, Charles Tait, in western Scotland and sent to myself and Smith along with the easting and northing from whence they were shot. Smith then compared these with his then current program. He concluded that the program was a good fit for the 50m data with the usual caveats: horizons within 50m of a site and lower than 50m would not be accurately portrayed with the most likely result being they would not exist on the 3D landscape.[15] Also, very narrow peaks might not be accurately drawn. However, as we were not testing for highly accurate astronomical alignments, the latter was certainly not an issue and the possibility of very close horizons would have to be checked in the future. Our current elevation data is more up-to-date: 50m horizontal data and 10m in the vertical, as is the software program (see www.Horizon.net for the manual). Smith has since tested his program with elevation data from other countries like the United States and Portugal and confirms its overall reliability and has joint publications in these areas.[16] For this paper, the 2011 version of the program was used to create the 3D landscapes.

As part of our preliminary investigations, the 3D landscapes were assessed using a typological method. That is, a classification system based upon physical characteristics which thus relies on a descriptive or morphological approach. The 3D landscapes were examined in detail in two ways: (i) printing out

---

to view the cosmic order', in Maravelia, ed., *Ad Astra per Aspera et per Ludu*, pp. 140–143.

[11] C. Ruggles, *Megalithic astronomy*.

[12] G. Higginbottom and R. Clay, 'Reassessment of sites in Northwest Scotland: A new statistical approach', *Archaeoastronomy* 24 (1999): pp. S1–S6.

[13] G. Higginbottom et al., 'A Re-creation of Visual Engagement and the Revelation of World Views in Bronze Age Scotland'; A. G. K. Smith, Horizon software manual, available at http://www.agksmith.net/horizon/ [accessed 18 July 2016].

[14] G. Higginbottom, *Interdisciplinary study of Megalithic monuments in Western Scotland* (PhD thesis, University of Adelaide, 2003): pp. 139–143.

[15] A. G. K. Smith, personal communication, 2000, 2002 and re-confirmed in 2015.

[16] A. G. K. Smith, personal communication, 2008, 2009, 2011, 2016; examples of his joint publications include: F. Pimenta, L. Tirapicos, and A. Smith, 'A Bayesian Approach to the Orientations of Central Alentejo Megalithic Enclosures', *Archaeoastronomy* 22 (2009): pp. 1–20; F. Pimenta, N. Ribeiro, A. Smith, and L. Tirapicos, 'The Sky and the Landscape of Rock Art in the Ceira and Alva Basins', in *Cosmology Across Cultures*, ed. J. A. Rubiño-Martín, J. A. Belmonte, F. Prada, and A. Alberdi, ASP Conference Series, Vol. 409 (San Francisco: Astronomical Society of the Pacific, 2009), pp. 359–63.

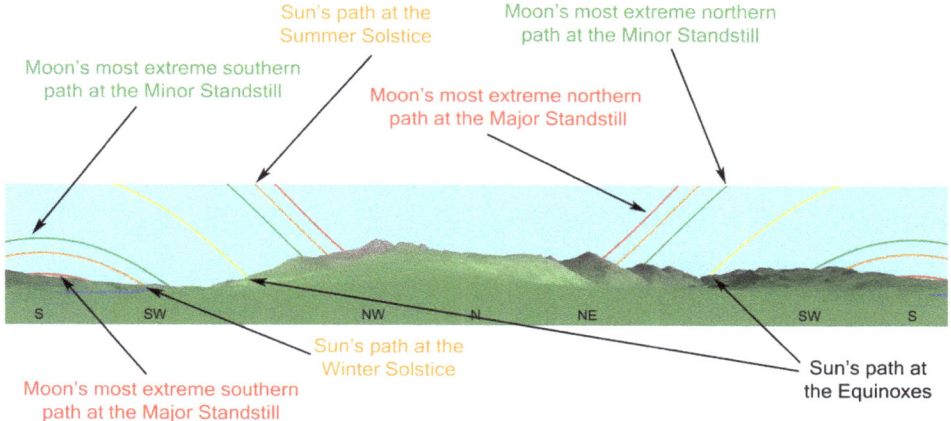

Fig. 15.1: This is the 3D rendering of the landscape around the classic site of Uluvalt on Mull along with a key to reading the paths of the sun and the moon on the other such figures below. N=north; S=south. Software created by Andrew Smith. Based upon the Ordnance Survey 1:50 000 Landform PANORAMA map with permission of the Controller of her Majesty's Stationery Office © Crown Copyright.

each landscape for an overview of the 360°-views and (ii) examining any detail via computer preview software. A 3D landscape was produced for every separate astronomical orientation. These were then all laid out together at the same time and arranged according to apparent horizon shape.

Significantly, in our previous work, we found two horizon landscape patterns, one that is basically the topographical reverse of the other. Only one site (1/41), that of Loch Seil or Duachy (LN22), was difficult to place into either category.

Classic sites
For our detailed regional studies to date, we have found that one or the other landscape pattern surrounds every site. For all the sites on Coll and Tiree (n=6/6), the majority of sites on the isle of Mull (n=9/16) and roughly half of the sites studied so far in mainland Argyll (with Lorn; (n= 10/21) there is a combination of usual visual cues, whether the sites are linear, single slabs, or small circular settings.[17] We called these 'classic sites' as they contained the first pattern we recognised. The usual dominant cues for classic sites are (Figs. 15.1 and 15.2a-b): 1. if water is seen, it is usually seen in the south as opposed to the north; 2. a northern horizon is closest, a southern most distant; 3. the northern horizon has a higher general profile or the highest vertical extent in the profile (apparent elevation); the southern horizon has a very distinct dip (concave) or a lower general profile than the northern or both; 4. the highest areas of the northern and southern horizons often focus around the four ordinal directions of NW, NE, SW and SE;

---

[17] G. Higginbottom et al, 'A Re-creation of Visual Engagement and the Revelation of World Views in Bronze Age Scotland', pp. 630–36; G. Higginbottom, 'The world begins here, the world ends here: Bronze Age megalithic monuments on the isle of Mull in western Scotland'. Draft available at https://www.academia.edu/22473630/The_world_begins_here_the_world_ends_here_Bronze_Age_megalithic_monuments_on_the_isle_of_Mull_in_western_Scotland_Draft_Only_ [accessed 17 July 2016]; G. Higginbottom, 'Megaliths in Argyll: integrating landscape formations and astronomical knowledge in the creation of place in the Bronze Age' (In preparation). Draft available at https://www.academia.edu/27237625/Megaliths_in_Argyll_integrating_landscape_formations_and_astronomical_knowledge_in_the_creation_of_place_in_the_Bronze_Age._Draft [accessed 25 July 2016].

occasionally the highest area is more generally northern if a single mountain or range fills the northern horizon; most commonly when sites have high ground near the ordinal points, it is usually found at all four points or at three out of the four; 6. the summer and winter solstitial sun and standstill moon tend to rise out of and set into these high ranges, hills, or ground; and 7. a site most often forms an alignment internally, or with another site, at a lunar or solar orientation (the majority of which fall within the statistically supported declination ranges). For the moon this is the Major or Minor Lunar Standstill (MajSS or MinSS), and for the sun it is at the winter solstice (WSol) or summer solstice (SSol). A few are aligned to the equinoctial sun at the horizon (n=3).

Reverse sites

As mentioned above, those sites that do not reveal this landscape pattern reveal a combination of reverse landscape traits (Mull (n=7/16) and Argyll/Lorn (n= 10/21), namely (see Fig. 15.2c): 1. if water is seen, it is usually seen in the north as opposed to the south; 2. a southern horizon is closest, a northern is most distant; 3. the southern horizon has the highest point(s) in profile; the northern horizon has a very distinct dip or overall lower horizon profile than the southern or both; 4. the highest areas of the northern and southern horizons often focus around the four ordinal directions of NW, NE, SW and SE; occasionally the apparent highest profile is more generally southern if a single mountain or range fills the southern horizon; most commonly when sites have high ground near the ordinal points, it is usually found at all four points or at three out of the four. Like classic sites, we find that: 1. the sun at the WSol and SSol and standstill Moon tend to rise out of and set into these high ranges, hills, or ground; and 2. a site most often forms an alignment internally, or with another site, at a lunar or solar orientation (the majority of which fall within the statistically supported declination ranges). For the moon this is the MajSS or MinSS, and for the sun it is at the WSol or SSol.

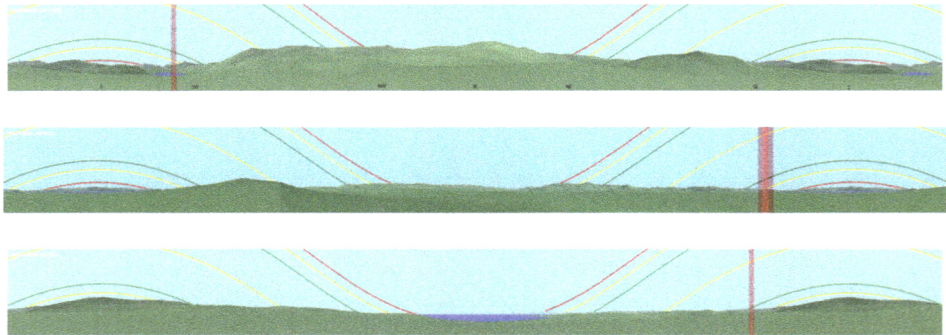

Fig. 15.2: 3D landscapes of slabs and single menhirs, where the centre of the landscape is north. The red, vertical lines indicate the direction of the alignment of the site where it touches the horizon. In order from the top: two classic sites of (a, top) Torran, Argyll NM87880488 (menhir, AR7; the alignment is looking towards Ford, AR8); (b, middle) Rowanfield, Argyll NM82059585 (Standing slab, AR16; the alignment is the axis of the megalith); as well as one reverse site: (c, bottom) Cillchriosd, Mull NM37735348 (menhir, ML7SE; the alignment is the axis of the megalith). Codes refer to Ruggles' 1984 site numbers. Created with the software Horizon by A. G. K. Smith, ©A.G.K. Smith. Based upon the Ordnance Survey 1:50 000 Landform PANORAMA map with permission of the Controller of her Majesty's Stationery Office © Crown Copyright. Image copyright © Andrew Smith & Gail Higginbottom (2013); http://www.ordnancesurvey.co.uk/docs/licenses/os-opendata-licence.pdf.

However, fascinatingly, at reverse sites it is not uncommon for the moon at the MajSS in the south to be completely or partially blocked by the horizon during its travels and this is occasionally true also for the WSol Sun and the moon at the MinLS in the south. Thus, our preliminary results so far reveal that in Argyll

& Lorn, 6/10 (60%) reverse sites altogether contain 22 southern blocking events and 39 events occur at 7/7 (100%) reverse sites on Mull, whereas we only find four such events at three out of ten (3/10) classic sites in Argyll.[18] Whilst 21 events were found at classic sites on Mull, this is only 35% of the total events on Mull and the majority of these are made up of small occlusions of part of the astronomical body for the first minutes when rising or setting (n=12/21) or short-term full body blocking, again at rising or setting, thus fractionally shortening the amount of daylight (WSol sun) or moonlight (MajLS or MinLS moon) by minutes to one hour (7/21). Blocking events are described as follows: where the horizon is used to cover the body for its entire travels in the south (reverse sites: Argyll=3, Mull=8; classic sites: Argyll =1, Mull=2), or temporarily at anytime in its travels along its path. Thus, for example if the moon is only seen to travel downwards towards the horizon after rising, clearly the moon's path has been blocked, shortening the time of its appearance above the horizon. We included all events where the body was totally blocked by the horizon for a quarter or more of an expected path at the rising or setting 'ends' as well as simple blocking instances where a body temporarily disappears behind a hill and then reappears, or partial blocking where part of the disc of the body is covered (all such events were counted). Detailed discussions of how some of these blocking events can play out at such sites is discussed in Higginbottom's paper on Mull.[19]

Astro-landscape qualities
The various astro-landscape qualities as a whole show us that the event of the SSol is likely just as firmly entrenched in the consideration of monument placement as those of the statistically-indicated astronomical events, set up via alignments. Such qualities include: occasional SSol monument alignments, the regular linking of the SSol sun rising and setting behind a hill in the NE or NW, and possibly a deliberate linking of this solstitial event to the MajSS in the south at the full moon – the direction of the majority of statistically supported orientations. The latter being more striking if we assume, like Ruggles and others, that there was a possible focus on the full-moon as a commanding display, which would have to be in summer in the few hours of darkness which occurred, as described by Higginbottom et al.[20] What is important to remember is that the sites discussed thus far are likely to be primarily from the latter BA, as discussed in the introduction.[21]

---

[18] Blocking events were counted at every different viewing location from where an alignment was measured by Ruggles within a site. For example, if there were three different standing stones each with their own easting and northing, the events seen at each were all counted. The reasoning behind this is linked to the research findings of Higginbottom, namely that each viewing position within a site clearly afforded a different but significant viewing of the astronomical phenomena. See G. Higginbottom, 'The world begins here, the world ends here: Bronze Age megalithic monuments on the isle of Mull in western Scotland'.
[19] G. Higginbottom, 'The world begins here, the world ends here: Bronze Age megalithic monuments on the isle of Mull in western Scotland'.
[20] C. Ruggles, *Astronomy in Prehistoric Britain and Ireland* (London: Yale University Press, 1999), pp. 118-123; G. Higginbottom et al., 'A Re-creation of Visual Engagement and the Revelation of World Views in Bronze Age Scotland', pp. 630–36, explains in detail how this visual display works at sites.
[21] See discussion in this paper's introduction regarding dating information and the coincident references in footnotes 2 and 3 above; G. Higginbottom et al., 'A Re-creation of Visual Engagement and the Revelation of World Views in Bronze Age Scotland', pp. 585–91, 598–99, 600–03.

Astro-landscape qualities of Late Neolithic sites: recent and new work
Site choices
As seen above, our earlier work was based upon Ruggles' 1984 database in order to conduct comparative analyses. In Higginbottom and Clay's paper of 2016, our aim was different. We wanted to discover the *earliest evidence* in Scotland of a clear interest in connecting monuments via their alignments to astronomical phenomena, as found in western Scotland so far. We also wished to discover if we could find any evidence at these earlier sites for same the landscape-astronomical patterns found for the majority of the likely BA sites (n=42/43). For these reasons we chose Callanish and Stenness, for they are grand monuments at possible regional centres with clear LN dates, and are two of the earliest standing stone monuments in Scotland, possibly even the earliest of their kind (though we recognise there are at least two complex StS sites with contemporary dates).[22]

So, with the availability of clearer dates of Callanish from Ashmore, the Bayesian analyses of Schulting et al. of Stenness, along with our newly developed statistical test that allowed quantitative assessments of the internal alignments of stone circles, we were able to conclude that the megaliths of these two earliest dated great circles of Scotland were likely deliberately arranged with astronomical phenomena in mind.[23] Having statistically established this, along with the interest in the same astronomical phenomena as the latter BA sites, we wanted to establish whether or not these same two LN sites also shared the same topographic-astronomical patterns as the latter BA sites. We also wished to establish whether or not other possible LN great circles in Britain shared these landscape setting qualities. So, along with the two earliest known great circles of Callanish and Stenness, we chose three other sites from the west coast of Britain, as this area has a secure megalithic tradition that appears to have lasted for the same length of time in Cumbria and Wales, as it did in western Scotland.[24] The sites were chosen according to probable age and size, ensuring that they shared similarities for comparison in this paper. Firstly, they were large stone circles (either in diameter or stone height or both). Specifically each diameter was greater than, or equal to, 13m – the size of Callanish, and they had to have at least 1 stone greater than 2m. Secondly, they were dated to the LN by scientific dating or by typological associations in accordance with long-term archaeological tradition.[25] The three sites were Castlerigg and Swinside in Cumbria and Druids Circle in Gwenydd.

3D-landscape reconstruction of large Neolithic circles in Scotland, England and Wales
When we observe Figures 15.3 and 15.4 of the LN stone circles it is striking how similar the horizon shapes, and the positioning of the rising and setting astronomical bodies in relation to these, are to those of the seemingly simpler monuments of the LBA seen in Figures 15.1 and 15.2.

---

[22] G. Higginbottom and R. Clay, 'Origins of Standing Stone Astronomy in Britain: New quantitative techniques for the study of archaeoastronomy', *Journal of Archaeological Science: Reports* (2016), http://dx.doi.org/10.1016/j.jasrep.2016.05.025; G. Higginbottom et al. 'A Re-creation of Visual Engagement and the Revelation of World Views in Bronze Age Scotland'.

[23] G. Higginbottom and R. Clay, 'Origins of Standing Stone Astronomy in Britain'; P. J. Ashmore *Calanais Survey and Excavation 1979-88*; R. Schulting et al., 'Revisiting Quanternes'.

[24] A. Burl, *From Carnac to Callanish: The Prehistoric Stone Rows and Avenues of Britain, Ireland, and Brittany* (New Haven, CT: Yale University Press, 1993).); A. Burl, *Stone Circles* (2000).

[25] A. Burl, *From* Carnac (1993); A. Burl, *The Stone Circles of Britain, Ireland and Brittany* (New Haven, CT: Yale University Press, 2000); C. Richards and S. Griffiths, 'A time for stone circles', pp 281–91.

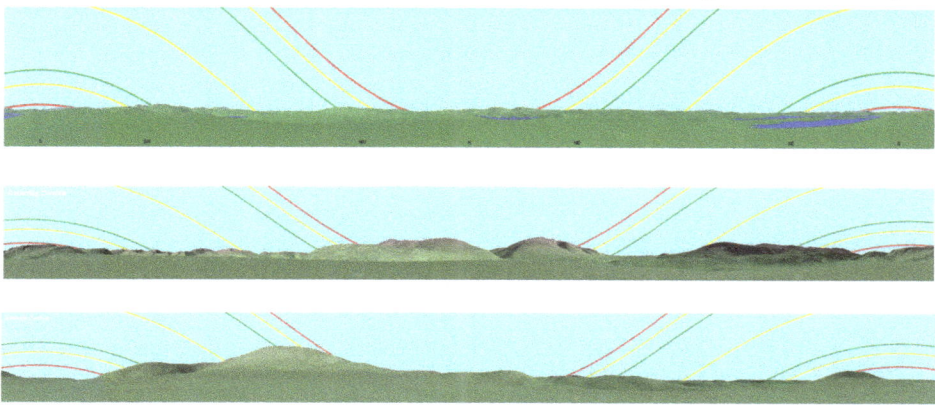

Fig. 15.3: 3D landscapes of Late Neolithic stone circles, classic sites. In order from the top: (a, top) Callanish, Isle of Lewis, Scotland; NB21303300; (b, middle) Castlerigg, Cumbria, England; NY29142363 and (c, bottom) Swinside, Cumbria, England; SD17168817. Note, due to the flatness of the horizon of Callanish, the z co-ordinate is multiplied by 1.5 for viewing at such a small scale here. Created with the software Horizon by A. G. K. Smith, © A. G. K. Smith. Created with Terrain 50. Contains Ordnance Survey data Crown copyright and database right (2012). http://www.ordnancesurvey.co.uk/docs/licenses/os-opendata-licence.pdf. Images copyright © Andrew Smith and Gail Higginbottom (2013).

Whilst these similarities are easy to see, we shall point out a few points of interest, focusing on the sites of Castlerigg and Swinside. Firstly we can see that, for both sets of figures, whether the overall landscape is mountainous or flat, prominent hills or peaks are chosen within the ordinal directions where possible and associated with astronomical events. For instance, at Castlerigg (Fig. 15.3a), like at the BA site of Uluvalt on Mull (Fig. 15.1), the northern astronomical phenomena rise out of the prominent hill's or peak's flanks in the NE and set into the top or the flanks of those in NW, and these peaks are clearly higher in the north than the south.

In the southerly direction there is the usual dip in the horizon with the phenomena's path rising and setting either-side. Swinside (Fig. 15.3b) is an example where there is only one major peak in the north being higher than the southern as we have found elsewhere, like on Coll and Tiree; nevertheless, there are still clear separate hills associated with the rising phenomena in the north and water in the SE (both much more obvious in a full-size version of the figures). Again there is the clear dip in the horizon profile in the south and prominent peaks in the SE and SW, into which all phenomena set into in the SW and which the MajLS rises out of in the SE. The MinLS and WS actually rise out of small hills but this is not clearly evident in the small figure here. In the SW there is an amazing view of the moon rolling along the horizon for about 12°. In actual fact, the moon at the MajLS actually partially sets and moves partially behind the horizon for about 5°, fully reappears for 2°, and then almost disappears completely again before reappearing fully and finally disappearing for the last time shortly after that. There are other strong similarities to seen between the LBA (Fig. 15.2c) and LN reverse sites (Fig. 15.4).

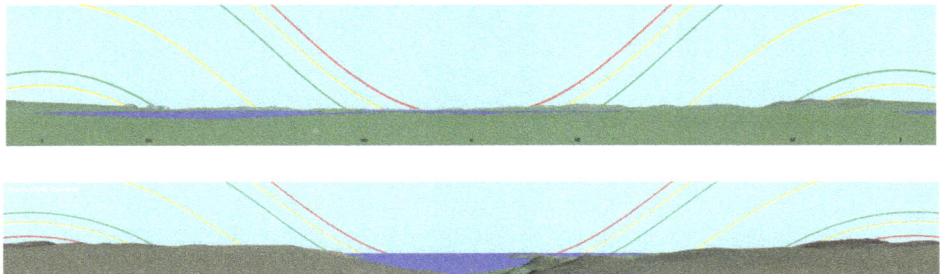

Fig. 15.4: 3D landscapes of stone circle reverse sites. In order from the top: (a) Stenness, Orkney, Scotland; HY30671252 and (b) Druids Circle, Gwenydd, Wales; SH72297466. Note due to the extreme flatness of the horizon of Stenness, the z co-ordinate is multiplied by 1.5 for viewing at such a small scale here. These hills and cliffs are easily viewed in person at the site, as Higginbottom has done. Created with the software Horizon by A. G .K. Smith, © A. G. K. Smith. Created with Terrain 50. Contains Ordnance Survey data Crown copyright and database right (2012). http://www.ordnancesurvey.co.uk/docs/licenses/os-opendata-licence.pdf. Images copyright © Andrew Smith and Gail Higginbottom (2013).

Water is clearly in the north and the northern horizon is distinctly lower than the southern for all sites. Figure 15.2c, of the menhir Cillchriosd, Argyll, shows strong likeness to Stenness, with the southern horizon too high to show the path of the MajLS moon even though the sites are 2.3° different in latitude: 56.6°N for Cillchriosd and 58.9°N for Stenness, emphasising further the deliberate nature of the location. Note that the 3D landscapes of Stenness and Cillchriosd also contain the full blocking of the moon at the MajLS in the south, more common at the BA reverse sites than at the classic. Overall, then, we can observe a consistency between the LN F-SS sites and those of the latter BA.

Concluding remarks

In this paper, we have explained that over 40 LBA sites share very complex topographic and astronomical features, and that there are some distinct features differentiating classic and reverse sites, in particular a greater amount of phenomena are blocked in the south at reverse sites. This blocking not only prevents one looking at the phenomena itself but effectively shortens the amount of full local daylight or significantly reduces the strength of moonlight at night. Narrative re-creations of how such events play out in real time at the time the sites were erected can be found in Higginbottom et al.'s paper of 2015 for the isles of Coll and Tiree and in Higginbottom 's work on Mull.[26] Our recent work demonstrated that the astronomical alignments of LBA Scottish F-SS monuments, along with their combined topographic and astronomical locational choices, were likely first 'set in stone' with the erection of the LN Scottish great circles.[27] Due to the complexity of the sites and their connection to the landscape and the sky we suggest that such locations had been known for some time before the setting up of the stones. It is not known when this occurred, though Higginbottom and Smith's report gives hints that it could be the Mesolithic. The astronomical alignments of Ruggles likely LBA sites, along with those of the LN great circles of Callanish and Stenness, are statistically supported. The 3D landscapes on the other hand, have been investigated through a typological approach.

What is newly presented in this paper is the possibility that the great circles along on the west coast

---

[26] G. Higginbottom et al., 'A Re-creation of Visual Engagement and the Revelation of World Views in Bronze Age Scotland'; G. Higginbottom, 'The world begins here, the world ends here: Bronze Age megalithic monuments on the isle of Mull in western Scotland'.

[27] G. Higginbottom and R. Clay, 'Origins of Standing Stone Astronomy in Britain'.

of Britain, not just Scotland, may share in these traditions of connecting the landscape and the sky to the stones in the same manner: a possibility that must be tested further.

Bibliography

Ashmore, P. J. 'Radiocarbon dating: Avoiding errors by avoiding mixed samples'. *Antiquity* 73, no. 279 (1999): pp. 124–130.

Ashmore, P. J. *Calanais Survey and Excavation 1979-88* with contributions by T. Ballin, S. Bohncke, A. Fairweather, A. Henshall, M. Johnson, I. Maté, A. Sheridan, R. Tipping, and M. Wade Evans. In press.

Barber, J. 'The excavation of the holed-stone at Ballymeanoch, Kilmartin, Argyll'. *Proceedings of the Society of Antiquaries of Scotland* 109 (1977–78): pp. 104–11;

Bradley, R. ed. *The Good Stones: A new investigation of the Clava Cairns.* (Edinburgh: Society of Antiquaries of Scotland, 2000).

Burl, A. *From Carnac to Callanish: The Prehistoric Stone Rows and Avenues of Britain, Ireland, and Brittany.* New Haven, CT: Yale University Press, 1993.

Burl, A. *The Stone Circles of Britain, Ireland and Brittany.* New Haven, CT: Yale University Press, 2000.

Duffy, P. 'Excavations at Dunure Road, Ayrshire: A Bronze Age cist cemetery and standing stone'. *Proceedings of the Society of Antiquaries of Scotland* 137 (2007).

Higginbottom, G. 'Megaliths in Argyll: integrating landscape formations and astronomical knowledge in the creation of place in the Bronze Age.' In preparation. Draft available at:
https://www.academia.edu/27237625/Megaliths_in_Argyll_integrating_landscape_formations_and_astronomical_knowledge_in_the_creation_of_place_in_the_Bronze_Age._Draft.

Higginbottom, G. 'Perception creates worlds: meaning and experience in the erection of the standing stones of western Scotland'. In *Yachay Wasi: a collection of papers in honour of Ian S. Farrington*, edited by Lisa Solling and Tom Knight. Oxford: Hadrian, in preparation. Draft available at https://www.academia.edu/6367035/Perception_creates_worlds_-_the_phenomenological_origin_of_the_spatiality_of_Nature_Husserl_-_Draft.

Higginbottom, G. 'The world begins here, the world ends here: Bronze Age megalithic monuments on the isle of Mull in western Scotland'. Draft available at
https://www.academia.edu/22473630/The_world_begins_here_the_world_ends_here_Bronze_Age_megalithic_monuments_on_the_isle_of_Mull_in_western_Scotland_Draft_Only_.

Higginbottom, G. 'Interdisciplinary study of Megalithic monuments in Western Scotland'. PhD thesis, University of Adelaide, 2003.

Higginbottom, G., A. Smith, K. Simpson, and R. Clay. 'Incorporating the natural environment: investigating landscape and monument as sacred space'. In Martin Gojda and Timothy Darvill, *Landscape archaeology: new approaches to field methodology and analysis*, British Archaeological Reports S987 (Oxford: Archaeopress, 2001).

Higginbottom, G., A. Smith, K. Simpson, and R. Clay. 'More than orientation: placing monuments to view the cosmic order'. In *Ad Astra per Aspera et per Ludum*, edited by Amanda-Alice Maravelia, BAR International Series S1154, pp. 39–52. Oxford: Archaeopress, 2003.

Higginbottom, G., A. G. K. Smith, and P. Tonner. 'A Re-creation of Visual Engagement and the Revelation of World Views in Bronze Age Scotland'. *Journal for Archaeological Theory and Method* 22, no. 2 (2015): pp. 584–645 (2015). First released online 12 December 2013, http://link.springer.com/article/10.1007/s10816-013-9182-7.

Higginbottom, G., and R. Clay. 'Reassessment of sites in Northwest Scotland: A new statistical approach'. *Archaeoastronomy* 24 (1999): pp. S1–S6. Higginbottom, G., A. Smith, K. Simpson, and R. Clay. 'Gazing at the horizon: sub-cultural differences in western Scotland?'. In César Esteban and Juan A. Belmonte, *The Oxford VI International Conference on Archaeoastronomy and Astronomy and Culture*, pp. 43–49. Tenerife: Organismo Autónomo de Museos del Cabildo de Tenerife, 2001.

Higginbottom, G., and R. Clay. 'The Origins of Standing Stone Astronomy in Britain'. *Journal of Archaeological Science: Reports* (2016), http://dx.doi.org/10.1016/j.jasrep.2016.05.025.

Martlew, R. D., and C. L. N. Ruggles. 'Ritual and landscape on the West Coast of Scotland: An investigation of the stone rows of Northern Mull'. *Proceedings of the Prehistoric Society* 62 (1996): pp. 125–129.

Richards, C. 'Interpreting Stone Circles'. In *Building the Great Stone Circles of the North*, edited by Colin Richards, pp. 2–30. Oxford: Windgather Press, 2013.

Richards, C. 'Wrapping the hearth: constructing house societies and the tall Stones of Stenness, Orkney'. In *Building the Great Stone Circles of the North*, edited by Colin Richards, pp. 64–89. Oxford: Windgather Press, 2013.

Richards, C., and S. Griffiths. 'A time for stone circles, a time for new people'. In *Building the Great Stone Circles of the North*, edited by Colin Richards, pp. 281–91. Oxford: Windgather Press, 2013.

Pimenta, F., L. Tirapicos, and A. Smith. 'A Bayesian Approach to the Orientations of Central Alentejo Megalithic Enclosures'. *Archaeoastronomy* 22 (2009): pp. 1–20

Pimenta, F., N. Ribeiro, A. Smith, and L. Tirapicos. 'The Sky and the Landscape of Rock Art in the Ceira and Alva Basins'. In *Cosmology Across Cultures*, edited by José Alberto Rubiño-Martín, Juan Antonio Belmonte, Francisco Prada, and Antxon Alberdi, ASP Conference Series, Vol. 409, pp. 359–363. San Francisco: Astronomical Society of the Pacific, 2009.

Ritchie, J. N. G. 'The Stones of Stenness, Orkney'. *Proceedings of the Society of Antiquities of Scotland* 107 (1976): pp. 1–60.

Ruggles, C. L. N. *Astronomy in Prehistoric Britain and Ireland*. London: Yale University Press, 1999.

Ruggles, C. L. N. *Megalithic astronomy: A new archaeological and statistical study of 300 Western Scottish Sites*. Oxford: British Archaeological Reports British Series 123, 1984.

Schulting, R., S. Sheridan, R. Crozier, and E. Murphy. 'Revisiting Quanterness: New AMS dates and stable isotope data from an Orcadian chamber tomb'. *Proceedings of the Society of Antiquaries Scotland* 140 (2010): pp. 1–50.

Sheridan, J. A. 'The National Museums' Scotland radiocarbon dating programmes: results obtained during 2005/6'. *Discovery and Excavation in Scotland* 6 (2006).

Sheridan, J. A. 'Towards a fuller, more nuanced narrative of Chalcolithic and Early Bronze Age Britain 2500–1500 BC'. *Bronze Age Review: The international journal of research into the archaeology of the British and European Bronze Age* 1 (2008): pp. 57–78.

WINTER SOLSTICE AT THE IBERIAN CAVE-SANCTUARY OF LA NARIZ

César Esteban and José Ángel Ocharan Ibarra

ABSTRACT: We present results of an archaeoastronomical study of the Iberian cave-sanctuary of La Nariz (Moratalla, Murcia, Spain). The chronology of the site has been established between the third century BCE and the first century CE; however, there are material evidences of its use from the Late Bronze Age. The cave has a striking symmetrical morphology, with two main almost parallel long cavities of similar size and proportions. Both cavities have water springs, and carved basins to collect water at their innermost areas. We have found that the northerly cavity is facing the point of the horizon where the winter solstice sunset takes place, producing a striking illumination phenomenon onto the carved basin. On the other hand, the southerly cavity is slightly tilted to the south with respect to the northerly one, facing the moonset at the major southern lunastice or Venus at its southernmost setting.

Introduction

The Iberians were a group of peoples who inhabited the Iberian Peninsula from the sixth century BCE up to practically the change of the era, after the Roman conquest of their territory. They occupied part of the centre and the Mediterranean facade of the Iberian Peninsula, as well as the French Languedoc region.[1] The main Iberian deity was apparently a fertility goddess that also had strong funereal associations. Her most common iconography is a reflection of the aristocratic Iberian female image, but sometimes represented with attributes of Eastern or Greek goddesses such as Astarte, Tanit, Artemis or Demeter.[2] Iberian sanctuaries were usually in locations that favour the manifestation of the sacred, such as on the top of mountains, within caves or in proximity to springs.[3] They mostly consist of open-air deposits or temples housing a statue of the divinity and a large number of offerings. The chronology of the known Iberian sanctuaries ranges from the sixth century BCE to the first century CE.

Recent archaeoastronomical studies are revealing the link between the Iberian religious world and celestial bodies, mainly the Sun.[4] To date, equinoctial markers (or most probably, the half day between solstices) have been the most common astronomical relationships found in Iberian sanctuaries.[5] Most of them are markers of the sunrise or sunset over remarkable topographic features of the horizon. Another ritual use of equinoxes is found in the cave-sanctuary of Cueva de La Lobera, where the light of sunset illuminates the innermost area of the cave producing a striking phenomenon.[6] On the other hand, there are some Iberian shrines that may be related to the sunrise or sunset on the summer solstice, such as La

---

[1] For an introduction to Spanish protohistory see: Richard J. Harrison, *Spain at the Dawn of History: Iberians, Phoenicians and Greeks* (London: Thames and Hudson, 1988).

[2] Teresa Moneo, *Religio Iberica: Santuarios, ritos y divinidades (siglos VII-I a.C.)* (Madrid: Real Academia de la Historia, 2003), pp. 427–439.

[3] José M. Blázquez, *Diccionario de las religiones prerromanas de Hispania* (Madrid: Ediciones Istmo, 1975), pp. 148–166.

[4] See compilation by César Esteban, 'Arqueoastronomía y religión ibérica', in *Santuarios iberos: territorio, ritualidad y memoria*, ed. Carmen Rísquez and Carmen Rueda (Jaén: Asociación para el desarrollo rural de la Comarca de El Condado, 2013), pp. 465–484.

[5] See César Esteban, 'Elementos astronómicos en el mundo religioso y funerario ibérico', *Trabajos de Prehistoria* 59, no. 2 (2002): pp. 81–100; Esteban, 'Arqueoastronomía y religión ibérica', pp. 465–484.

[6] César Esteban, Carmen Rísquez, and Carmen Rueda, 'An Evanescent Vision of the Sacred? The Equinoctial Sun at the Iberian Sanctuary of Castellar', *Mediterranean Archaeology and Archaeometry* 14, no. 3 (2014): pp. 99–106.

Fig. 16.1: Cave-sanctuary of La Nariz (Moratalla, Murcia). The insert on the left indicates its location in the Iberian Peninsula.

Escuera (San Fulgencio, Alicante), Ullastret (Girona)[7] or Cerro de las Cabezas (Valdepeñas, Ciudad Real).[8]

The cave-sanctuary of La Nariz is located in the highlands of the northwest of the Region of Murcia (Fig. 16.1), in an area known as Umbría de Salchite, on the southern slope of a mountain called Calar de la Cueva de la Capilla,[9] and hanging on a cliff about 40 meters high. The first archaeological reference about this site is by Pedro Lillo,[10] and since then, the sanctuary has always been associated with initiation rituals linked to the wolf.[11] This idea is based on the materials found at the site and especially a ceramic fragment of the second century BCE with the representation of a possible female deity whose arms were interpreted as wolf heads.[12]

One of the authors (Ocharan Ibarra) has carried out archaeological surveys and excavations at La Nariz since 2010.[13] The archaeological fieldwork included an archaeological excavation of the cave and a survey of the complete extension of the mountain where the sanctuary is located. The survey indicated that none of the numerous caves of Calar de la Cueva de la Capilla (apart from the cave-sanctuary of La Nariz) show artefacts of archaeological interest. Preliminary results of the excavations at La Nariz indicate a prolonged use of the sanctuary over time and that it should be understood as a rural sanctuary, linked to prehistoric and protohistoric roads and with a supraterritorial character.

While full use would be framed between the late third century and first half of the first century BCE, occasional use in earlier Iberian times and Late Bronze Age cannot be disregarded. The end-point of the cultic activity at the site came with Romanization. The most abundant offerings at the sanctuary are artefacts common in grave goods of Iberian women. This, coupled with the iconography of the ceramic

---

[7] Esteban, 'Elementos astronómicos', pp. 81– 100.
[8] César Esteban and Luis Benítez-de-Lugo-Enrich, 'Orientaciones astronómicas en el *oppidum* oretano del Cerro de las Cabezas (Valdepeñas, Ciudad Real)', *Trabajos de Prehistoria*, in press (2016).
[9] Its English translation is the 'Mountain of the Cave Chapel', perhaps as a memory of the sacredness of the site.
[10] Pedro Lillo, *El poblamiento ibérico en Murcia* (Murcia: Real Academia Alfonso X El Sabio, 1981), pp. 39–41.
[11] Pedro Lillo. 'Una aportación al estudio de la religión ibérica: La diosa de los lobos de la Umbría de Salchite, Moratalla (Murcia)', *Actas del XVI Congreso Nacional de Arqueología*, Zaragoza (1983): pp. 769-787.
[12] Lillo. 'Una aportación al estudio', pp. 769–787.
[13] Results of these excavations will be published elsewhere (Ocharan Ibarra, in preparation).

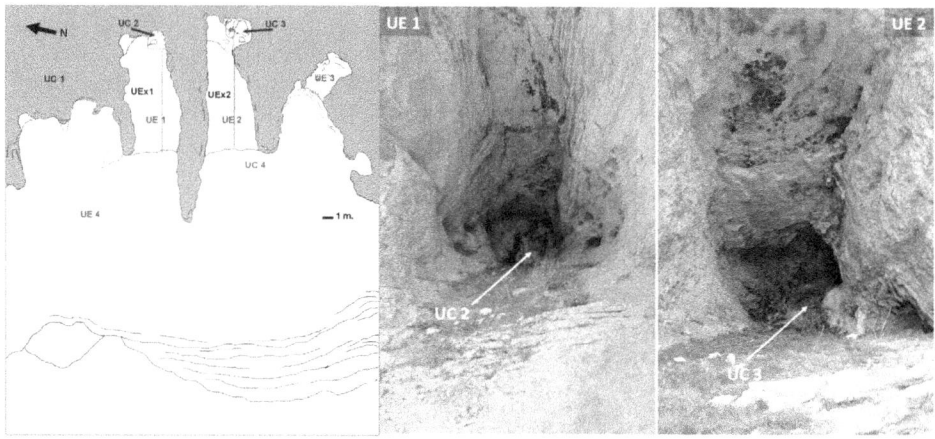

Fig. 16.2: Left, plan of the cave-sanctuary of La Nariz; UE 1 and UE 2 are the two main cavities. Right, photos of the interior of the cavities UE 1 and UE 2 obtained from their entrances; the location of the carved basins UC 2 and UC 3 are indicated with arrows.

fragment mentioned above, lead us to think that the cultic use of the cave would be related to a female deity, possibly linked to water, fire, conifers and birds.[14] Recently, González Reyero and collaborators[15] have discussed the similarity between that possible divine image and others at the Iberian settlement of Molinicos, located approximately 18.5 km west of La Nariz. Although Molinicos was abandoned in the mid-fourth century BCE there seems to be continuity in the kind of worship performed in both sites.[16] This is suggested because some early Iberian materials of about the fourth century BCE in the excavations of La Nariz have been found. Therefore, both spaces could have been used simultaneously during the fourth century BCE, solving the diachrony problem between the two sites alluded to by González Reyero et al.

La Nariz consists of two main large cavities that we called UE 1 and UE 2[17] (see Figs. 16.1 and 16.2) and two smaller ones (UE 3 and UE 4). UE 1 and UE 2 have very similar morphologies and dimensions, with a separation between axes of only 4 metres. The ensemble shows a curious and suggestive symmetrical appearance (see Fig. 16.1). In the innermost parts of UE 1 and UE 2 there are two natural but artificially modified basins carved on the rock (UC 2 and UC 3, see Fig. 16.2). The basins collect the water that filters through the inner walls of the cavities. UE 1 and UE 2 have a span of 11m high, 2.5 m wide and 12 m deep. The height of the cavities descends toward the interior reaching only 2 m at the basins.

---

[14] José A. Ocharan Ibarra, 'Aproximación al estudio de los santuarios rupestres ibéricos de la Región de Murcia; La Nariz (Moratalla, Murcia)', in *Santuarios iberos: territorio, ritualidad y memoria*, ed. Carmen Rísquez and Carmen Rueda (Jaén: Asociación para el desarrollo rural de la Comarca de El Condado, 2013), pp. 289–303; José A. Ocharan Ibarra, 'Santuarios Rupestres Ibéricos en la Bastetania Oriental. Aproximación a los posibles *loca sacra libera* de la Región de Murcia', *Orígenes y Raíces* 4 (2013): pp. 14-19.

[15] Susana González Reyero, F. J. Sánchez-Palencia Ramos, C. Flores Barrio, and I. López Salinas, 'Procesos de apropiación y memoria en el sureste peninsular durante la segunda Edad del Hierro: Molinicos la Umbría de Salchite en la construcción de un espacio político', *Zephyrus* LXXIII (2014): pp. 149–170.

[16] González Reyero et al.. 'Procesos de apropiación y memoria', pp. 149–170.

[17] To designate the different elements of the cave, we decided to use the same notation as in the report of archaeological excavation (UE 1, UE 2, etc.). This may lead to some confusion because the basin UC 2 is in the cavity UE 1 and UC 3 in UE 2.

Results

The orientation of the main axes of UE 1 and UE 2 and the horizontal coordinates of topographic elements of the surrounding horizon were measured with a precision compass and a theodolite in April 2012. The distant horizon visible from both cavities covers approximately the southwest quadrant from about 205° to 265°.

As it has been said above, both cavities, UE 1 and UE 2, have basins to collect the water that flows in their interiors. Both basins show carved drain canals, suggesting that the basins should be important elements in the ritual performed in the sanctuary. Standing on the basins and looking towards the entrance of the cavities we see that the portions of horizon visible are very narrow (especially from UE 1, see Fig. 16.3). The width of the portion of the horizon seen from UE 1 is only about 5.5° and about 14.5° from UE 2. The horizon seen from UE 1 covers the landmarks indicated as a and b; a coincides with the area occupied by the farthest mountain peaks visible from the site (Las Cabras and La Sagra, 2100 and 2380 metres above sea level and about 33 and 55 km away, respectively). After the measurements, we estimated that the sunset at winter solstice (WS), at the present date ($\delta = -23.5°$) or even the Iberian period ($\delta = -23.7°$ at the middle of the first millennium BCE) should take place at or very close to the landmark b.

Fig: 16.3: Left: Horizon visible from the carved basins of UE 1 (up) and UE 2 (down). Some landmarks are marker with letters. A circle indicates the size of the solar disc. Centre and right: Sunset at a date very close to winter solstice as seen from the interior of UE 1 and UE 2, respectively.

On the 18th December 2012, with the Sun located at $\delta = -23.4°$, we visited again the place to follow-up the illumination of the cave during sunset. Indeed, a few minutes before sunset, the sun illuminated the innermost zone of UE 1 (see Figs. 16.3 and 16.4). At 17:30 local time, sunlight, which from noon was penetrating into the cavity, reached the base of UC 2. From this moment and in the last twenty minutes before sunset, the illuminated area started a slow ascent by UC 2, not reaching the water contained in the basin until the last moments before the Sun was completely set (see Fig. 16.4). We estimated that this phenomenon is visible only about 15 days before and after the precise date of the WS.

Fig. 16.4: The four images on the top show the illumination of basin UC 2 (in the interior of UE 1) as the winter solstice sunset progresses. The lower photo shows the carved basin UC 2 just a few minutes before sunset. We can see that the light illuminates tangentially the water surface and the patch of light coincides very precisely with the borders of the basin

Let's come back to UE 2. The centre of the portion of horizon visible from the interior corresponds to the landmarks c and d and the extreme right (north) coincides approximately with landmark b. Our observations of 18 December 2012 demonstrated that in the last moments before sunset, the Sun penetrated just at the right end of the opening of UE 2 (only a portion of the solar disc). In this cavity, the illuminated area never reaches the water of basin UC 3 (nor in Iberian times). In UE 2, apart from this possible relationship with the sunset on WS, another interesting astronomical relation is that the moonset at southern major standstill (SMS, $\delta = -29.7°$ at the middle of the first millennium BCE) takes place approximately halfway between landmarks c and d, approximately at the centre of the opening.

Finally, due to its very difficult access and step floor, we could not take measurements from the cavity UE 3. In any case, its relative position inside the cave makes very unlike that it may present any astronomical relation of interest.

Discussion

As already mentioned in the introduction, the astronomical relations found in La Nariz are unusual in Iberian sanctuaries. A significant fraction of them show orientations or markers associated with the equinoxes or, more probably, with the half day in time between solstices.[18] The chronology of the sites showing that astronomical relation suggests that the onset of its use in Iberian sanctuaries occurred in the mid-fourth century BCE,[19] coinciding with a time of significant change in the ideology of the Iberian society and increased Punic influence. The striking visual relationship between the settlement of Molinicos and the mountain where La Nariz is located (Calar de la Cueva de la Capilla) found by Reyero González et al.[20] could also have an astronomical meaning. The line connecting the two places is very near the east-west line, so that from the urban shrine at room k of Molinicos, the sunset at the equinoxes would occur on Calar de la Cueva de la Capilla.

In the Iberian area, orientation towards the sunrise at WS are found in temple B of Illeta des Banyets (Campello, Alicante), the early Iberian necropolis of El Peñón del Rey (Sax, Alicante) and perhaps towards the sunset in the sanctuary of El Cigarralejo (Murcia).[21] The alignment of UE 1 to WS sunset inevitably makes us think that the rites celebrated in the sanctuary could be related to crop cycles, festivities associated with death and rebirth of nature or perhaps rites of passage[22] or initiation rituals.[23]

Although archaeoastronomical studies in Iberian cave-sanctuaries are limited to La Nariz and Cueva de La Lobera,[24] there are some similarities between both sites that deserve to be highlighted. First, both caves have their entrances open to the west (as it is the case in most Iberian cave-sanctuaries). On the other hand, the illumination phenomena that are produced inside both caves occur exactly at sunset, indicating that at least part of the celebrations should take place at the end of the day, when the sun returns to earth. Other remarkable aspects of the light phenomena are their short duration and the intense colour sensations they produce due to the reddish light of sunset. In addition, in both caves, some of the elements involved in the phenomenon (the west window of La Lobera that gives the shape to the patch of light and the basins of La Nariz) were artificially modified. All these facts suggest the willingness of the people charged with ritual to create an atmosphere of liminal character in the sanctuaries, even perhaps a dramatization of the sacred experience. In particular, at La Nariz, the brief illumination of the water surface by a tangent ray of twilight (that could be the central feature of the phenomenon) suggests a symbolic act of union of the two elements, water and sunlight, which only occurs in a unique time of the year, a moment of seasonal change. This seems to be full of meaning for a place of worship dedicated to a chthonic deity of fertility. This might be interpreted as a symbolic act of regular and seasonal fertilization that would closely relate the divinity with the celestial world and the vegetative cycle of nature. In the famous passage from Book III of his Geography, referring to a sanctuary at Cape St. Vincent, Strabo[25] tells how, according to Artemidorus, it was a native custom to turn up stones and offer a water libation at day-time but it was prohibited to offer sacrifices and spend the night in the sanctuary as it was thought to be occupied by the gods after sunset.

---

[18] Esteban, 'Arqueoastronomía y religión ibérica', pp. 465–484.
[19] Esteban, 'Arqueoastronomía y religión ibérica', pp. 465–484.
[20] González Reyero et al., 'Procesos de apropiación y memoria', pp. 149–170.
[21] Esteban, 'Arqueoastronomía y religión ibérica', pp. 465–484.
[22] Carmen Rueda, *Territorio, culto e iconografía en los santuarios iberos del Alto Guadalquivir (ss. IV a.n.e.-I d.n.e.)* (Jaén: Universidad de Jaén), p. 154.
[23] The difficult access and the seclusion of the cave seem compatible with the possibility of the celebration of such kinds of rites in La Nariz (we thank one of the referees for this suggestion).
[24] Esteban et al., 'An Evanescent Vision', pp. 99–106.
[25] Strabo, *Geography*, Book III, 3.1.4.

This reference is very suggestive, since it also relates the two main elements of the light phenomenon at La Nariz: water and sunset.

The possible lunar relation found in UE 2 also merits some discussion. There is a single known case of possible orientation toward the moonrise at the SMS in the Iberian world, the sanctuary of El Cigarralejo (Mula, Murcia); the building shows a double alignment towards the sunset at WS and moonrise at SMS due to the different height of the western and eastern horizons.[26] In a recent paper, González García et al.[27] have obtained archaeoastronomical results for three cultic sites in use at least since the Iron Age in the Celtic and Celtiberian areas of the northern half of the Iberian Peninsula. All these sites show horizon markers of the sunrise at WS and moonrise at SMS, the same astronomical relations found at the two main cavities of La Nariz, although in our cave-sanctuary orientations are towards settings. This similarity does not necessarily indicate a Celtic origin of the people who made La Nariz their sanctuary; perhaps, and this is just speculation, we may be facing spatial and temporally separate manifestations of a common ancestral astronomical tradition of the peoples of the Iberian Peninsula previous to the Iron Age. However, although it seems plausible that the orientation of UE 2 could have been related to rituals dedicated to the moon, we can raise another alternative astronomical explanation based on its possible relation with Venus. In a recent article, Esteban and Escacena Carrasco[28] presented a study of several Tartesian-Phoenician shrines in the south of the Iberian Peninsula, finding that many of the sanctuaries included in that work are orientated identically and that, to the west (average azimuth of 235°), they could be related to the setting of Venus at its southernmost declination ($\delta = -26.5°$ in the middle of the first millennium BCE) suggesting a cult of the Phoenician goddess Astarte, assimilated to Venus in Roman times. The setting of Venus at its southernmost declination would be observed between landmarks b and c (see Fig. 16.3), in particular about 2° to the right (north) of c. Assuming the Sun-Venus binomial as an interpretation for the ritual orientation of the two cavities of La Nariz, this site would be fairly similar to the emblematic and complex Tartesian-Phoenician sanctuary of El Carambolo (Camas, Sevilla), where two enclosures or temples have identical orientations to UE 1 and UE 2. For El Carambolo, Esteban and Escacena Carrasco[29] suggest a possible double cult of Baal/Melqart astronomically related to the Sun and Astarte as the personification of Venus. Therefore, assuming the binomial Sun-Venus in La Nariz, we could relate this sanctuary to the Tartesian-Phoenician binomial tradition that originated at least in the Orientalizing Period.[30] However, the chronology and the archaeological findings do not seem to support this parallelism for La Nariz.

Conclusions

We present results of archaeoastronomical research conducted in the Iberian cave-sanctuary of La Nariz (Moratalla, Murcia, Spain). The chronology of the site goes from the third century BCE to the first century CE, although there is material evidence of its possible use during the fourth century BCE and even the Late Bronze Age. The cave has a striking symmetrical morphology, with two almost parallel main cavities with similar sizes and proportions.[31] Both cavities have carved basins to collect the water flowing from

---

[26] Esteban, 'Elementos astronómicos', pp. 81–100.
[27] A. César González-García, Marco García Quintela, and Juan A. Belmonte, 'Landscape Construction and Time Reckoning in Iron Age Iberia', *Archaeoastronomy, The Journal of Astronomy in Culture*, in press.
[28] César Esteban and José L. Escacena Carrasco, 'Oriented for Prayer: Astronomical Orientations of Protohistoric Sacred Buildings of the South Iberian Peninsula', *Anthropological Notebooks* 19, supplement (2013): pp. 129-142.
[29] Esteban and Escacena Carrasco, 'Oriented for Prayer', pp. 129-142.
[30] Period of Phoenician colonization of the Iberian Peninsula at the beginning of the first millennium BCE.
[31] This morphological characteristic might be an aesthetic argument for the choice of this particular cave as a sacred place.

the inner walls. A very narrow part of the western horizon can be seen from the interior of the northern cavity (UE 1). The sunset at winter solstice (WS) takes place near the centre of this area and this was confirmed with observations at the site. We also found a striking phenomenon of illumination of the carved basin during the WS sunset. On the other hand, the southern cavity (UE 2) shows an orientation deviated about 4° south with respect to EU 1. The WS sunset also illuminates the interior of UE 2 but not the carved basin. The orientation of UE 2 is closer to the moonset at the southern major standstill and even the southernmost declination of Venus. This fact suggests the possibility of a cult with double astronomical implications in a cave formed by two almost identical cavities.

Acknowledgements
We thank the anonymous reviewer for his/her constructive comments.

Bibliography
Blázquez, José M. *Diccionario de las religiones prerromanas de Hispania*. Madrid: Ediciones Istmo, 1975.
Esteban, César. 'Arqueoastronomía y religión ibérica'. In *Santuarios iberos: territorio, ritualidad y memoria*, edited by Carmen Rísquez and Carmen Rueda, pp. 465–484. Jaén: Asociación para el desarrollo rural de la Comarca de El Condado, 2013.
Esteban, César. 'Elementos astronómicos en el mundo religioso y funerario ibérico'. *Trabajos de Prehistoria* 59, no. 2 (2002): pp. 81–100.
Esteban, César, and José L. Escacena Carrasco. 'Oriented for Prayer: Astronomical Orientations of Protohistoric Sacred Buildings of the South Iberian Peninsula'. *Anthropological Notebooks* 19, supplement (2013): pp. 129–142.
Esteban, César, and Luis Benítez-de-Lugo-Enrich. 'Orientaciones astronómicas en el *oppidum* oretano del Cerro de las Cabezas (Valdepeñas, Ciudad Real)'. *Trabajos de Prehistoria*, in press (2016).
González-García, A. César, Marco García Quintela, and Juan A. Belmonte. 'Landscape Construction and Time Reckoning in Iron Age Iberia'. *Archaeoastronomy. The Journal of Astronomy in Culture*, in press.
Esteban, César, Carmen Rísquez, and Carmen Rueda. 'An Evanescent Vision of the Sacred? The Equinoctial Sun at the Iberian Sanctuary of Castellar'. *Mediterranean Archaeology and Archaeometry* 14, no. 3 (2014): pp. 99-106.
González Reyero, Susana, F. J. Sánchez-Palencia Ramos, C. Flores Barrio, and I. López Salinas. 'Procesos de apropiación y memoria en el sureste peninsular durante la segunda Edad del Hierro: Molinicos la Umbría de Salchite en la construcción de un espacio político'. *Zephyrus* LXXIII, pp. 149–170.
Harrison, Richard J. *Spain at the Dawn of History. Iberians, Phoenicians and Greeks*. London: Thames and Hudson, 1988.
Lillo, Pedro. *El poblamiento ibérico en Murcia*. Murcia: Real Academia Alfonso X El Sabio, 1981.
Lillo, Pedro. 'Una aportación al estudio de la religión ibérica: La diosa de los lobos de la Umbría de Salchite Moratalla (Murcia)'. *Actas del XVI Congreso Nacional de Arqueología*, Zaragoza (1983) : pp. 769–787.
Moneo, T. *Religio Iberica. Santuarios, ritos y divinidades (siglos VII-I a.C.)*. Madrid: Real Academia de la Historia, 2003.
Ocharan Ibarra, José A. 'Aproximación al estudio de los santuarios rupestres ibéricos de la Región de Murcia; La Nariz (Moratalla, Murcia)'. In *Santuarios iberos: territorio, ritualidad y memoria*, edited by Carmen Rísquez and Carmen Rueda, pp. 289–303. Jaén: Asociación para el desarrollo rural de la Comarca de El Condado, 2013.
Ocharan Ibarra, José A. 'Santuarios Rupestres Ibéricos en la Bastetania Oriental. Aproximación a los posibles *loca sacra libera* de la Región de Murcia'. *Orígenes y Raíces* 4 (2013): 14–19.
Rueda, Carmen. *Territorio, culto e iconografía en los santuarios iberos del Alto Guadalquivir (ss. IV a.n.e.-I d.n.e.)*. Jaén: Universidad de Jaén, 2011.
Strabo. *Geography*, Book III, 3.1.4.

RAISING AWARENESS OF LIGHT POLLUTION BY SIMULATION OF NOCTURNAL LIGHT OF ASTRONOMICAL CULTURAL HERITAGE SITES

Georg Zotti[1] and Günther Wuchterl[2]

ABSTRACT: The 2007 La Palma declaration identified the preservation of intact and authentic night skies as part of natural and cultural heritage as a key vector towards enacting the 'right to starlight'. This paper describes steps towards understanding, communicating and protecting night-time authenticity in general and with the *Großmugl Starlight Oasis* near Vienna, Austria, as an example. To this end we introduce a new feature in the Stellarium desktop planetarium which can be used to visualize nocturnal illumination and light pollution with a degree of authenticity and accessibility that is novel.

A side effect of our modern civilization is the ever-increasing level of illumination in and around inhabited areas. Architectural and natural monuments were built and used in numerous cases in connection with celestial events. Many of them have already lost or are in danger of losing the original appearance[3] of the monument and/or the firmament that would allow experiencing them in the original context. That loss of context, as far as light and astronomical objects are relevant for it, makes it harder or impossible to infer the intention of the builders or users.

The La Palma Declaration of 2007 states that:[4]

> An unpolluted night sky that allows the enjoyment and contemplation of the firmament should be considered an inalienable right of humankind equivalent to all other environmental, social, and cultural rights.[5]

Many of the issues addressed by the La Palma Declaration cannot easily be visualised due to the existing technical limitations, both of 'sky simulators' (planetarium software) and to the intrinsic high dynamical range and special light-properties of night-time scenes. To communicate those to the public and provide a tool to site-managers and authorities administrating site protection we have augmented a large outreach popular planetarium software with a light and colour model that can handle key effects in real time on common personal IT-devices. Thus we are able to add environmental effects to planetarium scenery containing cultural heritage with direct or indirect connections to astronomy. Classically those were dominated by sunlit, moonlit or moonless sky conditions. With the increase of light pollution to levels above a small fraction of the natural light, a new light source started to shape and dominate the appearance

---

[1] Ludwig Boltzmann Institute for Archaeological Prospection and Virtual Archaeology, Vienna.
[2] Verein Kuffner Sternwarte, Vienna.
[3] We use 'appearance' to refer to the impression of a monument or landscape as depending on the light emitted by the astronomical objects and modulated by the atmosphere. The 'visibility' of an astronomical object or a remote monument or point of interest, e.g., a lighthouse, depends on similar factors. 'Appearance' is the consequence of the light falling onto the object and it is even modulated when the 'visibility' is zero as for the moon with complete overcast. The moon is not visible but its scattered light determines the appearance of a monument or landscape.
[4] Declaration in Defence of the Night Sky and the Right to Starlight (La Palma Declaration), International Conference in Defence of the Quality of the Night Sky and the Right to Observe the Stars, Canary Islands, 19–20 April 2007, available at http://www.starlight2007.net/pdf/StarlightDeclarationEN.pdf [accessed 21 December 2014].
[5] La Palma Declaration 2007, Art. 1.

of the night sky. Consequently it has to be included in visual representations if conformity and an easy connection to reality is the goal. Light pollution added regimes between the classical moonless nights, moonshine dominated skies and the twilight – e.g., nights with never-ending twilight-like conditions, skies subject to more light than that of the full moon - with severe consequences for the site appearance and technical challenges for the mapping and rendering.

> The intelligent use of artificial lighting that minimises sky glow and avoids obtrusive visual impact on both humans and wildlife has to be promoted. Public administrations, the lighting industry, and decision-makers should also ensure that all users of artificial light do so responsibly as part of an integral part of planning and energy sustainability policies,...[6]

If site managers decide to apply nocturnal artificial illumination to a monument, this usually completely changes the original appearance both of the natural lighting situation and the respective authentic lighting culture. Note that many sites likely have been used in moonlight. Many festival calendars connect cultural activities to the changing appearance and light of our moon and its ever-changing mixtures with other light from the firmament.

The astronomy and world heritage thematic study and the extended case studies on astronomy and world heritage, in particular the chapters on 'Windows to the Universe', point to the importance of core/buffer zone light management.[7] They essentially call for minimum light, with moonshine as reference point, in the *core zones* and downward directed, demand-based illumination in the *buffer zone*. National or provincial legislation should control light intrusion from an *external zone* of the *Starlight Reserve* concept or a *far zone* that may be hundreds of kilometres wide.[8]

The remoteness of light-protected areas produces a barrier for most people that live in bright urban areas from experiencing a reasonably intact night sky and night-time environment. The tool to resolve this conflict is the *Starlight Oasis*. While light levels in such a place are typically significantly above natural conditions, the most important features of light at night, such as its lunar illumination cycle and the culturally most important features of the night sky, such as the constellations of the respective culture, e.g. the ones of the Zodiac, are still intact. That leads to a typical zenithal visual limiting magnitude of 6^9 for moonless clear skies that also provides impressive views of the Milky Way.

An example is the *Großmugl Starlight Oasis* about 35km north of Vienna's centre.[10] Protected by three

---

[6] La Palma Declaration 2007, Art. 7.
[7] Clive Ruggles and Michel Cotte, eds., 'Heritage Sites of Astronomy and Archaeoastronomy in the context of the UNESCO World Heritage Convention - A Thematic Study'. *Paris: ICOMOS and the International Astronomical Union* (June 2010), available at http://www2.astronomicalheritage.net/index.php/thematic-study [accessed 25 December 2014]; Cipriano Marín, Richard Wainscoat, and Eduardo Fayos-Solá, 'Windows to the Universe': Starlight, Dark Sky Areas, and Observatory Sites', in Ruggles and Cotte, eds., 'Heritage Sites of Astronomy and Archaeoastronomy', Ch. 16.
[8] Marín et al., 'Windows to the Universe', p. 244. As proposed by Victor Reijs for the Bru na Boinne management plan.
[9] This is considered rather dark by some, following our daily experience. An exact specification of a culturally sufficiently 'starry' sky necessarily depends on the cultural context. For the western case we note that the Little Bear needs a solid 5th magnitude sky and a complete Zodiac, magnitude 6. A discussion is given by one of the authors in the context of the Astronomy and World Heritage Initiative: Günther Wuchterl, 'General considerations for relatively dark starlight/dark-sky areas with few or no direct cultural connections', 2012, available at http://www2.astronomicalheritage.net/index.php/show-theme?idtheme=21 [accessed 29 July 2016].
[10] Günther Wuchterl: 'Eastern Alpine Starlight Reserve and Großmugl Starlight Oasis, Austria', in Ruggles and Cotte, eds., 'Heritage Sites of Astronomy and Archaeoastronomy', Ch. 16.2, and available at http://starlightoasis.org [accessed 28 December 2014].

mountain ridges from the metropolitan light dome of Vienna, the visual limiting magnitude is a robust 6.3 and even the zodiacal light is still a visible phenomenon. The public observing site is next to a monumental Iron-age tumulus which attracts visitors as an ancient landmark. (It has no known archaeoastronomical relevance, though.)

The wider area contains a few of the even older monumental *Kreisgrabenanlagen* (Neolithic circular ditch systems)[11] which document a long-standing continuity in the permanent population and use of this landscape. A site like this allows fast access to very good skies (in modern civilisatory standards) to a large population. However, the low horizon towards Vienna is brightened up notably by astronomical standards, and also the surrounding villages cause some moderately bright spots visible along the horizon (Fig. 17.1). In and around *Großmugl*, there is continued activity around protecting the night sky and raising awareness with regard to the importance of good 'star-bright' sky conditions. (The natural clear sky should not be called 'dark' in this context, because of the word's negative connotations.) The community adopted the La Palma declaration in 2010 and effected a zero artificial light policy in the 'no building' zone that is in force in the framework of Austrian monumental protection legislation and which surrounds the tumulus.

Recently, a 'Star Walk' nature trail with information signposts has been installed, connecting the town with the observing site next to the tumulus, and occasional public 'starlight festivals' can attract many hundreds of people.[12] The Star Walk informally and informatively explains key facts and factors determining the authentic night-time appearance of an archaeological or natural landscape site. While proven or proposed alignments of built structures and celestial events like certain sun- or moonrises etc. are particular to only certain classes of prehistoric monuments, the night-time appearance generally dominated by natural light sources from the sky is an important and rarely addressed issue for all sites of prehistoric importance.[13] While the visitor leaves the last street-light near the beginning of the Star Walk, she or he experiences how progressing dark-adaption of the eye changes colour perception and field-of-view experience during a passage that ends with a 'step into starlight' at the exit of an overgrown *Kellergasse* (scenic road with a line-up of wine cellars) producing a natural tunnel of trees and bushes. The light at the end of the tunnel is starlight with a dash of moon at appropriate moments. Once in starlight, the visitor is introduced to colours of stars, and how the celestial sphere's daily motion interacts with the landscape. Once the end of the Star Walk is reached at the tumulus observing spot after at least twenty minutes, his or her perception is ready to perceive the Hallstatt-age giant tumulus in a way that is authentic for its period. Shadows and colours are seen as they are and were modulated and subdued according to the natural variations driven by sources like the moon and Milky Way. With the eyes having arrived in 'mesopic vision' (transition between day and night vision), the beauty of the Milky Way and the firmament is experienced in its authentic meld with the prehistoric remains. Arguably this relates to the eye's response to da Vinci's *sfumato* often made responsible for the charm of Mona Lisa's smile. The *sfumato* exploits the eye's changing resolution across its field of view to animate the face of the Gioconda

---

[11] Emília Pásztor, Judit P. Barna, and Georg Zotti, 'Neolithic Circular Ditch Systems ("Rondels") in Central Europe', in *Handbook for Archaeoastronomy and Ethnoastronomy*, ed. Clive Ruggles (New York: Springer, 2015), Ch. 113.

[12] Karoline Mrazek and Erwin Matys, 'A Star Walk for Everyone', *Sky&Telescope* (November 2014): pp. 34–37.

[13] Günther Wuchterl, 'In search of the perfect sky - towards a cultural and astronomical quantification', in *Proceedings of the Third International Starlight Conference Lake Tekapo, New Zealand June 2012* (The Starlight Conference Organising Committee, Feb. 2013), ed. John Hearnshaw, Karen Pollard, and Marilyn Head, available at http://www.phys.canterbury.ac.nz/Conferences/Third_International_Starlight_Conference [accessed 25 December 2014]; D. Brown, F. Silva, and R. Doran, 'Education, Archaeoastronomy and the Outdoor Classroom: Lessons from the Past', *Anthropological Notebooks* 19, supplement (2013): p. 525. We thank one of the reviewers for providing this reference to support the argument here.

Fig. 17.1: A fisheye photograph taken near the Großmugl public observing site on 5 November 2008 2:21CEDT (top) and simulation of the scene with the new illumination layer in Stellarium 0.13.1 (bottom). North is at top. The strongest source of light pollution is Vienna towards the south-southeast (bottom, slightly to the left). The illumination layer is hand-made and tries to recreate the view with the unaided eye. (Photo: G. Zotti)

and brings forward different expressions. In the nocturnal landscape the changing sharpness of day and night vision play together with the modulation by differences of perception across the human field of view to create a person's authentic experience of the *sky landscape system*.[14] If that system hosts or constitutes an archaeological site the effect contributes to an authentic experience of the site. It may be controversial what creates the unique experience of the Milky Way arching across an artefact, but certainly the appropriately low light levels and the adaptation of the eye are a necessary condition with broad consensus.

Visualisation

While we cannot yet visualise all aspects of authenticity – the focus here is the authenticity as considered by the world heritage convention – recent developments provide important steps towards the authentic visualisation of archaeological sites at night by introducing better light and colour management into state

---

[14] See also, Wuchterl, 'In search of the perfect sky'.

of the art planetarium software. Thus we aim to provide a better practical bridge between what people imagine of the night or view on their screens and what they may expect to experience when actually being there – a bridge between sites as the one in *Großmugl*, education and conservation, and visualisation.

To visualize sites with and without nocturnal illumination, at least in a qualitative way, a new feature has been introduced in the popular open-source desktop planetarium program Stellarium by one of the authors. One of the highly favoured features of this software is the ability to easily configure a photographically created landscape panorama as foreground behind which the starry sky is visible. The panorama photograph is usually taken during daytime and displayed with a brightness level that follows the brightness of the sky, i.e., the landscape is just displayed dimmed down during night time. Originally this feature had been introduced merely for illustration and decorative purposes, but since V0.10.6 the horizon imagery can be adjusted to high accuracy. This allows a meaningful assessment of potentially intended astronomical alignments of celestial objects with features on the horizon and foreground objects (built architecture) as they may have appeared in the past.[15] In addition to the previously existing global light pollution settings which simulate the artificially brightened sky with its reduced star visibility, we have introduced an optional site-dependent nocturnal light layer for landscape panoramas with the recent V0.13 release of Stellarium. This layer can illustrate the changed appearance of natural nightscapes introduced by human civilisation, like the annoying direct glare of foreground street lights, illuminated windows, floodlit buildings and also the direction-dependent sky glow along the horizon near larger cities. The light layer is added-in during the deeper phases of civil twilight, and can be switched off if needed.

The light layer is most easily created together with the horizon panorama. In case of a photography-based panorama, the most accurate procedure is to take a panorama in daytime and one during the night, and combine those in the same panorama project to align the 'daylight' photos with the 'nightlight' photos, for example in the free panorama maker software Hugin.[16] Then, two panoramas are exported from the day and night photos respectively, which are now properly aligned. As usual, the sky of the daylight panorama has to be set transparent in the postprocessing required to make a panorama usable with Stellarium, in image editing software like Photoshop or the Gimp, while the nightlight panorama should be based on a black background. Celestial objects visible in the night photographs have to be removed, so that only the artificial illumination remains.

Another option is to just paint an illustration of nocturnal illumination (street lights, bright windows, or city skyglow) on top of the existing daylight panorama. Both mentioned software titles are layer-aware image editing programs, so that images can be created and combined from a stack of sub-images (layers). Therefore the easiest way to create this light layer is to use a semi-transparent dark layer (to darken the daylight panorama during the paint process) and at least one new transparent layer to paint the lights on. Filter effects can be applied to illustrate glare halos around excessively bright lights. Creating city skyglow is made easy with layer masks which cover the ground and prevent disturbing foreground artefacts. The creator of a skyglow layer may be tempted to use a simple vertical brightness gradient, however sites may show uneven illumination or bright spots on the horizon in the direction of villages nearby or larger cities farther away, so it may be useful to even work with several layers which can be combined accordingly in additive mode. The final landscape then again requires two PNG exports for day and night as mentioned above.

---

[15] Georg Zotti, 'Visualization Tools and Techniques', in *Handbook for Archaeoastronomy and Ethnoastronomy*, ed. Clive Ruggles (New York: Springer, 2015), Ch. 29. See also a discussion about skyscape experiences and about a few known limitations of a previous version of Stellarium in Daniel Brown, 'Exploring Skyscape in Stellarium'. *Journal of Skyscape Archaeology* 1, no. 1 (2015): pp. 93–111.

[16] Hugin panorama stitcher available at http://hugin.sourceforge.net/ [accessed 1 January 2015].

The light layer is available for all image-based 'landscape' types as of V0.13.1 of Stellarium. In 'old-style' landscapes light layer tiles share the same geometry as the underlying daylight horizon tiles and are therefore most useful to accentuate the nocturnal landscape in natural-sky areas with a few distinct extra light dots. On the other hand, in 'fisheye' and 'spherical' landscapes, the light layer is geometrically mostly independent from the daylight panorama, which makes these landscape types better suited for skyglow simulation. Note that the fisheye landscape type is geometrically not accurate and thus useful for decoration only.

This light pollution layer is currently not based on a physical simulation and does by itself not change visibility of celestial objects, but serves as illustration of distracting lights. It also does not include animation like blinking tower lights or moving sky beamers. For a future development it may be interesting to attempt creating a skyglow layer based on a physics-based light pollution model[17].

To provide a sample for this new landscape feature, a landscape for the *Großmugl Starlight Oasis* has been added to the Stellarium standard landscapes that come included with the main program download. Its configuration file provides more complete insight into the technical details of setting up the various parameters, and full description of all landscape setup parameters is given in the Stellarium wiki.[18]

This new feature may help site managers and proponents of light pollution measures to better visualize their tasks and concerns. By preparing several different 'painted' night scenarios it is easy to illustrate the likely appearance of historically important sites in a plausible representation of original illumination, likely by fire or candles, and also to present simulated scenarios that may involve the visual effects of present or planned floodlighting. In the latter case, the panoramas may obviously also be prepared by artificial renderings of architectural 3D models.

Fig. 17.2: A simulated night scene of Valletta Harbour, Malta, created by combination of day and night photographs. Screenshot from Stellarium 0.13.0 with the new nocturnal illumination layer. The gain in realism when simulating the sky over a contemporary light-polluted site over a scene in previous versions with only a darkened daylight foreground is evident.

---

[17] E.g., Roy H. Garstang, 'Model for Artificial Night-Sky Illumination', *Publications of the Astronomical Society of the Pacific* 98 (1986): pp. 364–375.
[18] http://www.stellarium.org/wiki/index.php/Customising_Landscapes [accessed 31 December 2014].

Bibliography

Brown, Daniel. 'Exploring Skyscape in Stellarium'. *Journal of Skyscape Archaeology* 1, no. 1 (2015): pp. 93–111.

Brown, D., F. Silva, and R. Doran. 'Education, Archaeoastronomy and the Outdoor Classroom: Lessons from the Past'. *Anthropological Notebooks* 9, supplement (2013): p. 525.

Declaration in Defence of the Night Sky and the Right to Starlight (La Palma Declaration), International Conference in Defence of the Quality of the Night Sky and the Right to Observe the Stars, Canary Islands, April 19-20, 2007, available at http://www.starlight2007.net/pdf/StarlightDeclarationEN.pdf.

Garstang, Roy H. 'Model for Artificial Night-Sky Illumination'. *Publications of the Astronomical Society of the Pacific* 98 (1986): pp. 364–375.

Großmugl Starlight Oasis. http://starlightoasis.org.

Hugin panorama stitcher. http://hugin.sourceforge.net.

IAU/UNESCO Portal to the Heritage of Astronomy. http://www.astronomicalheritage.net.

Marín, C., Richard Wainscoat, and Eduardo Fayos-Solá. 'Windows to the Universe': Starlight, Dark Sky Areas, and Observatory Sites'. In Clive Ruggles and Michel Cotte, eds. 'Heritage Sites of Astronomy and Archaeoastronomy in the context of the UNESCO World Heritage Convention - A Thematic Study'. *Paris: ICOMOS and the International Astronomical Union* (June 2010), Ch. 16.
http://www2.astronomicalheritage.net/index.php/thematic-study

Ch.16 of Ruggles and Cotte, 2010 and available at http://www2.astronomicalheritage.net/index.php/show-theme?idtheme=21

Mrazek, Karoline, and Erwin Matys. 'A Star Walk for Everyone'. *Sky&Telescope* (November 2014): pp. 34–37.

Ruggles, Clive and Michel Cotte, eds. 'Heritage Sites of Astronomy and Archaeoastronomy in the context of the UNESCO World Heritage Convention - A Thematic Study'. *Paris: ICOMOS and the International Astronomical Union* (June 2010).
http://www2.astronomicalheritage.net/index.php/thematic-study

Pásztor, Emília, Judit P. Barna, and Georg Zotti. 'Neolithic Circular Ditch Systems ("Rondels") in Central Europe'. Ch. 113 In *Handbook for Archaeoastronomy and Ethnoastronomy*, edited by Clive Ruggles. Ch. 113. New York: Springer, 2015.

Stellarium website. http://www.stellarium.org.

Wuchterl, Günther. 'Eastern Alpine Starlight Reserve and Großmugl Starlight Oasis, Austria'. In Clive Ruggles, and Michel Cotte, eds. 'Heritage Sites of Astronomy and Archaeoastronomy in the context of the UNESCO World Heritage Convention - A Thematic Study'. *Paris: ICOMOS and the International Astronomical Union* (June 2010), Ch.16.2.

Wuchterl, Günther. 'In search of the perfect sky - towards a cultural and astronomical quantification'. *In Proceedings of the Third International Starlight Conference Lake Tekapo, New Zealand June 2012* (The Starlight Conference Organising Committee, Feb. 2013), edited by John Hearnshaw, Karen Pollard, and Marilyn Head.
http://www.phys.canterbury.ac.nz/Conferences/Third_International_Starlight_Conference.

Wuchterl, Günther. 'General considerations for relatively dark starlight/dark-sky areas with few or no direct cultural connections'(2012). http://www2.astronomicalheritage.net/index.php/show-theme?idtheme=21.

Zotti, Georg. 'Visualization Tools and Techniques'. In Handbook for Archaeoastronomy and Ethnoastronomy, edited by Clive Ruggles. New York: Springer, 2015, Ch. 29.

NEW FINDINGS AT THE 'PETRE DE LA MOLA' MEGALITHS

L. Lozito, F. Maurici, V. F. Polcaro, and A. Scuderi

ABSTRACT: The Croccia Cognato archaeological site is sited at 1150m above sea level on a mountain belonging to the Lucania Dolomites range. Traces of human frequentation go back to Neolithic Age (12,000–8,000 BCE) and during the Bronze Age the area was surely frequented. Later (6th century BCE), a Lucani settlement was established on the top of the mountain. The site was abandoned in the 4th century BCE and was never further frequented. At a distance of about 200m east of the main gate of this settlement, an imposing group of rocks is sited on a small, rather flat area of the mountain slope, at a height of 1049m. These rocks, named in local dialect *Petre de la Mola* (*Grindstone Rocks*), are natural outcrops of the limestone bedrock, and cracked into various boulders because of the rain and wind erosion. However, these rocks were deeply modified by human work, in order to generate alignments to the meridian and to winter solstice sunset, when the sun sets in a small, artificial gallery. A further survey of the site, performed by us at the winter solstice of 2013, revealed a number of previously undetected details: a petroglyph carved in the rock at the observation point clearly shows the meridian and the solstice sunset directions and other observation points, marked by basins carved in the rocks and by petroglyphs, were discovered. Finally, a Bronze Age ceramic fragment was found inside a basin carved on the top slab of the solstice-oriented gallery. These findings strongly support the hypothesis that the *Petre de la Mola* were used for calendric and cultural purpose during the Bronze Age.

1. The 'Petre de la Mola' megalithic complex

The 'Petre de la Mola' megalithic complex (40°33'02"N, 16°11'39"E) is a natural outcrop of calcarenithic rock that has been artificially modified, in order to show an impressive hierophany on the winter solstice.[1] It is sited near the top of Monte Croccia, a 1150m high mountain sited in Basilicata (Italy), belonging to the Lucania Dolomites range and dominating the Basento Valley and the upper flow of the Cavone river. It is surrounded by an uncontaminated and mostly unpopulated forest, now a National Biological and Archaeological Park. Traces of human frequentation of the summit area of Monte Croccia go back to the Neolithic Age (12,000–8,000 BCE); also during the Bronze Age the area was likely frequented. The first archaeological investigations of the area were performed in 1887 by Michele Lacava;[2] later, Vittorio Di Cicco performed five excavation campaigns, uncovering an osco–samnite settlement, dated between the sixth and the fourth centuries BCE, with a double surrounding wall and various structures sited on the acropolis, whose building technique was similar to the Hellenic one.[3] A number of graves of Proto-Lucani epoch (eight century BCE) were found during excavations performed in 1919 near to the megalith, which is sited at about 200m east of the settlement.[4]

A survey of the site was performed in 1950 and the last excavations were carried out in 1998 by the Basilicata Superintendence: the southern side of the fortification was explored for a length of about 60m and the southern and the eastern part of external wall, joining the acropolis one, was reconstructed[5].

---

[1] E. Curti, M. Mucciarelli, V. F. Polcaro, Witte C. N. Prascina, 'The "Petre de la Mola" megalithic complex on the Monte Croccia (Basilicata)', in the Proceedings of the SEAC 17th annual meeting *From Alexandria to al-Iskandariya, astronomy and culture in the ancient Mediterranean and beyond*, 25–31 October 2009, Alexandria, Egypt, ed. M. Shaltout and M. Rappenglück (London: British Archaeological Reports, in press).
[2] M. Lacava, 'Accettura. Avanzi di città', *Notizie dagli Scavi* (1887): p. 332.
[3] V. Di Cicco, 'Accettura. Cinta muraria', *Notizie dagli Scavi* (1896): p. 53.
[4] V. Di Cicco, 'Oliveto Lucano. Prima relazione sugli scavi a Monte Croccia-Cognato', *Notizie dagli Scavi* (1919): pp. 243–260.
5 A. Russo, 'Le prime tracce dell'uomo in Basilicata. Dal Paleolitico al Neolitico', in *Conoscere la Basilicata: Cultura,*

The summit of Monte Croccia was definitely abandoned at the end of the fourth century BCE because of Roman pressure, and there are no traces of further human frequentation until present days.

To date, there is thus no firm dating of the frequentation of the megalith, though Bronze Age pottery fragments were found in the surrounding area during a surface survey performed in 2009. During the same survey, two clearly astronomically aligned foresights were recognized (see Fig. 18.1).[6]

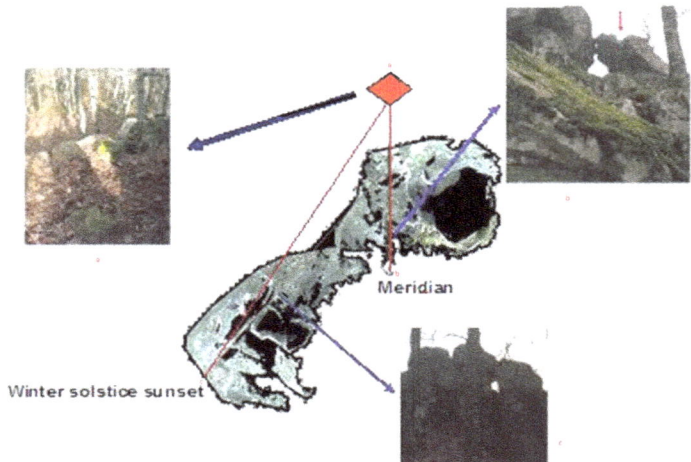

Fig. 18.1: Laser scan section of the megalith, performed by Faber Srl in 2008: the observing point (a), the viewfinder oriented to the meridian (b) and the gallery pointing to the winter solstice sunset (c) are marked. Blue arrows show pictures of these artificial features.

The first one is an artificial notch showing the meridian direction. The second one is a gallery, obtained by superimposing a slab to a natural split of the rock, pointing to the winter solstice sunset over the local horizon. Both foresights point to the abovementioned directions when viewed from the same observing point: this is a break of about 1m length and 0.3m wide in the soil platform on the North side of the megalith.[7] This observing point is lit by the sun's rays, passing through the meridian foresight, exactly at noon on the winter solstice.[8]

At sunset of the same day, the sun appears in the solstitial gallery immediately before it sets (see Fig. 18.2).

New findings

We performed a further survey of the site at the winter solstice of 2013. This survey revealed a number of new interesting findings.

The most important one is a petroglyph carved on a flat rock, adjacent to the observing point. It was masked by a deep fossil and living moss coverage. After cleaning, it was found to be shaped as a sort of

---

Itinerari archeologici (1999), available at http://www.old.consiglio.basilicata.it/conoscerebasilicata/cultura/archeologia/itinerari/russo_001.pdf [accessed 3 July 2016].

[6] Curti et al., 'The "Petre de la Mola" megalithic complex on the Monte Croccia (Basilicata)'.

[7] This platform is probably artificial, obtained by landfill, as suggested by a dry-stone wall closing its only side not delimited by natural outcrops of the bedrock.

[8] Curti et al., 'The "Petre de la Mola" megalithic complex on the Monte Croccia (Basilicata)'.

Y, cut by a horizontal bar (see Fig. 18.3). Measurements performed by a precision compass shown that, taking into account the local magnetic declination (3.3° ±0.3°), the arms of the petroglyph indicate the meridian and the winter solstice sunset directions (see Fig. 18.4).

Fig. 18.2: Light effects at the «Petre de la Mola» at winter solstice: A) noon B) sunset. Both pictures have been taken from the same observing point: the break on the platform on the North side of the megalith shown in Fig. 18.1.

Fig. 18.3: The meridian and the winter sunset directions indicated by the petroglyph.

The same petroglyph, with the same orientations, was found on a further flat rock, sited at about 50m from the main observation point, suggesting that this rock too was used as an observing point. Furthermore, on a third flat rock, where a clearly artificial basin, with a circular section of about 40cm diameter and a deep of about 60cm, is present, a cross is carved, indicating the cardinal directions (see Fig. 18.5). The basin shows two outlets, as seen on the ones present on the basins sited on the top of Petre de la Mola.

A further observation point was identified in another natural outcrop of the bedrock, at about 60m from the main observing site, which northwest side was artificially flattened and oriented to the winter solstice sunset.

Last, a further basin carved on the artificially positioned slab acting as a roof of the winter solstice oriented gallery of the Petre de la Mola was discovered. Inside this basin, a pottery fragment of the Early Bronze Age was found.

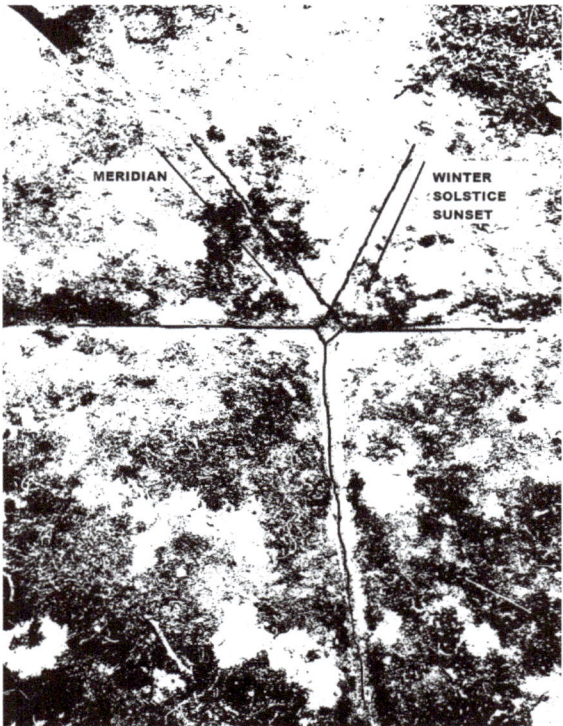

Fig. 18.4: The petroglyph indicating the cardinal direction and the basin sited in the third modified flat rock.

Fig. 18.5: The artificial basin and carving on the third flat rock.

Conclusions

The presence of multiple observing points, of artificially carved basins and of petroglyphs indicating cardinal and winter sunset directions strongly support the hypothesis of a religious use of the megalithic complex and complex ceremonies, involving many people. The in situ finding of an Early Bronze Age indicates the use of the complex in this epoch.

Astronomically oriented megalithic monuments are relatively common in Cilento and Basilicata (see, e.g. the 'Preta 'ru Mulacchio' on Monte Stella,[9] the 'Velo della Madonna' on Monte Gelbison,[10] the megalith of Cannalicchio near to Castelgrande[11]). All these monuments, often built by modifying natural rocks on the slopes of high mountains, show meridian and solstitial alignments. Though it is possible to hypothesize a first use of these monuments at the beginning of the Bronze Age, this crucial information can be proven only by archaeological excavation on these sites. Astronomically oriented megalithic monuments dated to the Bronze Age, sharing similar characteristics to the ones of Cilento and Basilicata, are found in Apulia and, possibly, in Calabria.[12] Astronomically oriented megalithic structures dated to the Bronze Age are also present in Sicily (e.g. the megaliths of the Alto Belice Valley).[13] However, these structures show a completely different and more sophisticated technology: this fact suggests that they are products of a different culture, though with similar practical and social needs.

Bibliography

Curti E., M. Mucciarelli, V. F. Polcaro., Witte C. N. Prascina. 'The "Petre de la Mola" megalithic complex on the Monte Croccia (Basilicata)'. In the Proceedings of the SEAC 17th annual meeting *From Alexandria to al-Iskandariya, astronomy and culture in the ancient Mediterranean and Beyond*, 25–31 October 2009, Alexandria, Egypt. Edited by M. Shaltout and M. Rappenglück . London: British Archaeological Reports, in press.

Di Cicco, V. 'Accettura. Cinta muraria'. *Notizie dagli Scavi* (1896): p. 53.

Di Cicco, V. 'Oliveto Lucano. Prima relazione sugli scavi a Monte Croccia-Cognato'. *Notizie dagli Scavi* (1919): pp. 243–260.

Lacava, M. 'Accettura. Avanzi di città'. Notizie dagli Scavi (1887): p. 332.

Polcaro, V. F., and D. Ienna. 'The Megalithic Complex of the "Preta 'ru Mulacchio" on the Monte della Stella'. In *Cosmology Across Cultures*, proceedings of the conference held 8-12 September, 2008, at Parque de las Ciencias, Granada, Spain. Edited by J. A. Rubiño-Martín, J. A. Belmonte, F. Prada, and A. Alberdi, pp. 370–374. San Francisco: ASP Conference Series 409, 2009.

Polcaro, V. F., M. Mucciarelli, S. Pizzimenti, and A. Polcaro. 'Il megalite di Cannalicchio di Castelgrande (PZ) ed i suoi analoghi nell'antica Lucania'. In *Cielo e cultura materiale. Recenti scoperte di archeoastronomia nel bacino del Mediterraneo*, proceedings of the XIII Borsa mediterranea del turismo archeologico, Paestum (Salerno), 20

---

[9] V. F. Polcaro and D. Ienna, 'The Megalithic Complex of the "Preta 'ru Mulacchio" on the Monte della Stella', in *Cosmology Across Cultures*, proceedings of the conference held 8–12 September, 2008, at Parque de las Ciencias, Granada, Spain, ed. J. A. Rubiño-Martín, J. A. Belmonte, F. Prada, and A. Alberdi (San Francisco: ASP Conference Series 409, 2009), pp. 370–374.

[10] V. F. Polcaro, M. Mucciarelli, S. Pizzimenti, and A. Polcaro, 'Il megalite di Cannalicchio di Castelgrande (PZ) ed i suoi analoghi nell'antica Lucania', in *Cielo e cultura materiale. Recenti scoperte di archeoastronomia nel bacino del Mediterraneo*, proceedings of the XIII Borsa mediterranea del turismo archeologico, Paestum (Salerno), 20 November 2010, ed. L. Lozito and F. Pastore (Salerno: Salternum, 2011) pp. 19–36.

[11] V. F. Polcaro et al., 'Il megalite di Cannalicchio di Castelgrande (PZ) ed i suoi analoghi nell'antica Lucania'.

[12] A. M. Tunzi, M. Lozupone, E. Antonello, V. F. Polcaro, and F. Ruggieri, 'The "Madonna di Loreto" Bronze Age Sanctuary and its Stone Calendar', in *Cosmology Across Cultures*, proceedings of the conference held 8–12 September, 2008, at Parque de las Ciencias, Granada, Spain, ed. J. A. Rubiño-Martín, J.A. Belmonte, F. Prada, and A. Alberdi (San Francisco: ASP Conference Series, 409, 2009), pp. 375–380.

[13] A. Scuderi, F. Vitale, S. Tusa, V. F. Polcaro, R. Scuderi, and F. Maurici, *The Pulpito f the King*, this volume.

November 2010. Edited by L. Lozito and F. Pastore, pp. 19–36. Salerno: Salternum, 2011.

Russo, A. 'Le prime tracce dell'uomo in Basilicata. Dal Paleolitico al Neolitico'. In *Conoscere la Basilicata: Cultura, Itinerari archeologici.* http://www.old.consiglio.basilicata.it/conoscerebasilicata/cultura/archeologia/itinerari/russo_001.pdf.

Scuderi, A., F. Vitale, S. Tusa, V. F. Polcaro, R. Scuderi, and F. Maurici. 'The Pulpit of the King'. This volume.

Tunzi, A. M., M. Lozupone, E. Antonello, V. F. Polcaro, and F. Ruggieri. 'The "Madonna di Loreto" Bronze Age Sanctuary and its Stone Calendar'. In *Cosmology Across Cultures*, proceedings of the conference held 8–12 September, 2008, at Parque de las Ciencias, Granada, Spain. Edited by J. A. Rubiño-Martín, J.A. Belmonte, F. Prada, and A. Alberdi, pp. 375–380. San Francisco: ASP Conference Series, 409, 2009.

ASTRONOMY IN CULTURE IN HISTORICAL TIMES

SIRIUS (AL-'ABŪR) PROPER MOTION AS RECORDED IN THE ARABIC STAR MYTHOLOGY

Flora Vafea

ABSTRACT: In their attempt to describe the relative positions of the stars, ancient peoples invented the constellations and created various mythologies. Arabic star mythology is preserved by certain authors, such as al-Ṣūfī and al-Bīrūnī. This paper focuses on a myth of Orion (*Jawzā'*), Canopus (*Suhaīl*), Sirius (Yemenite Sirius or *al-'Abūr*) and Procyon (Syrian Sirius).

According to the myth, the Yemenite and Syrian Sirius were sisters of *Suhaīl*, both of them situated north of the Galaxy. *Suhaīl* got married to *Jawzā'* (Orion), but he fell on her and broke her vertebrae and back. He was afraid that he would be held accountable for harming *Jawzā'*, so he escaped to the south, so as not to be visible in the sky. The Yemenite Sirius crossed the Galaxy so as to be with him, and therefore is called *al-'Abūr*, the one that traversed, while the Syrian Sirius remained on the north side of the Galaxy, crying for *Suhaīl*, until her eyes became bleary, and therefore is called *al-Ghumaīṣā'* (the bleary). The important element of that myth is the movement of the stars portrayed in it.

Using the value of the proper motion of Sirius and the astronomical software Voyager, it can be confirmed that Sirius was north of the Milky Way galaxy 42,000 years ago, having a galactic latitude of +1°36', while it is now situated south of the galaxy, having a galactic latitude of -8°53'39". Due to the rotation of the axis of the Earth, the declination of Canopus was decreasing approximately between 49,850 and 36,360 BCE, and also between 24,170 and 12,080 BCE; this fact can confirm that Canopus moved to the south and became invisible.

In their attempt to describe the positions of the stars and the shapes they form, ancient peoples invented the constellations and created mythologies to relate the stars and the constellations to each other.

Sirius, which is the brightest star in the night sky, always had a special position in the various mythologies. For the ancient Egyptians, Sirius was associated with the Goddess Isis; for instance, Sirius was depicted as Isis (3st/spdt) in a boat on the ceiling of Senmut's tomb in Luxor.[1] Its heliacal rising, near the summer solstice, was connected with the flooding of the Nile and the beginning of the new year,[2] although the duration of the year, estimated as 365 days, caused a difference between the celebrations of the new year and the heliacal rising of Sirius.[3]

In the Ptolemaic conception of the sky, Sirius belongs to the constellation of the Great Dog, 'Κύων', southeast of the constellation of Orion.[4] The figures of the constellations Orion, Canis Major and Canis Minor are preserved, among others, on a miniature celestial globe dating from the second century CE.[5] According to the Greek mythology, the hunter Orion is followed by two dogs, represented by the constellations Κύων – Canis Major and Προκύων – Canis Minor; Sirius is the brightest star of the former,

---

[1] Marshall Clagett, *Ancient Egyptian Science: Calendars, clocks and astronomy*, Vol. 2 (Philadelphia, PA: American Philosophical Society, 1995), pp. 113–14, 226, 582.
[2] Ptolemy, *Tetrabiblos*, II.10, trans. F.E. Robbins (Cambridge, MA: Harvard University Press, 1940), pp. 196–97; Clagett, *Calendars, clocks and astronomy*, pp. 28–37, 327, 490–91.
[3] Clagett, *Calendars, clocks and astronomy*, pp. 5–6.
[4] Ptolemy, *Syntaxis mathematica* (*Almagest*), in *Claudii Ptolemaei opera quae exstant omnia*, Vols. 1.1-1.2, ed. J. L. Heiberg (Leipzig: Teubner, 1898 and 1903), Vol. 1.2, VIII.1, H.132–47.
[5] Ernst Künzl, 'Der Globus im Römisch-Germanischen Zentralmuseum Mainz: Der bisher einzige komplette Himmelsglobus aus dem Griechisch-Römischen Altertum' / 'The globe in the „Römisch-Germanisches Zentralmuseum Mainz". The only complete celestial globe found to-date from classical Graeco-Roman antiquity', *Der Globusfreund* 45/46 (1997/98), pp. 7–153, figures 45, 34, 35 on p. 10.

and Procyon the brightest one of the latter.⁶

The Persian astronomer of the tenth century al-Ṣūfī, in his *Book of the Stars*,⁷ presented the constellations described in Ptolemy's *Almagest*, after he had examined all the stars by himself, correcting their coordinates, and adding figures representing the constellations with their stars, as seen in the sky and as drawn on the celestial globes.⁸ Al-Ṣūfī also connected the Ptolemaic names of the stars and constellations with those used by the Arabs, preserving the Arabic mythology of the stars, as well.

The Arabic myth concerning the constellation of the Orion (*Jawzā'*), and the stars Canopus (*Suhail*), Sirius (Yemenite Sirius or *al-'Abūr*) and Procyon (Syrian Sirius or *al-Ghumaiṣā'*) is investigated below. Sirius was called Yemenite Sirius, because it sets in the south, on the side of Yemen, while Procyon was called the Syrian Sirius, because it sets in the north, on the side of Syria. This myth is preserved by al-Ṣūfī (903–986 CE) and al-Bīrūnī (973–1048 CE). ⁹

According to the myth, the Yemenite and the Syrian Sirius were sisters of *Suhail*, both of them situated north of the galaxy. *Suhail* got married to *Jawzā'* (Orion), but he fell on her and broke her vertebrae and back. He was afraid that he would be held accountable for the life of *Jawzā'*, so he escaped to the south, so as not to be visible in the sky. The Yemenite Sirius crossed the Galaxy to be with him, therefore it is called *al-'Abūr*,¹⁰ the one that traversed, while the Syrian Sirius remained on the north side of the Galaxy, crying for *Suhail*, until her eyes became bleary, and therefore it is called *al-Ghumaiṣā'* (the bleary); the latter is said to explain why Syrian Sirius is fainter than Yemenite Sirius. The important element of that myth is the movement of the stars portrayed in it.

The question that arises is whether this myth describes a real motion of Sirius, so it must be investigated whether Sirius was ever on the north side of the Milky Way. For our époque, the distance of Sirius from the plane of the Milky Way, namely its galactic latitude, is approximately 8.89° to the south.

The first step will be to find the proper motion of Sirius and to examine whether this proper motion could vindicate a position of it north of the Galaxy. Sirius is one of the closest stars to our solar system. Although its mass is approximately double the mass of our Sun and its absolute visual magnitude is M_V=1.45, it has a visual magnitude of -1.44, being the brightest star, because of its proximity to us (trigonometric parallax π=379.21 milliarcsec).¹¹ Indeed, Sirius is among the 150 stars in the Hipparcos Catalogue with largest proper motions; its total proper motion is $|\mu|$=1339.42 milliarcsec per year and it can be analysed in -546.01 milliarcsec per year in right ascension and -1223.08 milliarcsec per year in declination.¹² This means that the direction of its motion is such that it had to be north of the Milky Way some thousands years ago.

---

[6] Eratosthenes, *Catasterismi*, Chapter 1, Sections 32, 33 and 42D, in *Pseudo-Eratosthenis Catasterismi*, ed. A. Olivieri (Leipzig: Teubner, 1897).

[7] Al-Ṣūfī, *The Book of the Fixed Stars*, in *Description des étoiles fixes*, trans. Schjellerup (St. Petersburg: Académie Impériale des sciences, 1874).

[8] Well drawn figures of the constellations by al-Ṣūfī's son al-Ḥusayn are preserved in al-Ṣūfī: *The Book of Constellations, Kitāb Ṣuwar al-kawākib, facsimile of MS Marsh. 144, Bodleian Library, Oxford* (Frankfurt am Main: Institute for the History of Arabic-Islamic Science, 1986).

[9] Al-Ṣūfī in Schjellerup, *Description des étoiles fixes*, pp. 220–23; Al-Bīrūnī, *The Book of Instruction in the Elements of the Art of Astrology*, ed. and trans. R. Ramsay Wright (London: Luzac, 1934), section 163, pp. 80–81.

[10] The name *al-'Abur* (العبور) for Sirius is very common in the Arabic star-literature and it appears on astrolabes and celestial globes. It comes from the verb *'abara* (عبر) that means cross, traverse.

[11] James Liebert, Patrick Young, David Arnett, J. B. Holberg, and Kurtis Williams, 'The Age and Progenitor Mass of Sirius B', *The Astrophysical Journal* 630 (2005) The American Astronomical Society: L69–L72..

[12] European Space Agency (ESA), *The Hipparcos and Tycho Catalogues*, 17 vols, Vol. 1: *Introduction and Guide to the Data*, (Noordwijk, The Netherlands: ESA Publications Division, c/o ESTEC, 1997), section 3.6, pp. 482, 485.

I have used the software Voyager 4.5 to calculate the variance of the galactic latitude of Sirius. The results are presented in Table 19.1 and the graph in Fig. 19.1. According to these results, Sirius was on the plane of the Milky Way about the year 33,000 BCE, while previously it was to the north of it. It must be taken in consideration that the Milky Way has a width that varies along its longitude. Sirius was moving for thousands of years crossing the Milky Way, as depicted in Figures 19.3 to 19.5, which represent various positions of Sirius, from the year 49,850 BCE up to the year 815 CE.

year	g.l. α CMa	year	g.l. α CMa
-52,000	4.059	-16,000	-4.029
-50,000	3.666	-12,000	-5.063
-48,000	3.266	-80,00	-6.124
-44,000	2.448	-40,00	-7.211
-40,000	1.603	-1000	-8.044
-38,000	1.171	150	-8.366
-36,000	0.732	500	-8.465
-32,000	-0.166	964	-8.596
-28,000	-1.090	1000	-8.606
-24,000	-2.042	2014	-8.894
-20,000	-3.022		

Table 19.1: Variance of the galactic latitude of Sirius (α CMa) within 54 millennia, as calculated with the software Voyager 4.5.[13]

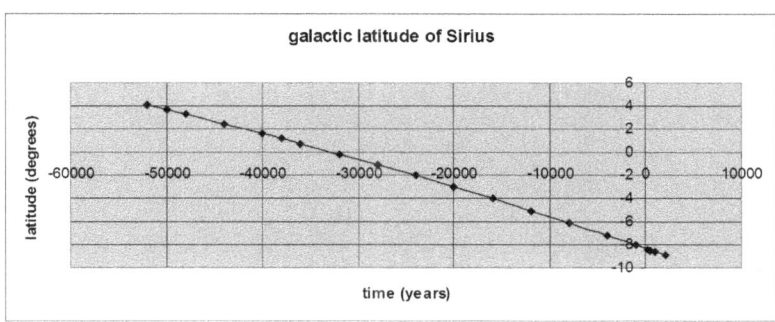

Fig. 19.1: Graph showing the variance of the galactic latitude of Sirius (α CMa) within 54 millennia, as calculated with the software Voyager 4.5.

But the myth also speaks about *Suhaīl*, the brother of Sirius, which 'escaped to the south, so as not to be seen in the sky'. *Suhaīl* is identical to Canopus, α Carinae, the second brightest star of the night sky, after Sirius.

---

[13] In the tables and graphs I use the astronomical notation for dates, as the software *Voyager* does. In that notation: 2 CE=2, 1 CE=1, 1 BCE=0, 2 BCE=-1, 3 BCE=-2, etc.

Although Canopus is less bright than Sirius, having a visual magnitude of -0.62, its absolute visual magnitude is M_v=-5.53 and it is relatively far from the sun with a trigonometric parallax of π=10.43 milliarcsec, and total proper motion of $|\mu|$=30.98 milliarcsec/year.[14] About the year 52,000 BCE, the galactic longitude of Canopus was -25.664905, while in our days it is -25.292021.[15] Thus, the proper motion of Canopus cannot explain its disappearance.

The visibility of a star from a certain location depends on its declination, so the variance of the declination of Canopus, due to the wobble of the axis of the Earth, should be taken in consideration. The main factors that constitute this wobble are the following:

1. The **precession** is the conical motion of the earth's axis of rotation with a period of ~25,800 years, in the opposite sense of the rotation of the Earth's axis. Due to this motion, the north and south celestial poles, which are the points of intersection of the axis of the earth with the celestial sphere, trace circles on the celestial sphere. The positions of the celestial equator and the circles parallel to it also change; this affects the celestial coordinates of the stars: the right ascension and the declination.
2. The **nutation** is the periodic oscillation observed in the precession of the earth's axis, with a period of ~18.6 years. The nutation has similar effects as the precession but to a lesser extent.
3. The **change in the obliquity of the ecliptic** affects the direction of the axis of the Earth, as well.

Next step will be to study the variance of declination of the stars involved in the myth, namely for *Suhaīl* – Canopus (α Car), *al-'Abūr* – Sirius (α CMa) and *al-Ghumaīṣā'* – Procyon (α CMi), plus Betelgeuse (α Ori) and Rigel (β Ori) as representative stars of *Jawzā'* – Orion, for a period of 54 millennia. The results, as calculated with the software Voyager, appear in Table 19.2 and the graph in Fig. 19.2.

The extreme values of the declination of each star are not always the same, since the declination is affected by all of the above mentioned factors and the proper motion of the star.

The graph in Fig. 19.2 shows that depending on the (terrestrial) latitude of the place of observation, Canopus was invisible for periods of millennia. For example, for a latitude of 25° N (Riyadh), Canopus was invisible between 42,000 BCE and 30,000 BCE, also between 18,000 BCE and 6200 BCE; for a latitude of 30° N (Cairo), it was invisible between 43,600 BCE and 28,200 BCE, also between 20,100 BCE and 4600 BCE.

From the graph in Fig. 19.2, it is obvious that the declinations of the rest of the involved stars are such that allow the stars to be always visible from latitudes less than 30° N during the examined period. It is also remarkable that about the year 52,000 BCE, the declination of Procyon was 24.37; this means that it culminated near the zenith of places of latitude ~25°N.

Next, the positions of the stars of the myth for a course of 52 millennia, at the times of the extreme declination values of Canopus, are examined. In the following figures, generated by the software Voyager 4.5, the position of the Milky Way is taken stable, while the positions of the celestial equator and its poles are changing, due to the wobble of the axis of the Earth. This affects the polar distance and the declination of each star and, subsequently, its visibility.

---

[14] ESA, *The Hipparcos and Tycho Catalogues*, Vol. 1: *Introduction and Guide to the Data*, p. 488.
[15] These values are calculated with the software *Voyager 4.5*.

year	α Car	α CMa	α CMi	α Ori	β Ori
-52,000	-51.64	1.18	24.37	6.13	-10.75
-50,000	-50.33	0.68	20.80	10.04	-5.33
-48,000	-51.18	-4.50	12.51	8.78	-4.35
-44,000	-58.79	-24.55	-10.28	-6.93	-15.57
-40,000	-71.64	-46.21	-27.49	-28.94	-37.07
-36,000	-80.14	-49.58	-24.61	-39.16	-52.97
-32,000	-70.07	-31.93	-6.15	-26.39	-43.97
-28,000	-59.57	-14.86	9.90	-5.77	-23.27
-24,000	-55.81	-11.86	8.81	4.61	-10.42
-20,000	-60.15	-25.55	-9.31	-4.07	-14.57
-16,000	-70.46	-47.73	-30.46	-24.84	-33.06
-12,000	-78.24	-60.61	-36.41	-40.91	-53.15
-8000	-69.82	-46.70	-21.06	-34.46	-52.17
-4000	-58.47	-26.42	-0.81	-12.97	-31.27
-1000	-53.37	-17.18	7.58	1.42	-16.26
150	-52.55	-15.88	8.02	4.98	-12.12
500	-52.44	-15.75	7.83	5.77	-11.10
1000	-52.39	-15.81	7.27	6.63	-9.86
2014	-52.71	-16.74	5.19	7.41	-8.19

Table 19.2: The variance of declination of the stars Canopus, Sirius, Procyon, Betelgeuse and Rigel, as calculated with the software Voyager 4.5.

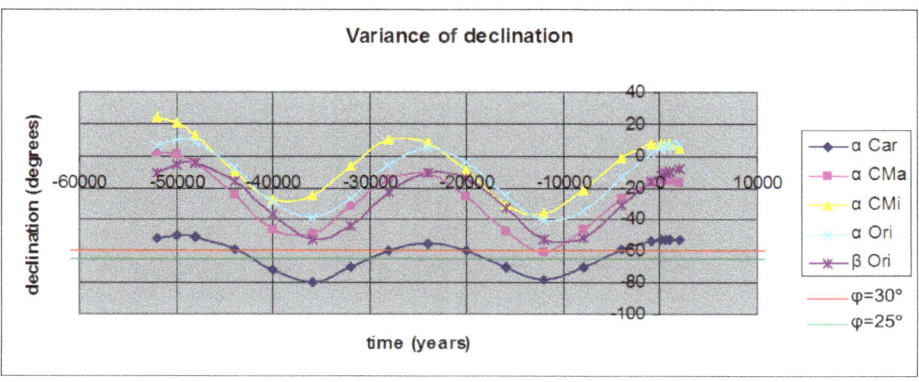

Fig. 19.2: Graph showing the variance of declination of the stars Canopus, Sirius, Procyon, Betelgeuse and Rigel, as calculated with the software Voyager 4.5. The values come from Table 19.2.

218   Flora Vafea

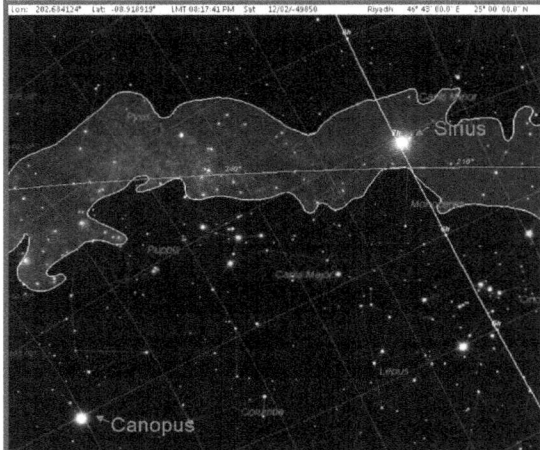
Fig. 1: Year -49850

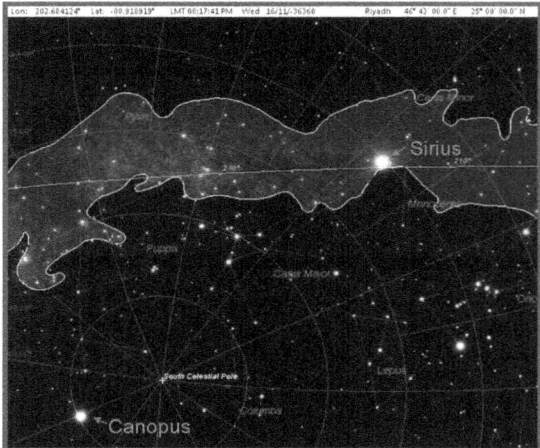
Fig. 2: Year -36360

Fig. 19.3.

In year 49,850 BCE (Fig 19.3, top), Canopus was in its maximum declination (δ=-50.318°) and maximum distance from the south celestial pole and visible from latitudes less than 39° north. Sirius was inside the Milky Way zone, north of the galactic equator; its galactic latitude was +3.636°. In year 36,360 BCE (Fig. 19.3, bottom), Canopus was in its minimum declination (δ=-80.231°) and minimum distance from the south celestial pole and invisible from latitudes greater than 10° north. Sirius was inside the Milky Way zone very close to the galactic equator, having a north galactic latitude of +0.811°.

Sirius (al-ʿAbūr) Proper Motion as Recorded in the Arabic Star Mythology 219

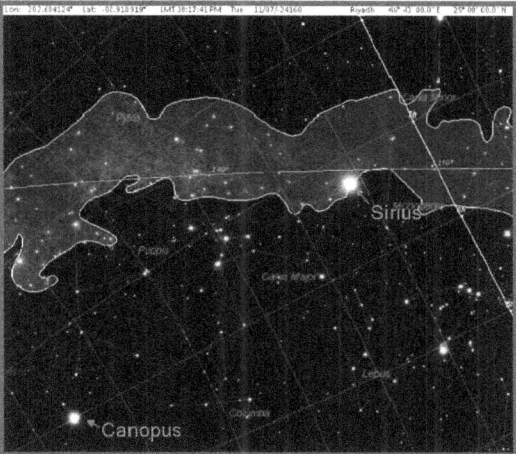

Fig. 3: Year -24168

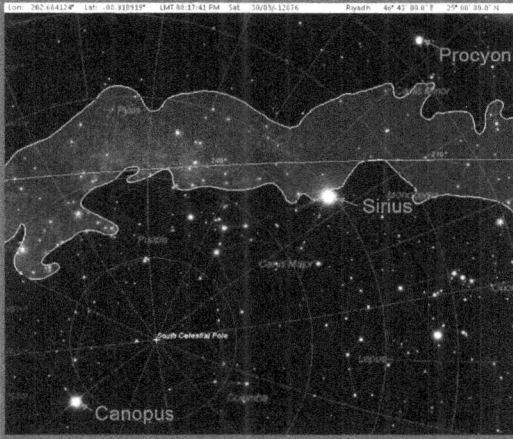

Fig. 4: Year -12076

Fig. 19.4.

In the year 24,168 BCE (Fig. 19.4, top), Canopus was at its maximum declination (δ=-55.802°) and maximum distance from the south celestial pole and visible from latitudes less than 34° north. Sirius was inside the Milky Way zone, south of the galactic equator; its galactic latitude was -2.002°. In the year 12,076 BCE (Fig. 19.4, bottom), Canopus was in its minimum declination (δ=-78.241°) and minimum distance from the south celestial pole and invisible from latitudes greater than 11.5° north. Sirius exited the Milky Way zone; its south galactic latitude was -5.043°. Procyon was approaching the Milky Way, its galactic latitude was +17.262°.

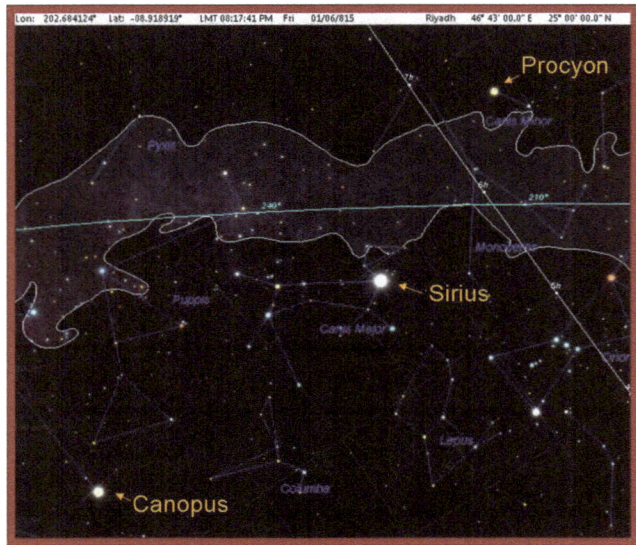

Fig. 19.5: Year 815.

In year 815 (Fig. 19.5), Canopus was in its maximum declination (δ=-52.389°) and maximum distance from the south celestial pole and visible from latitudes less than 37.5° north. Sirius was south of the Milky Way zone having a south galactic latitude of -8.554°. Procyon was closer to the Milky Way; its galactic latitude was +13.383°.

Conclusions
The Arabic myth about *al-'Abūr, Suhail, al-Jawzā'* and *al-Ghumaīṣā'* contains truths about the motion of the stars.

In the introduction to *The book of the stars*, al-Ṣūfī criticizes Abū Ḥanīfa al-Dīnawarī, who wrote astronomical books following the Arabic tradition, because the latter mentions that the stars change positions, the disposition of the stars is not always the same, and the movement of the so called fixed stars is very slow in comparison with the planets.[16]

Which period may this myth describe? It may be 42,000 BCE: the disappearance of Canopus for φ=25°, Sirius moves inside the Milky Way, north of the galactic equator with a galactic latitude of +2.04°. Or, it may be 18,000 BCE: the disappearance of Canopus for φ=25°, Sirius moves inside the Milky Way, south of the galactic equator with a galactic latitude of -3.5. Other periods could also be possible, depending on the latitude of observation.

The problem that arises now is how this knowledge has been preserved in a myth? It is impossible, within a human lifetime, to see any observable difference in the galactic latitude of a star, also to sense a dramatic change in its declination. Usually myths are transmitted orally through generations. This myth refers to the positions of the two brightest stars of the night sky.

Were these stars used as reference points for some repeated processes (e.g. travel orientation) and after some thousand years they could not longer play this role?

Was there a very ancient depiction of the stars Sirius and Canopus that was compared to a later image

---

[16] Al-Ṣūfī in Schjellerup, *Description des étoiles fixes*, pp. 32–33.

of the sky and made obvious that Canopus was not visible and Sirius was on the other side of the Galaxy? These are only speculations. These questions cannot be answered with certainty.

Bibliography

Al-Bīrūnī. *The Book of Instruction in the Elements of the Art of Astrology*. Edited and translated by R. Ramsay Wright. London: Luzac, 1934.

Clagett, Marshall. *Ancient Egyptian Science: Vol. 2 Calendars, clocks and astronomy*, Vol. 2. Philadelphia: American Philosophical Society, 1995.

Eratosthenes. *Catasterismi*. In *Pseudo-Eratosthenis Catasterismi*, edited by Alexander Olivieri. Leipzig: Teubner, 1897.

European Space Agency (ESA). *The Hipparcos and Tycho Catalogues*, 17 vols, Vol. 1: *Introduction and Guide to the Data*. Noordwijk, The Netherlands: ESA Publications Division, c/o ESTEC, 1997.

Künzl, Ernst. 'Der Globus im Römisch-Germanischen Zentralmuseum Mainz: Der bisher einzige komplette Himmelsglobus aus dem Griechisch-Römischen Altertum' / 'The globe in the „Römisch-Germanisches Zentralmuseum Mainz". The only complete celestial globe found to-date from classical Graeco-Roman antiquity'. *Der Globusfreund* Nr. 45/46. Wien (1997/98): pp. 7–153.

Liebert, James, Patrick Young, David Arnett, J. B. Holberg, and Kurtis Williams. 'The Age and Progenitor Mass of Sirius B'. *The Astrophysical Journal* 630 (2005): L69–L72.

Ptolemy. *Syntaxis mathematica* (*Almagest*). In *Claudii Ptolemaei opera quae exstant omnia*, Vols. 1.1-1.2, edited by J. L. Heiberg. Leipzig: Teubner, 1898 and 1903.

Ptolemy. *Tetrabiblos*. Translated by F. E. Robbins. Loeb Classical Library 435. Cambridge, MA: Harvard University Press, 1940.

Al-Ṣūfī. *The Book of Constellations, Kitāb Ṣuwar al-kawākib, facsimile of MS Marsh. 144, Bodleian Library, Oxford*. Frankfurt am Main: Institute for the History of Arabic-Islamic Science, 1986.

Al-Ṣūfī. *The Book of the Fixed Stars*. In *Description des étoiles fixes*, translated by Schjellerup. St. Petersburg: Académie Impériale des sciences, 1874.

THE STONES OF PENAS DE RODAS: CAN THE 'SPELL OF ARCHAEOASTRONOMY' CREATE A CONTEMPORARY SACRED PLACE?

Benito Vilas Estevez

ABSTRACT: The aim of this paper is to explore the role of archaeoastronomy in the contemporary creation of a sacred site out of a natural rock formation, focusing on Penas de Rodas in Galicia (Spain). Through archaeoastronomical measurement, fieldwork and photography, as well a limited number of semi-structured interviews and archaeological data, this paper explores how this place was reinterpreted as a Roman observatory and became a place where people now gather to observe the summer solstice sunset. The results reveal that as a consequence of the 'spell of archaeoastronomy', meaning research work that is conducted by people with no archaeoastronomical knowledge, the common idea that this was a natural monument has started to change.

Introduction

The aim of this project is to explore and compile information about how Penas de Rodas, a rock formation in the village of Outeiro do Rei, situated in Galicia (Northern Spain) in the province of Lugo, may or may not be seen as a contemporary sacred site through the influence of archaeoastronomy. The aim is to see how Penas de Rodas, a rock formation, can be reinterpreted as having been a Roman observatory through the work of amateur archaeoastronomers and analysis, and also to discover what people from the village and elsewhere think about this new vision; whether they think that this place was sacred in the past and whether it is now.

To determine these objectives, I used archaeological data from the Galician Cultural Heritage Institution. I measured and corrected the orientations of the place with a compass, a clinometer, and a GPS unit.[1] Finally using a questionnaire, I conducted semi-structured interviews that surveyed seven people so as to know their opinions on the site; I present a discussion of the different visions and ideas about sacred places in the Findings and Discussions section. Photographs of the site are included in Appendix A.

Pena de Rodas: Situation, Rock Formation, and Archaeology

Penas de Rodas is situated at latitude 43° 7′ 12.67″ N and longitude 7° 40′ 47.65″ W. The rock formation has two massive bowling stones that are almost spherical, which are supported by other stones at a minimum point (Fig. 20.1). Bowling alterations are produced under the surface, where through different processes of erosion are reduced to their most resistant parts. These kinds of formations are natural and very typical in Galicia and other parts of the world.[2] Furthermore, according to the Galician Cultural Heritage Institution, the council of Outeiro do Rei last reported (on 1 June 2012) a total of 63 archaeological fields, including 24 mounds, 12 hill-forts, 4 Roman fields, 4 medieval fields, and 16 undetermined sites that correspond to isolated fragments, mostly ceramic remains. So, in agreement with the updated archaeological records from the Galician Cultural Heritage Institution, there is no proof of the existence of any archaeological site or remains at this place or of the artificial collocation of the stones by human beings.

---

[1] Clinometer and compass SILVA, tandem. GPS model eTrex 30.

[2] Juan R. Vidal Romani and Charles Rowland Twidale, *Formas y paisajes graníticos* (Universidade da Coruña: Servizo de Publicacións, 1998).

Fig. 20.1: The stones of Penas de Rodas.

In fact, the nearest archaeological fields to Penas de Rodas are O Castro de Gaioso at a distance of 1.18 miles northeast of Pena de Rodas, and the archaeological site of Paus de Medorras at a distance of 0.413 miles southwest of Pena de Rodas.[3]

However, according to Carlos Sánchez-Montaña, the author of the first blog describing the site of Penas de Rodas as a possible Roman sanctuary, 'Penas de Rodas was projected according to a complex ritual function; besides functioning as an observatory place… (it is) a stone construction of possible Roman origin'.[4] Although he recognizes that the stones were a product of natural phenomena, he states that their positioning over the base of the altar was man-made: 'In the two spherical stones, the marks that a strong chain left are perfectly visible' (Fig. 20.2),[5] although, according to Galician Cultural Heritage Institution, they are natural grooves made by water over thousands of years. Furthermore, he also bases his idea on the 1943 work *Lugo en los Tiempos Prehistóricos* by Manuel Vazquez Seijas (1884–1982), who identified Penas de Rodas as an emblematic monument related with the Celts.[6]

---

[3] The distances were obtained through the use of the SixPac programme available on Internet. http://emediorural.xunta.es/visorsixpac/ [accessed 3 January 2015].

[4] My translation from Carlos Sánchez Montaña, 'El conjunto de "Penas de Rodas" fue proyectado de acuerdo con una compleja función ritual además de como lugar de observación astronómica….una construcción en piedra de posible origen romano.', at http://aspenasderodas.blogspot.com.es/ [accessed 3 January 2015].

[5] My translation from Sánchez-Montaña, 'Es perfectamente visible en las dos bolos las huellas que una fuerte cadena dejó en cuatro caras de su superficie.', at http://aspenasderodas.blogspot.com.es/2008/08/elementos.html [accessed 3 January 2015].

[6] Manuel Vazquez Seijas, *Lugo en los tiempos prehistóricos* (Lugo: Junta del Museo Provincial, 1943), pp. 29–30.

Fig. 20.2: Natural mark at the top of the stone, marked in red colour.

Sánchez-Montaña also describes the shape of the other stones situated at the front of Penas de Rodas as 'a semicircular theatre… in the shape of a stair'.[7] These stairs are the work of stonemasons through a system of wedges, who in living memory used the site as a quarry, and there is evidence of this in many of the stones around the complex (Figs. 20.3 and 20.4).

The interpretation of some stone configurations, eroded by nature over thousands of years, as being worked or built by human action, is identified by Xurxo Ayán and Manuel Gago, as 'o Mal das Pías (the illness of the cupmarks). This name is due to the fact that most reinterpretations of natural rock formations in Galicia are linked to cupmarks, although any reinterpretation of a stone as being worked by human action is included.[8]

The 'Spell of Archaeoastronomy' in Penas de Rodas

By the 'spell of archaeoastronomy', I refer to one of the biggest problems within the discipline of archaeoastronomy: research that is conducted by people with no or little (although it could be good on a few occasions) archaeoastronomical knowledge who make highly speculative claims. Archaeoastronomy attracts the attention of many people, and people think that it is easy to work within the discipline. As Juan Antonio Belmonte and Michael Hoskin state:

> Unfortunately the mutual incomprehension over decades between archaeologists and archaeoastronomers has allowed the uncontrolled proliferation of a series of "experts" in archaeoastronomy, with little or no formation in the field… and they have proposed a number of theories each more fantastic about the most diverse civilizations.[9]

---

[7] My translation from: Sánchez-Montaña, 'Los restos de una grada semicircular… en forma de escalera', at http://aspenasderodas.blogspot.com.es/2008/08/elementos.html [accessed 3 January 2015].

[8] Xurxo Ayán and Manuel Gago, *Herdeiros Pola Forza. Patrimonio Cultural e sociedade na Galicia do século XXI* (O Milladoiro, Ames: 2.0 Editora, 2012), pp. 167–173.

[9] My translation from: 'Desgraciadamente, la mutua incomprensión imperante durante décadas entre arqueólogos y arqueoastrónomos ha permitido la proliferación incontrolada de una serie de "especialistas" en arqueoastronomía, con

Fig. 20.3: 'Supposed' semicircular theatre.

Fig. 20.4: Marks of the wedges in other stones.

The 'spell of archaeoastronomy' causes alternative histories, related to what Eric Hobsbawm called invented tradition: 'Invented tradition is taken to mean a set of practices, normally governed by tacitly accepted rules and of a ritual or symbolic nature, which seek to inculcate certain values and norms of behaviour by repetition... In fact where possible, they normally attempt to establish a continuity with a suitable historic past'.[10]

The following demonstrates how the 'spell of archaeoastronomy' functions in Penas de Rodas: Sánchez-Montaña reports, 'Some modest astronomic research ... suggests that these [stones] are aligned to the sunset at the summer solstice and to the sunrise at the winter solstice'.[11] He also identifies two markers at the site for viewing the solstices: 'At the west there is a stone with a triangular shape that points to the exact position of the sunrise at the winter solstice. And in the east direction, the semicircular theatre suggests the place where the assistants could see the sunset at the summer solstice'.[12]

According to the descriptions of the markers, I took measurements with a compass and a clinometer, and after correcting the measurements, the results were the following:

AZ	ALT	LAT	DEC	MARKER
118	0	43.7	-19.8°	From the triangular shape
299	2.5	43.7	22.4°	From the "semicircular theatre"

The results show that the measurements are very close to aligning with the solstices, which have a declination of ±23.5°. The semicircular theatre measurement are especially accurate, as there is no real marker and one can move around the stones to choose the perfect alignment. I took the measurements from the place where it is possible to see the biggest gap, between the two stones of Penas de Rodas. In the case of the triangular shape, it is not accurate but again it is also possible to move around and get a perfect alignment. The problem is that there are no real markers, and it is possible to get any declination, as can be seen in the views of other bloggers like Iván Blanco, who visited the site and concluded that the stones are also aligned to the equinox – the difference is that he does not regard this place as Roman, he considers that this place is situated on a mound that could be hollow.[13] In any case, their theories are an example of what I call the 'spell of archaeoastronomy'.

Methodology of the Interviews

I used qualitative methods to gather data by conducting semi-structured interviews with five people from the village of Outeiro do Rei, one foreign blogger, and one other foreigner, all of who were familiar

---

poca o ninguna formación en el campo que se han dedicado a propugnar una serie de teorías, a cuál más fantástica, sobre las más variadas civilizaciones', in Juan Antonio Belmonte Avilés and Michael Hoskin, *Reflejo del Cosmos: Atlas de Arqueoastronomía en el Mediterráneo Occidental* (Madrid: Equipo Sirius, 1992), pp. 13–14

[10] Eric Hobsbawm And Terence Rangers, eds., *The Invention of Tradition* (Cambridge: Cambridge University Press, 1984, p. 1.

[11] My translation from Sánchez Montaña, 'Modestas investigaciones astronómicas... apuntan a que estas se encuentran orientadas hacia la puesta del sol en el solsticio de verano y hacia la salida del sol en el solsticio de invierno', at http://aspenasderodas.blogspot.com.es/2008/08/solsticio.html [accessed 3 January 2015].

[12] My translation from, Sánchez-Montaña, 'Situada hacia el Oeste de las principales, una con forma triangular, que asemeja un fiel o mira, y que señala la dirección exacta de la salida del sol en el solsticio de invierno. Y en dirección este, los restos de una grada semicircular permite sugerir el lugar de la puesta del sol en el solsticio de verano.', at http://aspenasderodas.blogspot.com.es/2008/08/solsticio.html [accessed 3 January 2015].

[13] Ivan Blanco and Miguel Blanco, http://www.nearbing.com/es/knowledge/a-hombros-de-gigantes.html [accessed 3 January 2015].

with Penas de Rodas. As a group the average age was 53 years, the youngest being 35 and the oldest 70. Five of the interviewees were from Outeiro do Rei, including the mayor and the president of the Cultural Association of Penas de Rodas (PCAPR). Of the two foreigners, one was a blogger and the interview was carried out through mail, and the other was a man visiting the place.

I was persuaded to use qualitative research methods by the arguments presented by David Silverman. As he put it, 'If you want to discover how people intend to vote, then a quantitative method, like a social survey, may seem the most appropriate choice. On the other hand, if you are concerned, with exploring people's life histories or everyday behaviour, then qualitative methods may be favoured'.[14] I was looking for personal experiences and responses to capture in order to study what Alan Bryman calls 'the uniqueness of individual cases and contexts'.[15] I was working as an insider (as I am Galician and I was in direct contact with the people of the village) but was concerned that an outsider perspective was necessary to analyse the broad-based data. Although the interviews were structured, I left space to reply, to confront ideas, and even to talk about other facts that were interesting for the project.

In the following section, I present the information collected from my interviews with references to the construction of the sacred spaces.

Interview Findings

The seven people interviewed in this work knew about Penas de Rodas and had visited it. Five of them knew of the place because they were born in Outeiro do Rei, and the others knew the place through an article in the newspaper *El Mundo* in 2011 (the second most read newspaper in Spain), which is available on the Internet,[16] and from the blog of Sánchez-Montaña.

The first question of the questionnaire was, 'Do you think this place was man-made or natural? Why?'. The informants started to diverge: two of them clearly stated that it was natural, without a doubt, by the erosion of water; three of them (the foreigners and one villager) said that it was man-made; and the other two were undecided. One stated, 'I do not know. If the rocks were smaller, I would have no doubt that they were man-made, but I do not know'.[17] The other stated, 'I do not know. They have always been there. My grandfather remembered the stones always being there, but people say that they are natural'.[18]

The second question was, 'Do you know about the ritual of the solstice there? And have you ever watched it?'. Two of them said no, and the other five said yes. Two of them had watched the solstice *in situ*, and the others want to see it. The PCAPR clarified that, 'Here in Pena de Rodas, this idea [of viewing the solstice from the site] was not very successful among the natives, as there are those who believe that it was an invention. But, maybe in the future, it will be very well known among the neighbours'.[19] This could explain why two people from neighbouring areas said no.

The third question I only asked those who knew about the ritual of the solstice: 'Do you think there

---

[14] David Silverman, *Doing Qualitative Research: a Practical Handbook* (London: Sage, 2000), p. 105.
[15] Alan Bryman, 'The Debate About Quantitative and Qualitative Research: A Question of Method or Epistemology?', *The British Journal of Sociology* 35, no. 1 (March 1984): p. 77.
[16] Silvia Pena, 'Penas de Rodas, el misterio del solsticio', *El Mundo*, 22 June 2011, at http://www.elmundo.es/elmundo/2011/06/22/galicia/1308731748.html [accessed 3 January 2015].
[17] My translation from: 'Non o sei, se as rocas foran maís pequenas non tendría dúbida de que están feitas de forma artificial, pero non sei'.
[18] My translation from: 'Non sei, sempre estiveron ahí, meu avó xa lembraba as pedras alí, pero a xente dí que son naturais'.
[19] My translation from: 'Aquí en Pena de Rodas, a idea non tivo moito éxito entre os veciños, hai quen cre que foi unha invención. Pero quizaís nun futuro sexa maís coñecida entre os veciños'.

was an astronomic sanctuary here?'. The foreigners and one villager believed so, whilst the other two said that they did not know.

Concerning the fourth question about legends, everybody referred to the legend that one of the stones is full of gold and the other full of petrol, but nobody knows which of the stones is which. This legend was even described by the Galician poet Manuel María (1929–2004) and is on a plaque at the site. Two of the interviewees referred to a legend about a woman who sometimes appears to offer a large treasure, and the PCAPR referred to a legend in which Penas de Rodas was an altar where people performed blood offerings.

As for the fifth question, 'Do you think this place was sacred? Why?', the blogger said yes, 'because rituals were performed there'.[20] One of the villagers said that he did not know. The two persons who ignored the ritual of the solstice said no, and the mayor of the village said that it could have been. The PCAPR said that for some people, it should have been, because there exists the legend of a bloody altar. In response to the sixth question, whether they believed that this place is still sacred, the blogger said, 'Not exactly, but observing *in situ* the phenomena of the solstice, there is something magic, and you can imagine the rituals that people did'.[21] The rest of them said that the place is not sacred now, but they stated that it is a magic or particularly charming place.

Question seven, about the existence of and the reason for a pilgrimage (Romería) to Penas de Rodas, was answered by explaining that in the year 1979, a group of friends decided to start a pilgrimage the Sunday before the 12th of August every year, because it was the time when almost everybody was in the village and also because at the beginning of August there are no other feasts in the surrounding areas.

Concerning the last question about if they noticed more tourism to the area because of the solstice ritual, the two persons who did not know about the phenomena said no and believed that tourism was as usual. The two foreigners, the mayor, and one villager said yes. The blogger recognized that he went there because of the ritual solstice, and the PCAPR said:

> This place has always been visited. Only during the two last years, we created an itinerary with the help of Sánchez-Montaña [for people] to visit the parish of Gaioso and finish with the observation of the sunset during the summer solstice in Penas de Rodas. We did not do it this year [2013], because of the bankruptcy, and I do not know if people went there this year.[22]

Discussion

The research question was, 'Can the "spell of archaeoastronomy" create a contemporary sacred site?'. For now, it is well established that the site is not a human made construct, Roman or otherwise; however this does not mean that this site is not fulfilling the notion of sacred space, as can be seen through the interviews. Next, different views will be discussed with reference to the construction of a sacred space.

Marion Bowman described the position of the New Age religious movements in Glastonbury stating, 'They include… the revival of traditional customs; the search for hidden wisdom; perception of past lives, reincarnation and interconnectedness; and the importance of sacred landscapes, topophilia and pilgrimage'.[23] In the case of Penas de Rodas, an alignment of the stones with the solstice could be nothing

---

[20] My translation from: 'Rituais eran levados a cabo alí'.
[21] My translation from: 'Non exactamente, pero observar *in situ* o fenómeno do solsticio é algo máxico, e un pode imaxinar os rituais que a xente alí levaba a cabo'.
[22] My translation from: 'Este lugar foi sempre visitado. Só durante estes dous últimos ano, nós creamos unha camiñata coa axuda de Sánchez-Montaña para visitar a aldea de Gaioso e rematar coa observación da posta do sol no solsticio de verán en Penas de Rodas. Aínda que este ano (2013) non o fixemos, e non sei se foi xente'.
[23] Marion Bowman, 'More of the same? Christianity, Vernacular Religion and Alternative Spirituality in Glastonbury', in

more than just a coincidence; for this reason, Sánchez-Montaña uses the 'spell of archaeoastronomy' to create a sacred space and connect this phenomenon with the Romans by looking for hidden wisdom in these stones, envisioning how the Romans could have used the stones to control the harvest seasons, reliving the ritual, and making the place sacred by associating it with a sanctuary. From a New Age point of view, this place could be considered to be sacred; in fact, as is possible to see in the interviews that some people believe this, especially foreigners to the village. In fact, when one searches for Penas de Rodas on Google.es (the Spanish version of Google) the first two results are the blog of Sánchez-Montaña and the blog of Ivan Blanco, and the fifth result is the official tourism page for Lugo, which recognizes the idea of the Roman-astronomic-sanctuary as true.[24] In addition, the newspaper *El Mundo*, in an article from 2011[25] (mentioned before), supported this same information, and it was even discussed on TVG (Galicia's own public television station)[26] in 2011 and this year (2014) on the national public television show 'Aqui la Tierra'.[27] Moreover, Sánchez-Montaña have published a book in 2014 titled *El Bosque Sagrado* de Lugo (*Sacred forest of Lugo*) in which is included Penas de Rodas.

Gerardus Van der Leeuw (1890–1950) considered that some places become sacred due to their natural peculiarity, like Penas de Rodas, after revealing a sacred attribute, such as alignments to the solstices. Van der Leeuw states, 'A sacred space may also be defined as that locality that becomes a position by the effects of power repeating themselves there, or being repeated by man. It is the place of worship'.[28] In Penas de Rodas, some people repeat the observation of the sunset solstice that Romans supposedly did thousands of years ago, in the mistaken belief that this is what the Romans did, while others may go just because they want to capture the phenomenon of the solstice sun, but in any case both positions are born because of the idea of the solar sanctuary.

Due to the complexity of defining a sacred space, Belden C. Lane has proposed four different rules or axioms by which to understand the character of a sacred space – these are basic self-evident principles that underlie the way a landscape is moulded into the religious imagination, which could fit Penas de Rodas.[29]

First, a sacred place is not chosen, it chooses. A sacred place is a construction of the imagination that affirms the independence of the holy. In Penas de Rodas, as has been stated, the consecration of this place is a construction of the imagination: through the spell of archaeoastronomy it was created as a solar sanctuary.

Second, a sacred place is an ordinary place ritually made extraordinary. A place becomes recognized as sacred because of certain ritual acts that are performed there, setting it apart as unique. Penas de

---

Beyond the New Age: Exploring Alternative Spirituality, ed. Marion Bowman and Steven Sutcliffe (Edinburgh: Edinburgh University Press, 2009), p. 92.

[24] Deputación de Lugo, 'Pena de Rodas', http://www.descubrelugo.com/turismo/lugo/pena-de-rodas/ [accessed 3 January 2015].

[25] Silvia Pena, 'Penas de Rodas, el misterio del solsticio', *El Mundo*, 22 June 2011, at http://www.elmundo.es/elmundo/2011/06/22/galicia/1308731748.html [accessed 3 January 2015].

[26] 'Ruta polo bosque sagrado de Lugo', *Televisión de Galicia*, 1 November 2011, at 'http://www.crtvg.es/informativos/ruta-polo-bosque-sagrado-de-lugo-158353 [accessed 3 January 2015].

[27] 'Solsticio de verano en Penas de Rodas - "Aquí la tierra" La 1 TVE', *YouTube*, 20 June 2014, at https://www.youtube.com/watch?v=y858IcLBFLE [accessed 3 January 2015].

[28] My translation from: 'El espacio sagrado es un lugar que se transforma en sitio cuando se repite en él el efecto del poder o cuándo lo repite el hombre. Es el sitio de culto', in Gerardus Van der Leeuw, *Fenomenología de la Religión* (México: Fondo de Cultura Económica, 1975), p. 379.

[29] Belden C. Lane, *Landscapes of the Sacred: Geography and Narrative in American Spirituality* (Baltimore, MD and London: The Johns Hopkins University Press, 2001), p. 19–20.

Rodas was an ordinary place that has been made ceremonially significant through the ideation of it as an astronomic Roman sanctuary and the creation of a modern ritual.

Third, a sacred place can be trod upon without being entered. The identification of a sacred place is thus intimately related to states of consciousness. Penas de Rodas is an open-air sanctuary, although even the people who do not know about the existence of the sanctuary consider the place magical, as seen from the interviews.

Fourth, the impulse of a sacred place is both centripetal and centrifugal, local and universal. By connecting the phenomenon with the Roman world, the local phenomenon is converted into a universal phenomenon.

James Twyman proposes that a sacred space is a place in which to find enlightenment. He uses a mall as an example, where he spent five days researching. He proposes the creation of new sacred spaces even in unlikely places.[30] This idea is also related to Lane's, who notes that a McDonalds is a place without any distinctive sense of presence; if you have seen one of them, you have seen them all. However, if you made a marriage proposal to someone you loved in one, that McDonalds becomes sacred to you.[31] This assumption reveals that sacredness is a human construct and that any place can be made to be a sacred space.

Barbara Bender's theory of contested space could be applied to Penas de Rodas, although it is not the same argument. Bender suggested that one space, as in the case of Stonehenge, can have a succession of identities, either in sequence or simultaneously. In the case of Stonehenge, for each period, the evidence of its origin is very different, and the monument is claimed by different people and researchers. Even in the present, there are discrepancies in how people discuss it.[32] Penas de Rodas is suffering a succession of identities as well, which can be seen in what the PCAPR said about the ritual of the solstice, 'This idea was not very successful among the neighbours… But maybe in the future it will be'.[33] It is a very recent idea that is gaining ground.

Anthony Thorley and Celia Gunn have also proposed a basic definition of a sacred site: 'It is a place in the landscape, occasionally over or under water, which is especially revered by a people, culture or cultural group as a focus for spiritual belief and practice and likely religious observance'.[34] Moreover, to meet this definition, a sacred site must satisfy both the statement and at least one of the nineteen characteristics that are divided into four inter-related categories: descriptive, spiritual, functional, and other. If the site satisfies more than one of the characteristics, it does not follow that it is more powerfully sacred; rather, it simply reflects the variety and richness of its sacred qualities.[35] Penas de Rodas would fall into the functional category and the characteristic of having a significant relationship with astronomical order and/or calendrical phenomena.

Conclusion

Returning to the essay question, Can the 'spell of archaeoastronomy' create a contemporary sacred site? The answer might be yes. The idea of an astronomic Roman sanctuary is derived from the 'alignment' of the stones to the solstices, and without the alignment, this theory is invalid. From the recognition of this

---

[30] James F. Twyman, *Ten spiritual lessons I learned at the Mall* (Forres: Findhorn Press, 2001).
[31] Lane, *Landscapes of the Sacred*, p. 39.
[32] Barbara Bender, *Stonehenge: Making Space* (Oxford: Berg, 1998), pp. 98–100.
[33] My translation from: 'Esta idea non tivo moito éxito entre os veciños…. Pero en un futuro pode tela'.
[34] Anthony Thorley and Celia M. Gunn, *Sacred Sites: An Overview. A Report for the GAIA Foundation 2007* (London: The Gaia Foundation, 2007), p, 76.
[35] Thorley and Gunn, *Sacred Sites: An Overview*.

alignment comes the interpretation of the stones as man-made. Then, the vision of this place as a Roman sanctuary is made by misinterpreting some stones as worked by stonemasons into a semicircular theatre. Thus, an all-natural monument has become a complex project developed by the human imagination. The once-common idea that Penas de Rodas is a natural monument has started to change – the people of the village who know about the alignment have their doubts of its verity, but to people from outside the village who know of this monument through the Internet or newspapers, the idea of it as a complex Roman sanctuary and therefore a sacred place is convincing. Through the different theories and definitions of what creates sacred spaces, it has been shown that Penas de Rodas could now be considered sacred, because it is becoming revered as a place of ritual.

Bibliography
Ayan, Xurxo, and Manuel Gago. *Herdeiros pola Forza: Patrimonio Cultural, poder e sociedade na Galicia do Século XXI*. O Milladoiro, Ames: 2.0 Editora, 2012
Belmonte Avilés, Juan Antonio, and Michael Hoskin. *Reflejo del Cosmos: Atlas de Arqueoastronomía en el Mediterráneo Occidental*. Madrid: Equipo Sirius, 1992.
Bender, Barbara. *Stonehenge: Making Space*. Oxford: Berg, 1998.
Bowman, Marion. 'More of the same? Christianity, Vernacular Religion and Alternative Spirituality in Glastonbury'. In *Beyond the New Age: Exploring Alternative Spirituality*, edited by Marion Bowman and Steven Sutcliffe, p. 92. Edinburgh: Edinburgh University Press, 2009.
Bryman Alan. 'The Debate About Quantitative and Qualitative Research: A Question of Method or Epistemology?'. *The British Journal of Sociology* 35, no. 1 (March 1984): p. 77.
Catalogue of the Galician Cultural Heritage Institutions.
Hobsbawm, Eric, and Ranger Terence, eds. *The invention of tradition*. Cambridge: Cambridge University Press, 1984.
Lane, Belden C. *Landscapes of the sacred: Geography and Narrative in American Spirituality*. Baltimore, MD and London: The Johns Hopkins University Press, 2001.
Vidal Romani, J. R., and C. Rowland Twidale. *Formas y paisajes graníticos*. Universidade da Coruña, Servizo de Publicacións, 1998.
Silverman, D. *Doing Qualitative Research: a Practical Handbook*. London: Sage, 2000.
Thorley, Anthony, and Celia M. Gunn. *Sacred Sites: An Overview. A Report for the GAIA Foundation*. London: The Gaia Foundation, 2007.
Twyman, James F. *Ten spiritual lessons I learned at the Mall*. Forres: Findhorn Press, 2001.
Van der Leeuw, Gerardus. *Fenomenología de la Religión*. México: Fondo de Cultura Económica, 1975.
Vazquez Seijas, Manuel. *Lugo en los tiempos prehistóricos*. Lugo: Junta del Museo Provincial, 1943.

THE SPHERE IN ANTIQUITY

Mª Pilar Burillo-Cuadrado[1]

ABSTRACT: This paper analyses the origin of the sphere, the result of the reflections of the Greek philosophers, who developed astronomical concepts that they used to shape instruments with which to continue their investigations and teaching. Their knowledge is reflected in the celestial sphere supported by the Farnese Atlas, a second century Roman copy of a Hellenistic sculpture, a paradigm of astronomical knowledge. The concept of the cosmic sphere led to the idea of the terrestrial globe that hung motionless in the centre of the spherical universe.

Both spheres became mythological symbols. To distinguish between them and understand what they mean it is necessary to analyse the signs that appear on their surfaces and the figures with which they are associated. And to pay particular attention to what they represent in the various myths that exist beyond the mere image and its evolution over time as beliefs gradually changed.

1. Introduction

The sphere or *sphaira* as a representation of our universe and of the world comes from Greek Philosophy, a broad field of knowledge covering all that was known at the time, including astronomy, mathematics and geometry, and it endures to the present day as a symbol and a didactic instrument. The sphere triumphed over other representations of the cosmos because it reflected totality, without beginning or end. A perfect, beautiful, mathematical figure, imbued with the magic it represents in itself. The product of reasoning that extends beyond astronomy to astrology and religious beliefs, becoming a divine entity that controlled destiny and took in the whole of known and imagined experience[2].

The sphere offers a view of the universe as seen from outside, from the position of Creator of the Cosmos. Its surface is decorated with representations of the stars and the circles into which the philosophers divided its surface in order to coordinate the stars and constellations. But the *sphaira* transcends iconography, and acts as a mirror of the beliefs and values projected onto it.

Both the cosmos and the earth are represented by the same geometric shape. The Celestial Sphere normally belonged to the gods so was associated with specific deities, unless it was the object of study by philosophers. But the *orbis terrarum* was ascribed to man, although there are cases in which it is depicted as an example of the divine domain.

So how, then, do we distinguish one sphere from the other? They can only be differentiated by applying the structuralist method and phenomenological and hermeneutical analysis.[3] In this deductive process we need to take into account not only the information provided to us by the iconography depicted on the surface of the sphere but also the context with which it is associated. The divine and human nature of the characters or their attributes will help us ascribe them to one type of sphere or the other and determine the meaning behind each symbol.

---

[1] University of Zaragoza. Departamento de Ciencias de la Antigüedad, Campus de Teruel. Spain. e-mail: pburillo@unizar.es.
[2] Otto Brendel, *Symbolism of the Sphere: A Contribution to the History of Earlier Greek Philosophy* (Leiden: E.J. Brill, 1977), pp. 50–51.
[3] Angela Ales-Bello, *Introduzione alla fenomenología*, (Rome: Aracne, 2009); Andres Ortiz-Osés, *C.G. Jung Arquetipos y Sentido* (Bilbao: Universidad de Deusto, 1988).

2. The Celestial Sphere, an Invention of the Greek Philosophers[4]

In the Greek world the idea that the universe is spherical arose from observing the movement of the stars and their appearance and disappearance on the horizon, as Ptolemy states in his *Syntaxis Mathematica*.[5]

It was Aristotle who, in his work *On the Heavens*, developed a cosmology that was used until Copernicus.[6] Using various arguments, he expounds the theory that the universe has existed since infinite time, is eternal and incorruptible. In contrast to the concept of infinity defended by some ancient philosophers, he claimed that the heavens rotated and moved in a circle in finite time, so the body of the universe is not infinite. The heavens, he said, must be in the form of a sphere, the first of the solids, without beginning or end, appropriate to its perfection. It moved together with the fixed stars in a uniform manner, while the earth remained at rest.

In ancient Greece the fixed stars and their arrangement in constellations, identified with animals and mythological characters, were depicted on the surface of the sphere with great precision, and in fact we have inherited the names and forms of the constellations fixed by the Greeks.

In order to measure the movements of the stars the Greek philosophers – there is some dispute whether these included Thales, Pythagoras, or his school – divided the sky into five parallel circles: the arctic circle, always visible, the summer tropic, the equator, the winter tropic and the invisible Antarctic circle. The Zodiac moves obliquely in relation to the middle three circles, corresponding to the constellations that appear in the band travelled by the sun's orbit.

An instrument for astronomers

Eratosthenes[7], who in 236 BCE was appointed chief librarian of the Library of Alexandria by Ptolemy III, is considered the founder of scientific astronomy. It was he who determined the accuracy of the equinoxes by measuring the ecliptic. The equator and the celestial ecliptic form an 'X', and the equinoxes occur at the two points where they intersect.

Scale models of the celestial sphere were made for educational and experimental purposes, an astronomical instrument used by surveyors, astronomers and metaphysicians in their study and veneration of the cosmos. An outstanding example can be seen in the Torre Annunziata mosaic dating to the first century BCE, preserved in the Museum of Naples.[8] The Seven Sages, whose identity is disputed, are seated in a circle fixing their gaze on a *sphaira*, with its meridians and parallels, carefully protected by a box. It is the subject of study and reflection, showing all the astronomical knowledge known up to that time. This scene must have been an archetypal subject for mosaics, because there is a similar depiction of the Seven Sages in another mosaic found in Sarsina that is now at Villa Albani in Rome. The only significant difference is that in the second case the celestial sphere is plain. A copy of a sphere similar in size to those depicted in these mosaics is curated in the German Museum of Mainz; in this case it shows the constellations. It is an exceptional survival of the spheres constructed for investigation and teaching.

In addition to these solid spheres Eratosthenes is credited with inventing the armillary sphere, used to reproduce the apparent motion of the stars around the earth. Archimedes wrote a treatise on

---

[4] Germaine Aujac, 'La sphéropée, ou la mécanique au service de la découverte du monde', *Revue d'histoire des sciences* 23, no. 2 (1970): pp. 93–107; 'L'image du globe terrestre dans la Grèce ancienne', *Revue d'histoire des sciences* 27, no. 3 (1974): pp. 93–210: 'Le ciel des fixes et ses représentation en Grèce ancienne', *Revue d'histoire des sciences* 29, no. 4 (1976): pp. 289–307; 'Sphérique et sphéropée en Grèce ancienne', *Historia Mathematica* 3 (1976): pp. 441–447.
[5] Ptolemy, *Syntaxis Mathematica*, 2 vols, ed. J. L. Heiberg (Leipzig: B. G. Teubner, 1898–1905, 2 vol.) 1, 3.
[6] Miguel Candel, 'Introducción a Aristóteles', *Acerca del Cielo* (Madrid: Biblioteca Clásica Gredos, 2008), p. 9.
[7] Giorgio Dragoni, *Eratostene e l'apogeo della scienza greca* (Bologna: Cooperativa Libraria Universitaria, 1979).
[8] Peter Sloterdijk, *Esferas II Globos Macrosferología*, (Madrid: Siruela, 2004), pp. 15–32.

Fig. 21.1: Bronze sphere found in Mainz, Germany. Photo by Pilar Burillo-Cuadrado.

the construction of spheres, but it has not survived. Cicero mentions in his *De re publica* that the consul Marcellus brought two mechanisms made by Archimedes showing the movements of the stars to Rome as spoils after the destruction of Syracuse, one of which was donated to the temple of Virtue. They were complex instruments like the one described by Cicero:

> … when Gallus moved the globe, the moon followed the sun through as many revolutions in this bronze invention as in the heavens themselves, so the sun was equally distant in the sky, and the Moon reached that position in which it cast its shadow on the Earth when it was in line with the Sun.[9]

The sphere of the Farnese Atlas
In Greek mythology Atlas was a Titan condemned by Zeus to carry the heavens on his shoulders so that they would not fall to the ground. The myth dates to ancient times when it was believed that the earth was a flat disk and the sky was a hemisphere that had to be held up so that it would not fall down. In ancient myths it was often supported by columns or mountains, but in the Greek world it was the Titan Atlas that held up the sky.[10]

---

[9] Cicero, *La república*, Ed. Akal (1989), Book 1, XIV, p. 22.
[10] Sloterdijk, *Esferas II Globos Macrosferología*, pp. 56–60.

Fig. 21.2: Farnese Atlas. *Detail of the intersection of the Celestial Equator with the Ecliptic. Photo by Pilar Burillo-Cuadrado.*

The most famous example is the *Farnese Atlas*, a Roman marble statue on display in the Meridian Room of the Museum of Naples of Atlas bearing a celestial sphere on his shoulders. The sphere has a diameter of 65 inches and very accurately depicts 42 constellations and the lines for measuring the cosmos in low relief. Bradley E. Schaefer concluded that the original dated to the Hellenistic era, specifically the second century BCE, in the time of Hipparchus, given the exact location of Aries at the intersection of the equinoctial colure with the equator and the ecliptic.[11] But a review by Dennis W. Duke notes that the position of Aries has a chronological range of ± 200 years and the sphere held by the Farnese Atlas depicts constellations not mentioned by Hipparchus, such as the wings of Pegasus or the Corona Australis, so we cannot be certain that he was the source.[12]

---

[11] Bradley E. Schaefer, 'The epoch of the constellations on the Farnese Atlas and their origin in Hipparchus' lost catalog', *Journal for the History of Astronomy* 36 (2005), pp. 167–196.

[12] Dennis W. Duke, 'Analysis of the Farnese Globe', *Journal for the History of Astronomy*, 37 (2006), pp. 87–100.

3. The Celestial Sphere, Symbol of Divinity

The celestial sphere is a symbol directly associated with religious beliefs, an attribute of divinity, but it is not always depicted with the details seen in the Farnese Atlas. There are cases where the surface is undecorated, it being taken for granted that the context associated with the deity in question identifies it as a celestial and not terrestrial sphere, although it is very probable that the statues that include these undecorated globes once had a polychrome surface that has not survived, and that the surface of the sphere was once blue to indicate the sky. In most cases the artist has removed all doubt by adding stars or the band of the zodiac.

However, there are examples in which doubts are now cast on the interpretation of signs that had an unequivocal meaning in their day, as in the case of the circles forming an 'X' that divide the sphere, which became a true archetype of astral symbolism. Whether this symbol is depicted with the circles crossing at an oblique angle or at 90° makes a significant difference.

We have to ask: why is the same symbol shown in different ways? Do all the variations mean the same thing or does the meaning change depending on the decoration and context in which it is depicted? In order to answer these questions it is necessary to reflect on the object we are studying, the sphere, in its relationship with the deity with which it is associated.

Zeus

Zeus is the supreme Greek god, father of the gods and man. In a fresco from Pompeii in the Museum of Naples, Zeus is shown seated majestically on his throne, holding a sceptre, with an eagle, one of his symbols, at his feet on the right and a celestial sphere on the left. The sphere has no other decoration apart from the colour blue, which removes any doubt about its symbolism. This association of Zeus with the celestial sphere implies that he is Creator of the Cosmos.

Helios

In Pompeii there is a fresco depicting a naked Helios with haloed and radiant head, symbol of the solar deity. In his right hand he carries a whip, a reference to controlling the chariot he rides each day, and in his left hand he holds a light-coloured globe crossed with an 'X'. If this schematic representation is compared with the globe supported by the Farnese Atlas it can be seen that the circles that intersect with an `X` coincide with the lines representing the Equator and the celestial ecliptic, whose intersection marks the equinoxes, solar dates of supreme importance in the Greco-Roman world, together with the solstices.

Fig. 21.3: Left, detail of the sphere held by Helios found in Pompeii, Italy; image from CEFIRE de Elda;[13] and on the right the sphere of the Farnese Atlas *showing the intersection of the equator and the ecliptic; photo by Pilar Burilllo-Cuadrado.*

---

[13] http://www.lavirtu.com [accessed 5 August 2016].

Aion

The lion-headed deity, or Aion, is sometimes depicted standing on a sphere. In Raffaela Bortolin's inventory of images of this deity, eight are standing on spheres.[14] Only one of them, No. 30, preserved in the Torlonia Museum in Rome, has two bands intersecting at right angles, not obliquely as in the case of Helios. From the interpretations given for the meaning of these bands, it is unanimously accepted that one of them represents the Zodiac, but the interpretation of the second is disputed. The fact that it is a band and not a line, and intersects the other at right angles, tends to support Raffaela Bortolin's proposal that it represents the Milky Way. Thus this figure is remarkable in that an element of the visible cosmos, the Milky Way, is shown in connection with an abstraction, the ecliptic.

This representation reflects the belief, already present in Pythagorean thought, that the Milky Way was the path taken by the soul on its journey to the beyond where it would become immortal, the gates to the other world standing at the intersection of the Milky Way with the celestial ecliptic, i.e., the solstices, which thus connect the two worlds, the earthly and the divine. This belief lived on in the Mithraic cult, where the lion-headed god takes souls to the world beyond, as reflected in the mithreum of the Palazzo Barberini by his position in the middle of the Zodiac, crossing the celestial sphere to the upper heavens.[15]

The Fates

The Fates, or *Moirai* in Greek mythology, were the goddesses who determined the destiny of men and the gods themselves, which even Zeus/Jupiter could not escape.[16] There are three of them, sisters, each shown with her own attributes, so that there could be no doubt about their identities. In Roman times they were given the names of: Nona, shown with a distaff and spindle, Decima with the thread of life and Morta with the globe, the wand and the book of life. These symbols indicate their power and influence over all phases of human existence, from birth to death. The images collected by Otto J. Brendel show the celestial sphere undecorated, with no symbols; as in the case of Zeus, this is an indication of their absolute power over the cosmos.[17]

Urania

Urania or *Ourania* is one of the nine muses who inspired the arts, sciences and literature. She is the muse of astronomy and astrology, so is shown with the cosmic sphere. She is also responsible for imparting celestial knowledge, so adopts a didactic pose with the sphere in her left hand or at her feet and holding the geometer's compass, used for making mathematical calculations, as can be observed in the Roman mosaic of the Wichten villa in the Luxembourg Museum.[18]

The surface of the celestial *Sphaira* can be decorated in different ways. In the case of the colossal Roman statue, a copy of the second-century Greek original, now in the Naples Museum, Urania holds the globe with all its measurement lines in her left hand, just like the sphere used for teaching the group in the Torre Annunziata mosaic. However, the Urania depicted with the Muses on a sarcophagus in the crypt of Palermo Cathedral holds a celestial sphere in her left hand with only the band of the zodiac marked on it.[19]

---

[14] Raffaella Bortolin, *Il Leontocefalo dei Misteri mitriaci. L'identità enigmatica di un dio* (Padova: Il Poligrafo, 2013), pp. 138-45.
[15] Mª. Pilar Burillo-Cuadrado and Francisco Burillo-Mozota, 'The swastika as representation of the Sun of Helios and Mithras', *Mediterranean Archaeology and Archaeometry* 14, no. 3 (2014): p. 33.
[16] Juan Ramón Carbó-García and Iván Pérez-Miranda, 'Daughters of the night (II): The fate of the Fates between past and present', *ARYS* 8 (Huelva: Universidad de Huelva, 2009-2010), pp. 141–54.
[17] Brendel, *Symbolism of the Sphere*, Plates XXIII and XXVI.
[18] P. Rodriguez-Oliva, 'Una escultura de musa sedente de Astigi (Écija, Sevilla). A propósito de una exposición celebrada en Málaga'. *Baetica* 30 (2008): Plate II, 4.
[19] The statue of Urania in the Museo Nazionale in Palermo and the sarcophagus of the muses in Palermo cathedral

Fig. 21.4: Left, detail of the Celestial Sphere held by the Muse Urania in the Naples Museum, Italy; and on the right Urania on the Sarcophagus of the Ostia Museum, Italy. Photo by Pilar Burillo-Cuadrado.

4. The *Orbis Terrarum*[20]

The shape of the Earth was also a subject for reflection among the Greek philosophers, whose reasoning took them beyond the empirical notions of the explorers. Aristotle notes that Thales was the first to argue that the earth was flat and held that it rested on water, while Anaximenes, Anaximander and Democritus asserted that it was supported by the air and that its flat shape held it in place.[21]

There is some debate about whether it was Parmenides or Pythagoras who first asserted that the earth was spherical, like a projection of the shape of the cosmos, of which it was considered the centre. But it was Aristotle who made the first scientific inferences about its spherical nature, observing the circular shadow of our planet on the moon and the way in which the stars rise and set. In his *Geography*, Strabo notes that the spherical nature of the Earth is a consequence of that of the Cosmos, since the same divisions will apply to it:

> The starting point therefore has to be the assumption that the sky has five zones, and the Earth likewise has five zones and that the zones here below and those above have the same name... Below each of the celestial circles is projected its earthly counterpart, and in the same way the celestial zone corresponds with the earth.[22]

Unlike Greek society the Roman world frequently used the globe or *Orbis Terrarum* as a symbol of the earth, especially on their coins from the time of the Emperor Augustus onwards, and it became an instrument of propaganda for Rome's dominion over the earth. The celestial sphere, as seen in Greece, continued to be associated with the realm of divinity.

The representations of the *Orbis Terrarum* appear on the reverse of coins in association with different figures such as Felicitas, Pax, the imperial eagle or the phoenix, thus emphasising the image of power and good government. Victory is depicted on the globe on an Augustan aureus; it is the so-called 'Nicephorus

---

were studied in person in August 2014.

[20] Aujac, 'L'image du globe terrestre dans la Grèce ancienne', pp. 193–210; Pascal Arnaud, 'L'image du globe dans le monde romain: (science, iconographie, symbolique)'. *Mélanges de l'Ecole française de Rome. Antiquité* 96, no. 1 (1984): pp. 53–116; Alberto Trivero R. Antvwala, 'L'Orbis Terrarum nella monetazione romana'. *Omni* 4, no. 2 (2010): pp. 51–63.
[21] Aristotle, *Acerca del cielo* (*On the Heavens*) (Madrid: Biblioteca Clásica Gredos, 2008), Book I, 294a, 30.
[22] Strabo, *Geography* (Madrid: Biblioteca Clásica Gredos, 1991), Book II, 2–3.

globe', example of the Roman conquest, which continued to appear on the coins of successive emperors until, with the triumph of Christianity, it was replaced with the 'globus cruciger', a globe surmounted by a cross, in a reinterpretation of imperial power. We also find this symbol associated with Christ and God the Father. In cases where the surface is smooth and coloured blue, it represents the universe.

However, this globe often has a 'T' on it. In this case it is a schematization of the Roman *Orbis Terrarum*, adapted to the idea of the divine creation of the Earth, which is divided into three parts in perfect symmetry, representing the three continents peopled by the three sons of Noah after the Great Flood: Africa, Asia and Europe. This schematisation appears in medieval cartography with Jerusalem at the centre[23].

5. Conclusions

The concept of the sphere derives from Greek philosophical thought. At first it was believed that the cosmos was a hemisphere covering a flat earth, but this idea evolved until the universe was conceived as being spherical, a concept that would later also be applied to the earth, or *Orbis Terrarum*.

Treatises were written on the celestial sphere and models made of it for investigation and teaching. The positions of the stars and constellations were established on its surface and the circles of the meridians and parallels labelled, together with the ecliptic or Zodiac, in order to determine their location and measurement. The globe supported by the Farnese Atlas was the paradigm that synthesised astronomical knowledge.

The celestial sphere was a symbol of deities (e.g. Zeus, Helios, Aion, the Fate Morta, the Muse Urania and the Christian God) while the *Orbis Terrarum* symbolised earthly power, in the case of the Roman emperors, but was also associated with Christ and God the Father. It is possible to determine which sphere is represented by studying the iconography on its surface, the context it is found in and the figure with which it is associated.

Fig. 21.5: The Orbis Terrarum. *Basilica S. Maria Sopra Minerva in Rome. Photo by Pilar Burillo-Cuadrado.*

---

[23] Erwin Raisz, *Cartografía* (Barcelona: Ediciones Omega, 1978), p. 26.

The cosmic sphere may appear plain and unadorned, and this in itself is important and of intrinsic value. When the colour has been preserved, it is blue.

When the sphere has two intersecting circles that cross obliquely, these correspond to the intersection of the equator with the ecliptic. But if they intersect at right angles they represent the Zodiac and the Milky Way, the path taken by souls to the upper heavens. The intersection of the two circles marks the solstices, interpreted as the gates to the Beyond.

The Romans, especially from the time of Augustus, minted coins with the *Orbis Terrarum* on the obverse, a symbol of power and good government reinforced by combining it with figures such as Felicitas, Pax, the imperial eagle and the phoenix with the sphere. When associated with Victory it is called the 'Nicephorus globe' which, with the coming of Christianity, was replaced by a sphere surmounted by a cross, the 'globus cruciger'. Christ and God the Father may appear with the celestial sphere surmounted by a cross. But if the sphere has a 'T', it symbolises the *Orbis Terrarum* divided into the three continents described in the Bible.

Acknowledgments

This study has been undertaken within the Iberus Research Group (H08), financed by the Government of Aragon and the European Social Fund, and R&D&i project: HAR2015-68032-P, "La Serranía Celtibérica y Segeda, el Patrimonio Histórico como motor de desarrollo rural", financed by the Spanish Ministry of Economy and Competitiveness and ERDF funds.

Bibliography

Ales-Bello, Angela. *Introduzione alla fenomenologia*. Roma: Aracne, 2009.

Aristotle, *Acerca del cielo* (*On the Heavens*), Madrid: Biblioteca Clásica Gredos, 2008.

Arnaud, Pascal. `L´image du globe dans le monde romain: (science, iconographie, symbolique)´. *Mélanges de l´Ecole française de Rome. Antiquité* 96, no. 1 (1984): pp. 53–116.

Aujac, Germaine. `La sphéropée, ou la mécanique au service de la découverte du monde´. *Revue d´histoire des sciences* 23, no 2 (1970): pp. 93–107.

Aujac, Germaine. `L`image du globe terrestre dans la Grèce ancienne´. *Revue d'histoire des sciences* 27, no. 3 (1974): pp. 193–210.

Aujac, Germaine. `Le ciel des fixes et ses représentation en Grèce ancienne´. *Revue d´ histoire des sciences* 29, no. 4 (1976): pp. 289–307.

Aujac, Germaine. `Sphérique et sphéropée en Grèce ancienne´. *Historia Mathematica* 3 (1976): pp. 441–447.

Bortolin, Raffaella. *Il Leontocefalo dei Misteri mitriaci. L'identità enigmatica di un dio*. Padova: Il Poligrafo, 2013.

Brendel, Otto. *Symbolism of the Sphere: A Contribution to the History of Earlier Greek Philosophy*. Leiden: E.J. Brill, 1977.

Burillo-Cuadrado, Mª. Pilar, and Francisco Burillo-Mozota. 'The swastika as representation of the Sun of Helios and Mithras'. *Mediterranean Archaeology and Archaeometry* 14, no. 3 (2014): pp. 29–36.

Candel, Miguel. 'Introducción´ a Aristóteles'. *Acerca del Cielo* (*On the Heavens*). Madrid: Biblioteca Clásica Gredos, 2008, pp. 9-39.

Carbó-García, Juan Ramón, and Iván Pérez-Miranda. 'Daughters of the night (II): The fate of the Fates between past and present'. *ARYS* 8. Huelva: Universidad de Huelva, 2009-10, pp. 141–54.

Cicero. *La República*. Madrid: Ed. Akal, 1989.

Dragoni, Giorgio. *Eratostene e l´apogeo della scienza greca*. Bologna: Cooperativa Libraria Universitaria, 1979.

Duke, Dennis W. `Analysis of the Farnese Globe´. *Journal for the History of Astronomy* 37 (2006): pp. 87–100.

Ortiz-Osés, Andres. *C.G. Jung Arquetipos y Sentido*. Bilbao: Universidad de Deusto, 1988.

Ptolemy. *Syntaxis Mathematica*, 2 vols. Edited by J. L. Heiberg. Leipzig: Lipsiae, in aedibus B. G. Teubneri,1898–1905.

Raisz, Erwin, *Cartografía*. Barcelona: Ediciones Omega, 1978.

P. Rodriguez-Oliva. `Una escultura de musa sedente de Astigi (Écija, Sevilla). A propósio de una exposición celebrada en Málaga´. *Baetica* 30 (2008): pp. 149–170.

Schaeffer, Bradley E. 'The epoch the constellations on the Farnese Atlas and their origin in Hipparchus´s lost catalogue'. *Journal for the History of Astronomy* 36 (2005): pp. 167–196.

Sloterdijk, Peter. *Esferas II Globos Macrosferología*. Madrid: Siruela, 2004.
Strabo. *Geography*. Madrid: Biblioteca Clásica Gredos, 1991,
Trivero R. Antvwala, Alberto. `L´Orbis Terrarum nella monetazione romana´. *Omni* 4, no. 2 (2010): pp. 51–63.

NORTH AND SOUTH AMERICA

HOUSES OF THE SUN
AND THE COLLAPSE OF CHACOAN CULTURE

J. McKim Malville and Andrew Munro

Abstract: The most monumental prehistoric masonry structures north of Mexico are the Great Houses of the Bonito Phase of Chaco Canyon. The largest was Pueblo Bonito with nearly 700 rooms. The Great Houses are associated with a regional system which dominated the southwest during the 11th century. During that time Chaco Canyon flourished as a pilgrimage and trade centre, apparently resulting in a concentration of wealth and political power in the hands of elite residents of the Great Houses. Following the drought of 1090–1100 CE the power of the residents of the Great Houses apparently waned. After 1100 the Great Houses built in the Late Bonito Phase are distinguished by their planned designs, relatively short construction period, and negligible middens. Unlike the earlier Great Houses, many of these are located at places that provide solstice sunrise or sunsets associated with notable horizon features. These Great Houses may have been intended to provide visitors and pilgrims with dramatic visual astronomical experiences. One interpretation involves an effort by the elite residents of the Great Houses to revive the pilgrimage system by new construction at astronomically significant locations. An alternate interpretation is that the Late Bonito Great Houses resulted from a spontaneous religious revival, in which the structures were built by pilgrims who visited the canyon to honour, worship, and/or make offerings to the sun. This movement may have presaged the rejection and eventual collapse of the hierarchal political system of Chaco. Animism and alternate ontologies are considered in interpreting the experiences of those viewing dramatic sunrises and sunsets in the canyon.

Introduction

During the Classic Bonito Phase of Chaco Canyon (1040–1100 CE), Great Houses became the most monumental prehistoric masonry structures in the Americas north of Mexico. The largest of these Great Houses, Pueblo Bonito, Chetro Ketl, Pueblo Alto, Pueblo del Arroyo, and Kin Kletso, are contained in a 2km wide area known as downtown Chaco.[1] The largest, Pueblo Bonito, was constructed over a 300 year period, contained nearly 700 rooms, stacked 4 or 5 stories high, covering an area of 0.8ha. Only the outer rooms had sunlight. Only a few families lived in this huge structure, perhaps having acquired wealth and power through the combination of pilgrimage and trade. Judging from the number of corn-grinding metates and large ovens, there were large festivals associated with these Great Houses that were likely pilgrimage and trade events occurring around the winter solstice.

The great drought of 1090–1100 CE appears to have changed the character of Chacoan society. After 1100 CE, in the Late Bonito Period, the majority of newly constructed Great Houses were located at sites that provide exceptional views of the solstice sunrise or sunset. Two of the Late Bonito Great Houses were contained in north-south alignments with earlier structures.

---

[1] Steven Lekson, ed., *The Archaeology of Chaco Canyon: An Eleventh-Century Pueblo Regional Center* (Santa Fe: School of American Research Press, 2006).

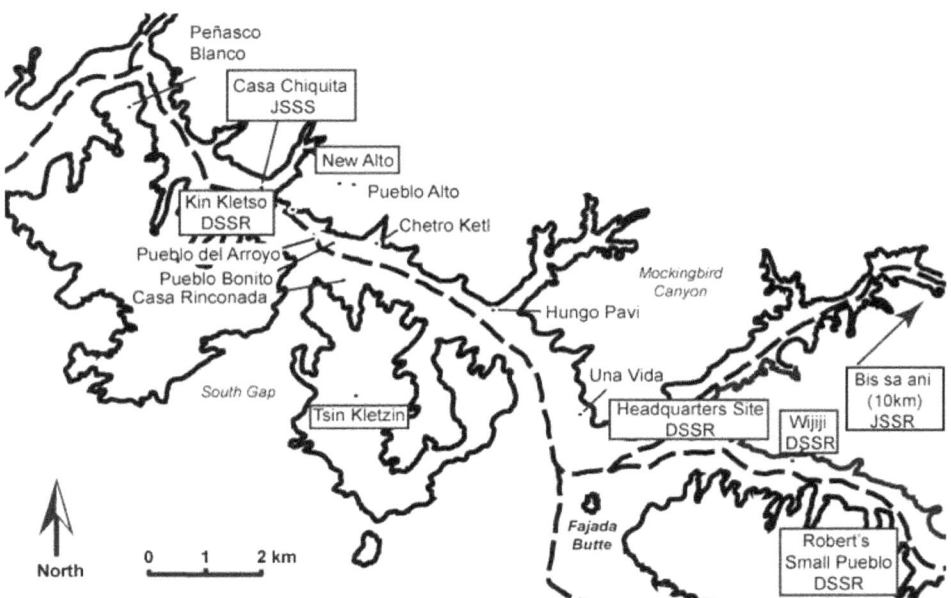

Fig. 22.1: Great Houses at Chaco Canyon. The Late Bonito Great Houses are identified in boxes, in which DSSR and JSSR indicate horizon markers for December solstice sunrise and June Solstice sunrise, respectively (adapted from Lekson).[2]

Figure 22.1 provides a map of Chaco Canyon in which Late Bonito Great House are labelled DSSR for the December solstice sunrise, JSSR for the June solstice sunrise, or JSSS for the June solstice sunset. Late Bonito Great Houses are distinguished by their pre-planned designs, relatively short construction period, and negligible middens, which indicates that they never fully functioned as residences. See Tables 22.1 and 22.2; those listed in Table 22.2 are the major Late Bonito Great Houses. We are currently investigating other less known cases. Lekson has suggested that they were administrative and storage buildings, an idea that has been contested by Van Dyke and others.[3]

Van Dyke suggested that the Late Bonito Great Houses were built at a time when Chaco was losing credibility as an efficacious ceremonial centre, because of the decrease in agricultural production in the Chaco basin brought about by the drought during the decade of the 1090s CE. Figure 22.2 shows the added storage area in the Great Houses estimated by Wilcox.[4] Note the gap in construction activity during the decade of 1090 CE. The burst of construction in 1110 CE and the following two decades was associated with the Late Bonito Great Houses.

---

[2] Lekson, *The Archaeology of Chaco Canyon, An Eleventh-Century Pueblo Regional Centre*. Andrew M. Munro, 'The Astronomical Context of the Archaeology and Architecture of the Chacoan Culture' (Unpublished PhD thesis, James Cook University, 2012); archived at the Hibben Center under NPS Permit CHCU-08-03; available at http://researchonline.jcu.edu.au/40277/.

[3] Ruth Van Dyke, *The Chaco Experience, Landscape and Ideology at the Center Place* (Santa Fe: School for Advanced Research Press, 2008).

[4] David Wilcox 'The Evolution of the Chacoan Polity', in *Chimney Rock: The Ultimate Outlier*, ed. J. McKim Malville (Lantham: Lexington Books, 2004), pp. 163–200.

Bonito Phase Great Houses	Late Bonito Phase Great Houses
Multi-phased construction, apparently unplanned, sometimes over centuries	Single stage, pre-planned construction
Hundreds of thousands of timbers	Radical reduction of number and size of timbers
Multiple orientations and varied floor plans	Most Great Houses are at locations for observing the solstice sun at notable horizon features and use a common floor plan
Inefficient construction using small pieces of tabular sandstone	Efficient construction utilizing large loaf sandstone blocks
Evidence of occupation in one-third or fewer of the rooms.	Negligible middens; little or no evidence for long-term occupation.

Table 22.1.

After 1100 CE, both Wijiji and Kin Kletso were built at locations that include foresights for the December solstice sunrise, as well as for anticipatory observations approximately two weeks prior to the solstice. Solstice sunrises visible from Wijiji and Kin Kletso are not architectural alignments of walls to significant azimuths; rather the buildings are located at observation sites for solstice horizon foresights.[5] Both contain anticipatory foresights, which may have been useful to prepare for ceremonies. The Great House of Wijiji has two solstice sunrise events: a horizon notch marks the December solstice sunrise and a short distance away the December solstice sun can be viewed dramatically rising over a prominent natural rock pillar.[6]

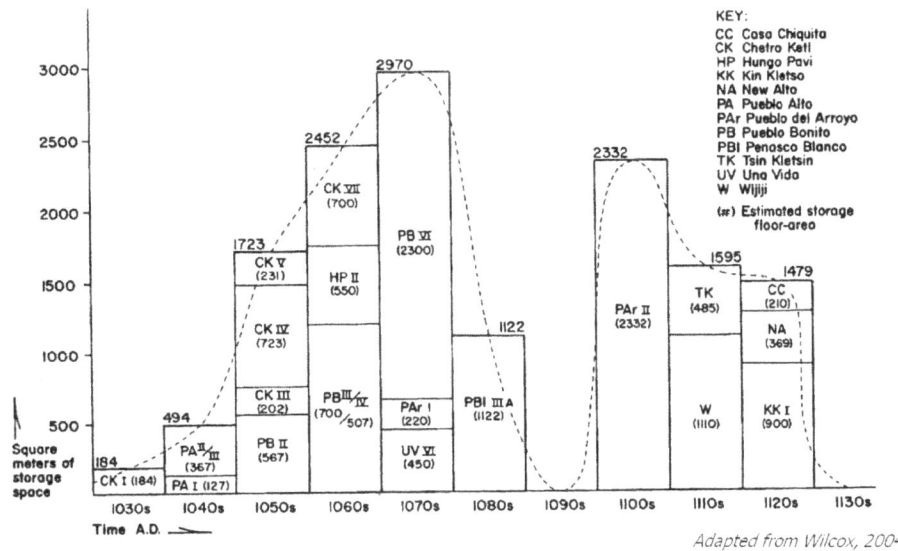

Fig. 22.2: Storage floor area added to Chaco Canyon Great Houses.

---

[5] J. McKim Malville, *A Guide to Prehistoric Astronomy in the Southwest* (Boulder, CO: Johnson Books, 2008), pp. 70–73.
[6] Michael Zeilik and Richard Elston, 'Wijiji at Chaco Canyon: a winter solstice sunrise and sunset station', *Archaeoastronomy* 6 (1983): pp. 66–73.

Fig. 22.3: Sunset observed from Casa Chiquita. Photo by G. B. Cornucopia.

Late Bonito Great House	Approximate Date	Foresight
Wijiji	1110-1115	DSSR
Kin Kletso	1125-1130	DSSR
Headquarters	1100s	DSSR
Robert's Small Pueblo	1100s	DSSR
Casa Chiquita	1100-1130	JSSS
Bis sa ai	1100s	JSSR
Tsin Kletsin	1110-1115	North to Pueblo Alto
New Alto	1100-1130	South to Casa Rinconada

Table 22.2.

By standing at the southeast corner of Kin Kletso, an anticipatory observation of the foresight may be made fifteen to sixteen days prior to solstice. On December solstice, sunrise is observable from the northeast corner, appearing the northern cliff meets the floor of the canyon (Fig. 22.4).

Fig. 22.4: Kin Kletso June Solstice Sunrise (photo by G. B. Cornucopia).

Robert's Small Pueblo was built 125m from a station with a December solstice sunrise horizon marker. The horizon at Casa Chiquita contains a June solstice sunset marker, similar to the December solstice sunrise foresight visible at nearby Kin Kletso but using a different horizon feature.

Fig. 22.5: December solstice sunrise near Roberts Small Pueblo.

The back-filled Late Bonito Great House known as Headquarters Site A has an eastern horizon feature that produced a dramatic December solstice sunrise light and shadow effect on most of the building's footprint. Sunrise on the December solstice occurs in a well-defined notch on the horizon.

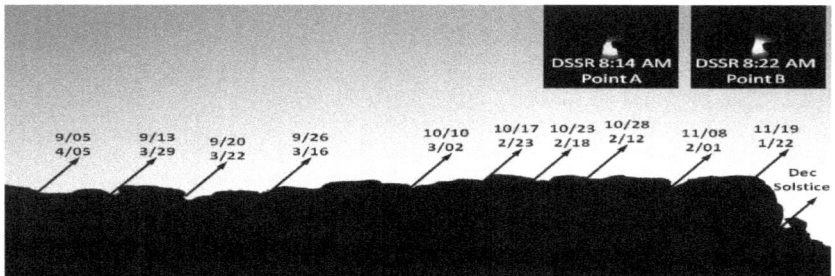

Fig. 22.6: *Horizon Survey Results at Headquarters Site A.*[7]

These two inter-site N-S alignments across the central canyon are especially interesting as possible demonstrations of the interaction of the north celestial pole with the human construct of a Great House. People at Tsin Kletsin could have watched the night sky rotate around the centre of the cosmos directly above Pueblo Alto. Similarly, people observing the night sky from Casa Rinconada could have watched the cosmos rotate over New Alto. Torchlight at the northern mesa-top sites could have increased dramatic visual demonstrations of the connections between the Chacoan Great Houses and the heavens.

Interpretation

Initially the power of the elite residents of the Great Houses may have been derived from control over trade associated with periodic regional festivals. The residents who organized these festivals must have acquired power, wealth, and prestige when rain making rituals appeared to be successful. They also should have acquired some measure of political or religious legitimacy through demonstrating their knowledge of astronomy and control of the calendar.

> The symbiotic linking of pilgrimage, periodic festivals, and entrepreneurial activity provides a means for integration of an extended population. If groups of people voluntarily visited Chaco Canyon to attend period festivals and religious ceremonies, the surrounding area could have become culturally integrated without any exercise of administrative control, force or political power.[8]

The suggestion by Van Dyke that the Late Bonito Great Houses were built after the drought of the 1090s to re-establish the reputation of Chaco as an efficacious ceremonial centre, implies the residents of the Great Houses acted in concert with each other or were dominated by a central authority in the canyon.[9] However, if the leaders of Chaco Canyon wanted to rebuild their legitimacy, it seems likely they would have enhanced the already dramatic and imposing Bonito Phase Great Houses. The monumental structures of Pueblo Bonito and Chetro Ketl already projected power. A case in point is Tiwanaku in the 11th century on the southern shore of Lake Titicaca. Vranich interprets Tiwanaku as a highly complex ceremonial centre, similar in many regards to the Chaco regional system.[10] It was dependent upon 'high-energy, popular ritual events' to attract resource bearing visitors from outlying communities. Massive new construction,

---

[7] Munro, 'The Astronomical Context of the Archaeology and Architecture of the Chacoan Culture', p. 160.
[8] J. McKim Malville and Nancy J. Malville, 'Pilgrimage and Periodic Festivals as Processes of Social Integration in Chaco Canyon', *Kiva* 66 (2001): pp. 327–344.
[9] Van Dyke, *The Chaco Experience, Landscape and Ideology at the Center Place.*
[10] Alexei Vranich, 'The Development of the Ritual Core of Tiwanaku', in *Tiwanaku*, ed. Margaret Young-Sanchez (Denver, CO: Denver Art Museum, 2009), pp. 11–34.

sometimes involving façades, was designed to impress the visitors. Tiwanaku had outlying communities, similar to Chacoan outliers in that their residents visited the centre for periodic festivals and constructed their own copies of the monumental structures of the centre.

Those pilgrimages that may have taken place during the Chaco florescence in the 11th century may fall under two general anthropological theories, characterized sometimes as Durkheimian or Turnerian.[11] Perhaps these Late Bonito Great Houses reveal a transition from a Durkheimian form of pilgrimage to one what can be described as Turnerian. From the viewpoint of Emile Durkheim, religion is not a spontaneous and inherent human creation but is the result of socio-political processes, sometime intentionally manipulative, in which an elite class creates a mythology and organizes participatory rituals to promote and maintain their own political authority.[12] Such pilgrimage systems 'legitimize domination and oppression' by an elite.[13] Construction of the Great Houses may have involved the corvée labour of pilgrims, consistent with the Durkheim model.

The alternative to such a 'top down' organization of pilgrimage by elites in the canyon is that of a spontaneous, self-organized religious revival. The return of the rains around 1100 may have caused pilgrims to visit the canyon to honour, worship, and/or make offerings to the sun and construct smaller versions of the Great Houses at places where the solstice sun could be seen joined on the horizon with the sacred landscape. This style of pilgrimage, proposed by Victor Turner (1969), involves a diametrically opposite relationship between pilgrims and political authority.[14] In his view, pilgrimage subverts established social order and is counter-hegemonic in that it challenges the authority of the state by setting up competing religious symbols and destinations. Turner argue that when pilgrims voluntarily embark on their journey they abandon the structures of their ordinary world and enter a landscape of 'anti-structure' where ordinary norms and differences of status are left behind. Pilgrims may have intentionally avoided the Great Houses and rejected the political power that they represented. Lekson suggests that historical egalitarian Pueblos reacted against the Classic Bonito political system, its palatial residences of leaders and hierarchal social structure, and that such a rejection may have occurred initially in the 1100s.[15]

We consider two primary alternate interpretations of Late Bonito Great Houses. They were constructed (1) to maintain the political power of leaders in the canyon or (2) as the result of spontaneous religious revival. We note that these interpretations need not be mutually exclusive. It is appropriate to ask what visitors to these Great Houses actually saw on the mornings or evenings of solstice. It is indeed likely that the experience was significantly different for them than it is for us. We see the sun as a hot ball of gas 92 million miles away fuelled by thermonuclear processes in its centre. For those watching a sunrise in the 1100s in Chaco canyon, it was a different universe. If we are to understand what ancient people saw and how they experienced the heavens, we have to go beyond judging other world views as 'fascinating but ultimately mistaken ways of knowing the world'.[16] As Fowles argues '... are we really to conclude that Native Americans 'see' Father Sun travelling across the heavens any less clearly than Anglo American

---

[11] John Eade and Michael J. Sallnow, eds., *Contesting the Sacred: The Anthropology of Christian Pilgrimage* (London: Routledge, 2001).

[12] Emile Durkheim, *The Elementary Forms of the Religious Life* (New York: Free Press, 1912).

[13] John Eade and Michael Sallnow, *Contesting the Sacred: The Anthropology of Christian Pilgrimage..*

[14] Victor Turner, *The Ritual Process: Structure and Anti-Structure* (Chicago: Aldine Publishing, 1969).

[15] Lekson, *The Archaeology of Chaco Canyon*.

[16] Benjamin Alberti and Yvonne Marshall, 'Animating Archaeology: Local Theories and Conceptually Open-ended Methodologies', *Cambridge Archaeological Journal* 19, no. 3 (2009): pp. 344–56; Benjamin Alberti and Tamara Bray, 'Animating Archaeology: Of Subjects, Objects and Alternative Ontologies', *Cambridge Archaeological Journal* 19, no. 3 (2009): pp. 337–343.

scientists see a stationary mass of hydrogen and helium?'[17] The growing interest among archaeologists about animism and alternate ontologies alerts us to consider the sacredness of both the sun and the natural horizon. Both may have been understood to be animate, powerful, and engaged in reciprocal relationships with humans.[18] These powerful and dramatic apparitions of the sun conjoined with unusual horizon features could have been theophanies, manifestations of powerful divine beings[19]. The places for such experiences may have been the Late Bonito Great Houses. The meaning and function of Great Houses may have evolved over time as the accumulation of memories of rituals, ceremonies, important events, and burials transformed residential structures into sacred realms. These post Bonito Great Houses appear to be uniquely Chacoan. They may have combined elements of the 'shrines, monasteries, temples, tombs, and cathedrals' of other cultures, but none encapsulates the fullness of their meaning. They may have become too sacred for ordinary uses.

Building these structures failed to preserve either the political system centred on Chaco Canyon or the tradition of pilgrimage to Chaco Canyon. For a variety of reasons including perhaps the greediness of its elites and reaction against its hierarchical polity, combined with an extended drought beginning around 1090 CE, the Chaco regional system failed. This failure may foreshadow the larger rejection of the Chaco Bonito Phase by modern Pueblos suggested by Lekson.[20] Portions of the political centre moved from Chaco Canyon to Aztec, where we have not yet found any evidence for an interest in solar theophanies. On the other hand, ontological continuity is suggested by similar juxtapositions of the sun and moon with unique features of the horizon at Chimney Rock, Yucca Great House and Cliff Palace of Mesa Verde.[21]

Bibliography

Alberti, Benjamin and Tamara Bray, 'Animating Archaeology: Of Subjects, Objects and Alternative Ontologies', *Cambridge Archaeological Journal* 19, no. 3 (2009): pp. 337–343.

Alberti, Benjamin, and Yvonne Marshall. 'Animating Archaeology: Local Theories and Conceptually Open-ended Methodologies'. *Cambridge Archaeological Journal* 19, no.3 (2009): pp. 344–56.

Eade, John, and Michael Sallnow, eds. *Contesting the Sacred: The Anthropology of Christian Pilgrimage*. London: Routledge, 1991.

Descola, Philippe. *Beyond Nature and Culture*. Chicago: University of Chicago Press, 2013.

Eliade, Mircea. *Patterns in Comparative Religion*. New York: New American Library, 1958

Fowles, Severin. *An Archaeology of Doings: Secularism and the Study of Pueblo Religion*. Santa Fe: School for Advanced Research Press, 2013.

Lekson, Stephen H., ed. *The Archaeology of Chaco Canyon, An Eleventh-Century Pueblo Regional Center*. Santa Fe: School of American Research Press, 2006.

Malville, J. McKim. *A Guide to Prehistoric Astronomy in the Southwest*. Boulder, CO: Johnson Books, 2008.

Malville, J. McKim, and Nancy J. Malville. 'Pilgrimage and Periodic Festivals as Processes of Social Integration in Chaco Canyon'. *Kiva* 66 (2001): pp. 327–344.

---

[17] Severin Fowles, *An Archaeology of Doings: Secularism and the Study of Pueblo Religion*, (Santa Fe: School for Advanced Research Press, 2013), p. 9.

[18] See, for example, Bill Sillar, 'The Social Agency of Things? Animism and Materiality in the Andes', *Cambridge Archaeological Journal* 19 (2009): pp. 367–377.

[19] Mircea Eliade. *Patterns in Comparative Religion* (New York: New American Library, 1958). Both the sacred sun and sacred landscape were alive, powerful, and vast beyond human imagining. They may best be understood as elements in Descola's ontology of analogism; see Philippe Descola, *Beyond Nature and Culture* (Chicago: University of Chicago Press, 2013).

[20] Lekson, *The Archaeology of Chaco Canyon*, pp. 29–31.

[21] Malville, *Prehistoric Astronomy*; John Ninnemann and J. McKim Malville, 'Using Photography to Test Hypotheses in Southwestern Archaeoastronomy', *Archaeoastronomy* 23 (2010): pp. 82–90.

Munro, Andrew M. 'The Astronomical Context of the Archaeology and Architecture of the Chacoan Culture'. Unpublished PhD thesis, James Cook University, 2012. Archived at the Hibben Center under NPS Permit CHCU-08-03: http://researchonline.jcu.edu.au/40277/.

Ninnemann, John, and J. McKim Malville. 'Using Photography to Test Hypotheses in Southwestern Archaeoastronomy'. *Archaeoastronomy* 23 (2010): pp. 82–90.

Sillar, Bill. "The Social Agency of Things? Animism and Materiality in the Andes'. *Cambridge Archaeological Journal* 19 (2009): pp. 367–377.

Turner, Victor. *The Ritual Process: Structure and Anti-Structure*. Chicago: Aldine Publishing, 1969.

Van Dyke, Ruth M. *The Chaco Experience, Landscape and Ideology at the Centre Place*. Santa Fe: School for Advanced Research Press, 2008.

Vranich, Alexei. 'The Development of the Ritual Core of Tiwanaku'. In *Tiwanaku*. Edited by Margaret Young-Sanchez, pp.11–34. Denver, CO: Denver Art Museum, 2009.

Wilcox, David. 'The Evolution of the Chacoan Polity'. In *Chimney Rock: The Ultimate Outlier*. Edited by J. McKim Malville, pp. 163–200. Lanham: Lexington Books, 2004.

Zeilik, Michael, and Richard Elston. 'Wijiji at Chaco Canyon: a winter solstice sunrise and sunset station', *Archaeoastronomy* 6 (1983): pp. 66–73.

ASTRONOMY AND THE CEQUE SYSTEM OF CUSCO

Steven R. Gullberg

ABSTRACT: With more than 328 huacas organized along 41 ceques, the Cusco basin has the highest concentration of huacas in the Inca world. Since the publication of Zuidema's monograph, *The Ceque System of Cusco: The Social Organization of the Capital of the Inca*, interest in the Cusco ceque system has grown such that it has become one of the most controversial and perhaps one of the least understood aspects of Andean culture. According to Zuidema's proposals, each huaca was part of a walk-through solar-lunar sacred calendar, which stopped functioning when the Pleiades were too close to the sun to be visible. In the 1970's Zuidema and Aveni proposed that some ceques functioned as straight sight-lines toward astronomical events on the horizon. Since then Rowe, Niles, and especially Bauer have shown that the ceques are not straight-lines, eliminating them as possible astronomical instruments. However, individual huacas on certain of the ceques do display astronomical meaning and intentionality. The ceques of Cusco and their huacas are analysed and correlated with astronomical and cosmological aspects such as direct sightlines of sunrises and sunsets over huacas, solar light and shadow effects, caves, light-tubes, springs, water channels, and carved non-functional stairs. Data is discussed regarding these astronomical and cosmological characteristics found with the huacas of certain ceques.

1. Introduction

Ceques were important features in the Cusco valley, and one of their symbolic roles may have been to affirm and supplement the inherent directionality of huacas. Ceques were organizational in intent, indicating sequences of ritual visits, responsibilities for individual panacas and ayllus, and assignment of territory and irrigation sources.[1]

Huacas were Inca shrines and prior to the Spanish invasion there were many hundreds of them. Major huacas required maintenance and caretaking and gifts were made to the powers of the shrines. Animals and produce were sometimes sacrificed to the huaca and used to support the attendants.

According to Bernabe Cobo the huacas were organized along ceques, partially to regulate their maintenance and to organize sacrifices at proper times.[2] Most ceques were arranged outward from the Coricancha and all were arranged in four basic groups within the four administrative quarters of Cusco.[3] There were nine ceques in Chinchaysuyu, nine in Antisuyu, nine in Collasuyu, and 14 in Cuntisuyu. The lines primarily did not overlap so that progression from one shrine to the next along a particular ceque was a straightforward matter. In all there were 42 paths, but two that are adjacent geographically were grouped together as one (Cu. 8), leaving 41 ceques in all. The huacas were not placed in a geometrically regular sequence along the ceques and they were typically within a half-day walk from the Coricancha. The ends of many ceques were reported to be where one would lose sight of Cusco.

Cobo described the ceques when he listed the 328 huacas:

From the Temple of the Sun, as from the center, there went out certain lines which the Indians call ceques; they formed four parts corresponding to the four royal roads which went out from Cuzco. On each one of those

---

[1] Panacas were royal kin groups and ayllus were kin groups made of non-royals. Both were given maintenance responsibilities for certain Cusco ceques.
[2] Bernabe Cobo, *Inca Religion and Customs,* trans. Roland Hamilton (1653; repr. Austin: University of Texas Press, 1990).
[3] Giulio Magli, 'On the astronomical content of the sacred landscape of Cusco in Inka times', *Nexus Network Journal 7,* no. 1 (2004): pp. 22–32.

ceques were arranged in order the guacas and shrines which were there in Cuzco and its region like stations of holy places, the veneration of which was common to all.[4]

Tom Zuidema described the ceques as comprising four groups corresponding with Cusco's four administrative quarters and ordered them sequentially.[5] John Rowe later referred to them by the suyu's abbreviation and a number (see Fig. 23.1).[6] The first ceque of Chinchaysuyu thereby became known as Ch. 1 and the last became Ch. 9. Ch. 1 separated Chinchaysuyu from Antisuyu and Ch. 9 divided Chinchaysuyu and Cuntisuyu. Similar delineations existed at each border of the four suyus. Even more should be discussed in a longer paper.

2. Huacas and Ceques

Huacas often were shrines to ancestors who, it was believed, could influence the living. The most powerful huacas required maintenance, care taking, and offerings. A huaca could be any material thing that manifested superhuman power, such as mountain peaks, springs, rock outcrops, hills, passes, and streams. Carving empowered outcrops and made them worthy of worship.[7] Water was also thought to empower huacas through a life-energizing force known as camay that provided sentience to the inanimate or renewed power in the living.[8] The world's water cycled through the heavens and earth in its journey down the Vilcanota River with return in the sky via the Milky Way.[9] Inca cosmology viewed the Milky Way as a river flowing across the night sky in a very literal sense. Incas saw earthly waters as being drawn into the heavens and then later returned to earth following a celestial rejuvenation. The earth was thought to float in a cosmic ocean.[10] When the celestial river's orientation was such that it dipped into that ocean, waters were drawn into the sky. 'The Milky Way is therefore an integral part of the continuing recycling of water throughout the Quechu universe'.[11] Frank Salomon and George Urioste suggest that huacas were understood to be living, energized beings brought to sentience by the earth's waters.[12] The circulation of running water and the pouring of offertory liquids could animate certain inanimate objects to become huacas, which were understood to be sentient beings with extraordinary and superhuman powers. Running water was located near many huacas, suggesting that this process was thought to vitalize the life within each of them.

---

[4] Cobo, *Inca Religion*, p. 51.
[5] R. Tom Zuidema, *The Ceque System of Cuzco: The Social Organization of the Capital of the Inca* (Leiden: E. J. Brill, 1964).
[6] John Rowe, 'Inca Policies and Institutions Relating to the Cultural Unification of the Empire', in *The Inca and Aztec States, 1400-1800*, ed. George A. Collier, Reneato I. Rosaldo, and John D. Wirth (New York: Academic Press, 1980), pp. 93–118.
[7] Maarten J. D. Van de Guchte, 'Carving the World: Inca Monumental Sculpture and Landscape', (PhD dissertation, University of Illinois at Urbana-Champaign, 1990).
[8] Frank Salomon and George L. Urioste, *Introductory Essay in The Huarochiri Manuscript: A Testament of Ancient and Colonial Andean Religion* (Austin: University of Texas Press, 1991); Tamara L. Bray, 'An Archaeological Perspective on the Andean Concept of Camaquen: Thinking Through Late Pre-Columbian Ofrendas and Huacas', *Cambridge Archaeological Journal* 19, no. 3 (2009): pp. 357–366; J. McKim Malville, 'Animating the Inanimate: Camay and Astronomical Huacas of Peru', in *Cosmology Across Cultures, ASP Conference Series, 409*, ed. Jose Alberto Rubino-Martin, Juan Antonio Belmonte, Francisco Prada, and Anexton Alberdi, (San Francisco: Astronomical Society of the Pacific, 2009), pp. 261–266.
[9] Gary Urton, *At the Crossroads of Earth and Sky: An Andean Cosmology* (Austin, University of Texas Press, 1981).
[10] Urton, *At the Crossroads*.
[11] Urton, *At the Crossroads*, p. 60.
[12] Salomon and Urioste, *Huarochiri Manuscript*.

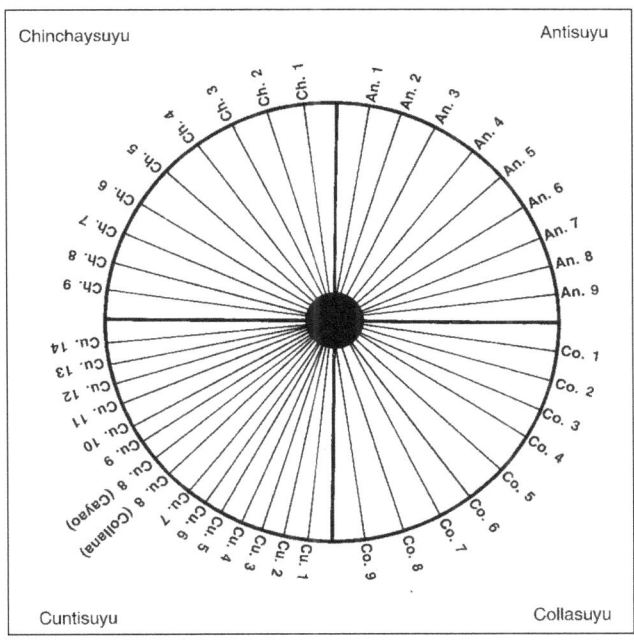

Fig. 23.1: Rowe's Cusco ceque numbering system.[13]

In 1539 the Spaniards began a campaign against the indigenous religion and proceeded to systematically extirpate huacas. The foundations of shrines were dug out, objects of worship were destroyed, anything flammable was burned, and finally a cross was often built over the site.[14] An unintended consequence of this campaign of destruction was that the names and locations of huacas were recorded so that they could be examined in the future to make certain no religious activity continued. Some of the huacas, such as large carved rocks, could not be eradicated and remain to this day.

Cobo published a comprehensive list and description of all 328 huacas of the Cusco ceque system.[15] Cobo was not the original author, however, and he presumably made use of a much older document, perhaps written by Juan Polo de Ondegardo.[16] Cobo mentions that the Coricancha, the primary sun temple, was counted as an extra huaca. John Rowe used Cobo's description to identify huacas by their sequence within the ceque that they lie upon. He classifies ceques and huacas in the order in which they were presented by Cobo. Kenko Grande, for instance, is the second huaca on Chinchaysuyu's first ceque and, as such, is designated Ch. 1:2. The Tired Stone at Sacsahuaman is referred to by Ch. 4:6. Lacco is An. 3:6.[17]

---

[13] Brian Bauer, *The Sacred Landscape of the Inca: The Cusco Ceque System*, (Austin: University of Texas Press, 1998), p. 8.
[14] Pablo Joseph Arriaga, *The Extirpation of Idolatry in Peru*, trans. L. Clark Keating (1621; repr. Lexington: University of Kentucky Press, 1968); Bauer, *Sacred Landscape*.
[15] Cobo, *Inca Religions*, pp. 51–84.
[16] Brian S. Bauer, 'Ritual Pathways of the Inca: An Analysis of the Collasuyu Ceques in Cuzco', *Latin American Antiquity* 3, no. 3 (1992): pp. 183–205; R. Tom Zuidema, 'Ceque System of Cuzco: A Yearly Calendar-Almanac in Space and Time'. In *Handbook of Archaeoastronomy and Ethnoastronomy*, ed. Clive L. N. Ruggles (New York: Springer, 2015), pp. 851–863.
[17] Bauer, *Sacred Landscape*; Cobo, *Inca Religions*; Rowe, *Inca Policies*.

Huacas were part of the officially organized worship of the Inca capital, and Cobo's account provides insight into what things were considered sacred. Sacrifices and offerings made at the shrines including such items as gold, silver, clothing, sea shells, and sheep. Extraordinary offerings involved the sacrifice of children, human and animal figurines, coca, llama blood, and firewood dressed as people.[18] Winding or zigzagged channels were carved into certain huacas, perhaps for the flow of chicha, a maize beer, or blood, either of which may have been used for divination.

Certain shrines were dedicated to the sun and exhibit light and shadow effects highlighting times such as solstices and equinoxes. These huacas express explicit orientations to sunrise or sunset at specific times of the year. Features provide vantage points for observing astronomical phenomena, for guiding the eye to solar horizon events or the rising of the Pleiades, and marking approximate dates by shadow or the casting of light.

Zuidema states that ceques marked the route of sequential offerings to adjacent huacas.[19] Some of the ceques may also have served as pathways and convenient routes of movement. Instead of simple trails for ritual-pilgrimage, ceques may have established administrative socio-political boundaries and may have at times followed existing roads.

Ceques have been interpreted both as social and administrative delineators and as having symbolic geometrical and astronomical orientations. Both appear to have merit. Certain ceques may have had astronomical elements, while others were used to organize and indicate the responsibilities for the care and maintenance of huacas provided by individual panacas and ayllus.[20]

Polo de Ondegardo and later Cobo related that every village had ceques.[21] While the ceque system of Cusco is the only one to have been extensively substantiated, the preponderance of huacas surrounding Machu Picchu and certain alignments suggest there may have been a ceque organization employed in this area as well. One such set of alignments connects the Sacred Plaza of Machu Picchu with the Sun Temple of Llactapata along the axis of June solstice sunrise and December solstice sunset.

3. Panacas and Ayllus

Kinship units from the city of Cusco cared for huacas on their assigned ceques and saw to the offering of established sacrifices at the proper times. The huacas surrounding Cusco were arranged on ceques which also helped to divide the region into the four suyus emanating from the city. Zuidema found the ceque system to be a significant component of social classification. He also proposed its usefulness for calendrical purposes that included the accomplishment of certain tasks at specific times throughout the year.[22]

The ceques were arranged in four groups corresponding with Cusco's four administrative quarters.

Within the suyus they were ordered primarily in groups of three with the names Collana, Payan, and Cayao assigned, one to each in a triad of ceques. The ceques of Chinchaysuyu were ordered counter-clockwise while all the others were clock-wise in progression.[23]

Collana, Payan, and Cayao refer to panaca and ayllu status. Zuidema stated that Collana was the most

---

[18] Cobo, *Inca Religions*, pp. 109–114.
[19] Zuidema, *Ceque System*.
[20] R. Tom Zuidema, 'The Astronomical Significance of Ritual Movements in the Calendar of Cuzco', in *Pre-Columbian Landscapes of Creation and Origin*, ed. John Edward Staller (New York: Springer, 2008), pp. 249–268.
[21] Juan Polo de Ondegardo, *A report on the basic principles explaining the serious harm which follows when the traditional rights of the Indians are not respected*, trans. A. Brunel, John V. Murra, and Sidney Muirden. 1571; repr. New Haven: Human Relations Area Files, 1965), p. 67; Cobo, *Inca Religion*.
[22] Zuidema, *Ceque System*.
[23] Zuidema, *Ceque System*.

prestigious and was made up of members of pure Inca descent.[24] Payan included subsidiary kin such as offspring born from a Collana man with a non-Collana woman. Cayao comprised the remainder of the population that were not of Inca descent.

A panaca was a patrilineal royal descent group established by a newly installed Inca. The panaca supported the emperor and took on increased importance upon the death of its patriarch as they then became responsible for the care and ceremonial functions of his mummy. No more than one panaca was assigned to each ceque cluster.[25]

An ayllu was a non-royal extended kinship group from the same patrilineal ancestor. It was the basic social unit beyond immediate relatives and provided structure for the regulation of marriage and inheritance.

Huacas required many prayers and offerings and the responsibility for such was given to specific panacas and ayllus. Panacas were assigned for the care and conduct of rituals at huacas on some ceques while ayllus were designated for others. Every ceque had a kin group assigned to manage its affairs.[26] By design, huacas were self-sufficient with large tracts of land and animals to fulfil the needs of their attendants. These caretakers belonged to the panacas and ayllus of Cusco and many thousands were employed and supported in this manner.[27] This system served as the state's method of allocating territory and irrigation sources.[28]

4. Ceque Astronomy

Inca rulers established a solar cult, which helped to facilitate some huacas on certain ceques exhibiting astronomical characteristics.[29] These often are orientations for the June solstice or December solstice sunrises. Sometimes light tubes or cave openings allow altars to be illuminated at specific times and in other cases there are orientations to guide the eye to the horizon on solar-significant dates. The Incas also placed pillars on the hills surrounding Cusco to calendrically mark the passage of the sun on the horizon.[30]

Zuidema described the ceque system as a means for counting days of the year with each of the 328 huacas used in order to mark the daily passage of time.[31] Ceques were grouped by threes and each of these groups represented one month. He suggested that the remaining 37 days in a 365 day tropical solar year equated to the approximate period the Pleiades are invisible by proximity to the sun between 3/4 May and 8/9 June. This calendar breaks into 12 sidereal lunar months of 27 1/3 days and the 41 ceques

---

[24] R. Tom Zuidema, 'Hierarchy and Space in Incaic Social Organization', *Ethnohistory* 30, no. 2 (1983).

[25] Bauer, *Sacred Landscape*; Zuidema, *Ceque System*.

[26] Zuidema, *Ceque System*.

[27] Brian Bauer and Charles Stanish, *Ritual and Pilgrimage in the Ancient Andes* (Austin: University of Texas Press, 2001).

[28] Susan Niles, *Callachaca: Style and Status in an Incan Community* (Iowa City: University of Iowa Press, 1987).

[29] David S. P. Dearborn and Brian S. Bauer, 'Inca Astronomy and Calendrics', in *Handbook of Archaeoastronomy and Ethnoastronomy*, ed. Clive L. N. Ruggles (New York: Springer, 2015): pp. 831–838.

[30] R. Tom Zuidema, 'The Astronomical Significance of a Procession, a Pilgrimage and a Race in the Calendar of Cuzco', in *Current Studies in Archaeoastronomy: Conversations Across Time and Space*, ed. John W. Fountain and Rolf M. Sinclair (Durham, NC: Carolina Academic Press, 2005): pp. 353–367; J. McKim Malville, Mike Zawaski, and Steven Gullberg, 'Cosmological Motifs of Peruvian Huacas', in *Astronomy and Cosmology in Folk Traditions and Cultural Heritage*, *Archaeologica Baltica 10*, ed. Jonas Vaiskunas (Vilnius: Klaipeda University Press, 2008), pp. 175–182.

[31] R. Tom Zuidema, 'Catachillay: The Role of the Pleiades and of the Southern Cross and α and β Centauri in the Calendar of the Incas', in *Ethnoastronomy and Archaeoastronomy in the American Tropic*, ed. Anthony Aveni and Gary Urton (New York: New York Academy of Sciences, 1982), pp. 203–229.

may then have represented 41 weeks of eight days each.[32]

Zuidema's ceque system thus became a walk-through ritual calendar, with each huaca honoured and worshipped on its own specific day of the year. The exact number of huacas, 328, is important in this system as it represents the lunar sidereal year, 12 x 27.3, the number of days in a sidereal lunar month, the remaining 37 days between this calendar and the tropical year accounted for by the approximately 37 days that the Pleiades are not visible. The Incas had no agricultural rituals during this 37-day period.

The astronomical and calendrical basis of ceques has been rigorously argued by Zuidema and Aveni, but Bauer and Dearborn state that they find less evidence for ceque celestial sightlines.[33] Bauer and Dearborn point out that the period of disappearance for the Pleiades can vary from year to year depending on the moon's phase, the horizon altitude, and atmospheric conditions.[34] They state that 'For the Pleiades, a disappearance from the night sky lasting forty to fifty days is quite reasonable'.[35] Even if the specific reason for the 37-day gap remains elusive, the ceque calendar as described would have been a most efficient tool for the Incas to use in the management of Cusco and their empire.

Niles, using descriptions by Cobo, place names, and Inca roads, found the ceques not to be straight and instead that they followed paths with changing courses (see Fig. 23.2).[36] Bauer's field research showed them to be indirect as well. Zuidema counters that ceques were not intended to be straight, but instead related huacas may have straddled a conceptually straight sight-line. Bauer continues that 'it is the specific locations of the huacas that define the course of the lines and not vice versa'. And that the considerable variance left or right was great enough to have made it unlikely that ceques could have formed sightlines to guide the eye to celestial horizon events.

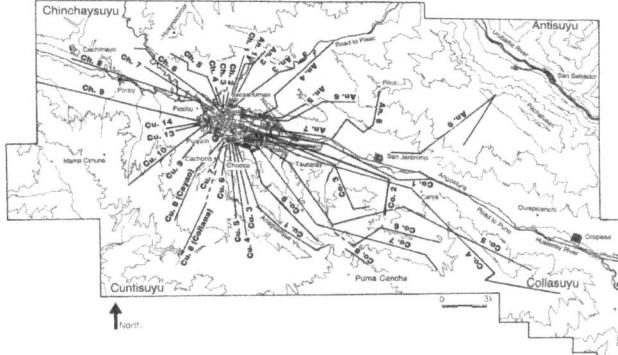

Fig. 23.2: Ceques of the four suyus of Cusco. The sun temple, Coricancha, is at the centre.[37]

Ziółkowski states that the 328-day/night cycle was never chronicled and he asserts that Cobo related there to be 350 huacas.[38] Ziółkowski also takes issue with the missing 37 days between the 328 days of

---

[32] Brian Bauer and David Dearborn, *Astronomy and Empire in the Andes: The Cultural Origins of Inca Sky Watching* (Austin: University of Texas Press, 1995), pp. 64–65; Zuidema, 'Catachillay'.
[33] Bauer and Dearborn, *Astronomy and Empire*, pp. 64–65; R. Tom Zuidema, 'The Inca Calendar', in *Native American Astronomy*, ed. Anthony Aveni (Austin, University of Texas Press, 1977); Zuidema, 'Catachillay'.
[34] Bauer and Dearborn, *Astronomy and Empire*, p. 65.
[35] Bauer and Dearborn, *Astronomy and Empire*, p. 130.
[36] Niles, *Callachaca*, pp. 177–179.
[37] Bauer, *Sacred Landscape*, p. 158.
[38] Mariusz Ziółkowski, 'Inca Calendar', in *Handbook of Archaeoastronomy and Ethnoastronomy*, ed. Clive L. N. Ruggles

Zuidema's ceque calendar and the full 365 day year.[39] He, as well as Bauer, Dearborn and Niles, have identified concerns of sufficient significance that cast doubt as to the use of the Cusco ceques for such a calendrical system. While certain assertions of Zuidema can have merit, the compelling arguments of others regarding the intended purposes for Cusco's ceques have called the calendrical concept into question, and due to their valid concerns have made the subject a matter of dispute.

5. Discussion of Research Data

Certain huacas exhibit astronomical alignments, frequently orientations for the rising sun at the time of the solstices in June or December. Light tubes and cave openings were found to allow altars to be illuminated at specific times (see Figs. 23.3 and 23.4) and intentional orientations guide the eye to the horizon on solar-significant dates. Additionally, solar pillars were constructed on hills to calendrically mark positions of the sun on the horizon. A symbolic staircase was found to illuminate ceremoniously and an animistic character was observed, having been created with light and shadow at the time of the June solstice. These are among many examples that demonstrate the Inca's astronomical passion. (Other huacas include topographical features such as hills, plains and flat places, trees, roads, fields, passes, and ravines).

Fig. 23.3: The opening in the northeast cave of An 3:4 (Lacco) is aligned for the June solstice sunrise.

(New York: Springer, 2015), pp. 839–850.

[39] Mariusz Ziólkowski, 'Knots and oddities: The quipu-calendar or supposed luni-sidereal calendar', in *Time and Calendars in the Inca Empire*, ed. Mariusz Ziólkowski and Robert M. Sadowski (Oxford: BAR International Series 479, British Archaeological Reports, 1989), pp. 197–208.

Fig: 23.4: The altar in the inner chamber of the southeast cave of An. 3:4 (Lacco) is illuminated through a light-tube at the time of the zenith sun.

	Astronomy	Structure	Cave	Spring	Other Water	Carved Rock	Other Stone	Steps/Seats	Other Huacas
Ch.	9	17	5	16	10	7	21	5	29
An.	7	6	5	29	6	6	26	4	23
N Tot	16	23	10	45	16	13	47	9	52
Co.		3		21	1	1	22	1	42
Cu.	2	5	2	22	1	2	28	1	33
S Tot	2	8	2	43	2	3	50	2	75
Total	18	31	12	88	18	16	97	11	127

Table 23.1: Astronomical and cosmological characteristics of huacas on ceques by suyu.

Table 23.1 lists counts of certain features found in huacas on ceques in each of the four suyus of Cusco. The rows are labelled Ch. (Chinchasuyu), An. (Antisuyu), N Tot (northern suyu subtotal), Co. (Collasuyu), Cu. (Cuntisuyu), S Tot (southern suyu subtotal), and Total (overall sum of each feature type). The feature categories are Astronomy (huacas with astronomical orientations), Structure (huaca structures with alignments), Cave (huacas with caves), Spring (huacas with springs), Other Water (huacas associated with water other than springs), Carved Rock (huacas with rock carvings), Other Stone (huacas with stone that is not carved), Steps/Seats (huacas with carved steps and seats), Other Huacas (huacas that have no features listed in any of the categories).

The table begins with astronomical orientations and structures with alignments. Caves have a significant place in Inca cosmology and sometimes are illuminated by solar orientations or light-tubes. Springs and water in general were felt to make huacas animate. It was thought that carving, including symbolic steps and seats, caused rock to become a hierophany to be worshipped. Non-carved rock is a common feature in other huacas.

The chart is divided into two groups, that of the two northern suyus (Chinchasuyu and Antisuyu) and also the two southern suyus (Collisuyu and Cuntisuyu). Since many huacas were destroyed during the Spanish extirpation, the data is derived both from field study and descriptions taken from Cobo's account of the 328 huacas on the 41 ceques. The counts represent features and not individual huacas since some of these shrines contain more than one.

As shown in Table 23.1, the preponderance of many of the features were found to the north of Cusco. All categories are related to Incan astronomy or cosmology with the exception of Other Stone and Other Huacas. Springs and Other Stone are distributed somewhat equally along the northern and southern ceques, but the astronomical and cosmological features (in addition to springs) were found more often in Chinchaysuyu and Antisuyu. Huacas without features related to any of these categories were counted in the Other Huacas column and they occurred more frequently to the south.

Of the nine ceques in Chinchasuyu, five of them each tally 10 or more of these features on their huacas. Ch. 1 has the most with 15, three of them categorized as astronomical, two with structure sightlines, and one with light and shadow effects. The rest of the five are Ch. 3, Ch. 4, Ch. 5, and Ch. 8. Adjacent to Ch. 1 is An. 1 (see Fig. 23.2), and Antisuyu has four ceques that meet or exceed a count of ten of the features. Among its 21 listed features, the huacas of An. 1 account for three that are astronomically related and three that are structural alignments. The other three in Antisuyu with 10 or more are An. 3, An. 5, and An. 8. There is an average of six features on the remainder of the ceques in each of the two northern suyus.

In the southern half of the ceque system only two ceques were found to have ten or more features, and they are also adjacent, Cu. 1 with 12 and Co. 9 with 11 (see Fig. 23.2). And while there are a total of ten features related to astronomy and sightlines in Collasuyu and Cuntisuyu, none of them were found on these two ceques. In contrast, there are 39 of these features in Chinchaysuyu and Antisuyu. The rest of the ceques in the south were found to average four features each with the large majority of southern features being springs and stones. There were also a large number of ceques there that exhibit no qualifying characteristics. Most features with astronomical association of any kind were found on huacas in Chinchasuyu. This was followed next by Antisuyu, with far fewer in Cuntisuyu and finally Collasuyu.

One possible factor regarding greater activity to the north was the large numbers of rock outcrops suitable for carving that were available. Secondary palaces built by the emperors Pachacuti, Topa Inca, and Huayna Capac were north and northwest of the city as well. Significant huacas such as Ch. 1:2 Patallacta (Kenko Grande) and An. 3:4 Chuquimarca (Lacco) are both north of and also close to Cusco. They each were extensively developed and had sightlines to the city below. Each was a ceremonial centre. Such purpose could give reason for the development of many of the astronomical and other features found in these shrines. The region north of Cusco exhibits a great astronomical and cosmological interest, even in huacas that are not located on the established ceques. One such example is that of a small, rather obscure huaca located between Ch. 1:2 Kenko Grande and An. 3:4 Lacco that has a large circle and a small circle carved into its rock in such a way that they exhibit orientations for six solar horizon events (see Fig. 23.5).

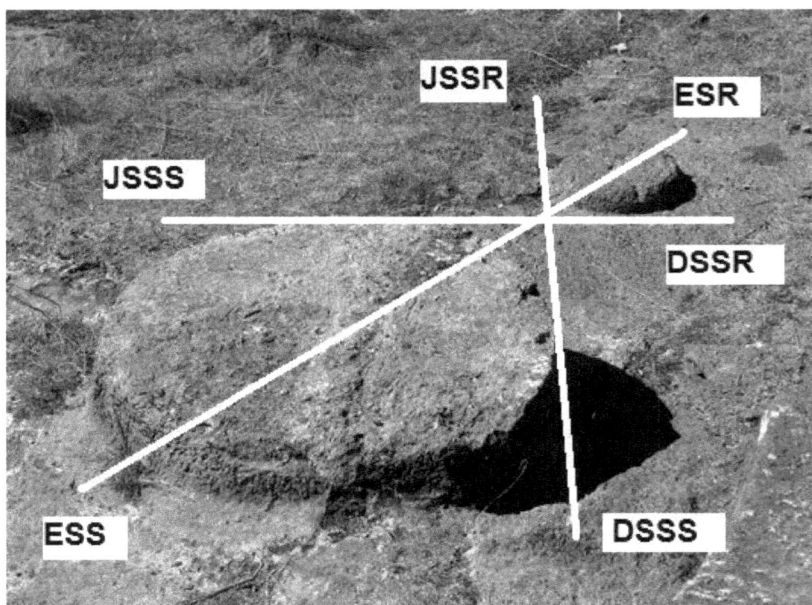

Fig. 23.5: Tangential lines at a small un-numbered huaca between Ch. 1:2 (Kenko Grande) and An. 3:4 (Lacco) guide the eye to the horizon for sunrises and sunsets at solstice and equinox.

The ceque system of Cusco served to organize Inca society and was complex with many meanings and uses, including those related to astronomy. Members of Cusco's panacas and ayllus cared for and worshipped huacas on their designated ceques. The ceques generally did not cross, but some contained dog legs. The reason for the dog legs remains elusive, but one suggestion is that they were possibly related to an order of worship. Ceques might also have served as pathways for worship with persons sometimes walking from one huaca to the next.

Ceques may have existed in other parts of the empire as well, and one such example has emerged in the area surrounding Machu Picchu. The rediscovery of Llactapata there has brought attention to sightlines that connect certain shrines. One such example is that of the June solstice sunrise – December solstice sunset axis that crosses Machu Picchu's Sacred Plaza, the River Intihuatana deep in the gorge below, and on to Llactapata's Sun Temple.[40]

The Cusco ceques were not in themselves astronomical, but many of their component huacas were, particularly to the north of the city. Ceques and astronomy were each important parts of Inca culture and thus it is reasonable to expect that we would find examples of them both together.

---

[40] Polo de Ondegardo, *A Report on the Basic Principles*; J. McKim Malville, Hugh Thomson, and Gary Ziegler, 'El Observatorio de Machu Picchu: Redecubrimiento de Llactapata y su templo solar'. *Revista Andina* 39 (2004): 9–40; J. McKim Malville, Hugh Thomson, and Gary Ziegler, 'The Sun Temple of Llactapata and the Ceremonial Neighborhood of Machu Picchu', in *Viewing the Sky Through Past and Present Cultures*, ed. Todd W. Bostwick and Bryan Bates (Phoenix: City of Phoenix Parks, Recreation and Library, 2006), pp. 327–339; Gary Ziegler and J. McKim Malville, *Machu Picchu's Sacred Sisters: Choquequiro & Llactapata* (Boulder: Johnson Books, 2013).

Bibliography

Arriaga, Pablo Joseph. *The Extirpation of Idolatry in Peru*. 1621. Translated by L. Clark Keating. Reprinted by Lexington: University of Kentucky Press, 1968.

Bauer, Brian S. 'Ritual Pathways of the Inca: An analysis of the Collasuyu Ceques in Cuzco'. *Latin American Antiquity* 3, no. 3 (1992): pp. 183–205.

Bauer, Brian S. *The Sacred Landscape of the Inca: The Cusco Ceque System*. Austin: University of Texas Press, 1998.

Bauer, Brian S., and David Dearborn. *Astronomy and Empire in the Andes: The Cultural Origins of Inca Sky Watching*. Austin: University of Texas Press, 1995.

Bauer, Brian S., and Charles Stanish. *Ritual and Pilgrimage in the Ancient Andes*. Austin: University of Texas Press, 2001.

Bray, Tamara L. 'An Archaeological Perspective on the Andean Concept of Camaquen: Thinking Through Late Pre-Columbian Ofrendas and Huacas'. *Cambridge Archaeological Journal* 19, no. 3 (2009): pp. 357–366.

Cobo, Bernabe. *Inca Religion and Customs*. 1653. Translated by Roland Hamilton. Reprinted by Austin: University of Texas Press, 1990.

Dearborn, David S. P., and Brian S. Bauer. 'Inca Astronomy and Calendrics'. In *Handbook of Archaeoastronomy and Ethnoastronomy*. Edited by Clive L. N. Ruggles, pp. 831-838. New York: Springer, 2015.

Magli, Giulio. 'On the astronomical content of the sacred landscape of Cusco in Inka times'. *Nexus Network Journal* 7, no. 1 (2004): pp. 22–2.

Malville, J. McKim. 'Animating the Inanimate: Camay and Astronomical Huacas of Peru'. In *Cosmology Across Cultures, ASP Conference Series, 409*. Edited by Jose Alberto Rubino-Martin, Juan Antonio Belmonte, Francisco Prada, and Anexton Alberdi, pp. 261–266. San Francisco: Astronomical Society of the Pacific, 2009.

Malville, J. McKim, Hugh Thomson, and Gary Ziegler. 'El Observatorio de Machu Picchu: Redecubrimiento de Llactapata y su templo solar'. *Revista Andina 39* (2004): pp. 9–40.

Malville, J. McKim., Hugh Thomson, and Gary Ziegler. 'The Sun Temple of Llactapata and the Ceremonial Neighborhood of Machu Picchu'. In *Viewing the Sky Through Past and Present Cultures*. Edited by Todd W. Bostwick and Bryan Bates, pp. 327–339. Phoenix: City of Phoenix Parks, Recreation and Library, 2006.

Malville, J. McKim, Mike Zawaski, and Steven Gullberg. 'Cosmological Motifs of Peruvian Huacas'. In *Astronomy and Cosmology in Folk Traditions and Cultural Heritage, Archaeologica Baltica 10*. Edited by Jonas Vaiskunas, pp. 175–182. Vilnius: Klaipeda University Press, 2008.

Niles, Susan. *Callachaca: Style and Status in an Inca Community*. Iowa City: University of Iowa Press, 1987.

Polo de Ondegardo, Juan. *A report on the basic principles explaining the serious harm which follows when the traditional rights of the Indians are not respected*. Translated by A. Brunel, John V. Murra and Sidney Muirden. 1571. Reprinted New Haven: Human Relations Area Files, 1965.

Rowe, John. 'Inca Policies and Institutions Relating to the Cultural Unification of the Empire'. In *The Inca and Aztec States, 1400-1800*. Edited by George A. Collier, Renato I. Rosaldo, and John D. Wirth, pp. 93–118. New York: Academic Press, 1980.

Salomon, Frank, and George L. Urioste. *Introductory Essay in The Huarochiri Manuscript: A Testament of Ancient and Colonial Andean Religion*. Austin: University of Texas Press, 1991.

Urton, Gary. *At the Crossroads of Earth and Sky: An Andean Cosmology*. Austin: University of Texas Press, 1981.

Van de Guchte, Maartin J. D. 'Carving the World: Inca Monumental Sculpture and Landscape'. PhD dissertation, University of Illinois, 1990.

Ziegler, Gary, and J. McKim Malville. *Machu Picchu's Sacred Sisters: Choquequiro & Llactapata*. Boulder: Johnson Books, 2013.

Ziólkowski, Mariusz. 'Knots and oddities: The quipu-calendar or supposed luni-sidereal calendar'. In *Time and Calendars in the Inca Empire*. Edited by Mariusz Ziólkowski and Robert M. Sadowski, pp. 197–208. Oxford: BAR International Series 479, British Archaeological Reports, 1989.

Ziólkowski, Mariusz. 'Inca Calendar'. In *Handbook of Archaeoastronomy and Ethnoastronomy* Edited by Clive L. N. Ruggles, pp. 839–850. New York: Springer, 2015.

Zuidema, R. Tom. *The Ceque System of Cuzco: The Social Organization of the Capital of the Inca*. Leiden: E. J. Brill, 1964.

Zuidema, R. Tom. 'The Inca Calendar'. In *Native American Astronomy*. Edited by Anthony Aveni, pp. 219–259. Austin: University of Texas Press, 1977.

Zuidema, R. Tom. 'Catachillay: The Role of the Pleiades and of the Southern Cross and α and β Centauri in the

Calendar of the Incas'. In *Ethnoastronomy and Archaeoastronomy in the American Tropics*. Edited by Anthony Aveni and Gary Urton, pp. 203–229. New York: New York Academy of Sciences, 1982.

Zuidema, R. Tom. 'Heirarchy and Space in Incaic Social Organization'. *Ethnohistory* 30, no. 2 (1983): pp. 49–75.

Zuidema, R. Tom. 'The Astronomical Significance of a Procession, a Pilgrimage and a Race in the Calendar of Cuzco'. In *Current Studies in Archaeoastronomy: Conversations Across Time and Space*. Edited by John W. Fountain and Rolf M. Sinclair, pp. 353–367. Durham, NC: Carolina Academic Press, 2005.

Zuidema, R. Tom. 'The Astronomical Significance of Ritual Movements in the Calendar of Cuzco'. In *Pre-Columbian Landscapes of Creation and Origin*. Edited by John Edward Staller, pp. 249–268. New York: Springer, 2008.

Zuidema, R. Tom. 'Ceque System of Cuzco: A Yearly Calendar-Almanac in Space and Time'. In *Handbook of Archaeoastronomy and Ethnoastronomy*. Edited by Clive L. N. Ruggles, pp. 851–863. New York: Springer, 2015.

THE TEMPLE OF THE INSCRIPTIONS IN THE SPIRITUAL LANDSCAPE AT PALENQUE

Stanisław Iwaniszewski

ABSTRACT: During the Late Classic period, Maya rulers were memorialized through monumental architecture built within cities and placed in accordance to local spiritual topography. Pyramids, shrines and ceremonial plazas were named to record local mythical histories and deified founders of royal lineages. The ritual landscape at Palenque combines sites associated with mythological beliefs with ceremonial locations linked to the seasonal procurement round.

The Temple of the Inscriptions was designed to represent the Underworld and to become K'inich Janaab' Pakal I's (615–683 CE) own memorial monument. His dead body was laid to rest in the burial chamber and covered with a huge sarcophagus lid that shows the ruler emerging from the Underworld. Pakal's passage into the afterlife is framed by the skyband and supervised by his ancestors and former rulers. Following the mythical path, his soul (or its animating entity) was supposed to enter the realm of the sun, the celestial realm, occupied by the exalted dead, his deified ancestors. The ascent of Pakal's soul (or its animating entity *k'in* – 'sunny', 'heated' – implied by the use of the k'inich title) was secured through the construction of a special rectangular conduit, commonly referred to as the 'psychoduct' connecting the burial chamber with the outside world. The long hieroglyphic inscription that covers the panels of the Temple of the Inscriptions established the complex web of different relationships existing between the Palenque's mythological patronal gods – who controlled time and space, meteorological events and celestial bodies – and Pakal himself, his ancestors and other of Palenque's former rulers.

The Temple of the Inscriptions (TI) was completed by Pakal's firstborn son Kan Balam II (684–702) around 690 CE when he was already finishing the construction of his own memorial monument, the Cross Group. This group was one of the most important ritual spaces at Palenque, hosting ceremonies linking the position of the divine ruler with the patron gods of Palenque. Due to its location on an elevated plaza, the erection of this architectural complex created new relationships between the sacred places and monuments in Palenque. To recreate the dynamic relationship between materialized mythical places associated with the spiritual landscape, the paper explores astronomical alignments and visual connections between the TI and the structures built at the end of the 8th century. The paper also evaluates recent research made by Klokočnik and Kokolecký.

Introduction

Palenque is best known for its Temple of the Inscriptions; a huge stepped pyramid built to house the tomb of K'inich Janaab Pakal I (603–683 CE), the greatest ruler of the Classic Maya kingdom of B'aakal (see Fig. 24.1). Both the tomb and the pyramid have long drawn scholarly attention, producing rich archaeological evidence, which in combination with intensive epigraphic, iconographic, architectural and environmental research, provide an effective framework for discussing the results of the ongoing archaeoastronomical investigation. In a recent paper in *Anthropological Notebooks* (*Proceedings of the 20th SEAC Conference*), Jaroslaw Klokočnik and Jan Kostelecký proposed an astronomical rationale for the orientation of Pakal's tomb, suggesting that its long axis was perpendicular to the position of the midsummer setting sun.[1]

---

[1] Jaroslav Klokočnik and Jan Kostelecký, 'Palenque: Astronomical-solar orientation of Pakal's tomb' in *Anthropological Notebooks, Proceedings of the 20th SEAC Conference, Ancient Cosmologies and Modern Prophets*, ed. Ivan Šprajc and Peter Pehani (Ljubljana: Slovene Anthropological Society, 2013), Vol. 19, supplement, pp. 305–317.

Fig. 24.1: The Temple of the Inscriptions, Structures XIII ('Red Queen') and XII.

Following a complete archaeoastronomical study of Palenque, I am now in a position to examine or re-examine some of their conclusions as well as to provide new data and interpretations regarding solar alignments of the Temple of the Inscriptions.

In the perspective that will be advocated in the present paper, astronomical alignments should be elicited within the context in which they once functioned. My point is that meanings cannot emerge just from astronomical alignments themselves. Instead, they should derive from complementary archaeological, architectural, epigraphic, historical, etc. evidence. Unfortunately, Klokočnik and Kostelecký's study does not provide any relevant evidence showing the reader some possibilities and constraints produced by local situations. To solve some practical problems it may be convenient to discuss astronomical and cultural issues separately, but the ultimate goal of our discipline is to study astronomical objects and phenomena as perceived, used and interpreted by local cultures. Therefore, the study of Pakal's tomb not only reveals certain patterns in burial orientations, but it also can produce a better understanding of various religious, ideological, political and social processes that mediated between the specific design of this funerary monument and 7th-century Palencan society.

I take this inquiry into the Pakal's tomb astronomical alignments as a starting point for further discussion of its possible meaning and significance. On astronomical and topographical grounds, Klokočnik and Kostelecký's proposed that the direction of the midsummer sunset could have motivated Palencan architects to orient the tomb perpendicularly to it.[2] This conclusion, however, disagrees with traditional scholarship[3] that associates the funerary symbolism of Pakal's tomb with the midwinter

---

[2] Klokočnik and Kostelecký, 'Palenque', pp. 314–316.
[3] Consult John B. Carlson, 'Astronomical Investigations and Site Orientation Influences at Palenque', in *The Art, Iconography & Dynastic History of Palenque, Part III. The proceedings of the Segunda Mesa Redonda de Palenque*, ed. Merle Greene Robertson (Pebble Beach CA: Pre-Columbian Art Research, The Robert Louis Stevenson School, 1976), pp. 107–122; Linda Schele, 'Palenque: The House of the Dying Sun', in *Native American Astronomy*, ed. Anthony F. Aveni (Austin: University of Texas Press, 1977), pp. 42–56; Linda Schele, 'Sacred Site and World-View at Palenque', in *Mesoamerican Sites and World-Views*, ed. Elizabeth P. Benson (Washington DC: Dumbarton Oaks Research Library and Collections, 1981), pp. 87–117.

sunset. Alternatively, the orientation of Pakal's tomb may be seen as being consistent with the local pattern of orientating the heads of the deceased to the north.[4] To produce sound interpretations of astronomical phenomena, we need to employ multiple sources of evidence.

The Problem

The authors used a magnetic compass and did a series of measurements of three of the sides of Pakal's sarcophagus, obtaining azimuths with a precision of about 2°, and determined that the long axis of the sarcophagus was oriented 24°±2° east of north. They hypothesized that a line which is perpendicular to this axis (i.e., with azimuths of 114°±2° in the east and 294°±2° in the west) roughly matched the positions of the sun at the solstices. The presence of the obscuring hill in the west, which impedes any precise observation of the midsummer sunset, suggests that this alignment was somewhat extrapolated. (Observe that at Palenque, the azimuth of 114.8°/294.8° and the true horizon produce a declination of ±23°.6, the value that is close to the solstices in 700 CE.)

At first glimpse, the idea that the orientation of Pakal's tomb depended on the observations of the sun at solstices is hardly controversial. Maya rulership was symbolically connected to the sun and the image of the rising sun represented the concepts of rebirth or resurrection. The image of the rising sun became a metaphor for the accession of a ruler, while certain animate entities (souls or spirits) representing dead rulers were believed to be transformed into celestial beings or even fused with the Sun God.[5] I do not cast doubt on the general significance of solar symbolism, but I would like to raise questions about the importance ascribed to the solstices.

Having determined that Pakal's sarcophagus was oriented at 24°±2° and that this orientation was hardly accidental and astronomically motivated, both authors found some supporting evidence in descriptions of solar hierophanies at Palenque. These hierophanies, in turn, increased their confidence in the solar solstitial alignment hypothesis.[6]

Solstitial hierophanies and alignments at Palenque are numerous. Numerous authors have sufficiently analysed them and there is no reason to discuss them all here.[7] Some buildings, such as Temple XIV

---

[4] Alberto Ruz Lhuillier, *El Templo de las Inscripciones, Palenque* (Colección Científica, 7) (México, DF: Instituto Nacional de Antropología e Historia, 1973); Alberto Ruz Lhuillier, *Costumbres funerarias de los antiguos mayas* (1968; México, DF: Universidad Nacional Autónoma de México, 1991); W. Bruce Welsh, *An Analysis of Classic Lowland Maya Burials* (Oxford: Archaeopress, BAR International Series 409, 1988).

[5] The Maya did not have one single concept of soul or spirit; instead, they believed that the spiritual energy was absorbed by the physical body in form of various soul-related entities (animate entities) responsible for the character, intelligence or material attributes of an individual. The head was the seat of one such entity. See Erik Velásquez García, 'Los vasos de la entidad política de 'Ík': una aproximación histórico-artística. Estudio sobre las entidades anímicas y el lenguaje gestual y corporal en el arte maya clásico' (PhD diss., Facultad de Filosofía y Letras, Universidad Nacional Autónoma de México, 2009).

[6] Klokočnik and Kostelecký, 'Palenque'.

[7] See Carlson, 'Astronomical Investigations and Site Orientation Influences at Palenque'; Horst Hartung, 'El espacio exterior en el centro ceremonial de Palenque', in *The Art, Iconography & Dynastic History of Palenque, Part III. The proceedings of the Segunda Mesa Redonda de Palenque*, ed. Merle Greene Robertson (Pebble Beach CA: Pre-Columbian Art Research, The Robert Louis Stevenson School, 1976), pp. 123–135; Schele, 'The House'; Schele, 'Sacred Site'; Anthony F. Aveni and Horst Hartung, 'Some Suggestions About the Arrangement of Buildings at Palenque', in *Tercera Mesa Redonda de Palenque*, ed. Merle Greene Robertson and Donnan Call Jeffers (Monterey CA:Pre-Columbian Art Research Center, Herald Printers, 1979), pp. 173–177; Neal S. Anderson, Moises Morales, and Alfonso Morales, 'A Solar Alignment of the Palace Tower at Palenque', *Archaeoastronomy, The Bulletin of the Center for Archaeoastronomy* 4, no. 3 (1981): pp. 34–36; Neal S. Anderson and Moises Morales, 'Solstitial Alignments of the Temple of the Inscriptions at Palenque', *Archaeoastronomy, The Bulletin of the Center for Archaeoastronomy* 4, no. 3, (1981): pp. 30–33: Alfonso Mendez,

and the Temple of the Sun, roughly face midwinter solstice, while the Temple of the Inscriptions is fully illuminated shortly after the midsummer sunrise. The most important among them is, however, the midwinter sunset observed above the tree-covered hill that stands behind the Temple of the Inscriptions.[8] This phenomenon can be observed from various locations within the Palace and the Temple of the Cross and during the days preceding or following the solstice. Obviously, this kind of solar hierophany cannot be effectively used to exactly mark the solstice, nonetheless, it could have provided a dramatic spectacle for Maya elite members placed in numerous locations within the Palace area or around the Temple of the Cross, showing that on that day the sun really entered the Underworld through Pakal's tomb.

Some Basic Facts
Palenque is a Classic Maya city located in the northern foothills of the Chiapas highlands. It became the most important political centre in the northwestern Maya region and had an estimated population of about 6,000 inhabitants at its population peak in the Late Classic (600–800 CE).[9] Palenque's urban core is on a narrow strip of a tableland that is about 3m long and 0.5km wide, and between 100 to 200m above the nearby flat swamp and savannah plains which extend further north towards the Usumacinta and Grijalva Rivers and the Gulf of Mexico. The area has a hot and wet tropical rainforest climate (Af) with most rain falling from May to November.[10] With regard to seasonal climatic changes the key months are: April (the driest month), May (the warmest month), September (the wettest month), and January (the coldest month). Current climatic patterns are similar to those of the Late Classic period (600–850 CE) when most of city buildings were erected.[11]

Ancient Palenque rulers carefully designed an intensive agricultural system to support its population. Remains of irrigation canals and agricultural terracing were identified by Liendo Stuardo just below Palenque's plateau and to the east and west of the city.[12] However, the plain below the urban core is considered as being not very appropriate for agriculture (low quality soils are here combined with the poor drainage system and seasonal floods).[13]

The city is located within a spectacular landscape. Several dozens of perennial springs and creeks (*arroyos*) run across the city from north to south causing seasonal inundations and erosion.[14] The abundance of running water pushed the inhabitants to construct subterranean aqueducts and

---

Edwin L. Barnhart, Christopher Powell, and Carol Karasik, 'Astronomical Observations from the Temple of the Sun, *Archaeoastronomy, The Journal of Astronomy in Culture* 19 (2005): pp. 44–73.

[8] Schele, 'The House'; Schele, 'Sacred Site'; Anderson, Morales, and Morales, 'A Solar Alignment of the Palace Tower at Palenque', pp. 34-36.

[9] Edwin L. Barnhart, 'Indicators of urbanism at Palenque', in *Palenque: Recent Investigations at the Classic Maya Center*, ed. Damien B. Marken (Lanham MD and Plymouth, UK: Altamira Press, 2007), pp. 107–121, here pp. 109–112.

[10] Jorge A. Vivó and José C. Gómez, *Climatología de México* (México, DF: Instituto Panamericano de Geografía e Historia, 1946), p. 10; Rodrigo Liendo Stuardo, *The Organization of Agricultural Production at a Classic Maya Center. Settlement Patterns in the Palenque Region, Chiapas, Mexico* (México, DF: Instituto Nacional de Antropología e Historia, University of Pittsburgh, 2002), p. 39.

[11] Liendo Stuardo, 'The Organization', p. 42.

[12] Liendo Stuardo, 'The Organization', pp. 124–151; Liendo Stuardo,' The Problem of Political Integration in the Kingdom of Baak', in *Palenque: Recent Investigations at the Classic Maya Center*, ed. Damien B. Marken, (Lanham MD and Plymouth, UK: Altamira Press, 2007), pp. 85–106.

[13] Liendo Stuardo, 'The Problem of Political Integration', pp. 89-93.

[14] See Kirk D. French, 'Creating Space through Water Management at the Classic Maya Site of Palenque', in *Palenque: Recent Investigations at the Classic Maya Center*, ed. Damien B. Marken (Lanham MD and Plymouth, UK: Altamira Press, 2007), pp. 123–132, here p. 125.

walled channels, pools, dams and drains. The city's ancient name was 'Lakamha', meaning 'Big Water', possibly denoting one of its biggest creeks, the Otulum River, which crosses through the main ceremonial precinct, separating the Palace and the Temple of the Inscriptions from the South and Cross Groups.[15] Palenque has many significant geographical features such as deep ravines, vertical cliffs, dark caves, pools, and springs that received diverse place names; some of them have been identified with modern place names. For example, *Yehmal K'uk' Lakam Witz*, 'Descending Quetzal Big Hill', has been identified with the mountain rising behind the Cross Group, known today as the Mirador Hill.[16] The terrain north of Palenque is absolutely plain, while tree-covered hills affect the skyline in all other directions, making any observation of the exact position of the sunrise or sunset difficult. Moreover, the series of steep escarpments hidden under the dense coat of tropical rainforest create divisions of the area into smaller units and isolated places, impeding the construction of a single observatory site.

The Temple of the Inscriptions

The Temple of the Inscriptions rests on the slope of a high terraced hill, where a few decades after Pakal's death the elite residential compound, known today as Blue Wood Group, was built.[17] The pyramid is stepped, with nine levels, possibly reflecting the concept of the nine levels of the Underworld in the Maya worldview, and rises almost 25m above the plaza floor (see Fig. 24.1). The single front stairway ascends at an angle of about 42°. Atop the pyramid is set a rectangular temple with a long frontal gallery and a rear central chamber, and crowned with a roofcomb. The summit temple measures 23.30 x 7.70m (approximately), and is orientated 22° northeast of north (see below). The pyramid had three constructive phases and a large stairway measured by Klokočnik and Kostelecký was added during the third phase.[18]

Archaeologists have long established that the funerary crypt was constructed before the pyramid.[19] First, a cavity was dug into the plaza to a depth of nearly 1.5m to make a place for a large limestone block from which Pakal's sarcophagus was made. The fish-like cavity was dug inside the block to accommodate the ruler's dead body while a limestone cover with stone plugs was carefully delineated to seal it. The whole sarcophagus was covered with a famous heavy lid (measuring 3.8 by 2.2m) and its four sides were carved, portraying Pakal's ancestors and B'aakal rulers as sprouting fruit trees.[20] Later,

---

[15] David Stuart and Stephen D. Houston, *Classic Maya Place Names* (Studies in Pre-Columbian Art and Archaeology, No. 37) (Washington DC: Dumbarton Oaks Research Library and Collection, 1994), pp. 30–32, p. 84; Simon Martin and Nikolai Grube, *Chronicle of the Maya Kings and Queens* (London: Thames & Hudson, Second Edition, 2008), p. 157; David Stuart, 'Gods and Histories: Mythology and Dynastic Succession at temples XIX and XXI at Palenque', in *Palenque: Recent Investigations at the Classic Maya Center*, ed. Damien B. Marken (Lanham MD and Plymouth, UK: Altamira Press, 2007), pp. 207–232, here p. 224.

[16] Stuart and Houston, 'Classic Maya Place Names', pp. 30–31; Stephen Houston, 'Symbolic Sweatbath of the Maya: Architectural Meaning of the Cross Group at Palenque, Mexico', *Latin American Antiquity* 7, no. 2 (1996): pp. 132–151, here p. 133; Stuart, 'Gods and Histories', p. 208; David Stuart and George Stuart, *Palenque: Eternal City of the Maya* (Thames and Hudson: London, 2008), p. 192.

[17] Ana Luisa Izquierdo y de la Cueva and Guillermo Bernal Romero, 'Los gobernantes heterárquicos de las capitales mayas del Clásico. El caso de Palenque', in *El despliegue del poder entre los mayas: nuevos estudios sobre la organización política*, ed. Ana Luisa Izquierdo y de la Cueva (México, DF: UNAM, 2011), pp. 151–192, here pp. 159–160.

[18] Lhuillier, *El Templo de las Inscripciones*, pp. 67–75; Klokočnik and Kostelecký, 'Palenque', p. 314.

[19] Lhuillier, *El Templo de las Inscripciones*, pp. 88–92.

[20] Linda Schele and Peter Mathews, *The Code of the Kings. The Language of Seven Sacred Maya Temples and Tombs* (New York: Scribner, 1998), pp. 119–123; Stuart and Stuart, *Palenque*, pp. 177–179; James L. Fitzsimmons, *Death and the Classic Maya Kings* (Austin TX: University of Texas Press, 2009), p. 209.

the tomb was sealed and enclosed within a vaulted chamber, forming a funerary crypt, showing nine figures modelled in stucco that represented former B'akaal rulers.[21] From that moment on the carved sarcophagus remained hidden to outside spectators. Finally, the pyramid was built with its interior stairway from the plaza floor up.

Unique to Pakal's tomb is the 'psychoduct', which leads from the tomb, up the stairway and through a hole in the stone covering the entrance to the burial. This construction is perhaps a physical reference to concepts about the departure of the soul at the time of death, corresponding to the phrase *ochb'ihaj sak ik'il* ('the white breath road-entered'), which is used to refer to the leaving of the soul.

Pakal's tomb orientation

It is probable that Pakal started to plan his own burial chamber himself a few years before his death.[22] His dead body was placed inside a coffin, extended on its back with the head toward the north-northeast.[23] One may expect that if Pakal oversaw the construction of his sarcophagus, he would probably follow the local rules of orienting the bodies and tombs of the dead and would supervise this alignment personally.

Already Ruz Lhuillier and Welsh have noticed that directional symbolism incorporated into Maya burial orientations reveals regional differences.[24] At Palenque, burials were generally aligned north-south with the head orientated to the north or northeast.[25] As stated above, the topographical configuration of the escarpment around the ceremonial area at Palenque privileges extensive views to the north where the horizon has a lack of impressive landmarks. Viewing orientation patterns from this perspective, it seems plausible to assume that the alignments found at Palenque may not be related to the local practice of orientating the tombs at a single particular yet not identified natural feature located somewhere north of the city, but rather appear to be more significant when the criteria governed by the Maya general directional symbolism are applied. I suggest that this pattern may recall the evidence from other Maya sites where north represents the heavens and south points to the underworld.[26] In other words, this rough north-south orientation may be attributed to the concepts of the afterlife, and a particular skeleton alignment may be ascribed to the practice of signalling the way to the celestial realm to an animate entity that left Pakal's body through the head after death.[27]

---

[21] Stuart and Stuart, *Palenque*, p. 180.

[22] Linda Schele and David Freidel, *A Forest of Kings: The Untold Story of the Ancient Maya* (New York: William Morrow, 1990), pp. 225–226; Schele and Mathews, *The Code of the Kings*, p. 97; Patricia A. McAnany, *Living with the Ancestors. Kinship and Kingship in Ancient Maya Society* (Austin TX: University of Texas Press, 1995); Vera Tiesler, 'Life and Death of the Ruler: Recent Bioarchaeological Finding', in *Janaab' Pakal of Palenque; Reconstructing the Life and Death of a Maya Ruler*, ed. Vera Tiesler and Andrea Cucina (Tucson AZ: The University of Arizona Press, 2006), p. 22; Markus Eberl, 'Death and Conceptions of the Soul', in *Maya: Divine Kings of the Rain Forest*, ed. Nikolai Grube (Potsdam: Hulmann, 2006), pp. 310–319, here p. 311; Stuart and Stuart, *Palenque*, p. 187.

[23] Lhuillier, *El Templo de las Inscripciones*, p. 201; Schele and Mathews, *The Code of the Kings*, p. 125; Tiesler, 'Life and Death', p. 32.

[24] Alberto Ruz Lhuillier, *Costumbres funerarias de los antiguos mayas* (1958; México, DF: Universidad Nacional Autónoma de México, 1991); Welsh, 'An Analysis'.

[25] Lhuillier, *El Templo de las Inscripciones*, p. 201; Lhuillier, *Costumbres funerarias*, p. 158–166, p. 196; Welsh, 'An Analysis' pp. 55, p. 63 Table 33, p. 229 Table 113, pp. 333–335, Table XV. Unfortunately, in those publications, all burial orientations are rounded off to the nearest cardinal direction (i.e. N, E, S, and W).

[26] Wendy Ashmore, 'Site-Planning Principles and Concepts of Directionality among the Ancient Maya' *Latin American Antiquity* 2, no. 3 (1991): pp. 199–226, here 216–217.

[27] The Classic Maya rulers were composite beings constituted of the human body and distinct animate entities ('souls', 'spirits') that once created their personal identities. Some of them were intrinsically associated with people's particular body parts and displayed certain agentive capabilities. One of them, called *b'aahis*, seems to be particularly

The overall message encoded in the Pakal's tomb is clearly associated with his death and rebirth. However, it is particularly dense and extremely elaborated, giving rise to different and slightly contradictory interpretations.[28]

The central figure displayed on sarcophagus lid is, of course, Pakal. He is dressed in the guise of the Maize God and depicted within the skeletal jaws of the Underworld (see Fig. 24.2).

Pakal's body is moving along the up-down axis represented by a type of a Cosmic Tree (often called World Tree in Mayanist literature).[29] It is covered with glyphic elements representing both jade and fruit. Its name is 'Shiny Necklace (or Jewelled) Tree'.[30] Stuart and Stuart proposed this tree represented one of the world directional trees, an eastern tree of jade, rather than a cosmic tree situated at the centre.[31] Atop the tree is perched a birdlike figure identified with *Muut Itzamnaaj*, 'Bird Itzamnaaj', an avian aspect of Itzamnaaj, the major god of the Classic Maya pantheon, associated with the heavens and the cosmic tree.[32] The bird indicates where the heavenly abode is.

The jaws of the Underworld represent a skeletal snake now recognized as *Sak B'ak Naj Chapat*, 'White Bone House Centipede', or 'First Centipede of White Bones', a mythological entity that represented a kind of threshold that connected the realm of the living with the abode of the ancestors.[33] Centipede images were sometimes associated with the sun *K'IN* glyph symbolizing either the travel of the night sun after the sunset through the Underworld or eclipses of the sun.

---

linked to a top or a forehead of the cranium. I am following here some of the concepts developed by William E. Duncan and Charles Andrew Hofling, 'Why the head? Cranial modification as protection and ensoulment among the Maya', *Ancient Mesoamerica* 22, no. 1 (2011): pp. 199–210. See also Stephen Houston and David Stuart, 'The ancient Maya self: Personhood and portraiture in the Classic period', *RES Anthropology and Aesthetics* 33 (1998): pp. 73–101; Fitzsimmons, *Death and the Classic Maya King*, pp. 47–48; Erik Velásquez García, 'Las entidades y las fuerzas anímicas en la cosmovisión maya clásica', in *Los mayas: voces de piedra*, ed. Alejandra Martínez de Velasco and María Elena Vega (México, DF: Ambardiseño, 2011), pp. 235–253.

[28] Simon Martin, 'The Baby Jaguar: An Exploration of Its Identity and Origins in Maya Art and Meaning', in: *La organización social entre los mayas prehispánicos, coloniales y modernos. Memoria de la Tercera Mesa redonda de Palenque*, ed. Vera Tiesler Blos, Rafael Cobos, and Merle Greene Robertson (México, D.F.: Instituto Nacional de Antropología e Historia and Universidad Autónoma de Yucatán, 2002), Vol. 1. pp. 49–78, here p. 56; Gerardo Aldana, *The Apotheosis of Janaab' Pakal* (Boulder CO: University Press of Colorado, 2007), pp. 100, 106; Stuart and Stuart, *Palenque*, pp. 174–175; Fitzsimmons, *Death and the Classic Maya Kings*, p. 57.

[29] See, for example, in Schele and Mathews, *The Code of the Kings*, p. 113.

[30] Stanley Guenter, 'The Tomb of K'inich Janaab Pakal: The Temple of the Inscriptions at Palenque', *Mesoweb*, 2007, pp. 26–27, available at http://www.mesoweb.com/articles/guenter/TI.pdf [accessed 12 January 2012]; Stuart and Stuart, *Palenque*, p. 176.

[31] Stuart and Stuart, *Palenque*, p. 255.

[32] Karl A. Taube, *The major gods of Ancient Yucatan* (Studies in Pre-Columbian Art & Archaeology, 32) (Washington, DC: Dumbarton Oaks Research Library and Collection, 1992), pp. 31–40.

[33] Nikolai Grube and Werner Nahm, 'A census of Xibalba: A Complete Inventory of Way Characters on Maya Ceramics', in *The Maya Vase Book*, ed. Justin Kerr (New York: Kerr Associates, 1994), Vol. 4, pp. 686–715, here p. 702; Schele and Mathews, *The Code of the Kings*, p. 113; Harri Kettunen, *Nasal Motifs in Maya Iconography* (Helsinki: Renvall Institute, University of Helsinki Printing House, 2006), pp. 307–308; Mercedes De la Garza, Guillermo Bernal Romero, and Martha Cuevas García, *Palenque-Lakamha'. Una presencia inmortal del pasado indígena* (México, D.F.: Fondo de Cultura Económica and El Colegio de México, 2012), p. 111.

Fig. 24.2: The Sarcophagus Lid (after Merle Greene Robertson, 'The Sculpture of Palenque. Volume 1: The Temple of the Inscriptions' (Princeton NJ: Princeton University Press, 1983), Fig. 99).

Now, moving downward to the bottom of the lid (see Fig. 24.2), we find a figure of the so-called Quadripartite Badge/Monster that supports a (sacrificial) solar bowl in which Pakal's body is resting.[34] The plate bears a *K'IN* glyph (a four-petaled flower) and denotes a type of an offering or sacrificial vessel that is also read as *EL*, meaning 'to burn, to cense'.[35] According to Stuart and Stuart the Mayan *el*

---

[34] Schele and Freidel, *A Forest of Kings*, p. 226, pp. 414–415; Schele and Mathews, *The Code of the Kings*, p. 113.
[35] David Stuart, *The Inscriptions from Temple XIX at Palenque* (San Francisco: The Pre-Columbian Art Research Institute, 2005), p. 164.

also means 'to exit' and 'to rise', possibly making a reference to the Mayan word *elk'in*, meaning 'east'.[36]

Earlier interpretations assumed Pakal's dead body represented the Maize God and it was depicted as moving downward, falling into the jaws of the Underworld[37]. Scholars supposed that, in his descent into the Underworld, Pakal is following the setting sun (the above mentioned *K'IN* glyph) making its journey across the Netherworld to approach the heavens from the eastern side, like the rising sun that appears each day on the east.[38] Finding support in the hierophany described above, this interpretation shows Pakal in the guise of the Maize God who is falling downward, together with the dying midwinter sun, into the jaws of the Underworld[39].

Recent interpretations assume the sarcophagus lid depicts the ruler's rebirth as the Maize God and his body is moving upward, emerging from the jaws of the Underworld.[40] The nasal motif attached to Pakal's nose probably represents the life essence (see below) meaning that he is considered to be alive, but instead of returning to the world of humans he is resurrected to be able to join his ancestors in the heavenly abode.[41]

Still others assume a different position, suggesting that Pakal is remaining in the Underworld, his body is lying on a sacrificial plate?? bearing a *K'IN* glyph, implying he would represent a newborn or rather a reborn young Maize God.[42]

Those interpretations emphasize Pakal's resurrection in the realm of his ancestors, whose figures are depicted on sarcophagus sides and walls of the funerary chamber, where they are shown as live persons. The upward rather than downward movement of Pakal's body seems to be much more probable. I concur with Fitzsimmons in seeing Pakal as emerging from the Underworld like the Maize God; the ruler (or rather his animate entity or soul) is kept alive to ascend into the celestial realm to join the animate entities (souls, spirits) of his ancestors.[43] In this light, the downward movement interpretation seems to be less convincing: after his death Pakal first must enter the Underworld, follow the sun and go to the Underworld, substantially delaying the moment of his rebirth. In any case, the sarcophagus was commissioned by Pakal, who believed in his resurrection and celestial apotheosis rather than in his descent and (temporal) residence in the Underworld.

Be that as it may, Pakal is shown falling into the Underworld of Xibalba, or emerging from it to join his ancestors in the sky, and the sarcophagus lid correctly displays the heavens in the north and the Underworld in the south (see Fig. 24.2). Even though the axis is misaligned by 24° from the North-South line, it still shows correct directions. In sum, the sarcophagus lid seems to embody two important visual metaphors for the rebirth of Maya rulership: the life-cycle of maize and the daily motion of the sun. Both metaphors were regularly used by the Maya elite to represent the idea of the resurrection and apotheosis of a particular dead ruler as well as to express the idea of the continuous dynastic succession: after the death of an old ruler, his son was expected to accede to the throne.[44]

---

[36] Stuart and Stuart, *Palenque*, p. 175.

[37] E.g. ,Schele, 'The House', p. 48; Schele, 'The Sacred', p. 98; Schele and Freidel, *A Forest of King'*, p. 412; Schele and Mathews, *The Code of the Kings*, pp. 115–116.

[38] Schele, 'The House', p. 48–49; Schele, 'The Sacred', p. 99;

[39] See arguments of Schele, 'The House'; Schele. 'The Sacred', and Oswaldo Chinchilla Mazariego, 'The stars of the Palenque sarcophagus', *RES, Anthropology and Aesthetics* 49/50 (2006): pp. 40–58, here pp. 51–54.

[40] Stuart and Stuart, *Palenque*, pp. 175–177; Aldana, *The Apotheosis*, pp. 100, 106; Fitzsimmons, *Death and the Classic Maya Kings*, pp. 53–57; and De la Garza, Bernal Romero, and Cuevas García, *Palenque-Lakamha'*, p. 27.

[41] Kettunen, *Nasal Motifs*, pp. 307–308.

[42] Martin, 'The Baby Jaguar', p. 55–56.

[43] Fitzsimmons, *Death and the Classic Maya Kings*, p. 57.

[44] For wider discussion of this change of royal power at Palenque, consult Schele and Freidel, *A Forest of Kings*, pp.

The Maya divine kingship was based on the idea of ancestor worship and rebirth. Legitimized ancestor rulers were apotheosized after death and believed to act for the benefit of their descendants. Because both the rulership and power were inherited, the Maya rulers used multiple strategies to make visible their connections to ancestors. Their actions ranged from monumental sepulchral architecture and iconographic representations of dead ancestors to detailed genealogical lists and calendrical and astronomical manipulations of dates. Very often the rulers installed shrines and temples at places tied to ancestors and mythological events to create the complex web of different relationships existing between the city's patronizing gods, who controlled time and space, meteorological events and celestial bodies, recent ancestors and current rulers. The role of such places was twofold: monumental royal tombs and shrines provided links to the sacred founders or to the city's patronizing gods (for example, in Palenque), but on the other hand they were *loci* providing access to the Underworld.[45]

As in many other agrarian societies, the ancient Maya kingship was closely associated with agriculture, and it was the ruler's major responsibility to conduct the rituals associated with maize cultivation and welfare of the kingdom.[46] During the Classic period, the life-cycle of maize, with its annual sequence of planting, sprouting, ripening and harvesting, was resumed in the story of the Maize God, who underwent a rhythm of events alternating between death and resurrection. The god was dead at sowing and reborn at plant sprouting. The Maya rulers were often portrayed in the guise of the young Maize God: their flattened head was akin to a maize cob and their dense hair represented a corn-silk. The descent of a dead ruler into the Underworld paralleled the descent of the dead Maize God into the Underworld, where the king was reborn in the form of his successor replicating the emergence of the Maize God from the Underworld.[47] The royal and maize life-cycles were metaphorically associated with each other.

The royal life-cycle was also associated with the daily motion of the sun. Each evening an aged K'inich Ajaw, the Maya Sun God, associated with the concepts of rulership, war, and sacrifice, was transformed into the nocturnal night sun, the Jaguar God of the Underworld. He was swallowed by the Starry Deer Caiman (the Milky Way) and travelled through the Underworld from west to east, to reappear in the form of a young K'inich Ajaw. As with the sun's motion, rulership was reborn when a new ruler re-emerged to accede to the throne shortly after the dead king descended into the Underworld.[48] The idea is clear: with the death of a current ruler, the ruling dynasty does not disappear, the continuity of the ruling lineage is assured.

Furthermore, many Late Classic Maya rulers added the title of the Sun God, K'inich Ajaw, to their royal names. The Mayan *k'inich*, 'hot', or 'the hot one', is derived from the adjective *k'ihnich* and denotes heat (or the state of 'hotness') as a vital quality ultimately derived from the sun, *k'in*; but it is also

---

234–261; Fitzsimmons, *The Death and the Maya Kings*, pp. 52–60; Martin, 'The Baby Jaguar', p. 61–62; Stuart and Stuart, *Palenque*, pp. 187–215; Aldana, *The Apotheosis*, pp. 128–141; Garza et al., 'Palenque-Lakamha'', pp. 117–148.

[45] See, for example, McAnany, *Living with the Ancestors*; Julia L. J. Sanchez, 'Ancient Maya Royal Strategies: Creating power and identity through art', *Ancient Mesoamerica* 16, no. 2 (2005): pp. 261–275, here pp. 268–272; Meghan O'Neil, 'Object, Memory, and Materiality at Yaxchilan: The Reset Lintels of Structures 12 and 22', *Ancient Mesoamerica* 22, no. 2 (2011): pp. 245–269, here pp. 247–248.

[46] Virginia F. Fields, and Dorie Reents-Budet, 'Introduction: The First Sacred Kings of Mesoamerica', in: *Lords of Creation: The Origins of Sacred Maya Kingship*, ed. Virginia M. Fields and Dorie Reents-Budet (Los Angeles: Los Angeles County Museum of Art, Scala Publishers, 2005), pp. 21–27.

[47] Nikolai Grube and Simon Martin, 'The Dynastic History of the Maya', in *Maya: Divine Kings of the Rain Forest*, ed. Nikolai Grube, with Eva Eggebrech and Mathias Seidel (Cologne: Könemann, 2006), pp. 148–171.

[48] Arthur G. Miller, *Maya Rulers of Time. A Study of Architectural Sculpture at Tikal, Guatemala* (Philadelphia PA: The University Museum, University of Pennsylvania, 1986), pp. 31–52.

accumulated with the advance of the age as well as with high offices of public or ritual functionaries.[49] During the Early Classic, those dead Maya rulers who were transformed into divine ancestors were sometimes apotheosized as the Sun, but during the Late Classic, this concept significantly evolved.[50] According to the Maya elite concepts of self, a human being had at least three animating entities roughly equivalent to the soul.[51] It seems that only the immortal animated entities of deceased and deified rulers called *b'aahis* travelled to the heavens after death, where they were fused with the Sun.[52] Strictly speaking, the strength of *b'aahis* entities increased when *k'in* essence was added.[53] At Yaxchilán, those animated entities of male rulers were fused with the sun, while those of their deified consorts were fused with the moon. At Palenque, the solar heat or 'hot breath' could have been represented by the 'Zip monster' squared snout, attached to the headdress of a ruler.[54] As stated above, the nasal motif attached to Pakal's nose probably represents this life essence and means that he is considered to be alive rather than dead in the realm of the Underworld.[55]

As noted above, the Maya burials at Palenque displayed a predominant north or northeast orientation. Pakal was the first ruler of Palenque who added to his name this special solar title *k'inich* and, in this, he was followed by all his successors.[56] Therefore, it is possible to assume that B'aakal rulers believed they received this vital essence from the sun. Since *b'aahis* was believed to reside in the head, then this particular head orientation pattern may mark the approximate direction of the celestial residence of all Pakal royal ancestors.[57]

When analysing the tomb's alignment, we must not overlook the fact that it is still placed within the 2° margin of error assumed by both authors. The presence of a smooth hill tends to suggest that the horizon might have been used for precise observations of the midsummer sunset. Due to the topographic location of Pakal's tomb, it is impossible to find a flat (true astronomical) horizon along the solstices. So, it is not clear whether the tomb's alignment is based on solstitial considerations. The Pakal's tomb appears to be orientated by prevailing local rules. Though the exact orientations of other graves at Palenque are not known at this moment, and were only rounded off to the nearest cardinal or inter-cardinal direction, the arguments made by both authors are weak.

On some astronomical alignments of the Temple of the Inscriptions
Klokočnik and Kostelecký also found that the stairway to the pyramid yielded an azimuth of 20°±2°, significantly deviating from that of the tomb (see Fig. 24.3).

---

[49] Søren Wichman, 'The Names of Some Major Classic Maya Gods. In: Continuity and Change in Mayan religious Practices in Temporal Perspective', in *Fifth European Maya Conference. University of Bonn, December 2000*, ed. Daniel Graña Behrens, Nikolai Grube, Christian M. Prager, Frauke Sachse, Stephanie Teifel, and Elizabeth Wagner (Acta Americana, Vol. 14) (Markt Schwaben: Anton Saurwein, 2004), pp. 77–86, here pp. 80–81; Velásquez García, 'Los vasos de la entidad política de 'Ik', pp. 542–543.
[50] Taube, *The major gods*, pp. 53–56; Carolyn Tate, *Yaxchilan: The Design of a Maya Ceremonial City* (Austin TX: University of Texas Press, 1992), pp. 59–62.
[51] Fitzsimmons, *Death and the Maya Kings*, pp. 120–122; Velásquez García, 'Los vasos de la entidad de 'Ik', pp. 560–561; Velásquez García, 'Las entidades y las fuerzas anímicas en la cosmovisión maya clásica', pp. 235–253, here p. 238.
[52] Stephen Houston and David Stuart, 'The ancient Maya self', pp. 73–101; Velásquez García, 'Los vasos de la entidad política de 'IK', p. 561.
[53] Velásquez García, 'Los vasos de la entidad política de 'Ik', p. 244.
[54] Stuart, *The Inscriptions from Temple XIX*, pp. 22–23; Velásquez García, 'Los vasos de la entidad política de 'Ik', p. 544.
[55] Kettunen, *Nasal Motifs*, pp. 307–308.
[56] Stuart, *The Inscriptions from Temple XIX*, pp. 150; Stuart and Stuart, *Palenque*, pp. 147–148.
[57] See note 29.

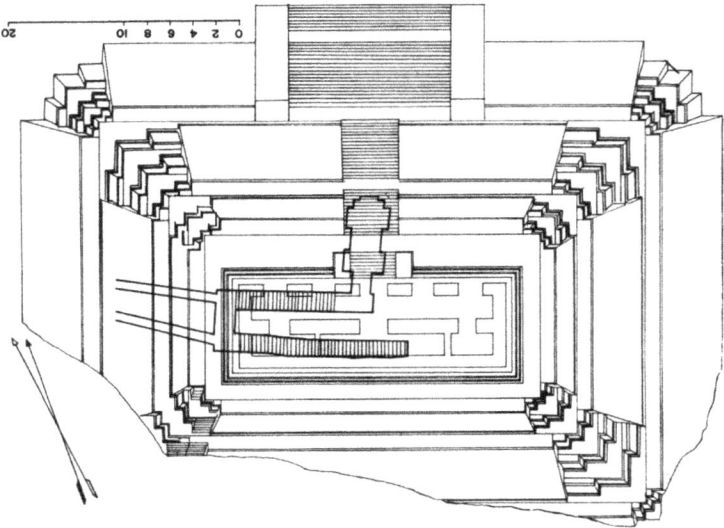

Fig. 24.3: The Plan of the Temple of the Inscriptions (after Ruz Lhuillier, 'El Templo de las Inscripciones', p. 240, Fig. 261g).

This discrepancy in azimuth values was already noticed and discussed by Ruz Lhuillier, who concluded that 'the builders' intention was that both the temple and the crypt would have had the same orientation'.[58] Naturally, diverse alignments may reflect subtly but deliberate arrangements of the tomb and the pyramid within the landscape, to emphasize diverse religious or political motivations. As Fleming once observed, tombs are erected for the dead as much as for the living.[59] On the other hand, however, though the tomb inside the pyramid remained invisible, the central doorway of the temple above has been carefully placed just above the entrance to the crypt, implying that precise architectural planning was intentional.[60]

Inside the pyramid, the narrow stairway leading to Pakal's funerary crypt is divided into two segments. The upper segment is built along the rear wall of the pyramid, reaches a descent, turns right at 180° and descends to the crypt. The rear wall of the temple and the inferior part of the upper segment were found slightly displaced from the temple's main axis. Today, the average dimensions of the temple are 23.38m east-west by 7.70m north-south by 23.30 west-east by 7.63m south-north, suggesting that all walls are of different length.[61] It is observed that the east and west facades are misaligned by 31'. The corners of the north façade show important deviations from a right angle: the northeast corner forms an 88°56' angle and the northwest corner forms an 89°27' angle (see Fig. 24.4).

---

[58] Lhuillier, *El Templo de las Inscripciones*, p. 84.
[59] Andrew Fleming, 'Tombs for the living', *Man* 8 (1973): pp. 177–193.
[60] Horst Hartung, 'Ancient Maya Architecture and Planning: Possibilities and Limitations for Astronomical Studies', in *Native American Astronomy*, ed. Anthony F. Aveni (Austin: University of Texas Press, 1977), pp. 111–129, here p. 122.
[61] Lhuillier, *El Templo de las Inscripciones*, pp. 76–77, 84.

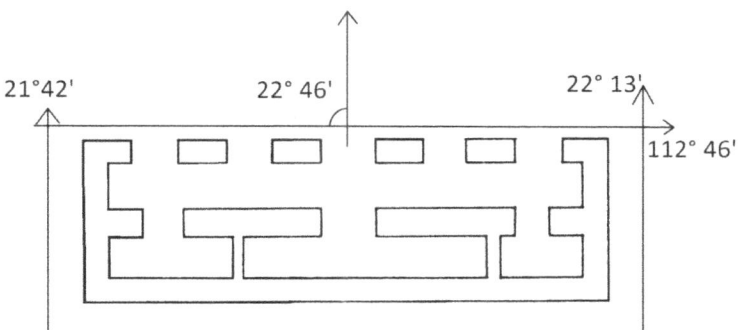

Fig. 24.4: Schematic plan of the temple atop the Temple of the Inscriptions.

Also, the rear or south façade is clearly misaligned: the wall was possibly displaced by the pressure of the slope of the nearby hill and was restored.[62] Ruz Lhuillier noticed that the main axis of the temple was skewed 15°-17° east of magnetic north while the funerary chamber itself displayed the direction 18°- 20° east of the magnetic north.[63] Thus, the long axis of the funerary crypt and the short axis of the temple would deviate from a right angle by about 3° (consult Fig. 24.3). Unfortunately, these measurements cannot be directly compared with those made by Klokočnik and Kostelecký, who measured the sides of Pakal's sarcophagus and the pyramid staircase and not the temple itself.[64]

My own measurements (see Fig. 24.4 and Table 24.1) indicate that the short temple's axes are skewed 21°42' - 22°13' of true north, and that the north façade has an azimuth of 112°46', implying that the perpendicular to this wall is orientated 22°46' east of true north. Now, adding 3° more, one arrives at 24° 42' - 25° 46' for the long axis of the funerary crypt, the value that stays very close to the azimuth of 24°± 2° given by Klokočnik and Kostelecký.[65] The addition of 3° to the recorded orientations displaces the azimuth from its strictly solstitial alignment but is still within the range of error proposed by the authors. The lack or more useful data makes any other conclusions speculative.

Wall	Azimuth (rounded to the nearest minute)	Horizon altitude	Declination (corrected for refraction)	Solar calendar dates	
North façade	112°46' eastward	8°00'	-18°55'	Jan 25	Nov 16
North façade	292°46' westward	0°50' estimated	21°47'	May 30	Jul 13
East wall	22°13'	-0°22'	61°19'		
West wall	21°41'	-0°22'	61°42'		
Perpendicular to the façade	22°46'	-0°22'	60°54'		

Table 24.1: Basic orientations of the temple located atop the Temple of the Inscriptions.

---

[62] Lhuillier, *El Templo de las Inscripciones*, p. 77, 80.
[63] Lhuillier, *El Templo de las Inscripciones*, pp. 84.
[64] Klokočnik and Kostelecký, 'Palenque: astronomical-solar orientation'.
[65] Klokočnik and Kostelecký, 'Palenque: astronomical-solar orientation'.

Now, a long time ago Aveni and Hartung suggested that Maya builders were more interested in precisely aligning the front and rear walls (long axis) of their buildings than in erecting short lateral walls (they were supposedly added later to the already standing walls).[66] It is, therefore, possible that the long axis of the temple shows the azimuth deliberately chosen by Maya architects. It may be speculated that the 'psychoduct' that connects Pakal's sarcophagus with pier no. 3 of the façade of the temple located atop the pyramid, depicting his mother, Lady Sak K'uk', signals the direction of the celestial abode. I suggest that this may be the religious-spiritual meaning of the line perpendicular to the façade.

A closer look at the tombs at Palenque shows that the north-northeast orientation was prevailing (see Table 24.2).

Monument	Burials	Orientations	Source[67]
Structure 18	Burial 1 Burial 2 Burial 3	15.5° (22°) 17.5° (24°) 12.75° (19.25°)	Lhuillier 1958: pp. 153, 181, Fig. 15.
Structure 18a	Burial 1 Burial 2	18.25° (20.25°) 17.75° (19.75°)	Lhuillier 1958: pp. 260–263, Fig. 5
Templo del Conde	Central burial Southern burial Northern burial	12.25° (17.75°) 14° (19.5°) 11.5° (17°)	Lhuillier 1958: pp. 204, 208, Fig. 3.
Group 1	Burial 1 Burial 2 Burial 3	8.5° 1(6.4°) 7° (14.9°) 8.5° (16.4°)	Ruz Lhuillier 1952: p. 54, Fig. 4.
Group 2	NW sarcophagus NE sarcophagus SW sarcophagus	9.25° - 14.50° (17.15° - 22.4°) 9.25° - 11.50° (17.15° - 19.4°) 8° (15.9°)	Ruz Lhuillier 1952: pp. 54–55, Fig. 5.

Table 24.2: Tomb orientations at Palenque. They all have different levels of confidence. The date for the tombs at Group 1 and 2 were corrected for a magnetic declination. The azimuths of the Temple of Conde are corrected by my own measurements of the façade of this building. The azimuths for Structure 18 and Structure 18a are extrapolated from my own measurements of the nearby Structure XVII. Corrected values are within brackets. These are not definite statements regarding tomb orientations at Palenque.

Though not all data can be adequately dealt with, several preliminary conclusions may be proposed. First, examples from the Temple of Conde, Structure XVIII and Groups 1 and 2 clearly show that the tombs are aligned with the façades of the buildings that house them. If this is a definite pattern to be found at Palenque, the deviation of the Pakal's tomb from the main axis of the Temple of the Inscriptions

---

[66] Anthony F. Aveni and Horst Hartung, 'Precision in the Layout of Maya Architecture', in *Ethnoastronomy and Archaeoastronomy in the American Tropics*, ed. Anthony F. Aveni and Gary Urton. (New York: Annals of the New York Academy of Sciences 385, 1982), pp. 63–80, here p. 69.
[67] The dates in Table 24.2 are compiled from the following reports by Alberto Ruz Lhuillier: 'Exploraciones arqueológicas en Palenque (1949)', *Anales del Instituto Nacional de Antropología e Historia* 4 (1952): pp. 49–60; 'Exploraciones arqueológicas en Palenque, 1954', *Anales del Instituto Nacional de Antropología e Historia* 10 (1958): pp. 117–184, pp. 185–240, pp. 241–299.

is unique. Second, combining the published plans of Palenque tombs with corrected azimuths or my measurements, I can now provide a more detailed description of alignments (see Table 24.2). It is shown that only tomb alignments from Structure 18 approach that of the Temple of the Inscriptions. These tombs may contain the bodies of elite members associated with a king who ruled between 721 and 734 CE. It can be speculated that after the death of Pakal, some elite members started to imitate his tomb orientation.

The Temple of the Inscriptions after Pakal's death

K'inich Kan Bahlam II (635–702) inherited the throne from his father, and we must bear in mind that the Temple of the Inscriptions was completed under his supervision.[68] This fact is supplied by the last passage of the West Tablet of the Temple of Inscriptions, which tells that K'inich Kan Bahlam acceded to the Palenque throne in 684 and 'took care of the Bal'un Eht Naah, the name of the tomb of K'inich Janaab Pakal, Divine Lord of Palenque'.[69] It is not clear whether the crypt or the pyramid were named. The name is not clear, it is alternatively translated as 'Nine Images House',[70] 'Nine Image House'.[71] 'Nine Works House',[72] 'Nine Figures House',[73] or 'House of Nine Companions'.[74]

Apart from the Temple of the Inscriptions, Kan B'ahlam focused his architectural activities on the buildings belonging to the Cross Group: the Temples of the Cross, of the Foliated Cross and of the Sun. Those three temples erected on stepped hills rising above a small plaza were dedicated to the Palenque patronizing gods, called the Palenque Triad, who were born in the remote mythological past (about 3000 years before the accession of Kan B'ahlam in 684 CE) in the place named *Matwil*. *Matwil* is the mythical place of origin of both the Palenque Triad and the B'aakal dynasty and a toponym that today is identified with the location of the Cross Group temples.[75] The whole assemblage was completed on 9.13.0.0.0 (692 CE) when *k'atun* ending was celebrated.

It's evident that Kan B'ahlam erected the Cross Group appropriating this mythological place for the benefit of his succession and of his B'aakal dynasty.[76] He also connected the Temple of the Inscriptions with the Cross Group by performing a series of ritual ceremonies around the same time. First, on 9.12.18.4.19 11 Kawak 17 Yaxk'in (3 July 690) he dedicated the Temple of the Inscriptions.[77] Later on 9.12.18.5.16 (20 July 690) he participated in an effigy-censers cooking ceremony held at the Cross Group, and on 9.12.18.5.17 (21 July 690) he conjured gods in the temple located on the summit of the Mirador Hill, or 'Big Hill of Descending Quetzal'. All those locations are observed from the temple erected on the top of the pyramid (see Fig. 24.5).

---

[68] Schele and Mathews, *The Code of the King*, p. 108; Martin and Grube, *Chronicle*, p. 168.
[69] Guenter, 'The Tomb of K'inich Janaab Pakal', p. 54–55.
[70] Schele and Mathews, *The Code of the King*, p.108.
[71] Guenter, 'The Tomb of K'inich Janaab Pakal', p. 55.
[72] Stuart and Stuart, *Palenque*, p. 180.
[73] Markus Eberl, 'Su aliento se separó. La muerte en el periodo Clásico', in *Los mayas: voces de piedra*, ed. Alejandra Martínez de Velasco and María Elena Vega (México, D.F.: Ambardiseño, 2011), pp. 255–263, here p. 259.
[74] Garza el al. 'Palenque-Lakamha", p. 18.
[75] Stuart, *The Inscriptions from the Temple XI*, pp. 79–83, p. 169.
[76] Peter Biró, 'Politics in the Western Maya Region (II): Emblem Glyphs', *Estudios de Cultura Maya* 39 (2012): pp. 33–66.
[77] Stuart and Stuart, *Palenque*, pp. 170–171.

Fig. 24.5: The Temple of the Cross as seen from the top of the Temple of the Inscriptions.

Above, we have seen that the Cross Group was linked to the Temple of the Inscriptions by solar hierophanies symbolizing the transfer of power from Pakal to his son and legitimate heir Kan B'ahlam. However, the sightline extending from the north façade of the Temple of the Inscriptions does not point to the central doorway of the Temple of the Cross as Hartung suggests.[78] This sightline may have been deliberately distorted to emphasize sunrise occurring along the horizon. It passes over the lower part of the platform on which stands the building and rises to an altitude of about 8° above the astronomical horizon. The dense forest cover rising on the northern slope of El Mirador today prevents a more distant view in this direction (see Fig. 24.5). This may be the case of a deliberate misalignment, or a kind of compromise between the northern orientation of Pakal's tomb and Kan B'ahlam's political exploitation of solar hierophanies and relationships with the mythical place of origin. Naturally, the dates of sunrise

---

[78] Hartung, 'El espacio exterior', pp. 129–131.

Fig. 24.6:. Western horizon as seen from the top of the Temple of the Inscriptions.

and sunset determined by alignments may also be significant (see Table 24.2). For example, it may be speculated that the date of the sunset, on 13 July, roughly coincides with the dates of ceremonies related to the dedication of the Temple of the Inscriptions (3 July 690 CE) and the consecration of incense burners in the Cross Group and ceremonies held atop El Mirador (20 and 21 July 690 CE) (see Table 24.2 and Fig. 24.6).[79]

In climatic-meteorological and agricultural terms, the sunset dates coincide with the rainy season;

---

[79] It has long been speculated that these ceremonies were made during a rare conjunction of Jupiter, Saturn, Mars and the moon, see, for example, Dieter Dütting, Anthony F. Aveni and Martin Schramm, 'The 2 Cib 14 Mol event in the inscriptions of Palenque, Chiapas, Mexico', *Zeitschrift für Ethnologie*, 57, no. 2 (1982): pp. 233–258. However, all texts that mention those dates do not mention the planets. Jupiter symbolism was adopted by Kan B'ahlam who probably invented the 819-day cycle. The Cross Group temples were dedicated together in 692 CE on *k'atun* ending ceremony.

nevertheless, since most rainfall occurs in the afternoon, one may expect that rain showers, mist, and dark clouds will make any direct sunset observation difficult, if not impossible. Both sunrise dates coincide with the ripening and harvesting of corncobs, the end of an agricultural season and the lowest average yearly temperature. Since, the Temple of the Inscriptions was not sealed immediately after Pakal's death, its inner staircase was used for restricted elite rituals and its outer stairway was exploited for public ceremonial gatherings, it may be further speculated that the eventual calendrical-agricultural meaning of eastern dates assured Pakal's important role in guaranteeing abundant crops and welfare to the city.[80] Though the Palenque Triad gods functioned as patronizing deities for Palenque, they were associated with both diurnal (rising) and nocturnal features of the sun (gods GI and GIII) and with rain, lightning, agriculture and royal succession.[81] Maya rulers acted as intercessors between gods and humans, performing rituals replicating the renewal of the universe. It should be reminded that Pakal's transit to the celestial abode and the sun was made in the guise of the maize god, revealing the ruler's power to generate maize, but this symbolism remained occult to the city commoners. Instead, public ceremonies that included visual connections of the Temple of the Inscriptions with the eastern horizon, where the mythic place of origin was located, could have provided arguments that the celestial power with which the ruling elite was once endowed would ensure crops for the entire community.

It is observed that by the Late Classic, during the seventh century, B'aakal rulers and their successors used a variety of strategies to legitimize their right to rule. The erection of the monumental Temple of the Inscriptions suggests that Pakal successfully solidified his position within Palenque elites, incorporating local burial traditions and beliefs in the afterlife in a highly elaborated ideological-cosmological symbolism assuring of his apotheosis and deification. Kan B'ahlam, his son and heir to the throne, tied the royal ancestry to the place of the mythical origins of the city's patronizing gods and founders of B'aakal dynasty. It may be expected that messages encoded in the Temple of the Inscriptions decoration and its astronomical alignments associated with the time of harvesting created the scenery for public rituals, reinforcing the idea of divine origins of the dynasty itself. According to Schele, Pakal initiated his master plan to offer a new interpretation of the local mythology of kingship by making arrangements of architectural structures.[82] The temples belonging to the Cross Group, raised under the reign of Kan B'ahlam, added new elements to this plan, creating an artificial or modified eastern horizon to accommodate the already existing alignments of Pakal's funerary monument.

Conclusions

In my opinion, claiming solstitial alignments in Pakal's sarcophagus is not definite and needs to be culturally sustained. The archaeological record examined at Palenque shows that the north-northeastern head orientation of deceased people was intentional. This evidence, in turn, has led to new interpretations of astronomical alignments of the Temple of the Inscriptions. However, it remains unclear whether the line that is approximately perpendicular to the axis of the funerary chamber and sarcophagus represents a collateral effect of orientating the tomb northerly or indicates a premeditated marking of the midsummer sunset alignment. Tombs orientations at Structures 18 and 18a may be indicative of this new trend attributable to the activities of Pakal. Naturally, in this stage of inquiry, any definite statement remains now premature.

---

[80] Garza et al., 'Palenque-Lakamha", p. 127.
[81] Stuart and Stuart, *Palenque*, pp. 189–190; Garza et al., 'Palenque-Lakamha", pp. 144–145.
[82] Schele and Freidel, *A Forest of Kings*, pp. 217–228; Schele and Mathews, *The Code of the Kings*, pp. 95–132.

Acknowledgements

This research was made on my sabbatical leave as part of the project 'Starry sky – animated sky'. This paper benefitted from my previous field studies and analyses that were funded by the Consejo Nacional de Ciencia y Tecnología (Project 25721-S). I appreciate greatly the constructive comments of Jesús Galindo Trejo, Erik Velásquez Garcían and Ricardo Moyano Vasconcellos. My grateful thanks to both anonymous reviewers whose suggestions resulted in the creation of Table 24.2. Responsibility for any errors and mistakes is solely mine.

Biography

Aldana, Gerard. *The Apotheosis of Janaab' Pakal*. Boulder CO: University Press of Colorado, 2007.

Anderson, Neal S., Moises Morales, and Alfonso Morales. 'A Solar Alignment of the Palace Tower at Palenque'. *Archaeoastronomy, The Bulletin of the Center for Archaeoastronomy* 4, no. 3 (1981): pp. 34–36.

Anderson, Neal S., and Moises Morales. 'Solstitial Alignments of the Temple of the Inscriptions at Palenque'. *Archaeoastronomy, The Bulletin of the Center for Archaeoastronomy* 4, no. 3 (1981): pp. 30–33.

Ashmore, Wendy. 'Site-Planning Principles and Concepts of Directionality among the Ancient Maya'. *Latin American Antiquity* 2, no. 3 (1991): pp. 199–226.

Aveni, Anthony F., and Horst Hartung. 'Precision in the Layout of Maya Architecture'. In *Ethnoastronomy and Archaeoastronomy in the American Tropics*. Edited by Anthony F. Aveni and Gary Urton, pp. 63–80.. New York: Annals of the New York Academy of Sciences 385, 1982.

Aveni, Anthony F., and Horst Hartung. 'Some Suggestions About the Arrangement of Buildings at Palenque'. In *Tercera Mesa Redonda de Palenque*. Edited by Merle Greene Robertson and Donnan Call Jeffers, pp. 173–177. Monterey CA: Pre-Columbian Art Research Center, Herald Printers, 1979.

Barnhart, Edwin L. 'Indicators of urbanism at Palenque'. In *Palenque: Recent Investigations at the Classic Maya Center*. Edited by Damien B. Marken, pp. 109–112. Lanham MD and Plymouth UK: Altamira Press, 2007.

Biró, Peter. 'Politics in the Western Maya Region (II): Emblem Glyphs'. *Estudios de Cultura Maya* 39 (2012): pp. 33–66.

Carlson, John B. 'Astronomical Investigations and Site Orientation Influences at Palenque'. In *The Art, Iconography & Dynastic History of Palenque, Part III. The proceedings of the Segunda Mesa Redonda de Palenque*. Edited by Merle Greene Robertson, pp. 107–122. Pebble Beach CA: Pre-Columbian Art Research, The Robert Louis Stevenson School, 1976.

De la Garza, Mercedes, Guillermo Bernal Romero, and Martha Cuevas García. *Palenque-Lakamha'. Una presencia inmortal del pasado indígena*. México, DF: Fondo de Cultura Económica and El Colegio de México, 2012.

Duncan, William E., and Charles Andrew Hofling. 'Why the head? Cranial modification as protection and ensoulment among the Maya'. *Ancient Mesoamerica* 22, no. 1 (2011): pp. 199–210.

Dütting, Dieter, Anthony F. Aveni and Martin Schramm. 'The 2 Cib 14 Mol event in the inscriptions of Palenque, Chiapas, Mexico'. *Zeitschrift für Ethnologie* 57, no.2 (1982): pp. 233–258.

Eberl, Markus. 'Death and Conceptions of the Soul'. In *Maya: Divine Kings of the Rain Forest*. Edited by Nikolai Grube, pp. 310–319.

Eberl, Markus. 'Su aliento se separó. La muerte en el periodo Clásico'. In *Los mayas: voces de piedra*. Edited by Alejandra Martínez de Velasco and María Elena Vega, pp. 255-263. México, DF: Ambardiseño, 2011.

Fields, Virginia F., and Dorie Reents-Budet. 'Introduction: The First Sacred Kings of Mesoamerica'. In *Lords of Creation: The Origins of Sacred Maya Kingship*. Edited by Virginia M. Fields and Dorie Reents-Budet, pp. 21-27. Los Angeles: Los Angeles County Museum of Art, Scala Publishers, 2005.

Fitzsimmons, James L. *Death and the Classic Maya Kings*. Austin TX: University of Texas Press, 2009.

Fleming, Andrew. 'Tombs for the living'. *Man* 8 (1973): pp. 177–193.

French, Kirk D. 'Creating Space through Water Management at the Classic Maya Site of Palenque'. In *Palenque: Recent Investigations at the Classic Maya Center*. Edited by Damien B. Marken, pp. 123–132. Lanham MD and Plymouth, UK: Altamira Press, 2007.

Grube, Nikolai, and Simon Martin. 'The Dynastic History of the Maya', in: *Maya: Divine Kings of the Rain Forest*. Edited by Nikolai Grube, with Eva Eggebrech and Mathias Seidel, pp. 148–171. Cologne, Könemann, 2006.

Grube, Nikolai, and Werner Nahm. 'A census of Xibalba: A Complete Inventory of Way Characters on Maya Ceramics'. In *The Maya Vase Book*. Edited by Justin Kerr, Vol. 4, pp. 686–715. New York: Kerr Associates, 1994.

Guenter, Stanley. 'The Tomb of K'inich Janaab Pakal: The Temple of the Inscriptions at Palenque'. *Mesoweb*, 2007.

www.mesoweb.com/articles/guenter/IT.pdf.

Hartung, Horst. 'Ancient Maya Architecture and Planning: Possibilities and Limitations for Astronomical Studies'. In *Native American Astronomy*. Edited by Anthony F. Aveni, pp. 111–129. Austin TX: University of Texas Press, 1977.

Hartung, Horst. 'El espacio exterior en el centro ceremonial de Palenque'. In *The Art, Iconography & Dynastic History of Palenque, Part III. The proceedings of the Segunda Mesa Redonda de Palenque*. Edited by Merle Greene Robertson, pp. 123–135. Pebble Beach CA: Pre-Columbian Art Research, The Robert Louis Stevenson School, 1976.

Houston, Stephen. 'Symbolic Sweatbath of the Maya: Architectural Meaning of the Cross Group at Palenque, Mexico'. *Latin American Antiquity* 7, no. 2 (1996): pp. 132–151.

Houston, Stephen, and David Stuart. 'The ancient Maya self: Personhood and portraiture in the Classic period'. *RES Anthropology and Aesthetics* 33 (1998): pp. 73–101.

Izquierdo y de la Cueva, Ana Luisa, and Guillermo Bernal Romero. 'Los gobernantes heterárquicos de las capitales mayas del Clásico. El caso de Palenque'. In *El despliegue del poder entre los mayas: nuevos estudios sobre la organización política*. Edited by Ana Luisa Izquierdo y de la Cueva, pp. 151–192. México, DF: UNAM, 2011.

Kettunen, Harri. *Nasal Motifs in Maya Iconography* (Helsinki: Renvall Institute, University of Helsinki Printing House, 2006.

Klokočnik, Jaroslav, and Jan Kostelecký. 'Palenque: Astronomical-solar orientation of Pakal's tomb'. In *Anthropological Notebooks, Proceedings of the 20th SEAC Conference, Ancient Cosmologies and Modern Prophets*. Edited by Ivan Šprajc and Peter Pehani, Vol. 19, supplement, pp. 305–317. Ljubljana: Slovene Anthropological Society, 2013.

Lhuillier, Alberto Ruz. 'Exploraciones arqueológicas en Palenque (1949)'. *Anales del Instituto Nacional de Antropología e Historia* 4 (1952): pp. 49–60.

Lhuillier, Alberto Ruz. 'Exploraciones arqueológicas en Palenque, 1954'. *Anales del Instituto Nacional de Antropología e Historia* 10 (1958a&b): pp. 117–184, pp. 185–240.

Lhuillier, Alberto Ruz. 'Exploraciones arqueológicas en Palenque: 1955'. *Anales del Instituto Nacional de Antropología e Historia* 10 (1958c): pp. 241–299.

Lhuillier, Alberto Ruz. *Costumbres funerarias de los antiguos mayas*. 1958. México, DF: Universidad Nacional Autónoma de México, 1991.

Lhuillier, Alberto Ruz. *El Templo de las Inscripciones, Palenque*, (Colección Científica, 7). México, DF: Instituto Nacional de Antropología e Historia, 1973.

Martin, Simon. 'The Baby Jaguar: An Exploration of Its Identity and Origins in Maya Art and Meaning'. In *La organización social entre los mayas prehispánicos, coloniales y modernos. Memoria de la Tercera Mesa redonda de Palenque*. Edited by Vera Tiesler Blos, Rafael Cobos, and Merle Greene Robertson), Vol. 1. pp. 49–78. México, DF: Instituto Nacional de Antropología e Historia and Universidad Autónoma de Yucatán, 2002.

Martin, Simon, and Nikolai Grube. *Chronicle of the Maya Kings and Queens*. London: Thames & Hudson, Second Edition, 2008.

Mazariego, Oswaldo Chinchilla. 'The stars of the Palenque sarcophagus'. *RES, Anthropology and Aesthetics* 49/50 (2006): pp. 40–58.

McAnany, Patricia A. *Living with the Ancestors. Kinship and Kingship in Ancient Maya Society*. Austin TX: University of Texas Press, 1995.

Mendez, Alfonso, Edwin L. Barnhart, Christopher Powell, and Carol Karasik. 'Astronomical Observations from the Temple of the Sun'. *Archaeoastronomy, The Journal of Astronomy in Culture*. 19 (2005): pp. 44–73.

Miller, Arthur G. *Maya Rulers of Time. A Study of Architectural Sculpture at Tikal, Guatemala*. Philadelphia PA: The University Museum, University of Pennsylvania, 1986.

O'Neil, Meghan. 'Object, Memory, and Materiality at Yaxchilan: The Reset Lintels of Structures 12 and 22'. *Ancient Mesoamerica* 22, no. 2 (2011): pp. 245–269.

Sanchez, Julia L. J. 'Ancient Maya Royal Strategies: Creating power and identity through art'. *Ancient Mesoamerica* 16, no. 2 (2005): pp. 261–275.

Schele, Linda. 'Palenque: The House of the Dying Sun'. In *Native American Astronomy*. Edited by Anthony F. Aveni, pp. 42–56. Austin: University of Texas Press, 1977.

Schele, Linda. 'Sacred Site and World-View at Palenque'. In *Mesoamerican Sites and World-Views*. Edited by Elizabeth P. Benson, pp. 87–117. Washington DC: Dumbarton Oaks Research Library and Collections, 1981.

Schele, Linda, and David Freidel. *A Forest of Kings: The Untold Story of the Ancient Maya*. New York: William Morrow,

1990.

Schele, Linda Schele, and Peter Mathews. *The Code of the Kings. The Language of Seven Sacred Maya Temples and Tombs*. New York: Scribner, 1998.

Stuardo, Rodrigo Liendo. *The Organization of Agricultural Production at a Classic Maya Center. Settlement Patterns in the Palenque Region, Chiapas, Mexico*. México, DF: Instituto Nacional de Antropología e Historia/University of Pittsburgh, 2002.

Stuardo, Rodrigo Liendo. 'The Problem of Political Integration in the Kingdom of Baak'. In *Palenque: Recent Investigations at the Classic Maya Center*. Edited by Damien B. Marken, pp. 85–106. Lanham MD and Plymouth, UK: Altamira Press, 2007.

Stuart, David. 'Gods and Histories: Mythology and Dynastic Succession at temples XIX and XXI at Palenque'. In *Palenque: Recent Investigations at the Classic Maya Center*. Edited by Damien B. Marken, pp. 207–232. Lanham MD and Plymouth, UK: Altamira Press, 2007.

Stuart, David. *The Inscriptions from Temple XIX at Palenque*. San Francisco: The Pre-Columbian Art Research Institute, 2005.

Stuart, David, and George Stuart. *Palenque: Eternal City of the Maya*. Thames and Hudson: London, 2008.

Stuart, David, and Stephen D. Houston. *Classic Maya Place Names* (Studies in Pre-Columbian Art and Archaeology, No. 37). Washington, DC: Dumbarton Oaks Research Library and Collection, 1994.

Tate, Carolyn. *Yaxchilan: The Design of a Maya Ceremonial City*. Austin TX: University of Texas Press, 1992.

Taube, Karl A. *The major gods of Ancient Yucatan* (Studies in Pre-Columbian Art & Archaeology, 32). Washington DC: Dumbarton Oaks Research Library and Collection, 1992.

Tiesler, Vera. 'Life and Death of the Ruler: Recent Bioarchaeological Finding'. In *Janaab' Pakal of Palenque; Reconstructing the Life and Death of a Maya Ruler*. Edited by Vera Tiesler and Andrea Cucina. Tucson AZ: The University of Arizona Press, 2006.

Velásquez García, Erik. 'Las entidades y las fuerzas anímicas en la cosmovisión maya clásica'. In *Los mayas: voces de piedra*. Edited by Alejandra Martínez de Velasco and María Elena Vega, pp. 235–253. México, DF: Ambardiseño, 2011.

Velásquez García, Erik. 'Los vasos de la entidad política de 'Ík': una aproximación histórico-artística. Estudio sobre las entidades anímicas y el lenguaje gestual y corporal en el arte maya clásico'. PhD diss., Facultad de Filosofía y Letras, Universidad Nacional Autónoma de México, 2009.

Vivó, Jorge A., and José C. Gómez. *Climatología de México*. México DF: Instituto Panamericano de Geografía e Historia, 1946.

Welsh, W. Bruce. *An Analysis of Classic Lowland Maya Burials*. Oxford: Archaeopress, BAR International Series 409, 1988.

Wichman, Søren. 'The Names of Some Major Classic Maya Gods'. In 'Continuity and Change in Mayan religious Practices in Temporal Perspective'. In *Fifth European Maya Conference. University of Bonn, December 2000*. Edited by Daniel Graña Behrens, Nikolai Grube, Christian M. Prager, Frauke Sachse, Stephanie Teifel, and Elizabeth Wagner (Acta Americana, Vol. 14), pp. 77–86. Markt Schwaben: Anton Saurwein, 2004.

EGYPT, THE MEDITERRANEAN AND ASIA

A COMPARATIVE STUDY OF MEGALITHIC MONUMENTS IN SARDINIA AND BEYOND

A. César González-García, Mauro P. Zedda and Juan A. Belmonte

ABSTRACT: We present a reappraisal of previous works on the megalithic structures of Sardinia together with new data collected in the last years for new monuments. In total we analysed more than 1500 megalithic structures on the island. We have performed two statistical cluster analyses on these data: first we used a Group analysis to search for similar clusters of orientation through a dendrogram; secondly we tried to group the orientations according with the shape of the orientation distribution by a K-means test. We confirm the existence of differences between the southern and northern halves of Sardinia since the appearance of the dolmens. We then compare the orientation of these groups in Sardinia with other groups of contemporaneous megalithic monuments in the central Mediterranean basin. The results of these analyses indicate that the differences observed between the North and South of Sardinia could be related to differences in the orientation customs in the North and South shores of the Western Mediterranean. In this sense we envisage Sardinia as a *carrefour* for orientation customs.

1. Introduction

Sardinia is without hesitation an open-air museum housing some of the most impressive megalithic monuments in the Western Mediterranean. Most of these monuments have already been investigated from the point of view of archaeoastronomy with varying results in the past. Those works resulted in a huge corpus of measurements which has been recently updated to implement the statistical analysis we performed recently and that is extended in this essay.[1]

A number of associations with different astronomical objects are pointed out in previous studies.[2] It is interesting to note that some results point towards an invisible dividing line across the central part of the island, indicating different traditions in orientation, especially with regard to the Tombe di Giganti.

Various megalithic or pseudo-megalithic structures appear quite early in Sardinian prehistory and have been one of the key elements to make Sardinia a case study for several archaeological hypotheses, like the Island laboratory.[3] Despite the caveats this hypothesis may have and the criticism it has received in the recent decades, it may be useful to study Sardinia in this frame, as the isolated nature of the island, together with its geo-morphological features, make it as an interesting laboratory to understand how the evolution of the megalithic structures might be both influenced by external and internal factors.

Sardinia is the second largest island in the Mediterranean, and although it is rather far from the continent, its proximity to other large islands, such as Corsica and the bridge of islands of Elba, made it not completely isolated. The geography of the island is marked by its variety, ranging from lowlands to rather high mountain ranges and volcanic terrain. Among the last it is important to note the presence of obsidian, as it was exploited and perhaps traded since the early Neolithic. Obsidian tools from Monte

---

[1] M. P. Zedda, *Astronomia nella Sardegna Preistorica* (Cagliari: Agora Nuragica, 2013); A. C. González-García, M. P. Zedda, J.A. Belmonte, 'On the orientation of prehistoric Sardinian monuments: a comparative statistical approach', *Journal for the History of Astronomy* 45 (2014): pp. 467–481.

[2] A. C. González-García, L. Costa Ferrer, and J. A. Belmonte, 'Solarist vs. Lunatics: modelling patterns in megalithic astronomy', in *Light and Shadows in Cultural Astronomy* ed. M. P. Zedda and J. A. Belmonte (Cagliari: Associazione archeofila sarda, 2007), pp. 23-30.

[3] Mark Patton, *Islands in time* (London: Routledge, 1996).

Arci are present in the coasts of northern Italy, southern France and even Northeast Spain.

While the predominant winds in the island are the northeastern ones (Mistral), their strength varies along the coasts or the interior and at different latitudes; for instance, there are important southern winds in the south of the island.

The oldest monuments we consider are the *Domus de Janas*. They are monumental burial tombs carved in the cliffs and rocky outcrops of the island, although it is likely that the larger ones also had a role as sanctuaries. The first *Domus* (like Cuccuru s'Arriu) could have been developed in the late phases of the Bonuighinu Culture (ca. 5800–3800 BCE), but the wide majority of them were sculpted during the late Neolithic period (3800–2900 BCE) within the frame of the so-called Ozieri Culture, although they continued to be built even at sub-Ozieri (2900–2700 BCE) and in the Copper Age (2700–2300 BCE). New *Domus de Janas* were seldom built during the Bronze Age, although the tombs already built in the Late Neolithic continued to be in use for several centuries.[4]

Dolmens were built during the Ozieri culture, but only in the northern half of the island. In contrast, dolmenic corridors are more widespread throughout the island, dating back to the period between the sub-Ozieri and the early Bronze Age (2300–1900 BCE). They are not very common but, it is highly probable that many of them were turned into tombs of giants in the Middle Bronze age. The *Tombe de Giganti* date back to the period between the Middle Bronze age (1900–1350 BCE) and the Late Bronze Age (1350–1150 BCE) and can be considered as dolmenic corridors built on a monumental scale to which an exedra and a frontis have been added. Nuraghe are towering cyclopean structures of different sizes and complexities, contemporaneous with the giant tombs. In our analysis, we consider the orientation of the entrance to the single-tower of simple nuraghe and to the central towers of complex nuraghe.

These central towers were probably the first elements to be built. The towers and ruins of the nuraghe dominate the Sardinian landscape, where the remains of some 7000 monuments have been identified. Of these, a number close to 500 are in a relatively good state of preservation (our data include ~540 of them).

The megara date back to the Bronze Age and extend especially to the Final Bronze Age (1150–900 BCE). The sacred wells should be contemporaries to the megara. Both types of cyclopean structures certainly had a strong ritual and religious component.

It is worth noticing that most of these monuments (and especially the *Domus de Janas*, nuraghe and *Tombe de Giganti* – hereafter TDG) remained in use for many centuries after the epoch in which they ceased to be built. Curiously, material culture shows substantial differences between the two halves of the island only during the Late Bronze Age (1350–1150 BCE), according to archaeological findings, although some small difference could be hinted upon a bit earlier.[5] This fact is also manifested in the architectural styles for both the nuraghe and the TDG.[6] However, some differences appear in the distribution of dolmens and the early presence of Bell Beaker items in the funeral spaces. The differences are self-evident in the cyclopean and megalithic monuments from the very beginning of their construction, but there are also suggestive differences in the way that the *Domus de Janas* were structured.

---

[4] G. Tanda, 'L'ipogeismo in Sardegna: arte, simbologia, religione', in *L'ipogeismo nel Mediterraneo Orogini, sviluppo, quadri culturali: atti del Congresso Internazionale*, 23-28 May 1994 (Sassari: 2000), pp. 399–426; G. Lilliu, *La civiltà dei Sardi, dal paelolitico all'età dei nuraghi* (Torino: Nuova Eri, 1988).
[5] A. Depalmas, 'Il Bronzo medio della Sardegna', in *Atti della XLIV Riunione Scientifica Istituto Italiano di Preistoria e Protostoria: La Preistoria e la Protostoria della Sardegna* (Florence: 2009), pp. 123–30, 131–40 and 141–60.
[6] Zedda, *Astronomia*; M. P. Zedda, *Archeologia del Paesaggio Nuragico* (Cagliari: Agorà nuragica, 2009).

2. Methodology

In the following we use the orientation data to build azimuth and declination histograms. We employ a smoothing of the histograms by a function called the kernel to produce the azimuth 'probability density function' (PDF). For this process, an Epanechnikov kernel is employed with a bandwidth of twice the estimated error in azimuth.[7] To be able to say whether a measurement is significant, we use a normalized relative frequency by dividing the number of occurrences of a given azimuth by the mean number of occurrences for that sample; this is equivalent to dividing or comparing with the results of a uniform distribution of the same size as our data sample, and with a value equal to the mean of our data.

To perform the comparisons between the statistics of the different monuments we use Multivariate techniques, commonly used in archaeology and used in the recent past for archaeoastronomical data. In this particular case, two techniques will be applied to the Sardinian data: the cluster analysis and a representation called the dendrogram on the one hand, and *k*-means clustering on the other.[8]

We have followed the same procedure as in previous works of our research team to characterize the different groups under study. The team has come to the conclusion that the kernel density distributions of each group are well represented by the following seven (7) numbers (see Table 25.1) which we have named 'genetic markers': the mean azimuth, the median azimuth, the standard deviation of the azimuthal distribution, the maximum and minimum azimuths of the distribution, and the azimuth of the first and second (if existent) maxima in the azimuth PDF diagrams of the group. In the rare case that the second maximum is non-existent, this is taken to be equal to the first one.

We have used IDL software to produce first the cluster analysis data and then the distances among groups besides the plotting procedures included in the package. In the present study, we have used a weighted pairwise average where the distance between two clusters is defined as the average distance for all pairs of objects between each cluster, weighted by the number of objects in each of the clusters. We have done tests with a nearest-neighbor algorithm and no significant changes are reported. The relative distance is given on the left side of the diagrams. This quantity will be used to find correlations between the different groups.

Finally, the *k*-means clustering links the groups of monuments into clusters by comparing the shape of the kernel density distribution of each data group with a given seed.[9] In each step, the method computes the distance of each distribution to the seeds and then does the grouping. The groups define a new seed by calculating the one in that cluster which is closer to the mean of the cluster, and the process is iterated until it reaches convergence.

3. Comparative Statistical Analysis: a discussion

3.1. *Conspicuous results in Sardinia*

We have calculated the seven 'genetic markers' for each of the 12 groups of monuments considered (see the first twelve rows of Table 25.1). The results are displayed in Figure 25.1 left as a dendrogram. There

---

[7] V. A. Epanechnikov, 'Non-parametric estimation of a multivariate probability density', *Theory of Probability and its Applications* 14, no. 1 (1969): pp. 153–58.

[8] M. Fletcher and G. Lock, *Digging Numbers* (Oxford: Oxford University Committee for Archaeology, 2005), p. 139; A. C. González-García and J. A. Belmonte, 'Statistical Analysis of Megalithic Tomb Orientations in the Iberian Peninsula and Neighbouring Regions', *Journal for the History of Astronomy* 41 (2010): pp. 225–38; A. C. González-García and J. A. Belmonte, 'Sacred architecture orientation across the Mediterranean: a comparative statistical analysis', *Mediterranean Archaeology and Archaeometry* 14, no. 2 (2014): pp. 95–113.

[9] B. S. Everitt, *Cluster Analysis* (London: Wiley, 1995). See also: González-García and Belmonte, 'Sacred architecture orientation across the Mediterranean'.

is a first division into two groups, with six elements in each of them and where they are clearly divided among north and south (with the *Pozi* in an intermediate position), with the notable exception of the *Domus de Janas*. The two groups, north and south, of *Domus de Janas* appear very close indeed in these computations, while the geographic division seems certain for the nuraghe (in both typologies) and the *Tombe di Giganti*. Dolmens, typical in the north, lie in the correct group accordingly.

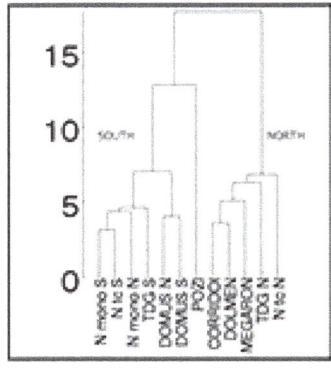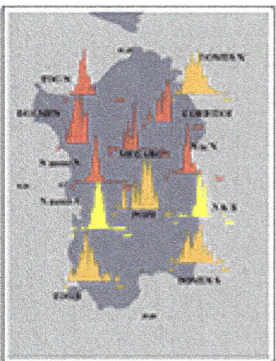

Fig. 25.1: Left, dendrogram for the orientation of megalithic monuments of Sardinia. Right, map showing the results of the k-means analysis. Adapted from González-García et al. (2014).

Figure 25.1 right shows the results for the *k*-means analysis. We have used three seeds for this computation, the *Corridoi*, the southern *Domus* (*Domus S*) and the southern simple nuraghe (*N mono S*). The results do not depend critically on the selected seed, as these change throughout the computation until convergence is reached. The results again present a cluster of mostly northern monuments (TDG North, Dolmen, Megaron, *Corridoi* and the two groups of northern nuraghe), a second cluster with the *Domus* of the north and the south (in agreement to the dendrogram) in the same group, plus some other southern monuments (*Pozi* and the southern TDG) and a third one with the southern nuraghe of both typologies.

Both analyses are independent and both produce similar results with a clear-cut division between the north and the south of the island, with the single exception of the *Domus de Janas*. The results are hence robust. Starting in the Chalcolithic and onwards there seems to be a division at least in the cultic buildings. Such division maintains a tradition closer to the orientation of the *Domus* in the south of the island, while the north seems to introduce a new custom with the dolmens that seems to be maintained later on.

3.2. *Comparison with other western Mediterranean cultural environments*

We have extended the analysis to include other megalithic monuments from the Western Mediterranean. First, we have included the monuments from neighbouring Corsica. These are divided into megalithic cists and dolmens. We have also included the megalithic monuments of the Balearic Islands and Malta, often typologically connected with the Sardinian ones in the literature, including the Maltese temples and the dolmens in two different groups. We have also considered the Languedoc L–dolmens and the BR–dolmens of Provence, and the megalithic monuments of the French departments of Herault and Aude, beside the ones at the department of Pyrenees Orientales. For completeness, we have also included monuments from the southeast of the Iberian Peninsula, such as Los Millares and Montefrio.[10]

---

[10] M. Hoskin, *Temples, tombs and orientations. A new perspective on Mediterranean Prehistory* (Bognor Regis: Ocarina Books, 2001), and references therein. Two exceptions are the Corsican Dolmens and the square talayots of Majorca, for these,

Finally, we have considered the megalithic monuments measured from the north of Africa divided into three groups (Tunisia, Algeria, and Foum al Rjam, in Morocco) together with the Tunisian Hawanat. Thracian dolmens are included in the analysis as a geographic and cultural interloper (see Table 25.1).[11]

Figure 25.2 presents the results of our multivariate analysis. The dendrogram presents a first complete separate section, which includes the Balearic monuments (megalithic monuments and navetas) plus those of southeast France, the Thracian dolmens – a clear geographic interloper – and the *Pozi*. All of them are predominantly orientated towards the setting part of the horizon.

The next division groups monuments from the Pyrenees (Spanish and French), the Corsican dolmens, and those in the north of Africa (Algeria, Tunisia, *Hawanat*) with those in the southern sector of the island of Sardinia, including all the *Domus de Janas*. A similar outcome to what the *k*-means computation yielded. A further division groups the rest of the island monuments (except the Corsican cists) with the related monuments of the Iberian NE and south of France. It is worth noting that the two Maltese groups appear closely related into this cluster. The last cluster finally groups the monuments of the northern part of Sardinia with those from the south of the Iberian Peninsula and Morocco.

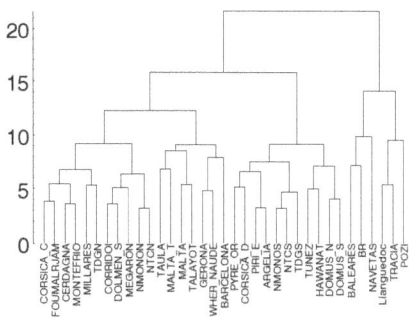

Fig. 25.2: Extended dendrogram for the orientation of megalithic groups in the Western Mediterranean.

---

see: G. Thury-Bouvet, J.-F. Santucci, E. H. Khoumeri, and A. Ottavi, 'Orientations of Corsican Dolmens', *Journal for the History of Astronomy* 37, no. 3 (2006): pp. 299–306; and X. Aramburu-Zabala and J.A. Belmonte, 'On the astronomical orientation of the square talayots of Mallorca', *Archaeoastronomy, The Journal of Astronomy in Culture* 33 (2002), pp. S67–74. See also the relevant chapters in J. A. Belmonte and M. Hoskin, *Reflejo del Cosmos, Atlas de Arqueoastronomía en el Mediterráneo antiguo* (Madrid: Sirius, 2002).

[11] J. A. Belmonte, C. Esteban, and J. J. Jiménez, 'Mediterranean Archaeoastronomy and Archaeotopography: pre-Roman tombs of Africa Procunsularis', *Archaeoastronomy, The Journal of Astronomy in Culture* 23 (1998): pp. S7–24; J. A. Belmonte, C. Esteban, L. Cuesta, M. A. Perera Betancort, J. J. Jímenez González, 'Pre-islamic burial monuments in Northern and Saharan Morocco', *Archaeoastronomy, The Journal of Astronomy in Culture* 24 (1999): pp. S21–34; J. A. Belmonte, M.A. Perera, R. Marrero, and A. Tejera Gaspar, 'The dolmens and *hawanat* of Africa proconsularis revisited', *Journal for the History of Astronomy* 34 (2003): pp. 305–20; and J-F. Santucci and E. H. Khoumeri, 'Orientations of Megalithic tombs in Algeria (1): Djebel Mazela and Rocknia Necropolises, and the Kabylian 'Allées couvertes'', *Journal for the History of Astronomy* 39 (2008): pp, 65–76. For data on the Tharcian dolmens, see: A. C. González-García, D. Z. Kolev, J. A. Belmonte, V. P. Koleva, and L. V. Tsonev, 'On the orientation of Thracian Dolmens', *Archaeoastronomy, The Journal of Astronomy in Culture* 22 (2009): pp. 21–33.

Group	N	<A>	Median	σ	Max A	Min A	1st Max	2nd Max
Corridoi	21	133.4	143.0	35.5	178.0	38.0	162.3	121.0
Dolmen Sardinia	27	137.4	136.0	34.9	198.0	64.0	155.3	131.0
Domus_North	435	151.9	148.0	55.9	356.0	5.0	172.7	124.0
Domus_South	208	164.6	168.0	68.4	356.0	5.0	201.3	113.0
TDG North	180	120.7	121.5	29.3	191.5	35.0	126.0	102.0
TDG South	125	152.4	145.0	64.6	353.5	24.5	136.4	163.0
Megaron	15	142.8	156.0	27.8	180.0	94.0	159.9	120.0
Pozi	20	190.8	209.0	83.0	325.0	27.0	238.0	207.0
Nuraghe Mono N	202	148.05	151.0	24.3	211.0	25.0	155.0	143.0
Nuraghe Mono S	122	167.7	170.0	38.4	356.0	10.0	151.2	171.0
Nuraghe Compl.N	101	148.5	148.0	29.2	234.0	25.0	152.0	122.0
Nuraghe Compl.S	116	164.7	167.0	37.6	329.0	26.0	153.5	171.0
Los Millares	48	105.9	101.0	28.9	216.0	62.0	98.0	111.0
Montefrio	40	107.3	110.0	15.9	138.0	74.0	115.3	105.0
Baleares	9	248.9	245.0	19.1	278.0	220.0	245.1	261.0
Taula	25	174.5	180.0	22.5	210.0	110.0	180.2	168.0
[Square] Talayots	28	159.0	144.0	37.0	242.5	122.0	144.4	129.0
Navetas	18	213.0	210.0	31.0	254.0	159.0	194.6	252.0
Barcelona	10	154.8	140.0	54.9	270.0	91.0	120.8	220.0
Cerdagna	22	117.6	121.0	24.8	161.0	73.0	121.6	85.0
Gerona	22	144.05	148.0	48.5	244.0	55.0	140.4	187.0
Pirineos Este	100	175.6	168.0	47.2	304.0	48.0	165.3	147.0
BR Dolmens	110	246.2	254.0	26.0	301.0	168.0	256.3	221.0
L in Languedoc	103	223.0	222.0	31.1	326.0	108.0	226.9	211.0
WHerault, NAude	36	163.1	165.0	52.1	250.0	20.0	148.7	170.0
Pyrenees Or.	72	166.3	164.0	53.7	293.0	38.0	135.8	111.0
Corsica_Cists	13	112.5	110.0	28.1	153.0	68.0	83.3	122.0
Corsica_Dolmens	28	148.9	147.0	60.4	288.0	59.0	129.8	150.0
Malta	15	150.9	144.5	43.8	264.0	105.5	110.9	159.0
Malta [Dolmens]	15	157.2	149.6	33.9	204.0	92.7	136.4	200.0
Tunez [Dolmens]	213	128.5	129.0	49.0	344.0	30.0	133.3	110.0
Algeria [Dolmens]	317	182.2	182.0	39.4	304.0	50.0	179.9	170.0
Foum al Rjam	41	109.9	106.0	22.6	156.0	33.0	96.1	127.0
Hawanat	89	129.6	136.0	67.3	347.5	0.5	151.6	81.0
Tracia	85	201.6	201.8	44.6	304.0	88.2	200.1	227.0

Table 25.1: Data considered for the cluster analysis. The first 12 rows are for Sardinian data, the rest are for the western Mediterranean groups. The first columns give the name of the different groups (identified either by typology or location) and the number of monuments in each group. The next seven columns include the so-called 'genetic markers': the mean azimuth, the median, the standard deviation, the maximum and the minimum azimuth and the principal and secondary maxima of the PDF histogram.

Broadly speaking these results could indicate the presence of two or three traditions or orientation customs in the western Mediterranean. A first one would include those monuments with orientations mostly towards the setting part of the horizon, basically including those of southern France and the Balearic Islands. A second family would include those groups with orientations mostly towards the SE quadrant of the horizon. This includes the monuments from the Iberian Peninsula and the south of France close to the Pyrenees. The Northern *Tombe di Giganti* present orientations compatible with this family and in general those from the northern sector of Sardinia. We could be talking of a 'northern group' custom. Finally, there are constructions mostly orientated towards the southern part of the horizon, which contain the monuments of Tunisia and southern Sardinia, including the *Domus de Janas* (North and South) and the *Hawanat*. Indeed, we are not arguing that these families are culturally related, at least not all the members of each group, but rather that the orientations of those monuments were possibly connected with similar astronomical events on the horizon.

4. Discussion: Is there any astronomy in all these?

Expanding the last sentence of the previous section we might try to search for an astronomical target that might explain the scheme described above. In order to do so, we include in Figure 25.3 the declination histograms of all the monuments considered in Sardinia, compared in each case with the distribution of a similar sample uniformly distributed in orientation throughout the horizon.

Domus de Janas present a clear distribution of orientation towards the southern half of the horizon, and the maximum is related to the accumulation point towards the south as compared with the position of the corresponding maximum of the uniform distribution to the south. It is interesting to find that a second maximum appears both in the north and the south of the island related to the winter solstice.

The mostly southern orientation is general to all the other monuments of the later periods. However, dolmens do not present a clear solstitial secondary maximum but a rather diffuse maximum in between the solar and lunar extremes.

Interestingly, the north and south dichotomy highlighted in the previous sections is apparent in the presence or absence of these maxima possibly related to the extreme positions of sun and moon. Consequently, the northern Tombe di Giganti and Nuraghe do present, apart from the south maximum (less clear in the Tombe di Gigati) two clear maxima towards both extremes, perhaps indicating both solar and lunar interests.

However the southern Tombe and Nuraghe mostly present the south peak and just a scarce trace of a lunar or solar interest.

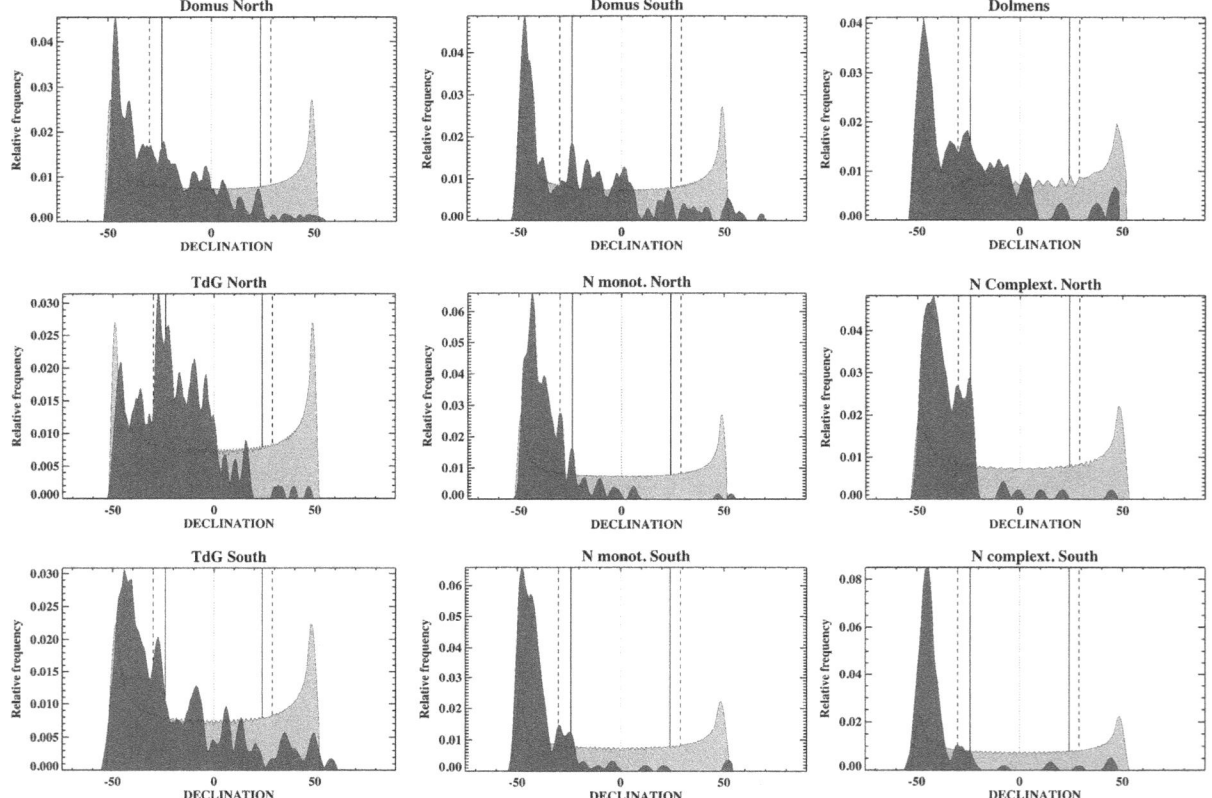

Fig. 25.3: Declination histograms for several groups of monuments in Sardinia (dark grey shade) compare with a homogeneous population of the same number of elements (light grey). TdG stands for Tombe di Giganti and N for Nuraghe.

Conclusions

During the Bonuighinu Culture, from the late 5th millennium BCE onwards, *Domus de Janas* started to be built in Sardinia with an orientating custom that possibly favoured the 'illuminated' southern half of the horizon. This tradition had certain parallelisms with the monuments later built or sculpted in other islands of the western Mediterranean Sea – such as Maltese temples and the taulas of Menorca – and the north of Africa – notably the *hawanat*. This custom spread throughout the insular territory with minor differences between the north and the south of the island. Noteworthy, an interest to the winter solstice also seems apparent.

The situation remained so during the period of maximum construction during the Ozieri Culture (ca. 3800–2700 BCE) but a new tradition of megalithic tombs (predominantly dolmens but also dolmenic corridors) came possibly from the sea and was established in the north of the island. These new burial buildings came with an orientating custom (southeast quadrant of the horizon: related to either the sun or the moon) that was loosely related with those present in other areas of the western Mediterranean region such as the Iberian Peninsula. Such a custom might have been related to a lunar phenomenon that mingled with the mostly solar custom in Sardinia.

During the Bronze Age, the Nuraghic civilization built literally hundreds of cyclopean monuments both for the living (nuraghe or megara) and the dead (*Tombe de Giganti*). The orientating custom was apparently copied from the earlier megalithic tradition, again possibly mixing solar and lunar traditions. However, when the nuraghe spread to the south of the island, the custom relaxed perhaps influenced by the original 'southern' tradition expressed in the orientation towards south of the *Domus de Janas*. Only the sacred wells (*Pozi*) do not fit to the proposed simple scheme and this could be related with an actual difference in the cult performed in those sacred structures – possibly related to the new crescent moon – or with a more prosaic statistical problem due to the small size of the sample.

Acknowledgements
This work has been partially financed within the framework of the projects P/309307 *Arqueoastronomía* of the Instituto de Astrofísica de Canarias, and *Orientatio ad Sidera* III (AYA2011-26756) of the Spanish MINECO. ACGG is a Ramón y Cajal researcher of MINECO.

Bibliography
Aramburu-Zabala, X. and J. A. Belmonte. 'On the astronomical orientation of the square talayots of Mallorca'. *Archaeoastronomy: The Journal of Astronomy in Culture*, 33 (2002): pp. S67–74.
Belmonte, J. A., and M. Hoskin, *Reflejo del Cosmos, Atlas de Arqueoastronomía en el Mediterráneo antiguo*. Madrid: Sirius, 2002.
Belmonte, J.A., C. Esteban and J. J. Jiménez. 'Mediterranean Archaeoastronomy: The Journal of Astronomy in Culture, and Archaeotopography: pre-Roman tombs of Africa Procunsularis'. *Archaeoastronomy: The Journal of Astronomy in Culture*, 23 (1998): pp. S7-24.
Belmonte, J. A., C. Esteban, L. Cuesta, M. A. Perera Betancort, and J. J. Jímenez González. 'Pre-islamic burial monuments in Northern and Saharan Morocco'. *Archaeoastronomy: The Journal of Astronomy in Culture*, 24 (1999): pp. S21–34.
Belmonte, J. A., M. A. Perera, R. Marrero, and A. Tejera Gaspar. 'The dolmens and hawanat of Africa proconsularis revisited'. *Journal for the History of Astronomy* 34 (2003): pp. 305–20.
Depalmas, A. 'Il Bronzo medio della Sardegna'. In *Atti della XLIV Riunione Scientifica Istituto Italiano di Preistoria e Protostoria: La Preistoria e la Protostoria della Sardegna* (Florence: 2009), pp. 123–30, 131–40 and 14–60.
Epanechnikov, V. A. 'Non-parametric estimation of a multivariate probability density'. *Theory of Probability and its Applications* 14, no. 1 (1969): pp. 153–58.
Everitt, B. S. *Cluster Analysis* (London: Wiley, 1995).
Fletcher, M., and G. Lock. *Digging Numbers*. Oxford: Oxford University Committee for Archaeology, 2005.
González-García, A. C., M. P. Zedda, J. A. Belmonte. 'On the orientation of prehistoric Sardinian monuments: a comparative statistical approach'. *Journal of the History of Astronomy* 45 (2014): pp. 467–481.
González-García, A. C., L. Costa Ferrer, and J. A. Belmonte. 'Solarist vs. Lunatics: modelling patterns in megalithic astronomy'. In *Light and Shadows in Cultural Astronomy*, edited by M. P. Zedda and J. A. Belmonte, pp. 23-30. Cagliari: Associazione archeofila sarda, 2007.
González-García, A. C., and J.A. Belmonte. 'Statistical Analysis of Megalithic Tomb Orientations in the Iberian Peninsula and Neighbouring Regions'. *Journal for the History of Astronomy* 41 (2010): pp. 225–38.
González-García, A. C. and J. A. Belmonte. 'Sacred architecture orientation across the Mediterranean: a comparative statistical analysis'. *Mediterranean Archaeology and Archaeometry* 14, no. 2 (2014): pp. 95–113.
González-García, A. C., D. Z. Kolev, J. A. Belmonte, V. P. Koleva, and L.V. Tsonev. 'On the orientation of Thracian Dolmens'. *Archaeoastronomy, The Journal of Astronomy in Culture* 22 (2009): pp. 21-33.
Hoskin, M. *Tombs, temples and their orientations. A new perspective on Mediterranean Prehistory*. Bognor Regis: Ocarina Books, 2001.
Lilliu, G. *La civiltà dei Sardi, dal paelolitico all'età dei nuraghi*. Torino: Nuova Eri, 1988.
Patton, Mark. *Islands in time*. London: Routledge, 1996.
Santucci, J-F., and E. H. Khoumeri. 'Orientations of Megalithic tombs in Algeria (1): Djebel Mazela and Rocknia Necropolises, and the Kabylian 'Allées couvertes''. *Journal for the History of Astronomy* 39 (2008): pp. 65–76.
G. Thury-Bouvet, J.-F. Santucci, E. H. Khoumeri, and A. Ottavi. 'Orientations of Corsican Dolmens'. *Journal for the History of Astronomy* 37, no. 3 (2006): pp. 299–306.

Tanda, G. 'L'ipogeismo in Sardegna: arte, simbologia, religione'. In *L'ipogeismo nel Mediterraneo Orogini, sviluppo, quadri culturali: atti del Congresso Internazionale*, 23-28 May 1994 (Sassari: 2000), pp. 399–426.

Zedda, M. P. *Archeologia del Paesaggio Nuragico*. Cagliari: Agorà nuragica, 2009.

Zedda, M. P. *Astronomia nella Sardegna Preistorica* Cagliari: Agora Nuragica, 2013.

ARCHAEOASTRONOMY IN SICILY: MEGALITHS AND ROCKY SITES

Andrea Orlando[1]

ABSTRACT: The first studies of archaeoastronomy in Sicily date back to the nineteenth century, when some orientations of the Greek temples were measured by German and British archaeologists.[2] After these pioneering studies we had to wait until the end of the twentieth century to find new archaeoastronomical analyses, which involved some important Sicilian prehistoric sites including the Sesi of Pantelleria and the rocky necropoleis of Roccazzo, Tranchina, Castelluccio and Cava Lazzaro.[3]

In this paper I present a review of the work already done on archaeoastronomy across Sicily since the nineteenth and twentieth centuries, as well as the first studies of the twenty-first century, which involves the Gurfa Grottos, the small island of Mozia and the Jato Valley.[4] In addition I present the preliminary analysis of a new campaign of archaeoastronomical measurements taken in 2012 by the author at different archaeological sites in Sicily, including the Argimusco plateau, the Rocca Pizzicata and the Temple of Diana at Cefalù.

Finally I briefly present the archaeological sites chosen for the future campaign of archaeoastronomical investigations in Sicily, which will take place between 2015 and 2018. Among them are sites whose potential archaeoastronomical value is quite interesting: the anaktoron of Pantalica, the pillar structure of Monte San Basilio, the shaft tomb necropolis of Thapsos, the small cromlech of Balze Soprane and the rocky necropoleis of the Alcantara Valley.

1. Introduction: nineteenth and twentieth centuries

The first archaeoastronomical study in Sicily dates from the second half of the 19th century, when the German historian Heinrich Nissen studied the orientations of some Greek temples.[5] At the end of the nineteenth century, other archaeoastronomical studies were undertaken on the Greek temples in Sicily.[6]

---

[1] Osservatorio Astrofisico di Catania (OACT/INAF), Laboratori Nazionali del Sud (LNS/INFN), and Istituto di Archeoastronomia Siciliana (IAS).

[2] H. Nissen, *Das Templum. Antiquarische Untersuchungen* (Berlin: Weidmann, 1869); R. Koldewey and O. Puchstein, *Die griechischen Tempel in Unteritalien und Sicilien* (Berlin: A. Ascher & Company, 1899); F. C. Penrose, 'On the Orientation of Greek Temples, being the Result of Some Observations Taken in Greece and Sicily in the month of May, 1898', *Proceedings of The Royal Society of London* 65, no. 413-422 (1899): pp. 370–375.

[3] S. Tusa, G. Foderà Serio, and M. Hoskin, 'Orientations of the Sesi of Pantelleria', *Journal for the History of Astronomy Supplement* 17 (1992): pp. S15–20; G. Foderà Serio and S. Tusa, 'Rapporti tra morfologia ed orientamento nelle architetture rituali siciliane dal IV al II millennio a.C', *Atti dei convegni lincei-Accademia Nazionale Dei Lincei* 171 (2001): pp. 297–323.

[4] C. Montagna, *Il tesoro di Minos: l'architettura della Gurfa di Alia tra preistoria e misteri* (Palermo: Officina di Studi Medievali, 2009); L. Nigro, 'L'orientamento astrale del Tempio del Kothon di Mozia', *Atti del VII Convegno Nazionale della Società Italiana di Archeoastronomia, Terme di Diocleziano* (2010): pp. 15–24; A. Scuderi, F. Mercadante, P. Lo Cascio and V. F. Polcaro, 'The Astronomically Oriented Megalith of Monte Arcivocalotto', *Anthropological Notebooks* 19 (Supplement) (2013): pp. 213–221.

[5] H. Nissen, *Das Templum. Antiquarische Untersuchungen* (Berlin: Weidmann, 1869); H. Nissen, 'Ueber Tempel-Orientirung II', *Rheinisches Museum für Philologie* (Bad Orb, J.D. Sauerländer Verlag, 1874); H. Nissen, 'Ueber Tempel-Orientirung IV', *Rheinisches Museum für Philologie* (Bad Orb, J.D. Sauerländer Verlag, 1885); and H. Nissen, 'Ueber Tempel-Orientirung V', *Rheinisches Museum für Philologie* (Bad Orb, J.D. Sauerländer Verlag), 1887.

[6] R. Koldewey and O. Puchstein, *Die griechischen Tempel*; Penrose, 'On the Orientation of Greek Temples'; and F. C. Penrose, 'Some Additional Notes on the Orientation of Greek Temples; Being the Result of a Journey to Greece and Sicily in April and May, 1900', *Philosophical Transactions of the Royal Society of London. Series A, Containing Papers of a*

It then took nearly 90 years for there to be a new study about archaeoastronomy in Sicily.[7] This study concerned the Sesi of Pantelleria, funerary megalithic monuments dating back to the Sicilian Ancient Bronze Age (about 2200–1400 BCE). The survey was intended to determine whether the orientations of the entrances of the funeral cells (26.1, on the left), situated within the Sesi, showed a deliberately targeted distribution.

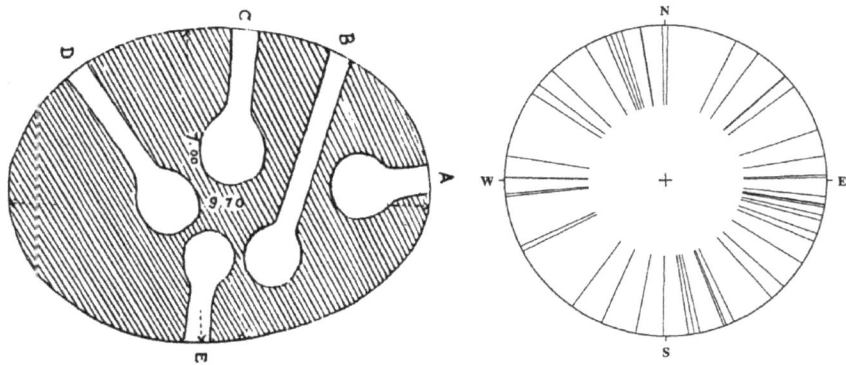

Fig. 26.1: *The plan of the Sese number 31 (on the left; from Orsi, 1899) and the orientations diagram of passages in the Sesi of Pantelleria (on the right; from Tusa et al., 1992).*[8]

The study was carried out in an incomplete manner; nevertheless, the data obtained showed a clear heterogeneous distribution of the orientations, in fact, as shown in Figure 26.1 (on the right) there are no preferred directions in the construction of the passages. This conclusion is very unusual for Bronze Age tombs in the Mediterranean (e.g., see papers by Hoskin and Morales Núñez in 1991; Hoskin in 1992; and Papathanassiou et al. in 1992),[9] but it supports the theory that the Mursia people came from Sicily where their tombs were cut into cliffs, inevitably without regard to orientations.[10]

2. Archaeoastronomical Research in Sicily on the twenty-first century

In 2001, Foderà Serio and Tusa carried out further studies on the orientations of the prehistoric monuments in Sicily. In particular they looked at pseudo-dolmenic structures, rock-cut tombs with dolmenic corridors, shaft tombs and rock-cut tombs built during the fourth to second millennium BCE. It was found that most

---

Mathematical or Physical Character 196 (1901): pp. 389–395.

[7] Tusa et al., 'Orientations of the Sesi of Pantelleria'.

[8] P. Orsi, 'Pantelleria', *Monumenti Antichi* 9 (1899): pp. 449–540; S. Tusa, G. Foderà Serio, and M. Hoskin. 'Orientations of the Sesi of Pantelleria', pp. S15–20.

[9] M. Hoskin and J. J. Morales Núñez, 'The Orientations of the burial monuments of Menorca', *Journal for the History of Astronomy* 22 (1991): pp. S15–S42; M. Hoskin, 'Orientations of megalithic sepulchres in Salamanca, Spain', *Journal for the History of Astronomy* 23 (1992): pp. 57–60; M. Papathanassiou, M. Hoskin, and H. Papadopoulou, 'Orientations of tombs in the late-Minoan cemetery at Armenoi, Crete', *Journal for the History of Astronomy Supplement* 17 (1992): pp. S15–S20.

[10] The protohistoric town of Mursia is dated to the late stages of the Sicilian Bronze Age and is framed within the Culture of Rodì-Tindari-Vallelunga, widespread especially in the northern and western Sicily between the thirteenth and fourteenth centuries BCE. The archaeological area of Mursia in Pantelleria is one of the most important archaeological sites for grade of conservation and scientific interest, as part of the Bronze Age of the central Mediterranean.

of these sites revealed intentional orientations.

With regard to the Eneolithic shaft and rock-cut tombs (fourth to third millennium BCE), the survey involved several necropoleis in the area of Trapani and Agrigento. From the analysis of the orientations of the cells it has emerged that in the underground tombs the cells were cut, in the great majority of cases, with a distribution of orientations in mind. Specifically, the axes of the cells' openings (inside to outside) are oriented within a specified range of directions, known as *sun rising – sun climbing – sun culmination*, namely the one between the sunrise at the summer solstice and the culmination at noon. So it appears that the orientation of the shaft tombs of Roccazzo and Tranchina are part of the typical model widespread in the Mediterranean.[11] However, around the middle of the third millennium BCE, the multi-chambered hypogeic tomb type spread to some Sicilian sites, such as Malpasso and Ciachea. These necropoleis are characterized by a shaft of access in which there are up to four cells, whose orientation is random, and therefore these funerary types no longer fall into the model to which the monosome shaft tombs belong.

With regard to the Bronze Age rock-cut tombs of the Castelluccio culture, Foderà Serio and Tusa's study examined some necropoleis located in Hiblaean territory (south-eastern Sicily).[12] The tombs are generally small in size with a circular or oval plan, convex section and small portal, square or sub-rectangular with rounded regularly edges.[13] The construction of the tombs in this period is strongly influenced by the territorial morphology, so the existence of intentionality orientation in the placement of the tombs is very problematic. The tombs are in fact always placed on almost vertical walls on the sides of river valleys or hills. For example, with regard to Cava Lazzaro, the rock wall where the rock-cut tombs were dug has a general east-west orientation; in this area there are also rock-cut tombs characterized by particular geometric decorations, such as the so-called Tomba del Principe.[14] The orientations of the tombs of Cava Lazzaro were therefore affected by this strong environmental quality and in fact the recorded azimuths are between 350° and 15°. However there are some exceptions to this strong north orientation and, moreover, it is also interesting to note that on the opposite wall to that used for the construction of the necropolis, there are no others tombs. But these considerations are not enough to support the intentionality of the orientations, especially given the fact that in Baravitalla the rocky wall is facing south and in Castelluccio the tombs are arranged on both walls of the Cava della Signora.

Finally it should be remembered that in south-western Sicily there appears in this period the rock-cut tomb with a dolmenic corridor, that is a tomb marked by a long entrance dromos. To this type belong, for example, the tombs of Pergole in the Belice Valley,[15] the area of maximum concentration

---

[11] M. Hoskin, E. Allan, and R, Gralewski. 'Studies in iberian archaeoastronomy: (1) orientations of the megalithic sepulchres of Almeria, Granada and Malaga'. *Journal for the History of Astronomy* 25 (1994): pp. S55–S82; M. Hoskin and M. Zedda, 'Orientations of Sardinian Dolmens', *Archaeoastronomy* 22 (1997): pp. S1–S16; M. Hoskin and T. Palomo y Perez, 'Studies in Iberian Archaeoastronomy: (4) the orientation of megalithic tombs of eastern Catalunya', *Journal for the History of Astronomy* 29 (1998): pp. 63–79; M. Hoskin, 'Studies in Iberian Archaeoastronomy: (5) orientations of the megalithic tombs of northern and western Iberia'. *Journal for the History of Astronomy* 29 (1998): pp. S39–S87.

[12] The culture of *Castelluccio* is an archaeological aspect dating back to Ancient Bronze Age of the prehistoric civilization of Sicily, originally identified by Paolo Orsi on the basis of a particular ceramic style, in the homonymous village located between Noto and Syracuse.

[13] P. Orsi, 'La necropoli sicula di Castelluccio', *BPI* 23 (1892): pp. 1–34, 67–84.

[14] G. Terranova, 'Maltese Temples and Hypogeism. New Data About the Relationships Between Malta and Sicily During the III and II Millennium BC', in *AA.VV., Exploring the Maltese Prehistoric Temple Culture, EMPTC 2003 Conference*, pp. 1–24. 2003.

[15] G. Mannino, 'La tomba di contrada Pergola', *Sicilia Archeologica* 15 (1971): pp. 52–56; S. Tusa, *L'insediamento dell'età del bronzo con Bicchiere Campaniforme di Marcita* (Trapani: Corrao Editore, 1997).

of the Bell-Beaker culture in Sicily (e.g., see papers by Bovio Marconi in 1963 and Tusa in 1987).[16] The archaeoastronomical study of Foderà Serio and Tusa into the orientations of this funerary type has allowed the measurement of fifteen tombs in the territory of Partanna and discovered how these tombs are placed in a different orientation type than the rest of the island. The orientations space out in fact in the second and third quadrant, between east and west; this orientation characteristic is linked to the presence of the Bell-Beaker culture, imbued with Sardinian-European elements.

So while in Malta and especially in Sardinia archaeoastronomical studies have been numerous over the past 50 years (e.g., see Agius and Ventura, 1981; Foderà Serio et al., 1992; Ventura et al., 1993; Hoskin and Zedda, 1997), only in recent years have new investigations involved Sicily, the largest of the Mediterranean islands.[17]

In the following paragraphs I present a review of the studies of archaeoastronomy realized in Sicily in the last 10 years. In the map shown in Figure 26.2 I have marked the sites involved in the archaeoastronomical studies in Sicily with coloured markers. In particular I marked with black marker the studies made at the end of the twentieth century and at the begin of the twenty-first century, while the orange marker indicates studies conducted in western Sicily in the last 10 years, the purple marker those ongoing studies carried out by the author, and the green marker indicates sites that will be studied in the near future.

Fig. 26.2: Map of the Sicily with the various markers that indicates the archaeological sites involved in the archaeoastronomical studies.

---

[16] J. Bovio Marconi, 'Sulla diffusione del bicchiere campaniforme in Sicilia', *Kokalos* 9 (1963): pp. 93–128; S. Tusa, 'The Bell Beaker in Sicily', in *Bell Beakers of the Western Mediterranean*, Vol. 331, ed. W. H. Waldren and R. C. Kennard (Oxford: British Archaeological Reports, 1987), pp. 523–550.

[17] G. Agius and F. Ventura, 'Investigation into the possible astronomical alignments of the Copper Age temples in Malta', *Archaeoastronomy: Bulletin of the Center for Archaeoastronomy* 4 (1981): pp. 10–21; G. Foderà Serio, G., M. Hoskin, and F. Ventura. 'The orientations of the temples of Malta'. *Journal for the History of Astronomy* 23 (1992): pp. 107–119; F. Ventura, G. Foderà Serio, and M. Hoskin, 'Possible tally stones at Mnajdra, Malta', *Journal for the History of Astronomy* 24 (1993): pp. 171–183; M. Hoskin and M. Zedda, 'Orientations of Sardinian Dolmens', *Journal for the History of Astronomy* 28 (1997): pp. S1-S16.

2.1 The Gurfa Grottos

In western Sicily, in the territory of Alia near Palermo, there is another interesting rocky site, the so-called Gurfa Grottos. This site is an outstanding example of rocky architecture; in fact Gurfa can be considered a sort of palace excavated in sandstone and is made up of six rooms located on two levels. On the lower level there are two entrance doors; one leads to a wide trapezoidal environment, with the ceiling at two slopings, called the 'tent room'. The second entrance, offset from the axis of the shaft plant, leads into a big room with an oval plan (about 12.85 x 13.30m), where the section is reminiscent of a *tholos*.[18] This spectacular cavity is one of the biggest rock *tholos* of the Mediterranean area, its size exceeds the famous so-called 'Greek Treasures', the magnificent *tholos* of Atreus in Mycenae (Argolida) and Minyas in Orchemenos (Boeotia).[19] The upper compartments of the complex are reached, from the outside, through a ladder carved into the rock that leads, by means of a reduced entrance, into a small room. On this level there are four rooms with a rectangular plan and that communicate with each other by means of short corridors. All these rooms are lit by large windows that are open on the southwest wall of the massive and looking towards the valley. The origins of the Gurfa are not so clear, nor its history, however it is important to note that upstream of the cavities there are some rock-cut tombs dating back to the Copper Age.[20]

The Gurfa Grottos have been the subject of study over many years by Carmelo Montagna, the Sicilian architect and art historian. Of particular interest is the archaeoastronomical approach to the bell room. It has been observed that the ray of light that filters through the hole (about 80cm in diameter) on the top of the *tholos* can act as a real sundial. Furthermore, on the southern side of the bell cavity, there is another hole, smaller in size, through which a ray of light hits the centre of the floor of the *tholos* at the equinoxes. But further studies are certainly needed to say that the Gurfa's *tholos* is a precursor of sundials in the darkroom that we still find today in churches or in ancient places of worship, first among them the Pantheon in Rome.[21]

2.2 Mozia and the Temple of Kothon

From many years a group of archaeologists of the University La Sapienza in Rome have studied the little island of Mozia in western Sicily.[22] Mozia was a Phoenician city, its foundation probably dates from the eighth century BCE, about a century after the foundation of Carthage in Tunisia. Particularly interesting is the study of the so-called Temple of Kothon.[23] The monumental entrance facing east and other installations along the median axis east-west have powerfully proposed the theme of the orientation of the Kothon's temple, which, although structurally connected with the artificial basin through a channel, did not have its own orientation. The Kothon, in fact, is oriented with the corners facing to the cardinal points (with the diagonal axis oriented in north-south direction), while the Temple presents a slight misalignment with respect to the cardinal points.

The study of the sky at Mozia in the antiquity by Nigro in 2010 showed that the facade of the temple

---

[18] The term *thòlos* refers to a burial structure characterized by its false dome created by the superposition of successively smaller rings of mudbricks or, more often, stones.

[19] C. Montagna, *Il tesoro di Minos: l'architettura della Gurfa di Alia tra preistoria e misteri*.

[20] S. Braida Santamaura, 'Le grotte della Gurfa', *Incontri e Iniziative, Memorie del centro cultura di Cefalù* I (1984): pp. 33–50.

[21] R. Hannah and G. Magli, 'The role of the sun in the Pantheon's design and meaning', *Numen* 58, no. 4 (2011): pp. 486–513.

[22] L. Nigro and G. Rossoni, eds., *"La Sapienza" a Mozia. Quarant'anni di ricerca archeologica (1964-2004)* (Roma: Missione Archeologica Mozia, 2004).

[23] The term *khoton* refers to the artificial dam reservoir characteristic of Phoenician ports.

and monuments erected in its central court were aligned with the constellation of Orion (*Baal* for the Phoenicians) when it rose at the winter solstice.[24] The temple was therefore oriented towards the point where the figure of the victorious god appeared, announcing the gradual return of the sunlight and the end of the winter night. The identification of Orion with Baal is considered very likely thanks to the extraordinary testimony offered from the accurate decoration in etching and punching of the Foroughi Cup, in which the god, dashed over the stars of the constellation, occupies the central medallion and insists on a ram, a likely indication of the winter solstice.[25]

2.3 The Jato Valley

In the Jato Valley, not far from Palermo, there are the so-called holed stones. These natural rocks, characterized by a central hole with a diameter of about 2m, have been studied by a team of Italian scholars in recent years. The Jato Valley, inhabited at least since the Neolithic period, strongly developed in the Eneolithic and in the Bronze Age. From the archaeoastronomical analysis it was discovered that the sun rises exactly on the centre of the hole of the Arcivocalotto's stone at the winter solstice, and this highlights the calendric purpose of the artefact.[26] At some kilometres from Monte Arcivocalotto it is possible to watch the sunrise at the summer solstice from the holed stone of Cozzo Perciata.[27] Unfortunately this stone, which is a similar size to that of Monte Arcivocalotto, is now broken on the top of the hill, because thirty years ago it was struck by lightning.

3. New research: preliminary analysis

Since 2012 I have carried out new studies of archaeoastronomy in Sicily, being aware of that few studies were carried out until a few years ago (Fig. 26.2) and being convinced of the great archaeoastronomical potential of numerous archaeological sites on the island. In the following paragraphs I present some of these places with their geological and archeological frameworks, their archaeoastronomical potential and the preliminary analysis. The method used in this first phase of the study are as follows: satellite images were analysed with GE's tool to get information about the azimuth, while for field measurements sophisticated technological instruments were used, including GPS, drones for photographs from the sky, total station, theodolite, magnetic compass and clinometer.

3.1 Argimusco Plateau

The naturalistic area of the Argimusco covers a wide plateau between 1100 and 1200m above sea level, in the centre of the Abacenino territory (in northern Sicily). The so-called *Argimusco Rocks* represent one of the biggest rocky natural sites of the whole of southern Italy. The Argimusco plateau is part of the natural reserve Bosco di Malabotta. As regards the geological framework in the Argimusco area, the most important outcrops are the Upper Oligocene-Lower Miocene siliciclastic turbidites of the Capo d'Orlando Flysch (COD) and secondarily the Floresta Calcarenites (CFL – Upper Burdigalian-Langhian) and the Antisicilide Complex (ASI – Upper Cretaceous-Lower Miocene).

The first historical information on the Argimusco date backs to the Middle Ages, with particular

---

[24] L. Nigro, 'L'orientamento astrale del Tempio del Kothon di Mozia'.
[25] The artefact is now housed in Shlomo Moussaieff Collection in London. L. Nigro, 'The Temple of the Kothon at Motya (Sicily): Phoenician Religious', in *All the Wisdom of the East*, ed. M. Gruber et al. (Fribourg: Academic Press Fribourg/ Vandenhoeck & Ruprecht Göttingen, 2012), pp. 293–331.
[26] A. Scuderi et al., 'The Astronomically Oriented Megalith of Monte Arcivocalotto'.
[27] A. Scuderi, V. F. Polcaro, and F. Maurici, 'New Archaeoastronomical findings in the alto Belice Valley (Sicily)', *Mediterranean Archaeology and Archaeometry* 14 (2014): pp. 93–98.

reference to an epistle (dated 16th July 1308) of Frederick III of Sicily addressed to James II of Aragon[28]. But it is the action of nature, mainly wind and water, which has shaped the huge rocks, creating rocks with particular human and animal shapes.[29] Among them, visitors are mesmerized by the charm of the so-called *Eagle's megalith*, a rock with the form of the bird of prey. There are also the so-called *Priest's rock*, a kind of human profile that has a hole that is reminiscent of an eye, and the so-called Praying's megalith, a female figure with hands clasped whose profile, 25m high, is enhanced by the soft light of sunset.

From surface surveys and archaeological finds in the surrounding area there was found pottery of the Roman and Byzantine age and also some dating back to the Bronze Age. Unfortunately there have never been any official archaeological excavations on the Argimusco plateau, so it is very difficult to date the inhabitation of the site. To discover its secrets we need to follow the signs that man has left on the plateau and on the surrounding area. The numerous anthropogenic signs on the Argimusco plateau allow us to frame the area in the pre-protohistoric age. On the so-called *Water's Rock*, for example, we find a tub or a rock tomb with a rectangular plan, and not far from it there is a sort of altar characterized by small ovals tubs and an arched corridor. Coming down from this cliff we find the so-called *Stone of the Seven Steps*, a privileged place for observations towards the western horizon.

Among the artefacts that show the unmistakable signs of man it is very important finally to remember the rock-cut tomb and the megalithic millstone, which includes a particular pentagonal tub.

From the Argimusco plateau is possible to observe all the four horizons, but the eastern one has a particular profile, suitable for creating what is generally called an horizon calendar. On the eastern mountainous profile is silhouetted the Rocca Novara (about 1340m), who serves still today as an indicator of the equinoxes. Originally I thought that the place where the observations at dawn were held was precisely at Water's Rock, and in particular the so-called *altar*. This is formed by a natural rock platform on which there are three steps cut into the rock, and where there are two small tubs used to collect water (Fig. 26.3) and a walkway that runs from east to west.[30]

Probably on this altar some ancient rites of worship took place; in fact, going up the three steps we meet the first tub immediately, then the second one and finally we are invited to walk the corridor from west to east. The corridor ends with a cliff, which appears to stretch out toward the Rocca Novara. However, as shown in Figure 26.3, if we consider the altar as a place to observe the eastern horizon, it should be noted that the azimuth of Novara Rocca is about 85°. Instead, from the rock called *Tower*, located in the northern part of the plateau, the azimuth of Rocca Novara is about 88°. The Tower is a very special rock, it has an almost cubic shape, and in front of it, on the eastern side, there is a kind of pseudo-natural dolmen. The Rocca Novara with its highest peak, still provides clear information on the alternation of the seasons, becoming a real equinoctial marker. [31]

---

[28] F. Giunta and A. Giuffrida (eds.), *Acta Sicula-Aragonensia II: Documenti sulla luogotenenza di Federico d'Aragona* (Palermo, 1972).

[29] G. Todaro, *Alla ricerca di Abaceno* (Messina: Armando Siciliano Editore, 1992); G. Pantano, *Megaliti di Sicilia* (Patti: Edizioni Fotocolor, 1994).

[30] These small tubs are very similar, for example, to those found in Basilicata, to 'Petre de la Mola', site of proven archaeoastronomical value. See E. Curti, M. Mucciarelli, V. F. Polcaro, C. Prascina and N. Witte, 'The "Petre de la Mola" megalithic complex on the Monte Croccia (Basilicata)', in *Proceedings of the SEAC 17th annual meeting From Alexandria to al-Iskandariya, astronomy and culture in the ancient Mediterranean and beyond, October 25th-31st 2009, Alexandria, Egypt*, ed. M. Shaltout and A. Maravelia (Oxford: British Archaeological Reports, 2009); and V.F. Polcaro and D. Ienna, 'The Megalithic Complex of the «Preta 'ru Mulacchio» on the Monte della Stella', in *Cosmology Across Cultures*, ASP Conference Series 409 (2009).

[31] A. Orlando, 'The Argimusco's Plateau and the Horizon's Calendar', in the *Proceedings of SEAC 2014*, held in Malta

Fig. 26.3: On the top: one of the rocky tubs present on the Water's Rock (photo by Massimo Calcagno); on the bottom: the azimuths of Rocca Novara measured from the altar (green line) and the Tower (red line).

3.2 Rocca Pizzicata

In the eastern Sicily, on the Alcantara Valley, there is Rocca Pizzicata, a site located at about 700m above sea level. This rocky site, known in the Medieval period as Petra Intossicata, is part of a wider system of rocky sites that involves the whole Alcantara Valley.[32]

The main structure of the Rocca Pizzicata is an altar, now surrounded by a little oak forest. The artefact shows itself as a complex cut in the sandstone rock, from which it was obtained a rectangular table surrounded on three sides by walls that leave exposed the east side. The altar is oriented toward the east, in fact the azimuth is about 90°, the direction in which the sun rises on the equinoxes.

(Lampeter: Sophia Centre Press, in press), this volume, specifically Argimusco's poster presented at the SEAC2014 meeting.

[32] G.L. Barberi, in G. Silvestri, ed., *I Capibrevi, Vol. II: I feudi del Val di Demina* (Palermo: 1886).

On the Rocca Pizzicata there are other artefacts, like a rock-cut tomb, crosses carved on different walls, and an inscription carved into the rock.[33] Moreover there is another interesting altar that sits on the top of Rocca Pizzicata. This one, however, is not a table but only carved walls dug into the sandstone. Lying on top of Rocca Pizzicata, with a southeast orientation, its function of observation towards the Alcantara Valley and the Etna volcano seems certain.

3.3 The Cefalù Rock and the Temple of Diana

The megalithic building called the Temple of Diana (Fig. 26.4) is located at about 150m above sea level on the imposing Cefalù Rock, the limestone massif that dominates the village of Cefalù in northern Sicily.

Fig. 26.4: The so-called Temple of Diana on the Cefalù Rock (from Houel, 1785).[34]

The megalithic temple seems to date back to the sixth and fifth centuries BCE, while the rocky tank with a dolmenic cover incorporated on the temple is considered of protohistoric age by many scholars.[35] The Cefalù Rock is linked to the myth of Daphnis,[36] and probably the sacred importance of the megalithic complex is linked to a worship of water.[37]

The first relief of the temple of Diana was performed by George Nott, an Anglican clergyman, scholar and lover of Italian culture, who lived between the eighteenth and nineteenth centuries[38]. At the beginning

---

[33] V. Platania and M. S. Scaravilli, 'Il complesso rupestre di Rocca Pizzicata', *Agorà* 45 (2013): pp. 123–125.

[34] J. Houel, *Voyages pittoresques des îles de Sicile, de Lipari et de Malte où l'on traite des antiquités qui s'y trouvent encore; des principaux phénomènes que la nature yoffre; du costume des habitants, & de quelques usages, vol. iv Paris* (Paris: Imprimerie de Monsieur, 1787), p. 92, tav. XLIX-LI, Parigi, 1785.

[35] P. Marconi, 'Cefalù (Palermo). Il cosiddetto «Tempio di Diana»', *Notizie degli Scavi* 5 (1929): pp. 273–395; V. Tusa, *Cefalù*, in J. W. Hayes, *Enciclopedia dell'Arte Antica II* (Rome: Rome: Instituto della Enciclopedia, 1959), p. 453; A. Tullio, *Memoria di Cefalù* (Palermo: Kefagrafica Lo Giudice, 1994).

[36] In Greek mythology Daphnis was a Sicilian shepherd who was said to be the inventor of pastoral poetry. According to tradition, he was the son of Hermes and a nymph, although Daphnis himself was mortal.

[37] A. Franco, *Le Radici e le Pietre* (Cefalù: Lorenzo Misuraca Editore, 2012).

[38] G.F. Nott, 'Avanzi di Cefalù', *Monumenti Inediti dell'Instituto di Corrispondenza Archeologica* (1831): pp. 270-287, tavv.

of 2014 an archaeoastronomical analysis on the Temple of Diana was begun. The preliminary study shows how the orientation of the front door of the megalithic temple has a well-defined direction: east-west. The azimuth of the entrance of the temple is about 270°, and this indicates perfectly the direction in which the sun sets over the period of the equinoxes (spring and autumn). So the megalithic structure is to be considered a real sun temple; moreover, considering the history of the building and the presence of a water source, we can consider the Greek temple to be an Artemision, a place where in antiquity rituals dedicated to the water were held.

4. Future Works

There are many archaeological sites that in the near future will be protagonists of new studies of archaeoastronomy in Sicily (Fig. 26.4). In the next paragraphs are a brief presentation of the archaeological sites chosen and their archaeoastronomical potential.

4.1 Alcantara Valley

Near Naxos, the first Greek colony in Sicily, there is Petra Perciata, a natural outpost place at the eastern entrance of the Valley.[39] Going westbound we meet other places, such as Rocca Perciata, Motta della Placa, Monti Cucco-Orgali, Pietra Marina, Monte Passo Moio and Serra Cinquonze, and pushing further into the western foothills of the Valley leads up to the so-called Rocca dei Giganti, not far from the extraordinary place of Balze Soprane where there are ancient lava flows frequented since Neolithic times.[40] The history of the Alcantara Valley is millenary but is yet to be written, because it has never been done an organic scientific project of inquiry that concerns its many anthropic rocky sites until now. The studies carried out in the area have involved only a few sites, such as the Sicilian necropolis of Cocolonazzo di Mola, some lava caves and some rock-cut tombs.[41]

The study of archaeoastronomy to be held in the Alcantara Valley will be aimed primarily at the study of the orientations of the rock-cut tombs. The study will then be compared with the broader work done to date on the rocky necropoleis in Sicily.[42] This study, once begun, will also survey the numerous rocky millstones present in the Alcantara Valley, artefacts that often are located just near the rocky necropoleis.[43]

4.2 Thapsos

Thapsos is the ancient name of the small Magnisi peninsula, located on the eastern coast of Sicily between the gulfs of Syracuse and Augusta. In Thapsos there developed a culture during the Middle Bronze Age (about 1400–1220 BCE) that had a vast expansion in Sicily.[44] Particularly sensitive in this period is the

---

XXVIII-XXIX.
[39] E. Gabrici, 'La Niobide di Taormina', *Monumenti Antichi* 41 (1951).
[40] B. Radice, *Memorie storiche di Bronte* (Bronte: Stabilimento tipografico sociale, 1936); F. Privitera, 'Necropoli tardo-neolitica in Contrada Balze Soprane di Bronte (CT)', *Atti della XLI Riunione scientifica IIPP: Dai Ciclopi agli Ecisti: società e territorio nella Sicilia preistorica e protostorica* (2011).
[41] For the Sicilian necropolis of Cocolonazzo di Mola, see P. Orsi, 'Taormina. Necropoli sicula al Cocolonazzo di Mola', *Notizie degli Scavi* 16 (1919): pp. 360–369; for the lava caves, see F. Privitera, 'Castiglione di Sicilia, contrada Marca. Grotta sepolcrale della tarda età del Rame del Bronzo Antico', *BCA Sicilia 1-2*, Palermo (1991–1992): pp. 21–25; and for the rock-cut tombs, see M. Privitera, 'Sepolture rupestri nella Valle dell'Alcantara', *Kokalos* 47-48, no. II (2009).
[42] Foderà Serio and Tusa, 'Rapporti tra morfologia ed orientamento nelle architetture rituali siciliane dal IV al II millennio a.C'.
[43] S. Puglisi, *La Valle dei Palmenti* (Messina: Armando Siciliano Editore, 2009).
[44] F.S. Cavallari, 'Thapsos. Appendice alla memoria: Le città e le opere di escavasene in Sicilia anteriori ai Greci',

influence of the Aegean civilization, moreover there are many Mycenaean imports.[45]

The prehistoric necropolis of Thapsos is split over three areas of the Magnisi peninsula. The necropolis of Thapsos is formed by rock-cut tombs with dromoi, shaft tombs and enchytrismòs tombs.[46] In particular the archaeoastronomical study will focus initially only on the shaft tombs, whose total number is about 80. The study of the orientations of the Thapsos' shaft tombs is very important because it will complete that achieved in western Sicily and will allow the realization of the first plan of the various shaft necropoleis located on the islet.[47]

4.3 Mura Pregne

Not far from Palermo, near the Tyrrhenian coast, there is the site of Mura Pregne. At the foot of Mount San Calogero, in the Brucato contrada, there are some important examples of megalithic structures, like a dolmen and some walls. Is certain that the use of the site started in prehistory (Neolithic) and ended in the Medieval period. The complex of Mura Pregne is probably the only example of village, in the hinterland, that predates the foundation of the Greek colony of Himera and survived even after the tragic end of the city.[48] Unfortunately, in the middle of last century, the Lambertini company used the archaeological site as a quarry for about 30 years, so what remains today is only a small part of the extraordinary area (Fig. 26.5). In particular the study will involve the orientation of the dolmen, whose measurements will be compared with the broader study on the orientations of the Sicilian and Mediterranean dolmens, including those of Hoskin and Zedda in 1997, González-García et al. in 2009 and Hoskin in 2011.[49]

Archivio storico siciliano Ser. NS, vol. 5 (1880): pp. 121-137; P. Orsi, 'Thapsos', *Monumenti Antichi* vol. VI (1885): pp. 89-150; L. Bernabò Brea, 'Thapsos (Augusta - Siracusa). Scavo di tombe nella necropoli dell'età del Bronzo (XIV- inizi XIII sec. a.C.)', *Bolletino d'arte* I-II I-II (1966); Bernabò Brea, L. 'Thapsos (Augusta - Siracusa). Scavo di capanne del villaggio dell'età del Bronzo (XIV- inizi XIII sec. a.C.)'. *Bolletino d'arte* (1966).

[45] S. Tusa, *La Sicilia nella preistoria* (Palermo: Sellerio, 1999).

[46] C. Veca, 'Contenitori "per i vivi" e contenitori "per i morti" a Thapsos (Siracusa): un approccio tecnologico a un problema interpretativo', *Rivista di Scienze Preistoriche* 64 (2014): pp. 203–225.

[47] Foderà Serio and Tusa, 'Rapporti tra morfologia ed orientamento nelle architetture rituali siciliane dal IV al II millennio a.C'.

[48] S. Vassallo, 'L'enigma del muro megalitico e dello pseudo-dolmen di Mura Pregne', in *From Cave to Dolmen. Ritual and symbolic aspects in the prehistory between Sciacca, Sicily and the central Mediterranean*, ed. D. Gullì (Oxford: Archaeopress, 2014), pp. 247–253.

[49] Hoskin and Zedda, 'Orientations of Sardinian Dolmens'; C. A. González-García, D. Z. Kolev, J. A. Belmonte, V. P. Koleva, and L. P. Tsonev, 'On the Orientation of Thracian Dolmens', *Archaeoastronomy* 23 (2009); M. Hoskin, 'Orientations of dolmens in Western Europe', *Proceedings of the International Astronomical Union* 5, no. S260 (2009): pp. 116-126.

Fig. 26.5: Planimetric sketch of the north wall (from Mauceri, 1896).[50]

4.4 Balze Soprane

In the territory of Bronte, on the western slope of Mount Etna, where there are millenary volcanic lavas, is situated one of the most enigmatic and least known structures of Sicily: the so-called megalithic spiral *of* Balze Soprane.[51] It consists of ten large lava stone slabs (average width 0.80/0.90m; average height 1.40/1.60m; average thickness of about 0.20/0.30m), only rough-hewn, placed vertically and aligned to form a sort of spiral. The maximum outer diameter is about 5m, while the interior is only 2.60/3.0m. Around its perimeter are several blocks, partly leaning against the slope which lay to the west, forming a sort of corridor connected to the megalithic spiral. The construction of Balze Soprane – as it seems similar to existing structures in Ragusa territory, like the burial enclosure of Contrada Paolina[52] or as a burial formed by slabs seats lining the walls of a pit in Monte Racello[53] – does not seem to be connected to the funerary sphere, given the lack of a proper burial chamber and a direct path to it from the outside. The

---

[50] L. Mauceri, *Sopra un'Acropoli pelasgica esistente nei dintorni di Termini Imerese* (Palermo, 1896).
[51] O. Palio and M. Turco, 'Strutture megalitiche nell'area etnea', *Notiziario di Preistoria e Protostoria* 2, no. 2 (2015): pp. 49–51.
[52] G. Di Stefano, 'Insediamenti e necropoli dell'Antico Bronzo dell'area iblea e Malta: contatti o influenze?', in *Malta in the Hybleans, the Hybleans in Malta*, ed. A. Bonanno and P. Militello (Palermo: Officina di Studi Medievali, 2008), pp. 49–54.
[53] P. Orsi, 'Miniere di selce e sepolcreti eneolitici a Monte Tabuto e Monte Racello presso Comiso', *Bullettino di Paletnologia Italiana* 24 (1898): pp. 165–206.

spiral structure, and the presence of a closed area around, seems more appropriate to a ritual destination, perhaps connected to initiatory ceremonies, given the complex access path to the central room.

The study will aim to determine the entrance's orientation of this megalithic structure, unique for its particular shape in Sicily and in the Mediterranean region.

4.5 Pantalica

Pantalica is a natural and archaeological locality of the province of Syracuse and is one of the most important prehistoric sites in Sicily, useful to understand the moment of transition from the Bronze Age to the Iron Age on the island. In 2005 the site was awarded, along with the city of Syracuse, the title of World Heritage Site by UNESCO for its high historical, archaeological, speleological and landscaping profile.

Pantalica is located on a plateau surrounded by canyons formed over millennia by two rivers, the Anapo and Calcinara. Pantalica is dotted with necropoleis across its vast territory, there are in fact about 5000 tombs; there are also many Byzantine churches and villages, and an acropolis with the remains of the Anaktoron, the so-called 'Prince's Palace'.[54] The acropolis of Pantalica is located on one of the highest point of the area, in a location where it dominates the view of the canyons and from where it controlled the possible arrival of enemies.

The study of archaeoastronomy in Pantalica will not be dedicated to the study of the thousands of rock-cut tombs, since it is more than obvious that it was the morphology of the place, or the canyons, that dictate their orientation characteristics; the study will instead focus on the orientation of the Anaktoron, located in a prime location and most likely also related to the observation of the stars to the horizons.

4.6 Monte San Basilio

The so-called Monte San Basilio is a hill feature that stands isolated near the town of Scordia, in the province of Syracuse. On this hill appear vestiges of considerable importance, both for their consistency and for the period to which they refer, which goes from Ancient Bronze Age to the late Middle Ages.

In the mid nineteenth century the first study of the site was achieved by Di Mauro, a Sicilian scholar, while the first archaeological excavations were conducted by Paolo Orsi in 1899 and from 1922–1924.[55]

Among the most significant discoveries were the walls and the pillar structure, already immortalized by Jean Houel in a famous painting in the eighteenth century. The construction excavated in the rock consist of a large rectangular room (about 18 × 16m) aligned to the cardinal points. The structure is characterized by thirty pillars, also cut into the rock, an access ladder and a cover with huge slabs of the same sandstone. This particular architecture, which is on top of the hill, was considered by Orsi as a water tank, while a recent hypothesis considers it to be a barn.[56]

The study will be dedicated to the relief of the rectangular structure and observations of the horizons.

5. Conclusions

In this work, I have reviewed the current status of research on archaeoastronomy in Sicily, summarising past and recent studies (see Section 1). Since the eighteenth century scholars have started to appreciate the archaeoastronomical potential of the Greek temples in Sicily, but until a few years ago, very few studies had been conducted on the island (see Section 2).

---

[54] Orsi, 'Miniere di selce e sepolcreti eneolitici a Monte Tabuto e Monte Racello presso Comiso'.

[55] M. Di Mauro, *Sul Colle di S. Basilio* (Catania, 1861).

[56] E. De Magistris, 'Granai pubblici di età romana', *La Parola del Passato-Rivista di Studi Antichi* 67 (2012): pp. 321–362; E. De Magistris, 'L'ipogeo di San Nicolò dei Cordari a Siracusa: fasi costruttive e funzione', *Journal of Ancient Topography* 24 (2015).

I have shown that there has been begun a new campaign of archaeoastronomical investigations that involve several sites in northern and eastern Sicily (see Section 3). This new research campaign, of which the preliminary data only have been presented, represents a new starting point for multidisciplinary studies carried out using the most modern topographic and geodetic survey techniques.

It is now a well-established fact in the archaeological domain that orientation studies are useful as a source of information that may shed light on a number of anthropological issues such as beliefs systems or landscape and territory acquisition by past cultures. In future, I aim to extend the analysis to others archaeological sites in Sicily that show archaeoastronomical potential (see Section 4), so that such activity can make an important contribution to the enhancement of all these archaeological sites, most of which are in a total state of disrepair and neglect.

Acknowledgements
I want to thank the members of the Institute of Sicilian Archaeoastronomy for their support, especially Dr Alessandra Alberici and Dr David Gori for their valuable collaboration and availability.

Bibliography
Agius, G., and F. Ventura. 'Investigation into the possible astronomical alignments of the Copper Age temples in Malta'. *Archaeoastronomy: Bulletin of the Center for Archaeoastronomy* 4 (1981): pp. 10–21.

Barberi, G.L., in G. Silvestri, ed., *I Capibrevi, Vol. II: I feudi del Val di Demina*. Palermo, 1886.

Bernabò Brea, L. 'Thapsos (Augusta - Siracusa). Scavo di tombe nella necropoli dell'età del Bronzo (XIV- inizi XIII sec. a.C.')'. *Bollettino d'arte I-II* (1966).

Bernabò Brea, L. 'Thapsos (Augusta - Siracusa). Scavo di capanne del villaggio dell'età del Bronzo (XIV- inizi XIII sec. a.C.)'. *Bollettino d'arte I-II* (1966).

Bovio Marconi, J.' Sulla diffusione del bicchiere campaniforme in Sicilia'. *Kokalos* 9 (1963): pp. 93–128.

Braida Santamaura, S. 'Le grotte della Gurfa'. *Incontri e Iniziative, Memorie del centro cultura di Cefalù* I (1984): pp. 33–50.

Cavallari, F. S. 'Thapsos. Appendice alla memoria: Le città e le opere di escavasene in Sicilia anteriori ai Greci'. *Archivio storico siciliano* Ser. NS, vol. 5 (1880): pp. 121-137.

Curti, E., M. Mucciarelli, V. F. Polcaro, C. Prascina and N. Witte. 'The "Petre de la Mola" megalithic complex on the Monte Croccia (Basilicata)'. In *Proceedings of the SEAC 17th annual meeting From Alexandria to al-Iskandariya, astronomy and culture in the ancient Mediterranean and beyond, October 25th-31st 2009, Alexandria, Egypt*. Edited by M. Shaltout and A. Maravelia. Oxford: British Archaeological Reports, 2009.

De Magistris, E. 'Granai pubblici di età romana'. *La Parola del Passato-Rivista di Studi Antichi* 67 (2012): pp. 321–362.

De Magistris, E. 'L'ipogeo di San Nicolò dei Cordari a Siracusa: fasi costruttive e funzione'. *Journal of Ancient Topography* 24 (2015).

Di Mauro, M. *Sul Colle di S. Basilio*. Catania, 1861.

Di Stefano, G. 'Insediamenti e necropoli dell'Antico Bronzo dell'area iblea e Malta: contatti o influenze?'. In *Malta in the Hybleans, the Hybleans in Malta*. Edited by A. Bonanno and P. Militello, pp. 49–54. Palermo: Officina di Studi Medievali, 2008.

Franco, A. *Le Radici e le Pietre*. Cefalù: Lorenzo Misuraca Editore, 2012.

Foderà Serio, G., M. Hoskin, and F. Ventura. 'The orientations of the temples of Malta'. *Journal for the History of Astronomy* 23 (1992): pp. 107–119.

Foderà Serio, G., and S. Tusa. 'Rapporti tra morfologia ed orientamento nelle architetture rituali siciliane dal IV al II millennio a.C'. *Atti dei convegni lincei-Accademia Nazionale Dei Lincei* 171 (2001): pp. 297–323

Gabrici, E. 'La Niobide di Taormina'. *Monumenti Antichi* 41 (1951).

Giunta, F. and A. Giuffrida, eds.. *Acta Sicula-Aragonensia II: Documenti sulla luogotenenza di Federico d'Aragona* (Palermo, 1972).

González-García, C. A., D. Z. Kolev, J. A. Belmonte, V. P. Koleva, and L. P. Tsonev. 'On the Orientation of Thracian Dolmens'. *Archaeoastronomy* 23 (2009).

Hannah, R., and G. Magli. 'The role of the sun in the Pantheon's design and meaning'. *Numen* 58, no. 4 (20110: pp. 486–513.

Houel, J. *Voyages pittoresques des îles de Sicile, de Lipari et de Malte où l'on traite des antiquités qui s'y trouvent encore; des principaux phénomènes que la nature yo ffre; du costume des habitants, & de quelques usages, vol. iv Paris*. Paris: Imprimerie de Monsieur, 1787. p. 92, tav. XLIX-LI, Parigi, 1785.

Hoskin, M. 'Orientations of dolmens in Western Europe'. *Proceedings of the International Astronomical Union* 5, no. S260 (2009): pp. 116-126.

Hoskin, M. 'Orientations of megalithic sepulchres in Salamanca, Spain'. *Journal for the History of Astronomy* 23 (1992): pp. 57–60.

Hoskin, M. 'Studies in Iberian Archaeoastronomy: (5) orientations of the megalithic tombs of northern and western Iberia'. *Journal for the History of Astronomy* 29 (1998): pp. S39–S87.

Hoskin, M., E. Allan, and R, Gralewski. 'Studies in iberian archaeoastronomy: (1) orientations of the megalithic sepulchres of Almeria, Granada and Malaga'. *Journal for the History of Astronomy* 25 (1994): pp. S55–S82.

Hoskin, M. and J. J. Morales Núñez. 'The Orientations of the burial monuments of Menorca'. *Journal for the History of Astronomy* 22 (1991): pp. S15–S42.

Hoskin, M., and M. Zedda. 'Orientations of Sardinian Dolmens'. *Archaeoastronomy* 22 (1997): pp. S1–S16.

Hoskin, M., and T. Palomo y Perez. 'Studies in iberian archaeoastronomy: (4) the orientation of megalithic tombs of eastern Catalunya'. *Journal for the History of Astronomy* 29 (1998): pp. 63–79.

Koldewey, R., and o. Puchstein. *Die griechischen Tempel in Unteritalien und Sicilien*. Berlin: A. Ascher & Company, 1899.

Mannino, G. 'La tomba di contrada Pergola'. *Sicilia Archeologica* 15 (1971): pp. 52–56.

Marconi, P. 'Cefalù (Palermo). Il cosiddetto «Tempio di Diana»'. *Notizie degli Scavi* 5 (1929): pp. 273–395.

Mauceri, L., *Sopra un' Acropoli pelasgica esistente nei dintorni di Termini Imerese*. Palermo, 1896.

Montagna, C. *Il tesoro di Minos: l'architettura della Gurfa di Alia tra preistoria e misteri*. Palermo: Officina di Studi Medievali, 2009.

Nott G.F., *Avanzi di Cefalù*, Monumenti Inediti dell'Instituto di Corrispondenza Archeologica, pp. 270-287, tavv. XXVIII-XXIX, 1831.

Nigro, L. 'L'orientamento astrale del Tempio del Kothon di Mozia'. *Atti del VII Convegno Nazionale della Società Italiana di Archeoastronomia, Terme di Diocleziano* (2010): pp. 15–24.

Nigro, L. 'The Temple of the Kothon at Motya (Sicily): Phoenician Religious'. In *All the Wisdom of the East*. Edited by M. Gruber et al., pp. 293–331. Fribourg: Academic Press Fribourg/ Vandenhoeck & Ruprecht Göttingen, 2012.

Nigro, L., and G. Rossoni, eds. *"La Sapienza" a Mozia. Quarant'anni di ricerca archeologica (1964-2004)*. Roma: Missione Archeologica Mozia, 2004.

Nissen, H. *Das Templum. Antiquarische Untersuchungen*. Berlin, 1869.

Nissen, H. 'Ueber Tempel-Orientirung II', *Rheinisches Museum für Philologie*, J.D. Sauerländer Verlag, Bad Orb 1874.

Nissen, H. 'Ueber Tempel-Orientirung IV', *Rheinisches Museum für Philologie*, J.D. Sauerländer Verlag, Bad Orb 1885.

Nissen, H. 'Ueber Tempel-Orientirung V', *Rheinisches Museum für Philologie*, J.D. Sauerländer Verlag, Bad Orb 1887.

Orlando, A. 'The Argimusco's Plateau and the Horizon's Calendar'. In the *Proceedings of SEAC 2014*, held in Malta (Lampeter: Sophia Centre Press, 2016), this volume.

Orsi, P. 'La necropoli sicula di Castelluccio'. *BPI* 23 (1892): pp. 1–34, 67–84.

Orsi, P. 'Miniere di selce e sepolcreti eneolitici a Monte Tabuto e Monte Racello presso Comiso'. *Bullettino di Paletnologia Italiana* 24 (1898): pp. 165–206.

Orsi, P. 'Pantelleria'. *Monumenti Antichi* 9 (1899): pp. 449–540.

Orsi, P. 'Taormina. Necropoli sicula al Cocolonazzo di Mola'. *Notizie degli Scavi* 16 (1919): pp. 360–369.

Orsi, P. 'Thapsos'. *Monumenti Antichi*, vol. VI (1885): pp. 89-150.

Palio, O., and M. Turco. 'Strutture megalitiche nell'area etnea'. *Notiziario di Preistoria e Protostoria* 2, no. 2 (2015): pp. 49–51.

Pantano, G. *Megaliti di Sicilia*. Patti: Edizioni Fotocolor, 1994.

Papathanassiou, M., M. Hoskin, and H. Papadopoulou. 'Orientations of tombs in the late-Minoan cemetery at Armenoi, Crete'. *Journal for the History of Astronomy Supplement* 17 (1992): pp. S15–S20.

Penrose, F. C. 'On the Orientation of Greek Temples, being the Result of Some Observations Taken in Greece and Sicily in the month of May, 1898'. *Proceedings of The Royal Society of London* 65, no. 413-422 (1899): pp. 370–375.

Penrose, F. C. 'Some Additional Notes on the Orientation of Greek Temples; Being the Result of a Journey to Greece and Sicily in April and May, 1900'. *Philosophical Transactions of the Royal Society of London. Series A, Containing*

Papers of a Mathematical or Physical Character 196 (1901): pp. 389–395.

Polcaro, V. F., and Ienna D. 'The Megalithic Complex of the «Preta 'ru Mulacchio» on the Monte della Stella'. *Cosmology Across Cultures*, ASP Conference Series 409 (2009).

Platania, V., and M. S. Scaravilli. 'Il complesso rupestre di Rocca Pizzicata'. *Agorà* 45 (2013): pp. 123–125.

Privitera, F. 'Castiglione di Sicilia, contrada Marca. Grotta sepolcrale della tarda età del Rame del Bronzo Antico'. *BCA Sicilia* 1-2. Palermo (1991–1992): pp. 21–25.

Privitera, M. 'Sepolture rupestri nella Valle dell'Alcantara'. *Kokalos* 47-48, no. II (2009).

Privitera, F. 'Necropoli tardo-neolitica in Contrada Balze Soprane di Bronte (CT)'. *Atti della XLI Riunione scientifica IIPP: 'Dai Ciclopi agli Ecisti: società e territorio nella Sicilia preistorica e protostorica'* (2012): pp. 543-556.

Puglisi, S. *La Valle dei Palmenti*. Messina: Armando Siciliano Editore, 2009.

Radice, B. *Memorie storiche di Bronte*. Bronte: Stabilimento tipografico sociale, 1936.

Scuderi, A., F. Mercadante, P. Lo Cascio, and V. F. Polcaro. 'The Astronomically Oriented Megalith of Monte Arcivocalotto'. *Anthropological Notebooks* 19 (Supplement) (2013): pp. 213–221.

Scuderi, A., V. F. Polcaro, and F. Maurici. 'New Archaeoastronomical findings in the alto Belice Valley (Sicily)'. *Mediterranean Archaeology and Archaeometry* 14 (2014): pp. 93–98.

Terranova, G. Maltese Temples and Hypogeism. New Data About the Relationships Between Malta and Sicily During the III and II Millennium BC. In AA.VV., *Exploring the Maltese Prehistoric Temple Culture, EMPTC 2003 Conference*, pp. 1–24. 2003.

Todaro, G. *Alla ricerca di Abaceno*. Messina: Armando Siciliano Editore, 1992.

Tullio, A. *Memoria di Cefalù*. Palermo: Kefagrafica Lo Giudice, 1994.

Tusa, V. *Cefalù*. In J. W. Hayes, *Enciclopedia dell'Arte Antica II*. Rome: Rome: Instituto della Enciclopedia, 1959, p. 453.

Tusa, S. *La Sicilia nella preistoria*. Palermo: Sellerio, 1999.

Tusa, S. *L'insediamento dell'età del bronzo con Bicchiere Campaniforme di Marcita*. Trapani: Corrao Editore, 1997.

Tusa, S. 'The Bell Beaker in Sicily'. In *Bell Beakers of the Western Mediterranean*. Vol. 331. Edited by W. H. Waldren and R. C. Kennard, pp. 523–550. Oxford: British Archaeological Reports, 1987.

Tusa, S., G. Foderà Serio, and M. Hoskin. 'Orientations of the Sesi of Pantelleria'. *Journal for the History of Astronomy Supplement* 17 (1992): pp. S15–20.

Vassallo, S. 'L'enigma del muro megalitico e dello pseudo-dolmen di Mura Pregne'. In *From Cave to Dolmen. Ritual and symbolic aspects in the prehistory between Sciacca, Sicily and the central Mediterranean*. Edited by D. Gullì, pp. 247–253. Oxford: Archaeopress, 2014.

Veca, C. 'Contenitori "per i vivi" e contenitori "per i morti" a Thapsos (Siracusa): un approccio tecnologico a un problema interpretativo'. *Rivista di Scienze Preistoriche* 64 (2014):, pp. 203–225.

Ventura, F., G. Foderà Serio, and M. Hoskin. 'Possible tally stones at Mnajdra, Malta'. *Journal for the History of Astronomy* 24 (1993):, pp. 171–183.

THE TALL GNOMON OF GUO SHOUJING: AN ASTRO-ARCHAEOLOGICAL ANALYSIS

Vance Tiede

ABSTRACT: The Tall Gnomon at Gaocheng, Dengfeng Township, Henan Province, China (34°24'8.73"N, 113°8'26.23"E) is described in scholarly literature as tracking changes in the gnomon's shadow at noon (meridian transit) between midsummer and midwinter. For example, the *Yuan History* (*Yuan Shih*) records that the sun's shadow cast near the winter solstice on 14 December 1278 CE measured 76.7400 Chinese feet (*chi*). However, the 128*chi* horizontal graduated template scale (*kuei piao*) of the Tall Gnomon is 40% longer than what is required to measure the sun's longest shadow at winter solstice. Using GIS and astro-archaeological computer modeling, nocturnal astronomical targets (i.e., midsummer full moon, Venus and bright stars) are investigated to account for the construction of the extra 51*chi* in length of the Tall Gnomon's template scale. Evidence is also presented confirming that the Tall Gnomon solstice observations in the *Yuan History* were made at Beijing, rather than Gaocheng.

Introduction

Ancient China's Astronomical Bureau validated imperial claims to rule under the Mandate of Heaven (*tiānmìng* 天命) and the Above Emperor (*Shangdi* 上帝) by monitoring the motions of celestial bodies, orienting the imperial urban grid to the cosmos, and preparing imperial calendar-almanacs.[1] An inherent limitation of calendars based on computed values is that over time they drift out of phase with empirical luni-solar observations, requiring periodic intercalation.[2] For example, the Yuan Dynasty Emperor Kublai Khan in 1276 ordered court astronomer (*Tai shi ling*) Wang Xun (王恂, d. 1282) and his assistant Guo Shoujing (郭守敬 Kuo Shou-Ching, 1231–1316) to correct the accumulated errors of the *Daming* Calendar prepared by Zhao Zhiwei in the previous Jin Dynasty. By combining the resolving power of the Chinese Shadow Definer ((*ying fu*) with the increased precision of the Persian Tall Gnomon of Nasir al-Din al Tusi (Maraghe Observatory, 1262),[3] Wang and Guo were able to calibrate the *Shoushi* Calendar of 1281 with empirical luni-solar observations and calculate the length of the tropical year 'to within 20 seconds of the correct figure'.[4] According to Guo Shoujing,

> [The ruler] has specifically commanded us to carry out a reform and set in order a new astronomical system. For this purpose, we have newly made a Simplified Instrument and a Tall Gnomon and have arrived at constants based on observations using them. We have carried out research (*k'ao-cheng*) on seven items: 1. Winter Solstice…; 2. Year Surplus…; 3. Tread of the sun…; 4. Travel of the moon…; 5. Crossing entry [*timing the intersection of the lunar nodes crossing the ecliptic – VT*]…; 6. Angular extensions of the lunar

---

[1] Joseph Needham, *Science and Civilization in China*, Vol. III (London: Cambridge University Press, 1959), pp. 189–191; Nathan Sivin, *Granting the Seasons: The Chinese Astronomical Reform of 1280* (New York: Springer, 2009), pp. 35–36.

[2] Christopher Cullen, *Astronomy and mathematics in ancient China: the Zhou bi suan jing* (Cambridge: Cambridge University Press, 1996), pp. 20–21.

[3] Needham, *Science and Civilization in China* III, p. 296; Vance Tiede, 'An Astro-Archaeological Analysis: The Tall Gnomon of Guo Shoujing' (presented at the SEAC Conference, Malta, 22–26 September 2014), slides 3–4, available at http://www.academia.edu/8604430/An_AstroArchaeological_Analysis_The_Tall_Gnomon_of_Guo_Shoujing_PPT_slides_ [accessed 1 April 2016].

[4] Hugh Thurston, *Early Astronomy* (New York: Springer-Verlag, 1994), p. 105; Y. Li and C. Z. Zhang, 'Secular Variation of Earth's Rotation: Inferred from the Chinese Ancient Shoushi Calendar (Ad 1281)', *Earth, Moon and Planets* 76, nos. 1 and 2 (1997): p. 11.

lodges...; and 7. Sunrise, sunset, day and night marks....⁵

This paper examines the Tall Gnomon's dimensions in order to investigate which astro-targets Guo Shoujing could theoretically have measured to derive such calendrical 'constants based on observations'. A second task is to resolve confusion in the scholarly literature regarding which of the two locations Guo Shoujing recorded his observations for the measurement of the sun's shadow, viz.,

> ... there is still the common impression, even with some serious historians, that the solsticial shadow observations listed by Guo Shoujing were made at Yangcheng [modern Gaocheng] in Henan province where there is extant an impressive tower, the 'Tower of Duke Zhou', with horizontal scale, apparently constructed by Guo Shoujing. Ironically, this impression is owing indeed to the work of Dong Zuobin [Tung Tso-Pin 1939] who wrote the definitive monograph on the tower, although, like Gaubil before him, Dong Zuobin made it clear that the observations listed by Guo Shoujing were made in Beijing [Yuan Dynasty capital of Dadu]. At a more popular level, Krupp presented a light account of the shadow measurements, where it is assumed that Guo Shoujing's measurements were made at the tower in Yang Cheng....⁶

*Fig. 27.1: From left to right, Guo Shoujing, Sun & Shadow at Meridian Transit, Grooved Template Scale (facing south).*⁷

Astro-Architectural Analysis

Only one of the two Yuan Tall Gnomons remains standing. It is located at Gaocheng (ancient Yangcheng), Dengfeng Township in Henan Province, China (34°24'8.73"N, 113°8'26.23"E; 266m above sea level (ASL) (Figs. 27.1 and 27.4 left). The second structure, now lost, was located at the Directorate of Astronomy in the vicinity of the Yuan capital at Dadu (modern Beijing, 39°54'N, 113°8'E, 49m ASL) (Fig. 27.4 right). 'Two were actually built, one at the Commission and one, not in a city, but in the ancient observatory at what is now Gaocheng, Henan. Archaeologists located the remains of the latter early in the twentieth century, and subsequently restored it'.⁸ Twin brick towers suspended an east-west horizontal gnomon 40*chi* (40

---

⁵ Sivin, *Granting the Seasons*, pp. 282–86.
⁶ Raymond Mercier, 'Solsticial observations in thirteenth century Beijing'. *SCIAMVS* 4 (2003): pp. 191–192; cf. Edwin C Krupp, *Echoes of the Ancient Skies: The Astronomy of Lost Civilizations* (New York: Harper & Row, 1983), pp. 60–61.
⁷ *http://hua.umf.maine.edu/China/astronomy/tianpage/0022MingGnomon6471w.html*; *http://whc.unesco.org/en/list/1305/video* [accessed 1 April 2016].
⁸ Sivin, *Granting the Seasons*, pp. 183–184 referring to Tung Tso-Pin (Dong, Zuobin, 1895-1963), Lin Tun-Chen, and Kao Phing-Tzu, 周公測景臺調查報告 [*Chou Kung Tshê Ching Thai Thiao Chha Pao Kao* [*Zhou gong ce jing tai diao cha bao gao*]

Chinese 'feet' ~ 9.75m) over a north-south 128*chi* (~31.19m) long horizontal stone Sky Measuring Scale (*liang thien chhih*).² The latter has a graduated template scale (*kuei-piao*) flanked by self-leveling grooved water channels.⁹

The primary advantage of the 40*chi* Tall Gnomon over the traditional 8*chi* gnomon is its five-fold increase in precision in plotting the focused gnomon bar shadow by sunlight at midday, and by moonlight at midnight. Meridian transit of planets and bright stars could have been marked at night with an Observing Table (*k'uei chi*), and perhaps also as reflections in the self-leveling water-filled grooves of the graduated horizontal template scale (Fig. 27.2, bottom). The higher precision achieved from the Tall Gnomon's large scale was possible only when the resulting enlarged but diffused shadow was focused with a portable pinhole lens or Shadow Definer (*jingfu*) (Fig. 27.2, top). For example, Guo's *Yuan History* records the length of the sun's shadow cast approaching the winter and summer solstices as:

> A shadow length near the summer solstice of year 16 of the Perfectly Great period, s.y. 16 (15 June 1279), on the 19th of month 4, s.d. 32 (30 May), was 12.3695 feet long. One near the winter solstice (14 Dec 1279), on the 24th of month 10, s.d. 35 (29 Nov), was 76.7400 feet long. {997}¹⁰

Astronomer Guo may not have been able to measure the gnomon bar's shadow to within one *hao* or 'about 0.03 mm' with a portable, angled, sheet-metal pin-holed Shadow Definer (*ying fu*) that slid (or floated) along the self-leveling water-filled channel grooves of the horizontal template scale (Fig. 27.1, right, and Fig. 27.2, top). Such high precision almost certainly resulted from mathematical interpolation of a series of symmetric empirical observations of the shadow cast by the horizontal gnomon bar bisecting a focused image of the solar disk on the template scale, not unlike the cross-hair of a surveyor's theodolite (Fig. 27.3, centre). 'The sun's light, passing through the hole, creates a bright spot of light on the scale, the size of a "grain of rice", and if the horizontal bar suspended 40 feet above the ground is in line with the Sun and the hole, a tiny image of the bar appears in the spot of light, as a black bar across it'.¹¹

*Fig. 27.2: Shadow Definer; Observing Table; Ying Fu Principle; Reflected starlight on water; Reconstructed Horizontal Scale.*¹²

Report of an Investigation of the Tower of Chou Kung for the Measurement of the Sun's (Solsticial) Shadow (Changsha: Com. Press Academia Sinica, 1939). See 1930s restoration photos at Needham, *Science and Civilization in China* III, p. 297–296.
⁹ Needham, *Science and Civilization in China* III, pp. 296–299; Fengxian Xu, 'Dengfeng Large Gnomon', in *Handbook of Archaeoastronomy and Ethnoastronomy*, ed. Clive L. Ruggles (New York: Springer-Verlag, 2014), p. 2112, available at http://link.springer.com/referenceworkentry/10.1007%2F978-1-4614-6141-8_218#page-1 [accessed 1 April 2016].
¹⁰ Sivin, *Granting the Seasons*, p. 189–190, 571; cf. Needham, *Science and Civilization in China* III, p. 299; and Mercier, 'Solsticial observations', pp. 197–198.
¹¹ Mercier, 'Solsticial observations', p. 195.
¹² Sivin, *Granting the Seasons*, pp. 188, 190; http://web.calstatela.edu/faculty/kaniol/a360/yingfu1.

Significance

Astro-archaeological analysis may on occasion resolve historians' conflicting interpretations of ancient documents and instruments, e.g., what is the reason for the Tall Gnomon horizontal template scale being longer than required to measure the sun's longest shadow at the winter solstice? Without accurate spherical trigonometric modeling or onsite observation, it is difficult to determine whether or not the length of the Tall Gnomon's horizontal template scale significantly exceeds the length required to observe Guo's primary research objective to measure the sun's shadow at the winter solstice (Fig. 27.3). In other words, the gnomon shadow maximum intercept point with the horizontal template scale determines whether the Tall Gnomon was designed exclusively as a diurnal solar scale, or was also intended for nocturnal lunar-planetary-stellar observations. For example, if Figure 27.3 (right) is accurate, then the Tall Gnomon was of use with only five of Guo's seven astronomical agenda items, viz.: 1. Winter solstice...; 2. Year Surplus...; 3. Tread of the sun...; 6. Angular extensions of the lunar lodges...; and 7. Sunrise, sunset, day marks (Fig. 27.2, top). Alternatively, if Figure 27.3 (left) is accurate, then the Tall Gnomon was also capable of plotting full moon shadows in support of the additional lunar research tasks, viz.: 4. Travel of the moon; 5. Crossing entry; and 7. Lunar-planetary-stellar 'night marks'.[13]

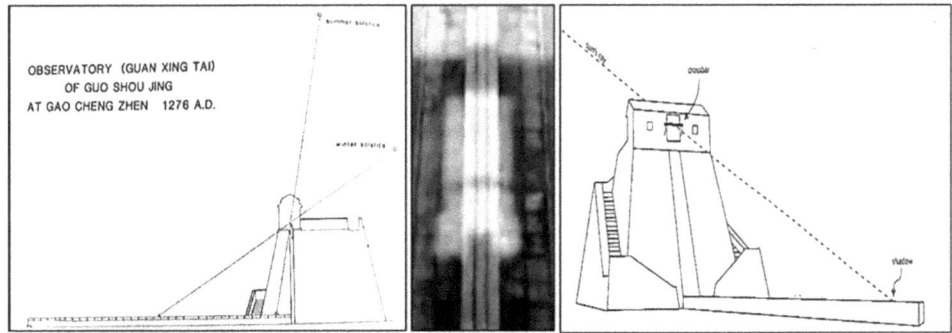

Fig. 27.3: Left and right, Solstice Altitudes; center, Shadow Cast on Template Scale at Meridian Transit.[14]

Purpose

The research objectives of this paper are threefold:

1. To refine the linear estimate of a Yuan *hau* as 'about 0.03mm' by using the Tall Gnomon dimensions to derive the limit of shadow resolution reported in the *Yuan Shih* in *chi* units as metric equivalents;
2. To determine the latitude of Guo Shoujing's solstice observations as either Gaocheng (34.4° N) or in the vicinity of Beijing (39.9° N); and
3. To account for Guo's Tall Gnomon's 128*chi* long horizontal template scale, which is roughly 51*chi* (40%) longer than required to measure the sun's (longest) winter solstice shadow.

---

htm; http://douglaslane.ie/photo8987452.html#photo; http://hua.umf.maine.edu/China/astronomy/tianpage/0022MingGnomon6471w.html [all sites accessed 1 April 2016].

[13] Tiede, 'An Astro-Archaeological Analysis', slides 8, 11–12.

[14] Left, see Krupp, *Echoes of the Ancient Skies*, p. 60 (Griffith Observatory); 'Historic Monuments of Dengfeng in "The Centre of Heaven and Earth"' video, available at http://whc.unesco.org/en/list/1305/video; Li Di, 'A Conjectural Construction of the Giant Chinese Gnomon in Dadu of the Yuan Dynasty', *Studies in the History of Natural Sciences* 20, no. 4 (2001): pp. 362–367, in Sivin, *Granting the Seasons*, p. 186, available at http://en.cnki.com.cn/Article_en/CJFDTOTAL-ZRKY200104008.htm [accessed 1 April 2016].

Methodology

The research designs regarding metrology, latitude and template scale are as follows:

1. Metric Equivalents. Ancient Chinese linear measurement units were converted to metric equivalents (Table 27.1) from architecture and historical records, viz.: the respective metric equivalents of 9.75m and 31.19m the Tall Gnomon height and length the *Yuan Shih* reference that the Tall Gnomon bar was placed 40*chi* above a 128*chi* long template scale [height/length = 40/128 = 5 (8/25.6)]; and Yuan powers of ten fraction (decimal) system of linear units (See Table 27.1).[15],

2. Latitude. To determine whether the 30 May and 14 December 1279 solar shadow observations were made at Gaocheng versus Beijing, an astro-architectural analysis was done with Google Earth-DigitalGlobe satellite imaging, Starry Night Pro Plus 6 digital planetarium software, and Program STONEHENGE.[16] The true azimuths for both Tall Gnomons is assumed to be 180° true azimuth, as solar noon shadows were so observed in Chinese historical records and the extant Gaocheng template scale appears so oriented. Latitudes and elevations (ASL) of Gaocheng's extant Tall Gnomon and the site of Dadu's Yuan Observatory were determined with Google Earth (Fig. 27.4).[17]

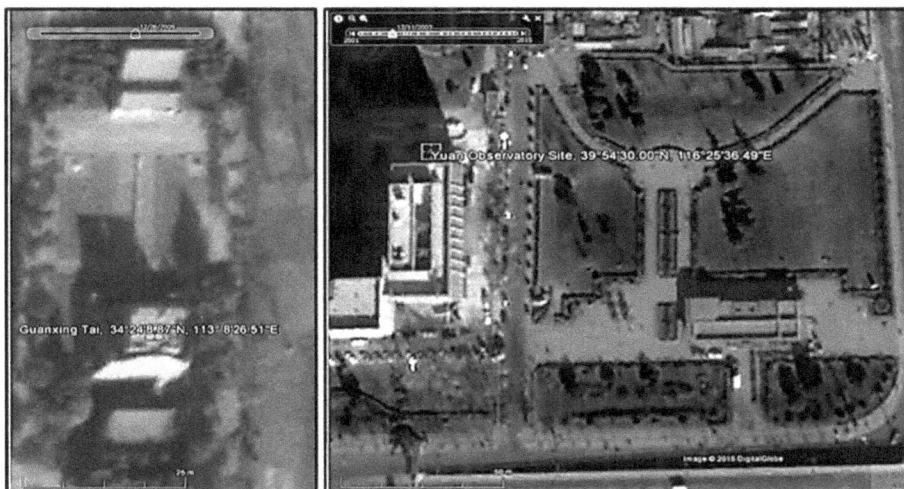

Fig. 27.4: Left – Tall Gnomon, Gaocheng (NB: Shadow on scale's mid-point at Midwinter, 26 Dec 2009 ≈ 11:00; Right – Site of Yuan Observatory, Dadu (Beijing), 11 Dec 2003 ≈ 11:00. North = top. Image©2015DigitalGlobe.

Solar declinations at meridian transit on 30 May and 14 December 1279 for both Gaocheng and Beijing were generated by Starry Night Pro Plus 6. The last step was to enter data from Table 27.1 into an Excel version of Program STONEHENGE to determine by trial and error what metric distance along the horizontal template scale generated the four respective declinations matching Guo's historical shadow lengths (Table 27.2). The derived horizontal template distance that more closely matches the historical shadow lengths will confirm the observation site as either Gaocheng or Beijing.

---

[15] Xu, 'Dengfeng Large Gnomon', pp. 2112; Mercier, 'Solsticial observations', p. 193.
[16] Gerald S. Hawkins, *Mindsteps to the Cosmos* (New York: Harper & Row, 1983), pp. 328–330.
[17] Mercier, 'Solsticial observations', Figure 6, p. 232.

3. Template Scale Length. To account for a template scale length beyond that required for marking the sun's longest shadow at the winter solstice, it was necessary to determine the minimum declination that could be plotted from the extreme end of the 128*chi* template scale. The theoretical minimum declination (longest template scale distance) was compared to that of the next brightest astronomical body capable of casting a longer shadow than the midwinter sun, e.g., the midsummer full moon at the major stand still (-29° declination). The lunar and solar minimum declinations were simulated in Starry Night Pro Plus 6, and entered into Program STONEHENGE. The respective shadow distances on the template scale were derived by trial and error (Table 27.3).

Limitations

Lacking historical documentation concerning the absolute length of the Yuan *chi*, the author constructed a mensuration conversion table based on the published metric dimensions of the Tall Gnomon's bar height and template scale length.[18] Therefore, the reader is cautioned that the metric dimensions of the Tall Gnomon published by Tung Tso-Pin et al. (1937) are assumed to have been preserved by the Ming dynasty (1368–1644) and 1930s restorations and accurately represent the original dimensions of the original Yuan masonry Tall Gnomon (Plan of 1279), which in turn was presumably based on the wooden prototype used to calibrate the *Shoushi* Calendar of 1281.[19] That said, three pre-1937 photographs of the Tall Gnomon with vegetation growing on the template scale support the assumption that at least the 1930s restoration accurately preserved the template length.[20]

Findings & Analysis

1. Metric Equivalents. In order to calculate astronomical declinations for the Tall Gnomon, it was necessary to convert Yuan *chi* to metric equivalents as inputs to Program STONEHENGE. Historically, the length of the Yuan *chi* is uncertain, as the Chinese linear units varied in length over time, e.g., the Han *chi* (ca. 100 BCE) varied 23.1 to 24.3cm, while the current (1984) *chi* is 33.3cm.[21] Based on the Tall Gnomon's metric dimensions, the *Yuan Shih* solar shadow lengths were recorded to the nearest *hao* (0.024mm = 24μm) (Table 27.1).[22]

2. Latitude. Because Table 27.2 shows that the angular error (delta d) in gnomon declination and solar declination is less than half of a degree at Beijing and the mean difference in solar declinations (~5.3° dec) observed at the two sites is equivalent to the difference in their latitudes (Δ 5.5° latitude), then the *Yuan Shih* observations were observable at Beijing, rather than Gaocheng.

3. Template Scale Length. While historians emphasize solstice ($\pm$23.5° declination) observations with the Tall Gnomon, Table 27.3 shows that the 128*chi* template scale was capable of measuring the longer shadow of the midsummer Full Moon (magnitude -12.81) at Major Standstill (-28.5° declination). However,

---

[18] Xu, 'Dengfeng Large Gnomon', p. 2112.
[19] Sivin, *Granting the Seasons*, pp. 181–183.
[20] Tso-Pin et al., *Chou Kung Tshê Ching Thai Thiao Chha Pao Kao*, pp. 53–54.
[21] K'ang-shen Shen, John N. Crossley, Anthony Wah-Cheung Lun, and Hui Liu, *The nine chapters on the mathematical art: companion and commentary* (Beijing: Oxford University Press, 1999), p. 8; cf. C. Y. Liu, 'A Research on the Implication of *Zhang-Chi* in Ancient Chinese Astronomical Records', *Acta Astronomica Sinica* 28, no. 4 (1987): p. 402, available at http://adsabs.harvard.edu/abs/1987AcASn..28..397L [accessed 1 April 2016].
[22] Xu, 'Dengfeng Large Gnomon', p. 2112; Sivin, *Granting the Seasons*, p. 271.

the real surprise is that a 128*chi* template scale is significantly longer than necessary for both midwinter solar and midsummer lunar shadow measurements (Table 27.3).

Yuan *znang/chi* - Metric Linear Unit Converter					Starry Night Declination (°′)	Pro Plus 6 Declination (decimal °)	Meridian Transit (LMT)
Feature	*chi*	meter	feet	Stone block			
Height: Gnomon Bar	40.0000	9.75	31.98	na			
Length: Template Scale	128.0000	31.19	102.30	36.00			
Length: Stone Block	3.5556	0.8664	2.84	1.00			
	4.1039	1.00	3.28	0.87			
Date, Latitude, Location	1.0000	0.24	0.80	0.28			
30 May +1279, +39° 54′	12.3695	3.01	9.89	3.48	+22° 41.68′	22.69466667	12:24:00
14 Dec +1279, Beijing	76.7400	18.70	61.33	21.58	−23° 32.21′	−23.53683333	12:27:15
Chinese Units	*zhang*	*chi*	*cun*	*fen*	*li*	*hao*	Meter
zhang	1	10	100	1,000	10,000	100,000	2.400000
chi	0.1	1	10	100	1,000	10,000	0.240000
cun	0.01	0.1	1	10	100	1,000	0.024000
fen	0.001	0.01	0.1	1	10	100	0.002400
li	0.0001	0.001	0.01	0.1	1	10	0.000240
hao	0.00001	0.0001	0.001	0.01	0.1	1	0.000024

Table 27.1: Yuan znang/chi *– Metric Linear Unit.*

Yuan Shih Dec & Tall Gnomon Locations OUTPUT: Program STONEHENGE (Hawkins 1983, 328-330)							
Chinese *chi*	Gno Distance (m)	mon Height (m)	Tall Gnomon Location N Latitude	Yuan shih Sun dec Date	#122 d Tall Gnomon Sun dec (deg & radian)	#127 delta d dec Sun MT - dec Gnomon (deg & radian)	
12.352739	3.01	9.75	Gaocheng 34.4° N	22.694666 30May+1279	17.2480 0.3010	5.4467 0.0951	
76.74293	18.7	9.75		−23.5368 14Dec+1279	−28.0586 −0.4897	4.5218 0.0789	
12.352739	3.01	9.75	Beijing 39.9° N	22.694666 30May+1279	22.2681 0.3887	0.4256 0.0074	
76.74293	18.7	9.75		−23.5368 14Dec+1279	−22.5552 −0.3937	−0.9817 −0.0171	

Table 27.2: Yuan Shih Dec & Tall Gnomon Locations.

Tall Gnomon Minimum Declinations at Gaocheng & Beijing						
OUTPUT: Program STONEHENGE (Hawkins 1983, 328-330)						
	Gno	mon	Tall Gnomon		#122 d	#127 delta d
Chinese *chi*	Distance (m)	Height (m)	Location N Latitude	Astro-Target	Declination on Skyline (deg/radian)	decl Target - decl Skyline (deg/radian)
84.7250	20.65	9.75	Gaocheng 34.4° N	Moon SMAJSS 15JUN1280	-29.3668 -0.5125	0.0007901684 0.0000137910
128.0006	31.19	9.75		?	-38.2401 -0.6674	6.4359772602 0.1123289938
111.2772	27.12	9.75	Beijing 39.9° N	Moon SMAJSS 15JUN1280	-29.3665 -0.5125	0.0004003563 0.0000069875
128.0006	31.19	9.75		Venus Absolute Min. decl.	-32.73672362 -0.571363613	0.9326069501 0.0162770619

Table 27.3: Tall Gnomon Minimum Declinations at Gaocheng & Beijing.

Is there an astro-target capable of casting a shadow at the 128*chi* mark (-32° declination) to account for the seemingly extra-long the template scale? In theory, Venus as Morning Star may cast a shadow near inferior conjunction at its theoretical absolute minimum declination of -32°.[23] Regardless of whether surviving historical records confirm that the Tall Gnomon was designed to measure Venus at its absolute minimum altitude, the fact remains that such an observation would require a 128*chi* long template scale at the latitude of the Astronomical Bureau in the imperial capital. Because of its proximity to the sun, the shadow of the third brightest celestial object, Venus (*Thai pei, Great White One*), would not have been easily observable, because the Tall Gnomon's north-south orientation (literally set in stone) faces away from any shadows cast by either the routinely observed the Morning Star at dawn or the Evening Star at dusk.[4] However, the noontime shadow of Venus, especially if focused by a Shadow Definer (used as a Helioscope with a pinhole lens) would have observable on the template scale during a midday solar transit of Venus. In fact, the *Shan Chǔ Hsin Hua* (*New Discussions From the Mountain Cabin*) written by Yang Yǔ in 1360, a Co-signatory Observer in the Bureau of Astronomy, states that 'And indeed only nine days later, the planet Venus 'crossed the meridian'…. c This was a very bad sign;… ' (Fig. 27.5).[24] Moreover, observing the projected image of the sun disk for black dots transiting the solar disk was a routine observation for Chinese astronomers who between 28 BCE and 1638 CE recorded 112 descriptions of sunspots.[25]

Venus itself may also visible during daylight at meridian transit during a rare solar eclipse or at

---

[23] Bruce McCurdy, 'Venus Vignette', *Journal of the Royal Astronomical Society of Canada* 99 (2005): pp. 187–189, available at http://adsabs.harvard.edu/full/2005JRASC..99..187M [accessed 1 April 2016]; cf. Brude McClure and Deborah Byrd, 'Venus brightest in mid-February for all of 2014', *EarthSky* (14 February 2014), available at http://earthsky.org/human-world/venus-brightest-greatest-brilliancy-greatest-illuminated-extent [accessed 14 July 2016]; Tiede, 'An Astro-Archaeological Analysis', slides 14–16.

[24] Needham, *Science and Civilization in China* III, p. 419.

[25] Needham, *Science and Civilization in China* III, pp. 434–436; Tiede, 'An Astro-Archaeological Analysis', slides 21–23.

maximum elongation (see Fig. 27.5).[26] Although 128*chi* Sky Measuring Scale is of a clearly excessive length for luni-solar observations, 128*chi* is the minimum length necessary for measuring Venus at its theoretical minimum altitude (-32° declination) from the Astronomical Bureau in Beijing. Given the considerable expense and labor involved in constructing the Tall Gnomons, it seems more probable than not that the doubling of the horizontal template scale length beyond that required for luni-solar observation was intentional, especially given that Venus observations played dual roles in both astrological prognostication and calendric computation.[27] As an example of the latter, Yuan astronomers recorded the Synodic Period of Venus as 583.90 days (vs. the modern value of 583.92 days). Planetary constants were used to calibrate luni-solar calendric cycles, e.g., 'The Era Epoch system (#71; in use 1106–27) went on to note the elongation of Great White [i.e. Venus], and to determine by observation after dusk or before dawn the position of the planet among the stars, thus deriving the tread [i.e., the position] of the sun'.[28]

Fig. 27.5: Transits of Venus: top left, 3 December +1360/1200 Local (-23° dec), Beijing; top right, 8 June +2004 (+23°dec) Helioscope, Yokohama Science Center (Itsuo Inouye, AP); bottom, 29 October +1244/1200 Local (-27° dec), Gaocheng (Starry Night Pro Plus-6).

Conclusions

We have explored the limits of Guo Shoujing's elegant astro-architectural solution to increasing the precision of luni-solar measurements by scaling up template scale and focusing the larger projected shadow with a pinhole lens. His innovative combination of the Tall Gnomon and Shadow Definer provided the increased precision required to refine by interpolation the observed shadow measurements to the nearest *hao* or 24μm. The 40*chi* Tall Gnomon solar observations of the *Yuan Shih* are shown to have been made at

---

[26] Tiede, 'An Astro-Archaeological Analysis', slides 17–20.
[27] Sivin, *Granting the Seasons*, pp. 524–525; Edward Schafer, *Pacing the Void: T'ang Approaches to the Stars* (Berkeley: University of California Press, 1977), pp. 214–215.
[28] Sivin, *Granting the Seasons*, pp. 518, 524–525, 570.

a second (now lost) Tall Gnomon site near the Yuan capital at Dadu (Beijing). The graduated 128*chi* Sky Measuring Scale and Shadow Definer were capable of measuring gnomon bar shadows on the horizontal template scale annually between the solstices ($\pm$ 23.5° declination), and the full moon between its major stand stills ($\pm$ 28.5° declination) every 18 or 19 years.

In addition, the combined instrumentation was theoretically capable of projecting shadows of sunspots daily at the solar meridian transit, and at the periodic noontime solar transits of Venus. Daytime measurement of meridian transits of Venus $\pm$32° declination) was theoretically possible in its Morning or Evening Star phase (-4.4 apparent magnitude) or during (very rare) solar eclipses, with either a sighting tube or Observing Table (Fig. 27.2 top). At night, bright planets and stars transiting the meridian at Beijing (-32° to +40° declination) and Gaocheng (-38° to +34° declination) may have been the targets recorded as 'night marks' mentioned by Guo Shoujing and presumably observed directly with a sighting tube or indirectly as reflections on the water-filled channels flanking the horizontal template scale.

Bibliography

Cullen, Christopher. *Astronomy and mathematics in ancient China: the Zhou bi suan jing*. Cambridge: Cambridge University Press, 1996.

Di, Li. 'A Conjectural Construction of the Giant Chinese Gnomon in Dadu of the Yuan Dynasty'. *Studies in the History of Natural Sciences* 20, no. 4 (2001): pp. 362–367, in Sivin, Granting the Seasons, p. 186. http://en.cnki.com.cn/Article_en/CJFDTOTAL-ZRKY200104008.htm.

Hawkins, Gerald S. *Mindsteps to the Cosmos*. New York: Harper & Row, 1983.

Krupp, Edwin C. *Echoes of the Ancient Skies: The Astronomy of Lost Civilizations* New York: Harper & Row, 1983.

Li, Y., and C. Z. Zhang. 'Secular Variation of Earth's Rotation: Inferred from the Chinese Ancient Shoushi Calendar (Ad 1281)'. *Earth, Moon and Planets* 76, nos. 1 and 2 (1997): pp. 11–17.

Liu, C. Y,. 'A Research on the Implication of *Zhang-Chi* in Ancient Chinese Astronomical Records'. *Acta Astronomica Sinica*, 28, no. 4 (1987): p. 402, http://adsabs.harvard.edu/abs/1987AcASn..28..397L.

McClure, Bruce, and Deborah Byrd. 'Venus brightest in mid-February for all of 2014'. EarthSky (14 February 2014). http://earthsky.org/human-world/venus-brightest-greatest-brilliancy-greatest-illuminated-extent.

McCurdy, Bruce. 'Venus Vignette'. *Journal of the Royal Astronomical Society of Canada*, 99 (2005): pp.187-189. http://adsabs.harvard.edu/full/2005JRASC..99..187M.

Mercier, Raymond. 'Solsticial observations in thirteenth century Beijing'. *SCIAMVS* 4 (2003): pp.191–232.

Needham, Joseph. *Science and Civilization in China*, Vol. III. London: Cambridge University Press, 1959.

Schafer, Edward. *Pacing the Void: T'ang Approaches to the Stars*. Berkeley: University of California Press, 1977.

Sivin, Nathan. *Granting the Seasons: The Chinese Astronomical Reform of 1280*. New York: Springer, 2009.

Shen, K'ang-shen, John N. Crossley, Anthony Wah-Cheung Lun, and Hui Liu. *The nine chapters on the mathematical art: companion and commentary*. Beijing: Oxford University Press, 1999.

Thurston, Hugh. *Early Astronomy*. New York: Springer-Verlag, 1994.

Tiede, Vance. 'An Astro-Archaeological Analysis: The Tall Gnomon of Guo Shoujing'. Presented at the SEAC Conference, Malta, 22–26 September 2014. http://www.academia.edu/8604430/An_AstroArchaeological_Analysis_The_Tall_Gnomon_of_Guo_Shoujing_PPT_slides_.

Tso-Pin, Tung, (Dong, Zuobin, 1895-1963), Lin Tun-Chen, and Kao Phing-Tzu, 周公測景臺調查報告 *Chou Kung Tshê Ching Thai Thiao Chha Pao Kao* [Zhou gong ce jing tai diao cha bao gao] *Report of an Investigation of the Tower of Chou Kung for the Measurement of the Sun's (Solsticial) Shadow*, (Changsha: Com. Press Academia Sinica, 1939).

Xu, Fengxian. 'Dengfeng Large Gnomon.' In *Handbook of Archaeoastronomy and Ethnoastronomy*, edited by Clive L. Ruggles, pp. 2111–2116. New York: Springer-Verlag, 2014.
http://link.springer.com/referenceworkentry/10.1007%2F978-1-4614-6141-8_218#page-1.

THE SOPHIA CENTRE FOR THE STUDY OF COSMOLOGY IN CULTURE

SCHOOL OF ARCHAEOLOGY, HISTORY AND ANTHROPOLOGY
THE UNIVERSITY OF WALES TRINITY SAINT DAVID

http://www.uwtsd.ac.uk/sophia

The Sophia Centre for the Study of Cosmology in Culture is a research and teaching centre within the School of Archaeology, History and Anthropology at the University of Wales Trinity Saint David. We define cosmology in the traditional sense as a view of the cosmos, and humanity's place within it. All cultures and all people therefore have a cosmology, and the Centre's remit is the consideration of how we live on planet Earth, with particular reference to the sky, stars, planets and cosmos as part of the wider environment. Our work includes exploration of the ways in which people have tried to live in harmony with the cosmos.

The Centre's academic and scholarly work is partly historical, partly anthropological and partly philosophical and draws strongly on recent developments in the study of religions. It has a wide-ranging mission to investigate the role of cosmological, astronomical and astrological beliefs, models and ideas in human culture, including the theory and practice of myth, magic, divination, religion, spirituality, architecture, politics and the arts. We deal with the modern world as much as indigenous or ancient practices, and our work ranges from the study of Neolithic sites to medieval cosmology, the history of astrology in the ancient world, India and the west, the nature of space and place on Earth as well as in the sky, and the ethics of modern space exploration. We have been instrumental in developing the concept of the skyscape as a vital feature of human cultures and can support work in any time period or culture.

The main qualification we teach is the distance-learning, online MA in Cultural Astronomy and Astrology, the only degree in the world which explores humanity's relationship with the sky. There is no need to live in the UK to study this MA and we have a global community of students and scholars in other MAs at the University, including the MAs in Ancient Religion, Body and Environment, and Ecology and Spirituality. We also contribute to the undergraduate module in Land, Sea and Skyscapes.

The Centre supervises PhD students, holds conferences, publishes books and articles, sponsors events and manages research projects. Aside from its annual conference, held every year since 2003 (at Bath Spa University from 2003 to 2007) the Centre has sponsored five other international conferences. In addition to the 2011 Heavenly Discourses conference these include the 2007, 2010 and 2015 conferences on the Inspiration of Astronomical Phenomena (INSAP) and the 2016 conference of the European Society for Astronomy in Culture (SEAC).

In addition to the volumes published by the Sophia Centre Press, our publications include the peer-reviewed history journal, *Culture and Cosmos*, and *Spica*, the online postgraduate journal written and edited by students. Our recent sponsored publications include: N. Campion, F. Pimenta, N. Ribeiro, F. Silva, A. Joaquinito and L. Tirapicos, eds, *Stars and Stones: Voyages in Archaeoastronomy and Cultural Astronomy – a Meeting of Different Worlds* (Oxford: British Archaeology Reports, 2015); Daniel Brown, ed., 'Modern Archaeoastronomy: From Material Culture to Cosmology', *Journal of Physics: Conference Series* 865, no. 1

(2016); Nicholas Campion, Barbara Rappenglück, Michael Rappenglück and Fabio Silva, eds, *Astronomy and Power: How Worlds are Structured* (Oxford: British Archaeology Reports, 2016); and Nicholas Campion and Dorian Gieseler Greenbaum, eds, *Astrology in Time and Place: Cross-Cultural Currents in the History of Astrology* (Newcastle: Cambridge Scholars Publishing, 2016). We also sponsor the *Journal of Skyscape Archaeology*, the peer-reviewed journal dedicated to examining the role and importance of the sky in the interpretation of the material record.

Our major research project of the moment is 'Welsh Monastic Skies', an investigation of the alignment of Welsh abbeys in relation to major features in the sky and land. We are also working on *Kepler's Astrology*, which hopes to make all Kepler's horoscopes available for study.

If you are interested in the way we use the sky to create meaning and significance then the Centre may be the best place for you to study. By joining the Sophia Centre you enter a community of like-minded scholars and students whose aim is to explore humanity's relationship with the cosmos.

For all academic inquiries and for further information, please contact Dr Nicholas Campion at n.campion@uwtsd.ac.uk.

> *'The work of the Centre is as broad as possible and the MA syllabus is groundbreaking, unique and innovative. We study the many ways in which human beings endow the cosmos with value and use the sky as a theatrical backdrop to tell stories and create meaning.'*
>
> — Dr Nicholas Campion

www.ingramcontent.com/pod-product-compliance
Lightning Source LLC
Chambersburg PA
CBHW042023100526
44587CB00029B/4279